MODERN ASPECTS OF ELECTROCHEMISTRY

No. 22

LIST OF CONTRIBUTORS

S. AMOKRANE
Laboratoire Structure et Réactivité aux Interfaces
Université P.M. Curie
75230, Paris Cedex 05
France

J. P. BADIALI
Laboratoire Structure et Réactivité aux Interfaces
Université P. M. Curie
75230, Paris Cedex 05
France

R. A. BATCHELOR
Department of Chemistry
The University
Newcastle-upon-Tyne NE1 7RU
United Kingdom
Present address: Pilkington Techology Centre
Lathom, Ormskirk, Lancashire L40 5UF, United Kingdom

BERNARD BEDEN
Laboratoire de Chimie 1
Electrochimie et Interactions
URA CNRS No. 350
Université de Poitiers
86022 Poitiers Cedex
France

STOJAN S. DJOKIĆ
Department of Chemistry
University of Ottawa
Ottawa, Ontario K1N 9B4
Canada
Present address: Sheritt Gordon Limited
Fort Saskatchewan
Alberta T8L 3W4
Canada

A. HAMNETT
Department of Chemistry
The University
Newcastle-upon-Tyne NE1 7RU
United Kingdom

CLAUDE LAMY
Laboratoire de Chimie 1
Electrochimie et Interactions
URA CNRS No. 350
Université de Poitiers
86022 Poitiers Cedex
France

JEAN-MICHEL LÉGER
Laboratoire de Chimie 1
Electrochimie et Interactions
URA CNRS No. 350
Université de Poitiers
86022 Poitiers Cedex
France

MIODRAG D. MAKSIMOVIĆ
Faculty of Technology and Metallurgy
University of Belgrade
Belgrade, Yugoslavia

BENJAMIN R. SCHARIFKER
Departamento de Química
Universidad Simón Bolívar
Caracas 1080-A
Venezuela

MODERN ASPECTS OF ELECTROCHEMISTRY

No. 22

Edited by

J. O'M. BOCKRIS
Department of Chemistry
Texas A&M University
College Station, Texas

B. E. CONWAY
Department of Chemistry
University of Ottawa
Ottawa, Ontario, Canada

and

RALPH E. WHITE
Department of Chemical Engineering
Texas A&M University
College Station, Texas

SPRINGER SCIENCE+BUSINESS MEDIA, LLC

The Library of Congress cataloged the first volume of this title as follows:

Modern aspects of electrochemistry. no. [1]
Washington, Butterworths, 1954-
v. illus., 23 cm.
No. 1-2 issued as Modern aspects series of chemistry.
Editors: no. 1- J. Bockris (with B. E. Conway, No. 3-)
Imprint varies: no. 1, New York, Academic Press. — No. 2, London, Butterworths.
1. Electrochemistry — Collected works. I. Bockris, John O'M. ed. II. Conway, B. E. ed. (Series: Modern aspects series of chemistry)
QD552.M6 54-12732 rev

ISBN 978-0-306-44061-8 ISBN 978-1-4615-3376-4 (eBook)
DOI 10.1007/978-1-4615-3376-4

Originally published by Plenum Press, New York in 1992

Preface

In this volume we start in the usual way with fundamentals, and the chapter by Amokrane and Badiali is a good one for putting forward the modern development in the theory of the double layer. Corresponding to this is the chapter on electrocatalysis by Beden, Léger, and Lamy, which deals with a central subject, oxidation of aliphatics on noble metals. Here all the problems of electrocatalysis and the intermediates are brought out, often with a spectroscopic tilt.

Now, changing to semiconductors, Batchelor and Hamnett have honored us with a chapter that deals with surface states in a very thorough way, perhaps more so than any work so far published. Surface states offer opportunities for semiconductor employment, particularly when photoelectrodes provide a component of electrocatalysis as well as the reason for loss of efficiency in the conversion of light to electricity.

The fourth chapter, by Djokić and Maksimović, on electrodeposition of nickel-iron alloys, represents a new study of an old subject in which the application of new techniques has been remarkably fruitful.

The last chapter, by Ben Scharifker, tells us about microelectrode techniques, which seem likely to replace rotating disk techniques in many applications.

This volume is illustrative of the present state of electrode process chemistry but omits our usual contribution from the applied side; two of these will come in our next volume.

J. O'M. Bockris
Texas A&M University
Brian Conway
University of Ottawa
Ralph White
Texas A&M University

Contents

Chapter 1

ANALYSIS OF THE CAPACITANCE OF THE METAL-SOLUTION INTERFACE: ROLE OF THE METAL AND THE METAL-SOLVENT COUPLING

S. Amokrane and J. P. Badiali

Chapter 3

SURFACE STATES ON SEMICONDUCTORS

R. A. Batchelor and A. Hamnett

Chapter 4

ELECTRODEPOSITION OF NICKEL–IRON ALLOYS

Stojan S. Djokić and Miodrag D. Maksimović

Chapter 5

MICROELECTRODE TECHNIQUES IN ELECTROCHEMISTRY

Benjamin R. Scharifker

1

Analysis of the Capacitance of the Metal–Solution Interface: Role of the Metal and the Metal–Solvent Coupling

S. Amokrane and J. P. Badiali

Laboratoire Structure et Réactivité aux Interfaces, Université P. M. Curie, 75230 Paris Cedex 05, France

I. INTRODUCTION

A specific role of the metal in many properties of the metal/solution interface has been known for a very long time. Insofar as the capacitance of the ideally polarized electrode is concerned, a possible contribution of the metal was suspected very early. However, the early attempts to include this idea in the modeling of the interface by using the theoretical methods available at that time were unsuccessful. For about the next 50 years, the so-called "traditional approach" was based on a macroscopic picture of the electrode surface. All the models for the capacitance proposed in this approach were based on two premises. The first is that the metal behaves like an ideal conductor. This means, for example, that a possible contribution to the capacitance related to a change of the surface potential of the metal with the applied potential is negligible. The second is that the electrode surface forms a rigid boundary on the solution side. This means that the distance of closest approach of solvent molecules or ions to the "electrode plane" is mainly determined by steric effects related to their size.

Modern Aspects of Electrochemistry, Number 22, edited by John O'M. Bockris *et al.*
Plenum Press, New York, 1992.

In such a macroscopic picture of the electrode, the influence of the nature of the metal could be accounted for only by invoking its influence on the polarization of the solvent via some "residual interaction." The influence of the structure of the electrode surface, for instance, the difference between liquid or solid electrodes or the effect of the crystallographic orientation of the surface, was discussed only in terms of steric effects. Thus, no coherent picture of the role of the metal emerged from this approach.

The impetus for an alternative analysis of the role of the metal did not come from new experimental facts which contradicted the traditional view but rather from the progress that was made in the understanding of the surface properties of metals on a microscopic level in the early seventies. These studies provide information precisely on those aspects for which the classical description of the electrode surface was insufficient. Their first important consequence is that a direct contribution of the metal to the capacitance cannot be neglected. In addition, this description of the metal surface on a microscopic level provides now a new insight into its interaction with the adjacent liquid phase and hence its indirect contributions. These ideas have now received sufficient support from theoreticians. Thus, it seems worth discussing their consequences in the models proposed in this past decade.

A comprehensive discussion of the properties of the metal/solution interface in the light of these developments requires that both sides of the interface be considered. Such a task is beyond the scope of this chapter. The main purpose of this work is to discuss the alternative analysis of the role of the metal which followed from the improved description of the surface properties of metals. Thus, important aspects of interfacial properties will not be covered here. A first limitation is that only capacitance curves will be discussed. Since both their shape and their magnitude depend on the nature of the electrode, they provide a good test for a new analysis of the metal at the interface. Insofar as the solution is concerned, its description in the classical models has been reviewed exhaustively in this series.[1,2] These models will be mentioned only in relation to their description of the metal. Recent progress[3,4] in the statistical mechanical description of the solution is also beyond the scope of this chapter. Other limitations of our discussion concern the kind of interfaces considered here. Interfaces showing a pronounced specificity related to the nature of the electrolyte or that of the solvent are naturally of the greatest interest. However, at the present stage in the process of reassessing the role of the metal, it is natural to test the theoretical models first on simpler situations such as, for example, interfaces without specific ionic adsorption. Quite generally, only interfaces where the interaction of the metal with the solvent and the electrolyte is weak will be discussed. We will leave out the most interesting case of specific ionic adsorption. Another important point is to select the metals to be discussed. In fact, some recent models have tried to describe the capacitance for a variety of metals, neglecting important questions such as are the capacitance measurements reliable for these metals, and are there sufficient theoretical studies and experimental data on their properties in vacuum? The discussion here will be restricted to cases which meet these requirements. Details of the models for the metal at the interface can be found in the chapter by Goodisman[5] in an earlier volume of this series. Since the latter chapter was written, the role of metal-solvent coupling has been emphasized in several papers. The present work, which is complementary to the chapter by Goodisman, will be mainly devoted to this aspect and its consequences for the analysis of the charge-capacitance curves of various metals.

Before proceeding further, the well-known difficulty one faces in the interpretation of interfacial properties should be recalled. It is common sense to state that most observed interfacial properties reflect contributions from the metal and the solution interacting with each other. From experimental data, one tries to obtain information on the interfacial structure and the interactions on a microscopic level. Information gleaned in this way concerns the interface as a whole. Therefore, the extent to which more specific information can be extracted from experiment depends on an ad hoc hypothesis on all possible contributions.[6] This is well illustrated by the example of the potential of zero charge $E_{\sigma=0}$, usually written as[7]:

$$E_{\sigma=0} = \Phi/F + [\delta\chi^m - g^s_{\text{dip}}]_{\sigma=0} - E_T(\text{ref}) \qquad (1)$$

This relation expresses the potential of zero charge in terms of the electronic work function Φ, the variation $\delta\chi^m$ of the electrostatic barrier at the metal surface χ^m induced by the solution, and the potential drop g^s_{dip} due to the orientation of the solvent dipoles. In this expression, $E_T(\text{ref})$ depends only on the nature of the reference electrode. From this expression for $E_{\sigma=0}$, it is possible to extract information only on the quantity in brackets. Because of the term $E_T(\text{ref})$, the information obtained in this way is only relative, generally referenced to Hg. Since a plot of $E_{\sigma=0}$ versus Φ gives a straight line with a slope less than unity for most *sp* metals, the dipolar term in brackets is metal-dependent.[7] Thus, no definite conclusion regarding g^s_{dip} alone can be obtained from $E_{\sigma=0}$. It follows that different estimates of the terms in $E_{\sigma=0}$ lead to different conclusions about the metal–solvent interaction (see the debate on the "hydrophilic" or "hydrophobic" character of silver in Refs. 8 and 9 and references therein).

In their turn, charge–capacitance curves relate macroscopic quantities. In order to extract information on a microscopic level from these curves, a model is always required. The link between the macroscopic and microscopic behavior being indirect, it is not possible to draw definitive conclusions on a microscopic level only from charge–capacitance curves.

In the analysis of the capacitance, the emphasis was placed for many years on the solution side, since it was assumed that the rearrangement of the electronic structure of the metal with the charge has no appreciable effect on the capacitance. Now, this assumption is challenged. In addition to its long known influence, in particular on the molecular polarization, the metal makes a non-negligible direct contribution to the capacitance. This raises the following question: To what extent will these improvements in our understanding of the properties of the metal lead to a different interpretation of experimental observations? In other words, will the separation of metal effects reveal qualitatively new aspects about the remaining parts of the interface?

This chapter is organized as follows. In Section II, some experimental data on the capacitance will be reviewed briefly. A formal statement of the problem together with a brief survey of the role of the metal in the traditional approach will then serve as an introduction to the more recent work. In Section III, some theoretical methods used for the study of metal surfaces will be presented. Some aspects of the theory which may be of interest in areas other than capacitance studies will also be pointed out. In Section IV a brief discussion of the results of experiments on adsorption from the gas phase and some theoretical studies concerning the interaction of adsorbed species with the surface will be presented. In the following section, the general structure of recent models of the capacitance will be discussed. Some models with fixed metal-solvent separation will be detailed in Section VI. A more general situation will be considered in Section VII. The last section will be devoted to the results of a semiempirical analysis of experimental data.

II. GENERAL ASPECTS

1. Some Experimental Facts

In the first part of this section, some classical data for the ideally polarized electrode have been selected as an illustration of the influence of the nature of the metal. In the original works, these data were analyzed in the framework of the traditional description of the electrode.

To begin with, it is useful to recall that the study of the electrical properties of the ideally polarized electrode started by considering the electrocapillary curve relating the interfacial tension γ to the electrode potential E.[10] The interfacial tension itself is known to differ little at the mercury/water interface from the surface tension of mercury.[10,11] Moreover, the variation of γ along the electrocapillary curve does not exceed 25% of its value at the potential of zero charge (pzc). Thus, the magnitude of the interfacial tension at the metal/solution interface is largely (75%) determined by the metal contribution. Accordingly, one may think that even a small variation of the metal contribution can be responsible for a large part of the observed variation. This situation holds also for gallium.[12,13] The electrocapillary curve and the capacitance are related through the Lippmann equation.[10,11] This link, which provides a possible route for studying the capacitance, suggests that the role of the metal should be considered carefully. This route has not been actually followed, mostly because the calculation of the second derivative of γ with respect to potential is difficult.

An important point to mention here is that the widely accepted analysis of the capacitance for nonadsorbing ions proposed by Grahame[14] will be followed throughout this chapter. In this analysis, the measured capacitance is separated into a contribution given by the Gouy-Chapman model[15,16] and a concentration-independent part $C_i(\sigma)$, as represented, for example, by Parsons-Zobel plots.[17] Therefore, in this chapter only $C_i(\sigma)$ curves will be discussed. $C_i(\sigma)$ is understood as the high-concentration limit of the measured capacitance at which the Gouy-Chapman contribution vanishes. The interpretation of this separation in terms of "diffuse layer" and "inner layer" contributions has been discussed in some recent work based on statistical mechanics.[2,3]

In the case of solid electrodes, an important aspect is that the $C_i(\sigma)$ curves are usually deduced from experiment by correcting the measured capacitance by a "roughness factor" R, since the real area of the electrode surface differs from the geometrical one. From capacitance data, the roughness factor can be determined either from Parsons-Zobel plots or from the criterion proposed by Hamelin and Valette that the capacitance should be monotonic near the pzc. The two methods may give slightly different estimates of R, so that the magnitude of the $C_i(\sigma)$ curves deduced from experiment may depend on the method employed to determine R. However, this variation of R does not dramatically affect their magnitude in the cases discussed below. Moreover, the shape of the $C_i(\sigma)$ curve is the same irrespective of which criterion is used to determine R. For a discussion of these points, see, for example, Ref. 18. Therefore, although an independent estimation of the roughness factor might be necessary for a quantitative comparison between theory and experiment, this is not strictly necessary for an analysis on a qualitative level, especially that which will be presented in the last section of this chapter.

In Fig. 1, $C_i(\sigma)$ curves obtained with four different metals are shown. They include results from the classical work on liquid electrodes by Grahame[14,19] (Hg/NaF) and by Frumkin *et al.*[12,13] (Ga/Na_2SO_4) and more recent work on monocrystalline solid electrodes by Hamelin *et al.*[20] ($Au/NaBF_4$) and Valette and Hamelin[21] (Ag/NaF). All these curves are for nonadsorbing electrolytes (note however that there is slight adsorption in the case of Ag/NaF,[22] and in the case of Ga/Na_2SO_4 the curve at 0.5 N is not corrected for the diffuse layer contribution[13]). Data on Ag/KPF_6[22] will also be discussed. The curves for solid electrodes are based on the real area. For a review on solid electrodes, see Ref. 23.

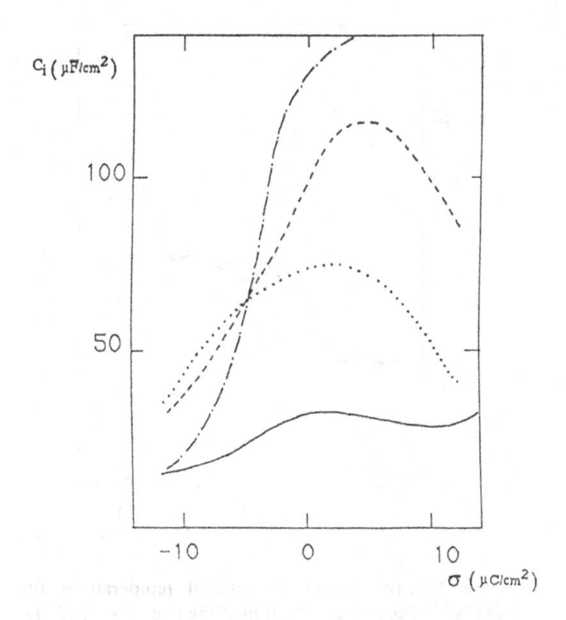

Figure 1. $C_i(\sigma)$ curves for Hg, Ga, Ag, and Au: ——, Hg/NaF[19]; —·—, Ga/Na_2SO_4 (Ref. 12); - - -, Ag/NaF[21]; · · ·, Au/$NaBF_4$.[20]

The curves in Fig. 1 exhibit appreciable variations both in shape (asymmetric, bell-shaped, or humped) and in magnitude with the nature of the metal. This last point is well illustrated by comparing the values of the capacitance at the pzc. All curves show either a maximum or a hump at a slightly positive charge. Figure 1 provides then a good illustration of the influence of the nature of the metal. This picture can now be completed by considering two other types of experiments.

The first one concerns the effect of temperature on $C_i(\sigma)$ curves. In Fig. 2, the classical results of Grahame for Hg/NaF[24] are shown. It is almost superfluous to recall the amount of work devoted to the interpretation of these curves and to the rather controversial origin of the hump in the low-temperature curves (see, for example, the reviews by Reeves,[1] Habib,[2] and Trasatti[6] in this series). We also mention very recent data on gold and silver obtained by Hamelin and co-workers.[25,26]

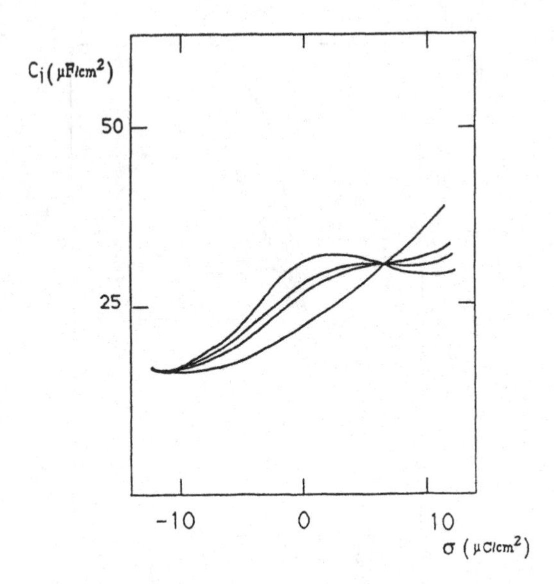

Figure 2. $C_i(\sigma)$ curves at different temperatures for Hg/NaF.[24] From top to bottom at the pzc: $T = 0$, 25, 45, and 85°C.

Among the effects related to temperature, important results concern the entropy of formation of the inner layer, S_{in} (see, for example, Fernando Silva[27]). In the case of the Hg/NaF interface, a well-known feature of the $S_{in}(\sigma)$ curve is that it exhibits a maximum at a negative charge.[28,29] In the models attributing the capacitance hump to dipoles, it was difficult to reconcile in a simple way the occurrence of the hump at a positive charge with the maximum of $S_{in}(\sigma)$ being at a negative charge. However, a dipolar origin of the hump was challenged in the work of Bockris and co-workers by invoking specific ionic adsorption. This was a long debated subject, and we refer the reader to Refs. 1, 2, and 6, cited above. Since none of the related interpretation considers a direct contribution of the metal, this raises the following question: What will happen if the effect of the metal can be separated from capacitance curves? This aspect will be discussed in Section VII.3.

In addition to the results shown in Figs. 1 and 2, there is an important feature observed in the case of solid electrodes. It is now well established that $C_i(\sigma)$ curves show a noticeable effect of the crystallographic orientation of the surface, especially for positive charging of the electrode.[23]

Finally, beyond the general shape of the $C_i(\sigma)$ curves, another important point which will be discussed later is the value of C_i at the plateau which appears at large negative charge. This value, usually denoted as K_{ion}, has been considered for a long time as metal-independent. K_{ion} is now considered to be weakly metal-dependent.[6] For solid silver it is higher than the classical value of about 17 $\mu F\ cm^{-2}$ pertaining to liquid Hg or Ga. Moreover, K_{ion} depends also on the orientation of the surface.[22,30]

From these data, it appears that the capacitance depends not only on the nature of the metal, but also on the crystallographic structure of its surface. A nontrivial feature of the $C_i(\sigma)$ curves is their strong dependence on temperature. It should be recalled that most of the data presented above have already been discussed in the classical models. However, because of the developments which have appeared in the past decade, these data should be reconsidered in the light of the new description of the metal. Thus, it is useful to discuss both classical models and more recent ones in a common framework if possible, at least from a formal point of view. This will allow us to point out why a more detailed description of the metal than the traditional one is really required.

2. Formal Expression of the Capacitance

For a planar interface, let $\Delta\phi_s^m$ be the potential drop between two planes, the first one located well inside the metal (at $z = -\infty$) and the second one in the bulk phase of the solution (at $z = +\infty$). Integrating Poisson's equation leads to[31,32,5]

$$\Delta\phi_s^m = \phi(-\infty) - \phi(+\infty)$$

$$= -4\pi\left[\int_{-\infty}^{+\infty} z\rho_m(z)\,dz + \int_{-\infty}^{+\infty} z\rho_s(z)\,dz + \int_{-\infty}^{+\infty} P(z)\,dz\right] \quad (2)$$

In this expression, $P(z)$ is the polarization due to the solvent molecules, in the direction z normal to the interface. It arises both from the permanent dipoles (orientational contribution) and from any induced dipoles (distortional contribution). The term $\rho_m(z)$ is the local density of charge at a position z in the interface, due to the spatial distribution of ionic cores and conduction electrons of the metal, averaged in the x-y plane. The net charge per unit area on the electrode is then $\sigma = \int_{-\infty}^{+\infty} \rho_m(z)\,dz$. The term $\rho_s(z)$ is the corresponding quantity due to the spatial distribution of the anions and the cations of the solution, so that $-\sigma = \int_{-\infty}^{+\infty} \rho_s(z)\,dz$. Note that the convention for the potential drop employed here is that adopted in classical textbooks of electrochemistry, while in recent models, the potential drop $\Delta\phi = -\Delta\phi_s^m$ is more often considered. The condition that the interface taken as a whole is electrically neutral implies that $\Delta\phi_s^m$ as defined in Eq. (2) is independent of the origin of the coordinates on the z axis. The normalized first moments Z_m of $\rho_m(z)$ and Z_s of $\rho_s(z)$ with respect to the same (and yet arbitrary) origin $z = 0$ are defined as

$$Z_m = 1/\sigma \int_{-\infty}^{+\infty} z\rho_m(z)\,dz \tag{3a}$$

$$Z_s = -1/\sigma \int_{-\infty}^{+\infty} z\rho_s(z)\,dz \tag{3b}$$

These quantities are often referred to as the "center of mass" or "centroid" of the charge distributions. By using the definitions of Z_m and Z_s in Eqs. (3), $\Delta\phi_s^m$ can be written as[32]

$$\Delta\phi_s^m = 4\pi\sigma(Z_s - Z_m) - 4\pi \int_{-\infty}^{+\infty} P(z)\,dz \tag{4}$$

and the differential capacitance can be written as

$$1/C = \partial\Delta\phi_s^m/\partial\sigma = 4\pi\partial/\partial\sigma[\sigma(Z_s - Z_m)] \\ -4\pi\partial/\partial\sigma \int_{-\infty}^{+\infty} P(z)\,dz \tag{5}$$

Let us define some other useful quantities. We first introduce a quantity $\chi^m(\sigma)$ defined as

$$\chi^m(\sigma) = -4\pi \int_{-\infty}^{+\infty} z\rho_m(z)\,dz = -4\pi\sigma Z_m \tag{6}$$

At the uncharged metal/vacuum interface where the charge distribution is $\rho_m^0(z)$, $\chi^m(\sigma)$ corresponds to the dipole barrier χ^m at the metal surface, referred to as the "surface potential of the metal" or "electron overlap potential" in the electrochemical literature. Note that, given the sign convention adopted, χ^m equals the electrostatic potential inside the metal minus the potential outside. At the metal/solution interface, $\rho_m(z)$ differs from $\rho_m^0(z)$ as a consequence of the metal-solution coupling. Thus, $\rho_m(z)$ reflects the perturbation due to the presence of the solution and the fact that it is possibly charged. The two effects may be formally considered separately by defining first the charge density $\Delta\rho_m(z) = \rho_m(z, \sigma) - \rho_m(z, \sigma = 0)$. This quantity accounts for the change in the local density of charges in the metal in the presence of the solution when the interface is charged. Second, one may consider the total variation $\delta\rho_m(z)$ of $\rho_m(z, \sigma)$ with respect to the uncharged metal in vacuum: $\delta\rho_m(z) = \rho_m(z, \sigma) - \rho_m^0(z)$. Then we may write

$$\chi^m(\sigma) = \chi^m + \delta\chi^m(\sigma) \tag{7a}$$

$$\delta\chi^m(\sigma) = -4\pi \int_{-\infty}^{+\infty} dz\, z\delta\rho_m(z) \tag{7b}$$

or, equivalently,

$$\chi^m(\sigma) = \chi^m(\sigma = 0) + \Delta\chi^m(\sigma) \tag{8a}$$

$$\Delta\chi^m(\sigma) = -4\pi \int_{-\infty}^{+\infty} dz\, z\Delta\rho_m(z) \tag{8b}$$

where, by definition, $\Delta\chi^m(0) = 0$, in contrast to $\delta\chi^m(0)$. From the definitions in Eqs. (7) and (8), we see that the "surface potential" $\chi^m(\sigma = 0)$ for the uncharged metal in the presence of the solution differs from the "surface potential" χ^m for the uncharged metal in vacuum. Accordingly, we may write

$$\Delta\chi^m(\sigma) = -4\pi\sigma Z_0(\sigma) \tag{9}$$

where $Z_0(\sigma)$ has the dimensions of length and is possibly nonvanishing at the pzc. By using Eqs. (9) and (8b), $Z_0(\sigma)$ can be written as

$$Z_0(\sigma) = (1/\sigma) \int dz\, z\Delta\rho_m(z) \tag{10a}$$

From the definition of $\Delta\rho_m(z)$, $Z_0(\sigma)$ given in Eq. (10a) is the equivalent at the metal/solution interface of the "center of mass" or "centroid" of the charge that an external electrical field induces on the metal surface for the case considered in the literature of the metal/vacuum interface.[33,34]. In that case, assuming that the distribution of the metal ions in unaffected by the field, $\Delta\rho_m(z)$ in Eq. (10a) must be replaced by $\delta n(z) = n_{\sigma=0}(z) - n_\sigma(z)$, where, following the conventions used, $n_\sigma(z)$ is the number density of electrons at a position z when the metal surface carries a charge σ per unit area and when atomic units ($e = 1$) are used for the electronic charge. It will also be convenient to define the differential quantity $X_0(\sigma)$:

$$X_0(\sigma) = \partial[\sigma Z_0(\sigma)]/\partial\sigma \tag{10b}$$

which can also be written as

$$X_0(\sigma) = -(4\pi)^{-1}\partial\chi^m(\sigma)/\partial\sigma \tag{10c}$$

Equation (10c) is obtained from Eq. (10b) by using Eq. (9) for $\Delta\chi^m(\sigma)$ and Eq. (8a) for $\chi^m(\sigma)$. In Eq. (10c), $X_0(\sigma)$, which has the dimensions of length, measures the rate of change with the surface charge of the "surface potential" of the metal. The physical interpretation of the length $X_0(\sigma)$ can be better seen from Eq. (10b), if the derivative is replaced by the ratio of infinitesimal variations and Eq. (10a) for $Z_0(\sigma)$ is used:

$$X_0(\sigma) = \int dz\, z[\rho_m(z, \sigma + \delta\sigma) - \rho_m(z, \sigma)]/\delta\sigma \tag{10d}$$

In Eq. (10d) the infinitesimal charge $\delta\sigma = \int dz\,[\rho_m(z, \sigma + \delta\sigma) - \rho_m(z, \sigma)]$ is related to a redistribution of the metal electrons across the interfacial region. Thus, $X_0(\sigma)$ has the meaning of the "center of mass" of the spatial distribution of the local excess or deficit of electrons corresponding, for example, to an infinitesimal variation of the applied potential. We may now define a similar quantity $X_s(\sigma)$ for the solution, with ρ_m being replaced by ρ_s in Eqs. (6)–(10). Then the capacitance may be alternatively written as

$$1/C = 4\pi[X_s(\sigma) - X_0(\sigma)] - 4\pi\,\partial/\partial\sigma \int_{-\infty}^{+\infty} P(z)\,dz \tag{11}$$

It is recalled here that $P(z)$ is the molecular polarization. This important relation will be the starting point of our theoretical analysis of the capacitance. It is important to note that, as a consequence of the coupling between the two sides of the interface, the different terms in this expression cannot be calculated from the properties of each side taken separately. The approximations used for these terms precisely define different models for the capacitance.

From Eqs. (4) and (9), another equivalent expression for $\Delta\phi_s^m$ can be obtained. By using Eq. (8a) for $\chi^m(\sigma)$, $\Delta\phi_s^m$ can be written as

$$\Delta\phi_s^m = \chi^m(\sigma = 0) + 4\pi\sigma(Z_s - Z_0) - 4\pi \int_{-\infty}^{+\infty} P(z)\, dz \qquad (12)$$

In Eq. (12) the contribution to $\Delta\phi_s^m$ of the surface potential $\chi^m(\sigma = 0)$ of the uncharged metal in the presence of the solution has been isolated.

3. Analysis of Classical Models

In order to understand the structure of classical models in the light of the previous analysis, we compare Eq. (12) to the classical expression for the potential drop (see, for example, Refs. 6, 10, and 30). Following the classical splitting, $\Delta\phi_s^m$ is written as

$$\Delta\phi_s^m = g_{\text{dip}}^m + g_s^m(\text{ion}) - g_{\text{dip}}^s \qquad (13)$$

In this expression, g_{dip}^m, $g_s^m(\text{ion})$, and g_{dip}^s are the potential differences due to the surface dipole of the metal, the free charges at constant dipole orientation, and the orientation of the permanent dipoles of the solvent molecules at the interface, respectively. This equation represents a particular separation of contributions to the potential drop $\Delta\phi_s^m$ and is based on the separation of the molecular polarization into a contribution related to the orientation of the permanent dipoles, P^{or}, and a contribution related to their polarizability, P^{pol}. The link between Eqs. (13) and (12) has been discussed in Ref. 5. Here, we proceed a little further. By definition, $g_s^m(\text{ion})$ vanishes at the pzc. Then, if we assume such a separation of $P(z)$ in Eq. (12), we may identify $[g_{\text{dip}}^m]_{\sigma=0}$ with $\chi^m(\sigma = 0) = \chi^m + \delta\chi^m(0)$ and g_{dip}^s with $4\pi \int_{-\infty}^{+\infty} P^{\text{or}}(z)\, dz$. At a charged interface the comparison of the various expressions of $\Delta\phi_s^m$ with the particular separation in Eq. (13) is less straightforward because all the terms depend on charge. There is no problem with g_{dip}^s and $4\pi \int_{-\infty}^{+\infty} P^{\text{or}}(z)\, dz$ since both arise from the permanent dipoles. The status of $g_{\text{dip}}^m + g_s^m(\text{ion})$ in Eq. (13) at the charged interface depends in the traditional approach on the models of the inner layer. It is generally considered[30] that these terms give rise to a linear potential drop (σ/K_{ion} + constant). In the simplest models, the inner layer in the absence of specific ionic adsorption consists of a monolayer of water located between the electrode plane and the outer Helmholtz plane. Then K_{ion} is the capacitance of a capacitor consisting of a dielectric medium of thickness e and dielectric constant ε between two fixed charged planes: $1/K_{\text{ion}} = 4\pi e/\varepsilon$. The dielectric constant ε is related to the distortional polarization of the solvent, assumed to be a linear function of the interfacial field. If, as usual, the "thickness" e of the capacitor is taken as equal to one molecular diameter, then from the value of the inner-layer capacitance at sufficiently negative charge, ε is approximately equal to 6. Note that in the model of Bockris and co-workers which was intended for the study of specific ionic adsorption, two water layers with different dielectric constants are considered as forming the inner layer. The important point here is that following the classical analysis the contribution of g^m_{dip} and $g^m_s(\text{ion})$ to the capacitance is a constant.

The formally exact expression for the potential drop given by Eq. (12) clarifies the approximations involved in the classical models. From Eq. (12) we have the following identification:

$$g^m_{\text{dip}} + g^m_s(\text{ion}) = \chi^m(\sigma = 0) + 4\pi\sigma(Z_s - Z_0) - 4\pi \int_{-\infty}^{+\infty} P^{\text{pol}}(z)\, dz$$

In these models Z_0 is the position of the "electrode plane," considered as fixed, that is, the geometrical boundary of the metal. The charge σ is assumed to be localized on the plane $z = Z_0$; that is, $\delta\rho_m(z)$ has the form $\delta\rho_m(z) = \sigma\delta(z - Z_0)$, where δ is the Dirac delta function. In actual models, Z_0 is taken as the origin of the coordinates ($Z_0 = 0$). Similarly, at high enough electrolyte concentrations and in the absence of specific ionic adsorption, the excess charge $\delta\rho_s(z)$ in the solution is assumed to be distributed on the outer Helmholtz plane located at Z_{oHp}: $\delta\rho_s(z) = -\sigma\delta(z - Z_{\text{oHp}})$. Then $(Z_s - Z_0)$ in Eq. (12) reduces to the geometrical distance $(Z_{\text{oHp}} - Z_0) = e$, considered as the "thickness" of the inner layer. Apart from the dielectric continuum in the diffuse layer, the molecular polarization $P(z)$ in these models is localized in the monolayer adjacent to the electrode. When the linear distortional polarization in this monolayer is taken into account, the quantity $(Z_{oHp} - Z_0)$ is divided by ε. The subsequent linear potential drop gives the classical expression of the capacitance K_{ion}. Obviously, in this case the "thickness" of the inner layer is charge-independent. However, in their well-known paper,[35] MacDonald and Barlow considered the possibility of a variation of the thickness of the inner layer with charge. Note that this aspect has a central importance even in the most recent versions of the traditional approach as it appears in the work of Guidelli and Aloisi.[36,37]

If we go back to Eq. (12), the results of the calculations show that in contrast to the assumption made in the classical models, $4\pi\sigma(Z_s - Z_0)$ is not a linear function of charge since both Z_s and Z_0 depend on σ. Although the potential drop $4\pi\sigma(Z_s - Z_0)$ is formally the same as that for two oppositely charged layers located at Z_s and Z_0, the distance $(Z_s - Z_0)$, that is, the "thickness" of the capacitor, is not constant with the charge. We stress that in Eq. (12), Z_s and Z_0 are not physical positions, but correspond to the centroids of charges distributed across the interface, which eventually change with the interfacial field. Physically, Z_0 being very small and independent of the interfacial field means that the induced charge on the metal surface has no spatial extension. This also means that an external electrical field does not penetrate the metal, being completely screened beyond Z_0. Although field penetration was conceptually accepted, this effect was, with very few exceptions (see below), considered as negligible in the calculation of the capacitance. In its turn, Z_s being constant means that the ions in the outer Helmholtz plane approach the metal surface always at the same distance, irrespective of the strength of the interfacial field. This means that the geometrical plane simulating "the surface" acts as a sharp boundary for the solvated cations.

The main task in the classical approach is then to determine the orientational contribution P^{or} to the polarization $P(z)$ in the layer adjacent to the electrode from a molecular model. It is there that the nature of the electrode enters the classical models. The role of the metal is then essentially indirect and appears mostly through a chemical interaction giving rise to some preferred orientations of the molecules. This aspect of the models has been extensively discussed in the literature. The reader is referred to Refs. 1, 2, and 6, for example. This chemical interaction may explain the shape of the curves and their variation with the nature of the metal (compare, for example, the work on the Hg/NaF interface by Parsons[38] and that of Valette on silver,[22] both of whom used the same model). Recall also the discussion of Trasatti[30] and of Parsons[39] on the relation between the strength of the metal-solvent interaction and the magnitude of the capacitance.

The interaction of the metal surface with the solution has also been discussed in terms of physical interactions. Electrostatic interactions including image forces on the ions or dipoles together with dispersion forces (in their *classical* formulation) were introduced in models of various sophistication, such as in the work of Watts-Tobin,[40] Bockris and co-workers,[2,10,41,42] Barlow and MacDonald,[35] Levine *et al.*,[43] Fawcett *et al.*,[44-45] Damaskin and Frumkin,[46] Parsons,[38] and Guidelli and Aloisi.[36,37] A vast amount of work has been done by these authors and many others in order to explain various experimental aspects, generally with appreciable success. However, we will now point out some aspects where this description of the role of the metal may not be sufficient.

4. Why a New Description of the Role of the Metal?

Despite the apparent success mentioned above, the classical models may be criticized on two different levels. First, these models use parameters which either are adjustable or sometimes correspond to quantities which are ill defined on a microscopic level. Just as an example of this last aspect, we mention the arbitrary values of the "dielectric constant" of the two water layers in the model of Bockris and co-workers or the value of the ratio of the dipole moment of the clusters to that of the monomer in the work of Parsons. The observed sensitivity of the results to the precise value of some of these parameters is a real drawback of these models. Second, except for very general considerations regarding the contribution of the electron overlap, with no treatment of it on a microscopic level, all these models do not take into account the new description of the electrode surface and its consequences for the analysis of properties as important as the capacitance.

Progress in the understanding of the electronic properties of metal surfaces which followed to a large extent from the work of Lang and Kohn[47] indeed provides answers to some of the questions raised by the use of the (over)simplified description of the electrode surface in the traditional approach. Just above, three assumptions which are implicit in the classical description of the electrode were mentioned. Among the problems encountered in the classical analysis of the interface is the fact that $\chi^m(\sigma)$ is considered differently depending on the state of charge of the interface. At the pzc, it is widely accepted that the presence of the solution may induce a rearrangement in the electronic distribution giving rise to $\delta\chi^m(0)$ in Eq. (7a). However, at a charged interface, the excess charge $\Delta\rho_m(z)$ in Eq. (8b) induced by an external field is assumed to be distributed on a geometrical plane whose position Z_0 is charge-independent. In other words, in the classical models, the electronic structure of the metal is sensitive to the presence of molecules but is practically not affected by an external field. Although from obvious physical considerations a possible effect of the applied field on the response of the electrons in the metal was accepted, this effect was considered as negligible or was not taken into account explicitly. In fact, the assumption that the induced charge $\Delta\rho_m(z)$ is sharply peaked at the geometrical surface is not correct: it is now recognized that $\Delta\rho_m(z)$ has a more diffuse character.[33,34,48] This means, for example, that there is an appreciable penetration of the electric field into the metal. As recalled above, this aspect of metal "nonideality" is in fact not a new concept. Indeed, it has been recognized since 1928[49] that the effect of field penetration at a metal surface may play a major role in the capacitance, eventually leading to what is sometimes referred to as the "Rice paradox" (see, for example, Refs. 154 and 170). Indeed, it follows from the estimation of the effect of field penetration made by Rice that the metal contribution gives the entire value of the capacitance, with little place for a contribution from the solution. The difficulties arise largely from the fact that in Rice's model the metal electrons were subject to an infinite potential barrier near the surface. It was only in the beginning of the seventies that a self-consistent treatment of the electron gas subject to a realistic potential barrier at a metal surface became successful. This point has been taken into account in the models proposed in the early eighties for the metal/electrolyte interface, which removed the difficulties of Rice's model. The relevant work will be discussed in the last two sections of this chapter. See also the chapter by Goodisman in this series.[5]

The modern description of the structure of the metal surface definitely clarifies the notion of "electrode plane." As shown in 1973 by Lang and Kohn,[33] in the case of the response of the metal to a weak field, the electrode plane is to be identified with the image plane, whose position X_0 is given by the centroid of the charge induced by a weak uniform field [see Eq. (10d) and Fig. 10]. This position X_0 of the "electrode plane" is approximately fixed only in the linear regime[33] close to the pzc. Away from the pzc, the shape of $\Delta\rho_m(z)$ changes noticeably with the value of the net (and non-infinitesimal) charge σ.[34,50] Its center of mass Z_0 [the integral of $\Delta\rho_m(z)$ is σ] is also charge-dependent. In the limit of weak surface charges, X_0 and Z_0 are of course identical. We shall see in the next section that X_0 (or Z_0) can be determined unambiguously with respect to the last ionic plane of the solid (and can also be determined from theory in the case of a liquid, at least in principle). For all electrical properties, this position X_0 should be considered as the *effective location* of the metal surface. This point is relevant to electrochemistry in many respects, and part of this chapter will be concerned with this point and related aspects. Let us already mention some important points. Its relevance in the determination of the capacitance can be visualized if one considers the metal in the presence of an assumed ideal charged plane located at a distance D far from the geometrical surface of the metal. The whole system is equivalent to a capacitor whose capacitance is given by $C^{-1} = 4\pi(D - X_0)$ rather than by the classical expression $C^{-1} = 4\pi D$. This aspect will be discussed in detail in Section III.4 (see Fig. 8).

More generally, a description of the metal surface which goes beyond its representation as a geometrical plane may be relevant in the estimation of interaction energies. For example, image forces in their classical formulation are common ingredients in many classical models. It is then important to recall that the classical formulation of image forces is incorrect near the surface. In classical electrostatics, the image force on a point charge at a distance Z from an ideal conductor ($Z = 0$ is the geometrical boundary of the metal) varies as $1/Z$. A microscopic description of the metal leads actually to a slower variation at small values of Z (see, for example, Refs. 51-54). A similar situation occurs for dispersion forces (see Section III). Since typical values of Z for ions or dipoles in contact with the metal are of the order of a few angstroms, the correction to the classical expression is indeed substantial. This may have serious consequences for the predictions of models where electrostatic forces (or dispersion forces) play an important role in the structuring of the particles in contact with the electrode. Then the consideration of the real structure of the electrode may be important for an accurate description of the interaction of adsorbed particles with the surface. For example, it is implicit in the classical view that the electrode acts as a sharp boundary representing the short-range surface-adsorbate repulsion. The actual metal-adsorbate interaction potential exhibits in fact a smoother variation with distance. This would mean that there is no reason to locate the position of the center of the water layer adjacent to the electrode *a priori* at one molecular radius from the "metal surface." A similar remark holds also for the position of the outer Helmholtz plane.

To sum up, a characteristic distance of the interface—the effective position of the electrode surface—can be determined now from the physical properties of the metal. Other important distances such as the distance of closest approach of the solvent molecules or the ions to the electrode should also be determined from the knowledge of their actual interaction with the metal surface by using some appropriate criteria (this aspect is discussed in Section IV.2). We thus expect a detailed description of the metal surface together with its interaction with adsorbed species to be a necessary step for a satisfactory description of the role of the metal in interfacial properties.

Before going further in this direction, a few words should be said about the description of the solution side of the interface. This field of investigation, which has been renewed in the last decade, will not be treated in this chapter (the interested reader is referred to the reviews by Carnie and Torrie[3,5] and by Henderson[4]). The major task is to calculate $P(z)$ and $\rho_s(z)$ in Eq. (2) for a given model of the solution, by using the tools of the modern statistical mechanical theory of inhomogeneous liquids (integral equations, for example). Some interesting results have already been obtained for model systems.[3-5] However, they have not yet been extended to real situations. An important aspect is that in most of these works the metal is replaced by a charged wall, which means $Z_0 = 0$. This group of methods includes computer simulations. We note here the interesting results obtained recently by Heinzinger and Spohr.[55]

See also references in their review paper.[56] The development of such techniques will certainly bring valuable information on the properties of the solution in the near future.

5. An Alternative to a Fully Microscopic Approach

We may now summarize the situation: the traditional approach has brought a large wealth of information, but also has its limitations and has left many questions open. The modern statistical mechanical description of the solution based on models which go beyond the "primitive" model of point ions in a dielectric continuum is still in a preliminary stage of development. The work of the last ten years based on a microscopic description of the metal seems more directly related to experimental aspects. However, all recent attempts toward an *ab initio* calculation of the capacitance which couple the recent view of the metal to models of the solution beyond the "primitive" model have failed to reproduce the experimental data, at least insofar as the charge-capacitance curves and their variation with the nature of the metal are concerned. In our opinion, this situation is understandable: besides the well-known limitations of the statistical mechanical methods presently available, such as the mean spherical approximation (MSA) or its variants,[57-59] the direct calculation of the capacitance requires that each term in Eq. (11) should be calculated with high accuracy since the *inverse* capacitance results from cancellation between terms of comparable magnitude. We hope that Section VII, devoted to our own work, will convince the reader on this point.

An alternative to the direct calculation of the capacitance by the evaluation of all terms in Eq. (11) may consist in calculating some of them as accurately as possible. Then if the calculated terms are separated from the experimental value of the (inverse) capacitance, a semiempirical estimate of the *remaining contributions* is deduced. Then one may have some hope that a simple interpretation of these terms which contain more limited information can be found *a posteriori*. In our recent work,[60] we found that this semiempirical approach proved to be rather fruitful. Practically, we define from Eq. (11) a contribution C_m. By using the experimental C_i, we isolate a contribution C_s by using the relation $1/C_s = 1/C_i - 1/C_m$. The precise meaning of C_m and C_s will be discussed in detail in Section

VII.2. We just indicate here that the calculation of C_m requires a model for the metal and for the metal-solution coupling. A review of the theory of metal surfaces and metal-adsorbate interactions will be the subject of the next section.

III. ELECTRONIC PROPERTIES OF THE METAL SURFACE

1. General Aspects

The preceding sections have shown that some properties of the metal/solution interface are determined to some extent by physical properties of the metal side. For example, we have seen that for mercury or gallium electrodes, the interfacial tension is determined largely by the surface tension of the metal/vacuum interface (Section II.1). Also, the potential of zero charge has been known for a very long time[61] to be related to the value of the electronic work function Φ. At a charged interface, a question of fundamental interest for the theory is to determine how these metal properties are affected by the interfacial field. A natural requirement that a model of the metal at the interface should fulfill is that it must first be satisfactory for typical properties of the bare surface such as the surface tension γ or the work function Φ. In the case of weak interfacial coupling, the models fulfilling this criterion may be used as a good starting point in the modeling of the interface.

From the experimental point of view, the surface tension is a more characteristic property of the metal than the work function. Indeed, while Φ varies roughly by a factor of two in the periodic table,[62] the surface tension of a light alkali such as Cs (40 dyn/cm) is an order of magnitude smaller than that of Hg (400 dyn/cm) or Al (1000 dyn/cm) and 50 times less than that of Fe (2000 dyn/cm). The surface tension should then provide a much more stringent test of theory than the work function.

Unfortunately, except in the case of the alkalies and a few polyvalent metals, all models proposed so far (apart from semi-phenomenological formulas) for a microscopic derivation of the surface tension for *sp* metals have failed to reproduce the experimental values of γ, especially for liquid Hg (see, for example, the recent work of Hasegawa[63] and references therein). For these metals, the difficulties originate to a large extent from the fact that an important contribution due to conduction electrons competes equally with that related to the ions. This cancellation makes the theoretical results very inaccurate. This in contrast with the case of the transition metals, which will not be discussed here; for these metals, only one phenomenon dominates, namely, the filling of the d band.[64] However, if only the variation of γ with the potential, which ultimately determines the capacitance through the Lippmann equation, is required, the actual magnitude of the surface tension might not be relevant. For this purpose, what seems important is to have a good description of electrostatic contributions to γ. The electrostatic energy depends on the average distribution of charges at the surface. In addition, a correct description of the electrostatic potential at the surface is also required for a satisfactory description of the work function. Thus, while a good account of electrostatics seems necessary for both γ and Φ, an accurate calculation of the magnitude of γ is not strictly required. This point is further illustrated by the following example.

For the uncharged surface, it is well known from the theoretical work of Lang and Kohn[47] that, except in the case of the alkalies, the jellium model, in which the discrete nature of the ions is ignored, predicts unreasonable negative values of γ (more precisely the surface energy U_s) while the values of Φ given by the same model are not in significant error when compared to experiment. The introduction of the discrete nature of the ions as a perturbation changes dramatically the values of U_s, which become much closer to the experimental values, while the values of Φ remain roughly in the same range. Much of the improvement is due to the appearance of additional terms in the expression of the surface energy. Among these new terms, the cleavage energy of the solid makes a very large contribution to U_s. However, this term is only related to the properties of the ionic lattice. Another effect comes from the change in energy related to the noncoulombic nature of the short-range interaction between the electrons and the ions; this term is calculated with the electronic profile corresponding to the jellium model in Ref. 47. These terms are important for determining the magnitude of U_s, but they do not dramatically change the values of Φ, that is, the potential drop at the interface. This ultimately means that a model which does not predict accurate values for the energy may still give reasonable values of Φ.

One may ask whether similar behavior occurs also in the case of a charged surface. For solids, the ions will contribute to the energy through terms not affected by an external electric field, since the latter is almost completely screened by the conduction electrons at the position of the ions.[33] The situation is then similar to that existing at the pzc. Then it will be sufficient to discuss a possible effect of the field related only to nonzero ionic size (in the electron–ion interaction energy) on the work function.

Due to the diffuse nature of the ionic profile in the case of liquid metals, screening of external fields by the free electrons might be less efficient than for solids. The ionic profile may then vary with charge, with a possible effect on both γ and χ^m. In this case, no definite conclusion can be reached prior to calculations.

To summarize, at the pzc and for solids it seems that reasonable values of the work function may coexist within the same model with somewhat less satisfactory values of the surface energy. At a charged surface, only the correct behavior of the electrostatic potential is needed. For all these reasons, in the following we will mainly present values of the work function in the case of solids. The case of liquids will be discussed only on a qualitative level.

2. Density Functional Theory

The density functional theory and its applications to the description of bulk and surface properties of metals have been the subject of numerous and extensive reviews. The interested reader is referred to Refs. 62, 65, and 66 for a detailed presentation of the theory. For a more specific application to electrochemistry, see, for example, Refs. 67 and 68 and the chapter by Goodisman in this series.[5] Here, only the aspects of the theory which will be useful for the discussion of recent models of the metal/electrolyte solution interface will be considered.

(i) *Ground State Energy of the Electron Gas*

The theoretical description of the electronic properties of metals requires the solution of a complicated quantum-mechanical many-body problem. For *sp* metals, including the alkalies and some polyvalent metals (e.g., Al, Mg), the electronic states in the metal can be separated into core states and conduction band states. For most properties of interest, one has to consider explicitly only states in the conduction band. Then, one is led to calculate the properties of a gas of conduction electrons immersed in a lattice of ions (the core electrons and the nucleus). It can be shown that the energy of these metals in the bulk phase can be calculated by considering the conduction electrons as nearly free electrons, weakly perturbed by the presence of a lattice of ions. This description is based on the so-called pseudo-potential theory.[69,70] In this approach, the electronic wave functions behave as plane waves outside the ionic core, while they are expanded in terms of core state functions near the nucleus. This leads to the replacement of the strong lattice potential by a weak pseudo-potential operator, which allows the use of perturbation theory. The description of bulk properties takes then great advantage of translational invariance and of the possibility to using, in general, very simple models for the pseudo-potentials.[70,71]

In the case of surface properties, the problem is further complicated by the loss of translational invariance introduced by the presence of a surface, in the vicinity of which the gas of conduction electrons is strongly inhomogeneous. This problem has been made tractable by Hohenberg and Kohn[72] and Kohn and Sham,[73] who introduced the density functional formalism (DFF). This formalism is specially devised for the study of strongly inhomogeneous systems like the electron cloud of atoms or molecules[66] or the inhomogeneous electron gas appearing when the surface of a solid is created. The main idea involved in the density functional formalism is that the problem should be greatly simplified if, instead of trying to determine the complex many-body wave function of the system, one focuses on the electronic density $n(\mathbf{r})$ at a given point $\mathbf{r}$ of the interface. Using the density as the basic variable, one constructs an appropriate functional $E[n(\mathbf{r})]$ which, when evaluated with the exact density, will correspond to the ground state energy. A fundamental property of this formalism is that the ground state energy of the electron gas in the presence of an external potential $V(\mathbf{r})$ is the minimum value, with respect to variations of $n(\mathbf{r})$ at a constant number of electrons, of $E[n(\mathbf{r})]$. The functional

$E[n(\mathbf{r})]$ can be written as:

$$E[n(\mathbf{r})] = G[n(\mathbf{r})] + \tfrac{1}{2}\int d\mathbf{r}\, d\mathbf{r}'\, n(\mathbf{r})n(\mathbf{r}')/|\mathbf{r} - \mathbf{r}'| + \int d\mathbf{r}\, n(\mathbf{r})\, V(\mathbf{r}) \quad (14)$$

where $G[n(\mathbf{r})]$ contains the kinetic energy of the electrons, the exchange energy related to the Pauli principle, and that arising from the correlated motion of the electrons. The second term is the electrostatic self-energy of the electron gas, and the last term is its interaction energy with the external potential $V(\mathbf{r})$, which may correspond, for example, to the potential of the semi-infinite lattice of ions. Note that one should add to Eq. (14) some constant terms independent of $n(\mathbf{r})$ in order to deal with finite quantities.

For a particular application, the main task is to devise some approximation for the unknown functional $G[n(\mathbf{r})]$, which contains all the complexity of the many-body problem. One major ingredient in the theory is the so-called "local density approximation" (LDA) for the exchange-correlation part, E_{xc}, of $G[n(\mathbf{r})]$:

$$E_{xc}[n(\mathbf{r})] = \int d\mathbf{r}\, n(\mathbf{r})\{e_x[n(\mathbf{r})] + e_c[n(\mathbf{r})]\} \quad (15)$$

where e_x and e_c are, respectively, the exchange and correlation energies per particle of a *uniform* electron gas of density $n(\mathbf{r})$. The idea involved in the LDA is to approximate the local properties of the electron gas by those of a gas of uniform density equal to the local density $n(\mathbf{r})$. While nonlocal treatments of E_{xc} have been proposed (see, for example, references in Ref. 66), most applications of the formalism are within the LDA. Frequently used expressions for exchange-correlation energies are the Hartree approximation for e_x: $e_x(n) = -\frac{3}{4}(3n/\pi)^{1/3} = -0.458/r_s$ for exchange; the Wigner interpolation formula for e_c: $e_c(n) = -0.44/(r_s + 7.8)$; or the analytically convenient form due to Pines and Nozières: $e_c(n) = \frac{1}{2}[-0.0115 + \ln(r_s)]$.[65,66,69] The parameter r_s, defined by $r_s = (3/4\pi n)^{1/3}$ where n is the bulk electron density, is a convenient way of characterizing the average free electron density of the homogeneous metal: r_s is the radius of the sphere containing on average, one electron.

In addition to the necessary approximations for $G[n(\mathbf{r})]$, the effect of the semi-infinite lattice of ions cannot be treated exactly.

One is then forced to construct some model of the surface in order to make the problem tractable. Some methods proposed for treating the functional $G[n(\mathbf{r})]$ and the related models of the surface are briefly presented below.

(ii) The Jellium Model

The simplest possible representation of the surface is found in the so-called "jellium" model. In this model, the discrete nature of the semi-infinite lattice of ions is ignored. Rather, it is replaced by a structureless uniform background of positive charge, terminating abruptly at the "surface" (Fig. 3) and which neutralizes the total negative charge of the electron gas. In this picture, the problem is reduced to that of an interacting electron gas subject to the coulombic potential $V(z)$ due to the semi-infinite jellium. Due to the structureless nature of the latter, the properties of the electron-jellium system are invariant in the direction parallel to the surface. This introduces a major simplification in that the problem is now reduced to a one-dimensional one. This feature is the reason for the wide popularity of the jellium model.

An approximate treatment of this problem was proposed by Smith in 1969.[74] A more exact treatment was published by Lang and Kohn in 1970.[47] Most of the subsequent developments in the theory of metal surfaces followed from the work of these authors. Since both the method proposed by Smith and the self-consistent calculations of Lang and Kohn (LK) have been used at the metal/solution interface, we briefly describe the two methods. Details can be found in Ref. 5.

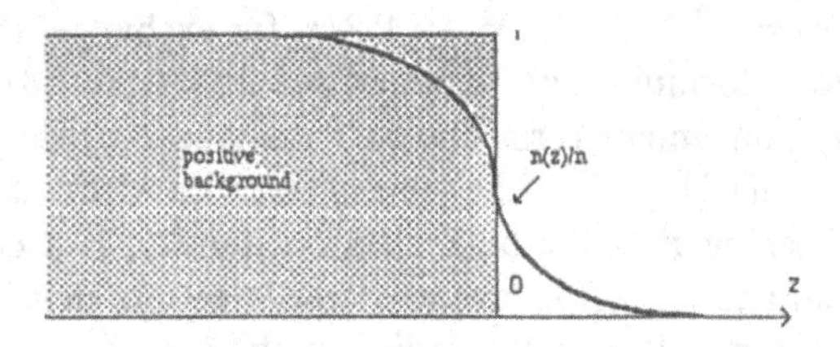

Figure 3. Schematic representation of the metal in the jellium model; $n(z)$ is the electronic density profile in the direction $0z$ normal to the surface, and n is the bulk density.

The LK approach is based on the fact that the minimization of $E[n(\mathbf{r})]$ given in Eq. (14) is equivalent to finding the solution of a Schrödinger-like equation for electrons in an effective potential V_{eff}:

$$\{(-\hbar^2/2m)\Delta + V_{\text{eff}}[n(\mathbf{r}), \mathbf{r}]\}\Psi_l = \varepsilon_l \Psi_l \tag{16}$$

where the first term on the left-hand side is the usual kinetic energy operator. In this equation, V_{eff} is the sum of $V(\mathbf{r})$, the coulomb potential of the electrons, and the exchange-correlation potential $\delta E_{xc}[n(\mathbf{r})]/\delta n(\mathbf{r})$, which in the LDA is equal to $d[n\varepsilon_{xc}(n)]/dn$ where $\varepsilon_{xc}(n)$ is simply $e_x[n(\mathbf{r})] + e_c[n(\mathbf{r})]$ as discussed above in Section III.2(i). The effective potential depends on the electron density $n(\mathbf{r}) = \sum_l \Psi_l^*(\mathbf{r})\Psi_l(\mathbf{r})$, so that the equations must be solved self-consistently. This procedure is similar to that followed in a Hartree–Fock calculation, with which the reader may be more familiar. The difference is that while Hartree–Fock treats exchange exactly but ignores correlations, DFF + LDA considers both in an approximate way.[66]

The approach of Smith is much simpler than that of LK in that it avoids having to find the complicated solution of Eq. (16) and the self-consistent calculation of V_{eff}. Instead, the functional $E[n(z)]$ is directly minimized in a given class of *trial functions* for the density profile $n(z)$. These functions are chosen in such a way as to reproduce the gross features of the exact profiles as obtained from an LK calculation. They depend on a set of variational parameters $\{\alpha_i\}$. Some of these are determined by general conditions of continuity of $n(z)$ and its derivative and by the requirement of charge conservation. The remaining free parameters are determined by the minimization of $E[n(z)]$. The practical procedure consists in minimizing the surface energy U_s per unit area A, obtained by subtracting from $E[n(\mathbf{r})]$ the contribution of an electron-jellium system whose electron density is uniform up to the jellium edge:

$$U_s = \{E[n(z)] - E(n)\}/A, \qquad \partial U_s/\partial\alpha_i = 0 \tag{17}$$

Note that the approach of Smith also uses the LDA for the exchange-correlation energy. An expansion of the kinetic energy $T[n(z)]$ in terms of density gradients is further made, in contrast to the approach of LK, which treats this quantity exactly. Then, $T[n(z)]$

is written as

$$T[n(z)] \approx \int dz\, n(z)\{t_0[n(z)] + t_1[n(z)] + t_2[n(z)] + \cdots\} \quad (18)$$

The first term in the integrand corresponds to a simple Thomas-Fermi approximation. Developments up to second gradient terms

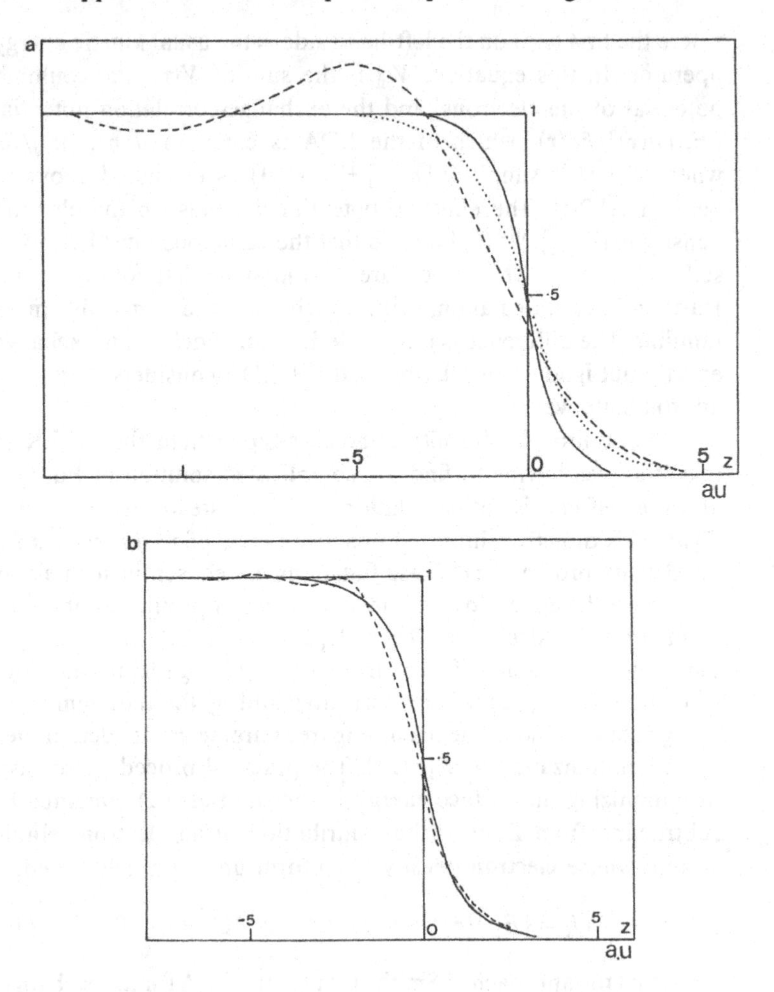

Figure 4. Calculated density profiles $n(z)/n$ for $r_s = 6$ (a) and $r_s = 2$ (b) in the jellium model. (——) from Smith[74]; (– – –) from Lang and Kohn.[47] In (a), (· · ·) is from the work of Ma and Sahni[75] and shows the effect of second gradient terms.

are now commonly used, following the work of Ma and Sahni,[75] who pointed out their importance.

As an illustration, we have plotted in Fig. 4 the density profiles obtained by the two methods for $r_s = 6$ and $r_s = 2$, corresponding roughly to Cs and Al.[47,74] The Smith profiles, obtained with an exponential type of trial function, are monotonic while in the exact profiles the Friedel oscillations are clearly visible in the case of Cs. Note that these oscillations are less marked when the bulk electron density n increases or equivalently when the value of r_s decreases. In general, the method of Smith is less accurate than that of LK, and the profiles are only approximations of the exact ones. This may have an important consequence in the calculation of the electrostatic potential. The latter can be in significant error when calculated directly by using in Eq. (6) the profiles obtained by the minimization of U_s. However, this feature can be circumvented by using an alternative determination of the potential, as detailed below. If such a procedure is followed, one can really take advantage of the simplicity of the method of Smith, compared to that of LK.

(iii) Calculation of the Work Function. Position of the Image Plane

By definition, the work function Φ is the minimum work required to extract from the metal one electron at the Fermi level and transfer it to the vacuum. Then it follows from the general formalism[33,65] that Φ is given by

$$\Phi = \chi^m - \bar{\mu}_e \tag{19}$$

where χ^m is the electrostatic energy barrier at the metal surface, and $\bar{\mu}_e$ is the bulk chemical potential of the electrons relative to the mean electrostatic potential in the metal interior and is calculated from the theory. Note that since atomic units for the electronic charge are used in Eq. (19), χ^m is also the surface dipole barrier as defined in Eq. (6). In the jellium model, the charge density is $\rho_m(z) = n\theta(-z) - n(z)$, where $n\theta(-z)$ is the background density, $\theta(-z)$ being the Heaviside step function (the metal occupies the $z \leq 0$ half-space). Then from Eq. (6) we have:

$$\chi^m = 4\pi \int dz\, z[n(z) - n\theta(-z)] \tag{20}$$

Equation (19) is sometimes referred as the "Koopman's theorem" expression of the work function. Note that Eq. (19) is the same as the expression used in the electrochemical literature.[6,7] Equation (19) shows that Φ results from a cancellation between the surface term χ^m and the bulk contribution $\bar{\mu}_e$. Especially in the case of low r_s, both terms of Eq. (19) must be calculated very accurately if this Koopman's theorem expression of Φ is used. As an extreme example, the recent work of Perdew and Wang[76] has shown that for very low r_s, values of Φ of a few electron volts are obtained from values of χ^m and $\bar{\mu}_e$ of more than 150 eV. In the case of a jellium density corresponding to Al, $\Phi = 3.78$ eV while $\chi^m = 6.59$ eV. If the desired accuracy is a few tenths of an electron volt, then the value of Φ in Eq. (19) obtained with χ^m calculated in Eq. (20) with the approximated profile $n(z)$ may not be accurate enough. If one does not use the LK scheme, this problem can still be avoided by using the alternative expression of the work function called the "change in self-consistent field" expression of Φ, denoted $\Phi_{\Delta SCF}$[77]:

$$\Phi_{\Delta SCF} = dU_s/d\Sigma|_{\Sigma=0} \tag{21}$$

where Σ is the number of electrons per unit area on the surface. This less familiar expression of Φ is a consequence of the general definition of the work function.[77] It may be understood as the change in energy (per unit area) of the metal when an infinitesimal charge is induced on its surface, due to the transfer of a few electrons from the metal to the vacuum. Since the induced charge is localized on the surface, this change in energy results from a change in surface energy. Because of the stationary property of $E[n(z)]$ (viz. U_s) with respect to a variation of the density profile, the ΔSCF expression is much less sensitive to errors in the profile than the direct Koopman's theorem expression.[77] Note that, in practice, one may obtain $\Phi_{\Delta SCF}$ from the change in surface energy when a charge $\sigma = e\Sigma$ per unit area is induced on the metal surface by an external field created by a charged plane located far away from the metal surface.

Equations (19) and (21) are formally equivalent, so that $dU_s/d\sigma|_{\sigma=0} = \chi^m - \bar{\mu}_e$. This relation can be generalized for an arbitrary surface charge, with χ^m being simply replaced by $\chi^m(\sigma)$.[78] Since only $\chi^m(\sigma)$ may change with charge, this relation provides an alternative to Eq. (20) for its calculation as a function of charge. In particular, one can obtain in this way the quantities $Z_0(\sigma)$ and $X_0(\sigma)$ defined in Eqs. (10a) and (10b) by using a relation derived from Eq. (21), as shown recently by Russier and Rosinberg.[78] The values of Z_0 and X_0 obtained by following this route are very accurate, even if Smith's method is used.

There exists also an important relation between the bulk energy per electron and the potential drop between the bulk of the metal and the jellium edge, called the Budd-Vannimenus sum rule.[79] These points will be further developed in the discussion of the results for the metal/solution interface.

If Eq. (19) were to be discussed in connection with experiments at the metal/vacuum interface, one faces again the problem that experiment gives information only on the global quantity Φ. Then, together with reasonable predictions for Φ, internal coherence of the model should be an indication that χ^m is itself not too badly calculated.

For the purpose of discussing the interfacial capacitance, the relevant quantity is $\chi^m(\sigma)$, which represents the generalization of χ^m for a charged interface, as discussed in Section II.2 (Eq. 7a). However, $\chi^m(\sigma)$, as well as χ^m, cannot be determined directly from experiment. However, by using a given model, they can be calculated from Eq. (6) or the generalization of Eq. (21). Then one may compare the theoretical results for $\delta\chi^m(0)$ to electrochemical estimates such as those made in Ref. 6. A more direct (and more demanding) comparison to experiment is the prediction of effects related to the crystallographic orientation of the surface of the electrode. For example, the predicted order of magnitude of the variations of $\delta\chi^m(0)$ with the exposed face may be compared with the observed change in the potential of zero charge.

(iv) Discreteness of Ions

(*a*) *Effect of the ionic lattice*

In the jellium model, the nature of the metal appears only through the value of the electronic bulk density *n*. While such a picture may be convenient at least as a first approximation for the surface energy and work functions of the alkalies, for which the discrete nature of the ions can be neglected, lattice effects must be reintroduced for a reasonable description of the surface properties of most metals used as electrodes in electrochemical cells.

In order to take into account the discreteness of the ionic cores, the electron–ion interaction at the surface is usually described in terms of *pseudo-potentials*, just as for bulk properties (the widely used "empty core model" potential introduced by Ashcroft[71] is shown in Fig. 5). In the treatment of Lang and Kohn,[47] the one-dimensional character of the jellium model is preserved. These authors treated the effect of the lattice by first-order perturbation theory. The perturbation of the jellium is taken as the difference $\delta V_{ps}(\mathbf{r})$ between the pseudo-potential of the lattice and the potential due to the jellium. The first-order correction to the energy is given by

$$\delta E_{ps} = \int n(z)\delta V_{ps}(z)\, dz \tag{22}$$

In this expression, $n(z)$ is the zero-order electronic density profile determined for the jellium, and $\delta V_{ps}(z)$ is the average in the plane parallel to the surface of $\delta V_{ps}(\mathbf{r})$. This first-order treatment has been criticized since δV_{ps} is in general not small compared to the unperturbed term. Perdew and Monnier[80] have shown that it is better to include this term in the expression of the functional of the energy $E[n(\mathbf{r})]$ given in Eq. (14). While, in principle, the problem is now unavoidably a three-dimensional one, the simpler treatments, such as the self-consistent variational approach of Perdew and Monnier,[80] still use one-dimensional wave functions (and hence density profiles) in Eq. (14). In the Smith-like approaches, the contribution δE_{ps} can also be included in the expression of the

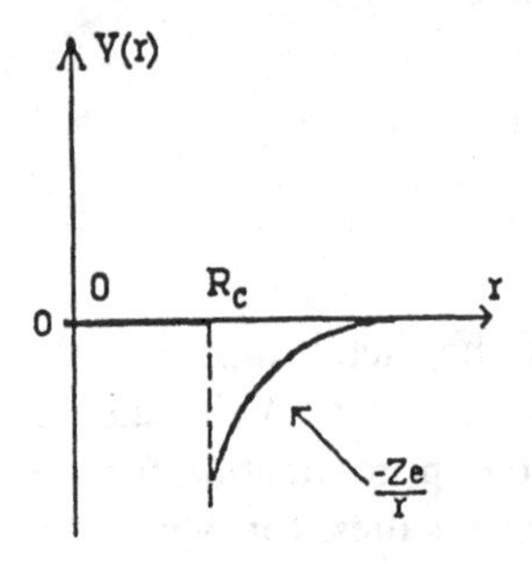

Figure 5. The one-parameter (R_c) Ashcroft pseudo-potential.[71] Z is the valence, and e is the electronic charge.

surface energy. Such models are sometimes referred to as "jellium-like" models. In this case, the electronic density profile differs from the jellium profile and can depend, for example, on the crystallographic orientation of the surface, as in the calculation of Perdew and Monnier. This is the procedure we followed in our own work.

More sophisticated calculations have attempted to incorporate the three-dimensional nature of the semi-infinite metal. Examples include the recent work of Aers and Inglesfield[81] or the calculation of Rose and Dobson.[82] An alternative approach is based on the simulation of the surface by a cluster of a few atoms. Such calculations undertaken with a view to studying the metal/solution interface can be found in Refs. 83 and 84, for example. Recently, Halley and Price[85] also used a three-dimensional version of the jellium without pseudo-potentials.

(b) Comparison between liquids and solids

By its very nature, the jellium model does not make any distinction between solids and liquids. Indeed, it has been often used as such in models for the capacitance of liquid mercury or gallium.[86,87] In this view, the step profile of the jellium background is either taken as the neutralizing background just as in the case of solids or is assumed to mimic a very steep ionic density profile at the liquid metal surface.

Since both liquid and solid electrodes are used in experiments, it is interesting to ascertain whether there are indications which support this viewpoint or whether, on the contrary, there are indeed some important differences. This can be done, at least on a qualitative level, on the basis of the classical work on solid electrodes and some more recent results on liquid metal surfaces. In Fig. 6, we have plotted the results of Perdew and Monnier for the electronic density profile $n(z)$ of the (110) face (denser face in the bcc structure) of cesium. The electronic density profile $n(z)$ and ionic density profile $\rho(z)$ from the work of Hasegawa[63] and a computer simulation by Harris *et al.*[88,89] for liquid Cs are also shown.

A first observation is that $\rho(z)$ bears little resemblance to the step profile of the jellium and exhibits pronounced oscillations (this

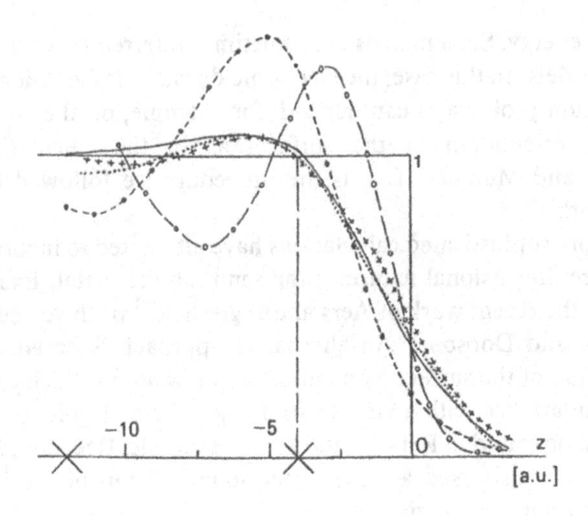

Figure 6. Normalized ionic $[\rho(z)/\rho]$ and electronic $[n(z)/n]$ density profiles for cesium: + + +, $n(z)/n$ for Cs(110) from Monnier and Perdew[80]; ——, $n(z)/n$, and - - . - -, $\rho(z)/\rho$ for liquid Cs from Hasegawa[63]; · · ·, $n(z)/n$, and —∘—, $\rho(z)/\rho$ for liquid Cs from Harris *et al.*[89] Crosses on the z axis indicate the position of ionic planes in the case of solid Cs. $z = 0$ is the position of the corresponding jellium background.

feature has been recently included in a model of the metal/solution interface by Goodisman[90]). A more important fact in our opinion is that while for solids $n(z)$ has non-negligible values far away from the last ionic plane, $\rho(z)$ and $n(z)$ have comparable decay lengths in the case of a liquid. This may be important when the metal comes in contact with the solution. While for solids one may neglect as a first approximation the short-range interaction of the metallic ions with the solution compared to that of the electron cloud, both should be considered in the case of liquids.

(*v*) *Extension of the Jellium Model to the Noble Metals*

Silver and gold electrodes are now commonly used in electrochemical cells.[23] Since the electronic properties of noble metals are dominated by the hybridization of states in the s band and in the d band, the formalism developed for sp metals cannot be used directly in this case. This issue was raised by Lang and Kohn in their seminal paper on the work function,[33] where they found that the values of Φ obtained from the formalism for *sp* metals were in error by about 0.5 eV (even 1 eV with more recent experimental determinations of Φ). The theory of bulk properties of *sp* metals has been extended to the case of noble metals mainly by Moriarty, who gave a synthesis of his work in Ref. 91, and by Wills and Harrison.[92] An essential feature of the theory is that the effective valence of noble metals differs from the nominal valence, as a result of the *s-d* hybridization. For the study of surface properties of Ag and Cu, Russier and Badiali recently proposed[32,93] an extension of the jellium model. The work function of Ag and Cu can be determined accurately by considering an effective simple metal contribution Φ_{ms}, which is the dominant contribution, and a potential drop $\Delta\phi_d$ due to the *d*-band states located on the last ionic sites of the semi-infinite lattice. In agreement with the *ab initio* calculation of Moriarty,[91] the simple metal part is calculated by attributing to the metal an effective valence $Z_s = 1.5$ instead of the nominal valence $Z = 1$. An approximate version of the tight binding theory[64] based on a sum rule due to Friedel can be used in order to compute $\Delta\phi_d$. Since the latter is much smaller than the potential drop arising from the simple metal part, one can neglect its variation with the charge at the metal/solution interface, as a first approximation.

3. Some Results for the Metal/Vacuum Interface

(*i*) *Work Functions*

We will not give here a detailed account of the experimental situation. Table 1 gives the work functions of four selected metals, including liquids (Hg, Ga) and solids (Ag, Al). The first three metals are those typically used in electrochemical experiments. Results for Al are included, since this polyvalent metal with a high electron density ($r_s = 2$) is often regarded as a good candidate for testing models. The values for solids are for monocrystalline surfaces.

For the metals in Table 1 the experimental values for the work function differ by less than 0.3 eV, and some of them even by less than 0.1 eV. For solids, an interesting indication is the sequence of values of Φ for different surface orientations. Some experimental

Table 1
Experimental Work Functions as a Function of the Bulk Electron Density n

Metal	$10^3 n$ (a.u.$^{-3}$)	Φ^{exp} (eV)
Hg[a]	12.7	4.48–4.52
Ga[a]	22.3	4.30
Ag(111)[b]	8.8	4.46
Al(111)	26.9	4.24,[c] 4.26[d]

[a] Values are those selected by Trasatti (Ref. 7).
[b] Ref. 94.
[c] From Ref. 95, cited in Ref. 77.
[d] From Ref. 96, cited in Ref. 77.

data for Ag and Al are shown in Table 2. For Ag, the spread of the experimental values is narrow. For Al, two sets of values, labeled *b* and *c*, may be distinguished. In values in footnote *c*, the work functions are ordered in the sequence $\Phi(111) > \Phi(100) > \Phi(110)$ as observed for Ag while in values in footnote *b* this sequence is not obeyed. This discrepancy has been analyzed in Ref. 77. Table 2 suggests in turn that the magnitude of the variation of Φ with the crystallographic orientation is at most 0.3 eV.

The information contained in Tables 1 and 2 provides some guidelines for selecting theories. In principle, the accuracy of calculated work functions should be better than 0.1 eV in order to describe correctly the variations in Φ with the nature of the metal. This can

Table 2
Effect of Crystallographic Orientation on Φ

	Φ^{exp} (eV)	
Face	Ag[a]	Al
111	4.46	4.24,[b] 4.26[c]
100	4.22	4.41,[b] 4.20[c]
110	4.14	4.28,[b] 4.06[c]

[a] Ref. 94.
[b] Ref. 95, cited in Ref. 77.
[c] Ref. 96, cited in Ref. 77.

hardly be fulfilled in practice, because Φ results from the cancellation of two relatively large quantities, as discussed previously. At present, a deviation in the range 0.1-0.2 eV should be satisfactory. For solids, when the order in the values of Φ for different orientations of the surface is well established experimentally, a more qualitative test is that the values predicted for different orientations should be properly ordered.

Calculated work functions are compared to experiment in Table 3. From the values in the table, we may draw the following conclusions:

1. In the case of solids, calculated values of Φ agree fairly well with experiment. For examples, the values of Monnier *et al.*[77]

Table 3
Comparison of Theoretical and Experimental Work Functions

Metal	Work function (eV) Theoretical	Experimental
Solids		
Al(111)	4.05,[a] 4.27[b]	4.24,[c] 4.26[d]
Al(100)	4.2,[a] 4.25[b]	4.41,[c] 4.20[d]
Al(110)	3.6,[a] 4.02[b]	4.28,[c] 4.06[d]
Ag(111)	4.40[e]	4.46[f]
Ag(100)	4.29,[e] 4.95,[g] 4.2[h]	4.22[f]
Ag(110)	4.10[e]	4.14[f]
Liquids		
Al	4.52,[i] 3.64[j]	4.19,[k,l] 4.28[k,m]
Hg	3.48[i]	4.50[n]
Ga	4.14[i]	4.30[n]

[a] Lang and Kohn (Ref. 33).
[b] Monnier *et al.* (Ref. 77).
[c] From Ref. 95, cited in Ref. 77.
[d] From Ref. 96, cited in Ref. 77.
[e] Russier and Badiali (Ref. 93).
[f] From Ref. 94.
[g] Aers and Inglesfield (Ref. 81).
[h] Arlinghaus *et al.* (Ref. 97).
[i] Badiali *et al.* (Ref. 98); jellium-like metal with step ionic profile.
[j] Goodisman (Ref. 90); jellium with oscillatory profile.
[k] Polycrystalline solid.
[l] Quoted in Ref. 33.
[m] From Ref. 99.
[n] From Ref. 6.

agree within less than one-tenth of an electron volt with the experimental values in footnote *d* of Table 3. However, deviations as large as 0.2 eV are observed with the data in footnote *c* of Table 3. In the case of silver, the calculated values of Φ reported by Russier and Badiali[93] are also in good agreement with experiment. However, in the case of Ag(100) the value of Aers and Inglesfield[81] obtained by a three-dimensional calculation is much larger than those of Arlinghaus *et al.*[97] (three-dimensional slab geometry) and Russier and Badiali[93] (one-dimensional jellium-like metal). We mention that the first two authors quote a value $\Phi^{exp} = 4.64$ eV from Ref. 100, much higher than that given in the table [see also the discussion of Valette[101] for Φ^{exp}(Ag)].

Concerning the effect of the surface crystallographic orientation, one can see that for Ag and Al (footnote *d* in Table 3), predicted values of Φ for the three low-index faces are correctly ordered, following the sequence $\Phi(111) > \Phi(100) > \Phi(110)$.

2. For liquids, due to an insufficient understanding of their surface structure, only rough estimates may be given. From the values in Table 3, it is clear that the situation is much less satisfactory than in the case of solids. As an example, a large spread of values can be observed for Al. More generally, predicted work functions differ from experiment for all metals by several tenths of an electron volt.

From these comparisons, one may conclude that theory appears to predict correct values of the work function for solids. However, it should be borne in mind that what is ultimately needed at the metal/solution interface is a good description of the variation with charge of $\chi^m(\sigma)$, which at the pzc is only a part of the work function. At this point, we may recall that similar values of Φ can be obtained with quite dissimilar values of χ^m,[67,77] largely because of the cancellations in Φ that we discussed previously. The only test at present is the comparison of different theoretical determinations of $X_0(\sigma)$, defined in Eqs. (10).

(ii) *Charged Surface*

At the uncharged surface, the position $X_0(0)$ of the image plane was obtained for the first time in the jellium model by Lang and Kohn.[33] They defined $X_0(0)$ as the center of mass of the infinitesimal charge $\delta n(z)$ induced by a weak uniform field, $X_0(\sigma) = \int z\delta n(z)\, dz/\int \delta n(z)\, dz$, this definition being equivalent to Eq. (10d). For typical metallic densities, Lang and Kohn found that $X_0(0)$ depends slightly on the average free electron density and ranges from 1.6 a.u. (Al, $r_s = 2$) to 1.2 a.u. (Cs, $r_s = 6$). These values of $X_0(0)$ are measured from the jellium edge, the latter being at half an interplanar spacing from the last ionic plane.[33]

Since the work of Lang and Kohn, the self-consistent calculation of $n(z)$ and $X_0(0)$ has been extended to the case of a charged surface by several authors (see the work of Gies and Gerhardt[34] and of Schreier and Rebentrost,[50] for example). These calculations may be used to test results obtained by approximate calculations as in the variational Smith-like approaches. In this case, $\chi^m(\sigma)$ calculated directly in Eq. (20) with trial density profiles may be inaccurate. An alternative route is to take advantage of the so-called Budd–Vannimenus sum rule,[79] which expresses part of the potential drop in terms of properties of the bulk metal. At the charged metal surface, this relation, sometimes referred to as the "half-moment condition," can be written as

$$\phi(0) - \phi(-\infty) = -2\pi\sigma^2/n - n\, d\varepsilon(n)/dn \tag{23}$$

where the term on the left-hand side is the potential drop between the jellium edge and the bulk of the metal, and ε is the energy per electron in the homogeneous metal. This condition allows $Z_0(\sigma)$ to be calculated from the alternative expression

$$Z_0(\sigma) = -\sigma/2n + 1/\sigma \int_0^\infty z[n_{\sigma=0}(z) - n_\sigma(z)]\, dz \tag{24}$$

This route has been followed by several authors.[102–104] The reader is referred to the recent review by Kornyshev for details and for a discussion of the proper use of this sum rule.[105] Without invoking the Budd–Vannimenus sum rule, Russier and Rosinberg[78] calculated $Z_0(\sigma)$ in the variational Smith scheme and compared their results to those obtained by Gies and Gerhardt,[34] who used an LK scheme. By using a special version of the ΔSCF route for Φ, they have derived an expression of $Z_0(\sigma)$ which exactly reduces to Eq. (24). Their results compare well with the self-consistent calculations. As an example, we have plotted in Fig. 7 $Z_0(\sigma)$ and $X_0(\sigma)$ deduced

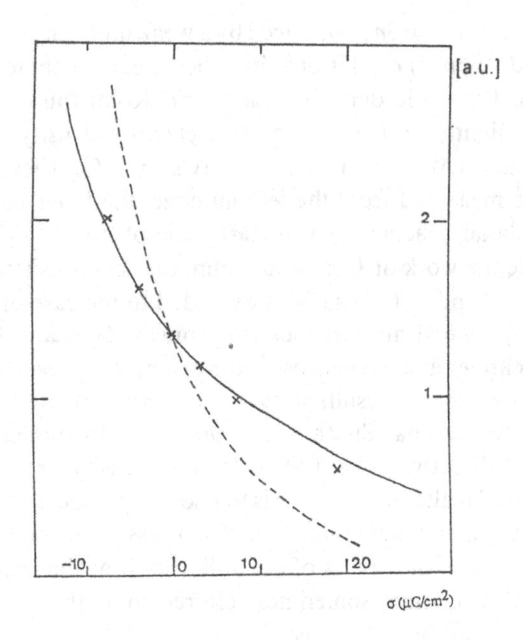

Figure 7. Center of mass $Z_0(\sigma)$ of a finite charge (——) and center of mass $X_0(\sigma)$ of an infinitesimal induced charge (– – –) for $r_s = 3$ calculated by Russier and Rosinberg.[78] Crosses show the results of Gies and Gerhardt[34] for $Z_0(\sigma)$.

from the results of Refs. 34 and 78 as a function of charge for $r_s = 3$. In general, $Z_0(\sigma)$ and $X_0(\sigma)$ are decreasing functions of the charge on the electrode. The larger increase at negative charging is generally attributed to the fact that it is easier to pull the electrons out of the metal than to push them back. Figure 8 is an illustration of the general behavior of $Z_0(\sigma)$ and $X_0(\sigma)$ for free-electron-like metals.

It can then be concluded that the weakness of the Smith-like approach can be corrected by using one of the previous alternative routes for evaluating $X_0(\sigma)$. This is important for applications to the metal electrolyte interface, since this method is much simpler than solving the self-consistent Kohn–Sham equations.

To our knowledge, no self-consistent calculations of $Z_0(\sigma)$ and $X_0(\sigma)$ of the Kohn–Sham type are available for a three-dimensional

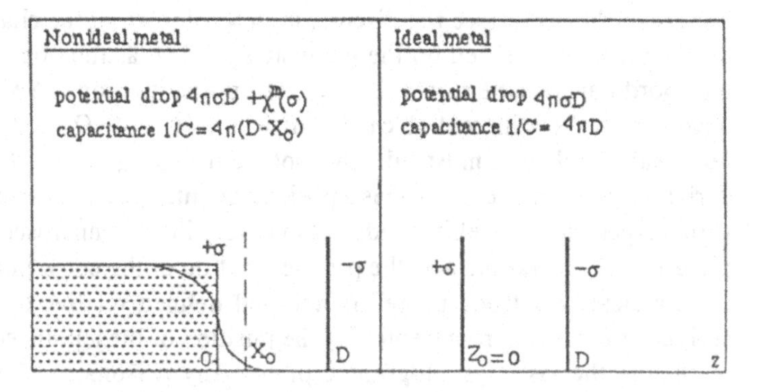

Figure 8. Schematic view of a model charged interface. Left panel: Microscopic view of the metal; right panel: representation of the metal surface by a charged plane located at Z_0. In both panels, D is the position of an ideal charged plane.

solid with a surface. A three-dimensional calculation based on a different formalism was performed by Aers and Inglesfield,[81] who calculated $Z_0(\sigma)$ for Ag(100). They suggested that the jellium calculation overestimates the variation of $Z_0(\sigma)$ by a factor of three on the basis of a comparison of their results to those of Gies and Gerhardt for $r_s = 3$ (silver with nominal valence $Z = 1$). We have checked that using the effective valence $Z = 1.5$ (higher electron density) and an electron-ion pseudo-potential indeed reduces the slope of $Z_0(\sigma)$, in agreement with the suggestion of Aers and Inglesfield.

4. Conclusion. Relevance to Electrochemistry

In order to illustrate how the ideas presented above will be useful at the metal/solution interface, some examples are considered below.

(i) *Capacitance of a Planar Capacitor*

Consider the charged interface represented in Fig. 8. An ideal charged plane with a charge $-\sigma$ per unit area is located at a distance D from the metal bearing an opposite charge σ. In the classical picture of the metal (see the discussion in Section II.3), the charge on the metal is localized on the plane at Z_0, taken as the origin of the coordinates on the z axis ($Z_0 = 0$). The potential drop is $\Delta\phi_s^m = 4\pi\sigma D$, and the differential capacitance is $1/C = 4\pi D$. For the nonideal metal, one must add the potential drop $\chi^m(\sigma)$, whose derivative is $-4\pi X_0(\sigma)$. $X_0(\sigma)$ is a positive quantity when measured with respect to the jellium edge. However, in this microscopic picture of the metal surface, the precise location of the corresponding classical "electrode plane" is not well defined relative to the surface of the metal, represented by the position of the jellium edge or that of the last crystallographic plane. Any reasonable choice for this classical value of Z_0 would locate it between the last ionic plane and the jellium edge. Thus, the classical value of $1/C$ will always be reduced by the presence of $\chi^m(\sigma)$. For example, if both D and X_0 are measured with respect to $Z_0 = 0$ taken as the jellium edge, the capacitance at the pzc is $1/C = 4\pi[D - X_0(0)]$. For a macroscopic capacitor, D is many orders of magnitude larger than X_0 so that the correction is truly negligible. The situation is now very different at the interface. There, one may imagine D as being the position Z_{oHp} of the outer Helmholtz plane. Since $X_0(\sigma)$ can be an appreciable fraction of Z_{oHp}, we expect this metal nonideality to have a substantial effect on the capacitance of the interface. Provided that $X_0(\sigma)$ is not dramatically reduced by the presence of the solution, the relaxation of the surface dipole of the metal with charge should then increase the inner-layer capacitance compared to that predicted by using the traditional picture of the electrode.

(ii) Predicted Slope of the Capacitance

Insofar as the variation of $X_0(\sigma)$ with charge will not be qualitatively changed by the presence of the solution compared to the metal–vacuum case, one may infer from Fig. 7 the resulting effect on the slope of the inner-layer capacitance. Quite generally, we may write $1/C_i = 1/\tilde{C}_i - 4\pi X_0(\sigma)$, where $\tilde{C}_i$ contains any other contribution to the capacitance (stemming from the effect of the interfacial field on the solvent polarization or the excess charge in the solution). The simplest way to picture the effect of $X_0(\sigma)$ is first to consider that $\tilde{C}_i$ is constant. In this case, the resulting C_i is

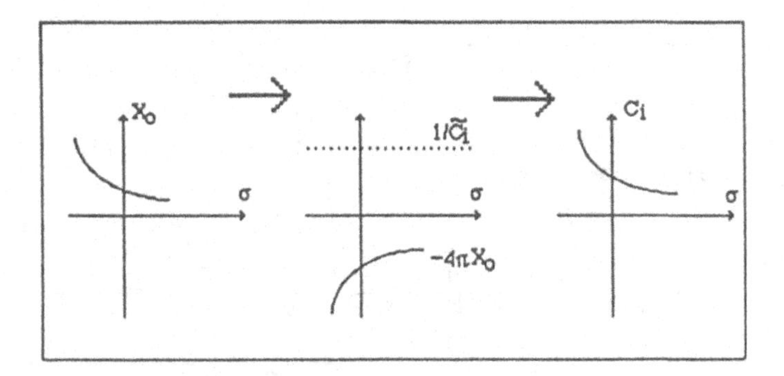

Figure 9. Predicted slopes of the capacitance C_i near the pzc assuming that (i) the metal contributes only via $X_0(\sigma)$, and (ii) all other contributions are constant.

a decreasing function of charge (Fig. 9). From this picture, one may infer that deviations of the experimental slopes from this behavior may be attributed to nonlinear effects in $P(z)$ or a possible variation with charge of the center of mass Z_s of the excess charge in the solution, as discussed by Kornyshev and Vorotyntsev.[106] This point will be discussed in detail in the last section of this chapter.

(iii) Physical Picture of the Electrode Surface

We may now draw the picture of the metal surface (Fig. 10) which emerges from the preceding discussion. The edge of the positive background of the jellium model, $z = 0$, is at half an interplanar spacing from the last ionic plane. The position $X_0(\sigma)$ of the image plane is then located in front of this jellium edge and is directly related to the spread of the induced charge. From $X_0(\sigma)$ one obtains, for example, the asymptotic expression of the image potential on a point charge far outside the metal as $U^{\text{im}} = 1/|Z - X_0(\sigma)|$, where both Z and $X_0(\sigma)$ are measured with respect to the jellium edge. A similar dependence on $Z - X_0(\sigma)$ appears in the expression of the image potential of a classical point dipole and also in the asymptotic expression of the van der Waals forces [in the latter case, $X_0(\sigma)$ is replaced by the position of the dynamic image plane[107,108]]. This picture of the metal surface may be particularly useful in the presence of adsorbed species.

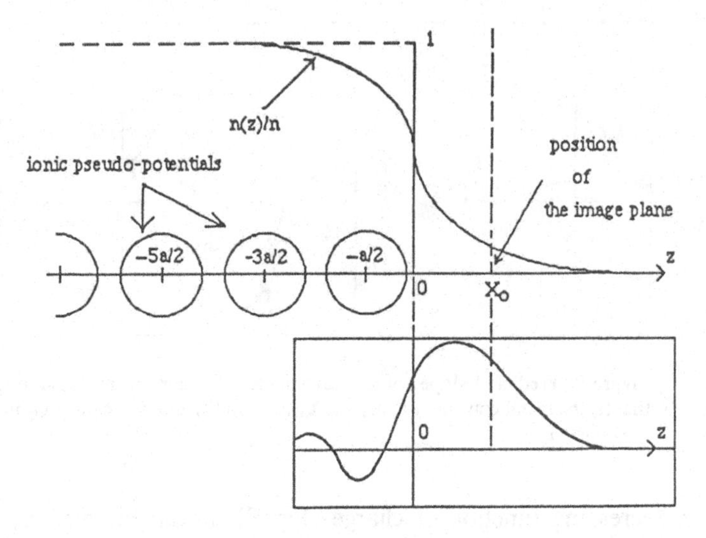

Figure 10. Schematic view of the metal, including ionic structure, with a the interplanar spacing in the direction normal to the surface. The inset shows the qualitative shape of the induced charge $\delta n(z)$.

IV. METAL–ADSORBATE INTERACTIONS

The present section is motivated by the fact that the interaction of the metal with the adjacent liquid is an important ingredient in models of the interface. In the traditional approach, the characteristics of this interaction are estimated on the basis of electrochemical data. As mentioned in Section I, the latter generally involve contributions from both sides of the interface. When available, independent studies which may provide more direct information on this interaction are of the greatest interest. Among such possible sources of information, some recent data from ultra-high-vacuum (UHV) experiments will first be described. Then some first-principle calculations and some related models for the metal-solvent interaction will be discussed.

1. Adsorption from the Gas Phase

Among the very many techniques developed recently in the field of surface science, some may be applied to the study of the electrochemical interface.[109] For example, adsorption experiments under UHV conditions such as thermal desorption spectroscopy (TDS) or work function change ($\Delta\Phi$) measurements may provide valuable information. A general review of water adsorption at metal surfaces has been presented recently by Thiel and Madey.[110]

TDS provides information on the metal-adsorbate bond strength. These experiments have been performed on various metals such as Ni, Rh, Ru, Ir, Pt, Cu, and Ag (see Refs. 110 and 111 and references therein). For these metals, the adsorption energy is estimated to be between 48 and 57 kJ mol^{-1}. For all metals studied in UHV, the adsorption energies are in the same range and correspond to the upper part of the physisorption range. Note that, as noted by Trasatti,[7] this is in line with the estimated variation of the adsorption energy by a few kilojoules per mole from Hg to Ga. It is also interesting to note that silver seems quite distinctive. For this metal, only one peak attributed to desorption of water from ice layers is observed in the desorption spectrum. This feature is interpreted as indicating that for silver, water-metal interactions are, at most, as strong as water–water interactions.[110,111]

Measurements of work function change upon adsorption[109] represent the second group of UHV experiments that we would like to mention. The observed initial decrease of the work function is usually related to the potential drop due to the orientation of the physisorbed molecules. The molecular dipole points toward the vacuum; that is, on average the oxygen atom is in contact with the metal. This interpretation requires some assumption about charge transfer and neglects any contribution from a charge redistribution in the metal,[7,109] similar to that giving rise to $\delta\chi^m(0)$ as discussed in Section II.2. The final saturation after about one layer of coverage indicates that the orienting effect induced by the surface is short-range. Finally, it is interesting to note that in the case of silver, $\Delta\Phi$ curves indicate that the orientation of water is less pronounced than in the case of copper.[109] This seems to confirm a weaker water–silver interaction.

2. Discussion

The strength of the metal–solvent interaction is a question of central importance in the classical analysis of electrochemical data. It was recalled in Section I that information on this strength was obtained to some extent from the analysis of the components of the electrode potential, that is, the term $\chi^m(\sigma = 0) - g^s_{\rm dip} = \chi^m + \delta\chi^m(0) - g^s_{\rm dip}$. From estimates of $\delta\chi^m(\sigma = 0) - g^s_{\rm dip}$, the metals have been classified on a "hydrophilicity scale."[6] By assuming that the variation of this quantity with the nature of the metal is mainly due to a metal-induced orientation of the solvent dipoles through $g^s_{\rm dip}$, it was inferred, for example, that water is much more oriented and thus more tightly bound on Ga than on Hg.[6] Thus, Ga is considered to be a hydrophilic metal whereas Hg should be hydrophobic. However, UHV experiments indicate, in agreement with other electrochemical estimates,[7] a relatively minor influence of the nature of these two metals on the water-metal bond strength. These seemingly conflicting findings were reconciled by considering that major variations of potential drop from metal to metal might be induced by relatively minor energy changes.[6,7] This point favored the interpretation of "hydrophilicity" in a relative sense.

Since these conclusions follow to a large extent from the assumption that $\delta\chi^m$ is nearly metal-independent, this point has been investigated by using a nontraditional description of the electrode. It is now known from theoretical calculations that $\delta\chi^m(0)$ is metal-dependent. Calculations indicate that $\delta\chi^m(0)$ may vary from metal to metal by a few tenths of an electron volt. Since this is of the same order as the estimated variation of $\delta\chi^m(0) - g^s_{\rm dip}$ for different metals, the presence of this metal contribution makes the interpretation of experimental data in terms of hydrophilicity only tentative, as suggested in the early work of Badiali *et al.*[98] At the present stage of development of theoretical calculations, it is not possible to give values of $\delta\chi^m(0)$ which are accurate enough to allow definitive conclusions on this point. One of the reasons for this situation is our insufficient understanding of the metal-solvent coupling. However, recent work in this field[106,112-115] suggests that a new concept be introduced in the analysis of the interface: *the effect of the metal-solvent bond length.*

Quite generally, substrate-adsorbate interaction may be characterized both by the *bond strength* and by the *bond length* (see Refs. 116 and 117). For a single adsorbed particle *at rest*, the two quantities are indeed correlated since the equilibrium surface-particle distance corresponds to the minimum of the potential energy curve. In the case of a chemisorbed molecule, the potential energy curve is expected to exhibit a pronounced minimum and rise rapidly with increasing distance between the surface and the molecule. Then the equilibrium position of the molecule at the surface is well defined and will be only slightly affected by, say, an external field or the presence of coadsorbed particles. This is just the view which underlies the classical analysis: it is implicit in this view that the water molecule has its center of mass rigidly fixed to the metal and only its orientation may change with the nature of the metal or under the effect of the interfacial field. Therefore, while "hydrophilicity" refers to the strength of the bond, in the underlying analysis the metal affects in fact only the orientation degrees of freedom of the molecule via g^s_{dip}.

While this picture might be valid for a truly chemisorbed solvent, it need not be true for water, which appears to be not so strongly bound on the metals analyzed in UHV, especially on silver. In the case of a weakly bound molecule, the potential energy curve is expected to have a less steep variation with the distance between the surface and the molecule than in the case of chemisorption. In a dense fluid, the average minimum metal-solvent distance will differ from the equilibrium value for the isolated particle at rest, since, in addition to the effect of other molecules or ions, the molecules in the liquid may cross an energy barrier of the order of their mean kinetic energy. Therefore, the average metal-solvent separation is expected to be associated with the repulsive part of the potential. The presence of the solvent at this average distance from the metal surface will give rise to the perturbation $\delta\chi^m$ of the surface dipole. The variation of $\delta\chi^m$ with the nature of the metal will thus be related in some ways to the variation of this metal-solvent separation, mainly through the repulsive part of the potential. Therefore, the precise value of the energy minimum for an isolated molecule will be much less relevant than in the case of a strongly bound solvent.

In contrast to the chemisorption situation, the binding of a weakly bound solvent should then not be associated with a rigid bond length. Then one may reasonably expect such weak binding to be also affected by the interfacial field. This does not exclude of course any orientational effect induced by the metal. These aspects will be discussed again in the sections of this chapter devoted to models. It should also be interesting to further investigate these points from the experimental point of view. To our knowledge, estimates of metal-solvent bond lengths from UHV experiments are rather scarce and have not been reported in the case of charged surfaces. Progress in this direction will clearly be of great help.

3. Theoretical Description of Metal–Adsorbate Interactions

The interaction of metal sufaces with various adsorbates has been studied from first principles both in chemisorption and physisorption situations. This aspect of the theory is a wide field of research in itself, and it is not the purpose of this section to present the state of the art in this field. Important aspects have been reviewed in Refs. 116 and 117. Here we will present briefly some studies on water adsorption on metal clusters and a typical physisorption calculation.

(i) *Cluster Calculations*

Theoretical studies of water adsorption on metal clusters are rather scarce (see the review by Müller[118]). This kind of calculation has been performed for one water molecule on clusters of Pt,[119] Al,[120] and Cu.[121] See also the recent work of Kuznetsov *et al.*[122] who also considered Hg, Cd, Zn, and Au along with the case of water clusters. The methods of quantum chemistry are used in order to study the interaction of the water molecule with a cluster of a few metallic atoms simulating the immediate environment of water adsorbed at the metal surface. From these calculations, the following findings emerge[118]: (1) water is adsorbed without dissociation or significant deformation of the molecule and with the oxygen atom toward the metal; (2) the binding involves mixing of the water lone pairs with molecular orbitals of the cluster; (3) the interaction potential involves a repulsion between metal electrons and the closed-shell orbitals of the molecule, similarly to the case of a rare gas; and (4) predicted binding energies are on the order of those estimated in UHV but depend on the crystallographic definition of the adsorption site.

As is well known, these calculations raise the question of the relevance of results obtained with a cluster containing a small number of atoms to the case of an extended surface. The latter situation involves effects related to delocalized electrons and to long-range electrostatic forces which may be absent in cluster calculations. For example, the total dipole moment of the water-cluster system leads to work function changes much greater than those estimated from UHV.[118,120] Another aspect specific to the metal/solution interface is that we are in fact interested in the effective interaction of a given molecule in the presence of the other molecules of the liquid. Since the molecule does not form a true chemical bond with the metal, one may expect that the presence of this dense phase will first modify the distribution of the free electrons on the metal surface so that the effective interaction will also reflect this effect. These considerations are clearly beyond the cluster approach. Then what one may do is to try to incorporate some results of these calculations in a model for the interaction at the metal/solution interface. The qualitative indications concerning the repulsion can be included in models of the potential in a rather simple way. In contrast, the attractive contribution is clearly not amenable to simple modeling. However, if one recalls the discussion given earlier on the equilibrium metal-solvent distance, the details of the attractive contribution may be not so important for a weakly bound solvent.

(ii) Metal–Rare-Gas Interaction

The interaction of a rare-gas atom with a metal surface has been studied from first principles by Zaremba and Kohn.[107,108] They have shown that, to a good approximation, the interaction potential $V(Z)$ of an atom located at a distance Z from the metal surface can be written as

$$V(Z) = V_{\mathrm{HF}}(Z) + V_{\mathrm{corr}}(Z) \tag{25}$$

The first term, $V_{\mathrm{HF}}(Z)$, is essentially a short-range repulsion due to quantum exchange between the conduction electrons of the metal and the closed-shell orbitals of the rare gas. It corresponds to a treatment in the Hartree-Fock approximation. Harris and Liebsch[123] have extended the treatment of Zaremba and Kohn in order to include correlations between the metal electrons. They introduced a pseudopotential V^{ps} which they treated by perturbation theory. They have shown that a first approximation is to consider V^{ps} as a local operator. Then, instead of $V_{\text{HF}}(Z)$, they obtain a repulsive term of the form

$$V_R(Z) = \int n(\mathbf{r}) V^{\text{ps}}(\mathbf{r}, \mathbf{R}) \, d\mathbf{r} \tag{26}$$

where $n(\mathbf{r})$ is the electronic density profile of the metal in vacuum, and $\mathbf{R}$ is the coordinate of the atom. Note that they mention that this local approximation may not be sufficient if one seeks the detailed description of $V(Z)$ required for interpreting the helium diffraction experiments for which the theory was initially devised. In our case, however, we may still use it for qualitative purposes.

In Eq. (25), the second term, $V_{\text{corr}}(Z)$, is an attractive contribution related to long-range electronic correlations between the metal and the atom. It corresponds to the presence of dispersion forces as in the van der Waals potential. Zaremba and Kohn have shown that the calculation of $V_{\text{corr}}(Z)$ requires the knowledge of the dynamic response function of the metal and the atomic polarizability. An expansion of the general expression of $V_{\text{corr}}(Z)$ in inverse powers of Z gives as the leading term the van der Waals energy:

$$U_{\text{vdw}}(Z) = C_3/|Z - Z_0^{\text{dyn}}|^3 \tag{27}$$

where the distance Z is measured with respect to the dynamic image plane position Z_0^{dyn}, which, as well as the strength C_3, can be calculated from the theory. The approach of Zaremba and Kohn presents two difficulties: on the one hand, the evaluation of C_3 and Z_0^{dyn} for a real solid is not straightforward. On the other hand, $U_{\text{vdw}}(Z)$ exhibits a nonphysical (artifactual) divergence for $Z \to Z_0^{\text{dyn}}$. Annett and Echenique[124] have derived an alternative expression which avoids these difficulties. The interested reader is referred to their paper for details. The important steps in the calculation are given in the appendix of this chapter. The input parameters for the metal are the surface plasmon frequency $\omega_s = (3/2r_s^3)^{1/2}$ and the static image plane position X_0, which enter the surface plasmon dispersion relation $\omega_s(q)$. For the molecule we need the dipolar oscillator strengths f_{n0} and the related frequencies ω_{n0}. In the formulation of Annett and Echenique, U_{vdw} remains finite at all metal-atom separations while being of course equivalent at large Z to the result of Eq. (27).

Finally, it should be mentioned that while the approach of Zaremba and Kohn considers the effect of bound electrons in the metal, a straightforward application of the formalism of Annett and Echenique is possible only in the jellium model. However, a possible heuristic extension of the formalism simulating the effect of the solid is to introduce the value of X_0 calculated for the solid and an appropriate value for ω_s, through r_s.[114]

(*iii*) *Effective Medium Theory*

An alternative approach to the theory of metal-adsorbate interactions is based on the use of the effective medium theory (EMT) (see the papers of Lang and Norskov,[125,126] for example). EMT is appropriate, in principle, both for chemisorption[125] and physisorption situations.[126] The main idea is that the adsorption energy of the adsorbate in the inhomogeneous electron gas at the surface can be determined from the corresponding energy $\Delta E(n)$ in a homogeneous host of uniform electron density n. To a good approximation, the adsorption position is such that the electron density is equal there to the density n for which $\Delta E(n)$ is minimum. EMT has been compared[125] to first-principle calculations[127,128] in the case of rare gases or simple atoms. The agreement found there supports the use of EMT, which indeed introduces a considerable simplification. To our knowledge, there is no rigorous justification for the use of EMT in a more general adsorption situation. At the metal/water interface, Goodisman found recently[90] that it leads to largely overestimated adsorption energies. Thus, the appropriate use of EMT in this case still awaits clarification.

We close this section with the following remarks: because of the complexity of a first-principles calculation of the interaction of water with an extended surface, the modeling of this interaction at the metal/solution interface is unavoidable for practical reasons. The simplest way to do this is to take metal/rare-gas-like interaction potentials. This is the route followed in most investigations of the metal/solution interface. This was done largely for simplicity. However, given the data (mostly from UHV) on the metal-water bond strength, it does not seem unreasonable—as a first approximation—to model the interaction potential by that of a rare gas, especially in the case of water on silver, for which it seems particularly weak. In addition, the truly important part of the potential for a relatively weakly bound solvent is the repulsive contribution, as discussed above. A final check of the adequacy of a given model potential may consist in verifying that the results for the capacitance show a weak dependence on details of this model potential. We will show later that this is ultimately the case by comparing the results obtained with weak adsorption potentials.

V. GENERAL STRUCTURE OF RECENT MODELS FOR THE CAPACITANCE

The models proposed recently for calculating the capacitance are naturally based on models for the metal, for the solution, and for their coupling. An important aspect in most of these models is that the three-dimensional nature of the interface is discarded, mostly for simplicity. The properties of the interface are assumed to vary only in the direction normal to the metal surface. However, the surface structure of the solid is usually taken into account approximately by the introduction of ionic pseudopotentials averaged in the direction parallel to the interface. A few exceptions are a recent work by Halley and Price,[85] where solvent molecules distributed in a two-dimensional array interact with a three-dimensional jellium, and the work on small metal clusters,[83,84] where the solution is absent. While in other contexts a one-dimensional picture may be considered as an oversimplified view of the interface, especially in the case of solids, the presently available three-dimensional calculations have not introduced features which may lead one to consider one-dimensional models as qualitatively inadequate for describing the capacitance. For example, Halley and Price found that the capacitance curves obtained with the three-dimensional calculation do not differ much from those obtained in the one-dimensional case. This behavior seems related to the fact that the capacitance, as a macroscopic quantity, smooths details of the interfacial structure in the direction parallel to the surface (see, for comparison, the definition of the capacitance in the work on small clusters[83,84]). In another respect, three-dimensional calculations may be required in the future for a more quantitative discussion than the one we intend to present in this chapter.

1. The Metal Side

Most models describe the metal on the basis of the jellium model, in some cases including electron–ion pseudo-potentials. As discussed in Section III.2(iv), the jellium model is used in some models in an indentical manner for solids and liquids.[86,87] In the latter case, the step profile also mimics the ionic profile, whose exact form is still a subject of debate,[63,88,129–131] as discussed previously. Using the step profile as an approximation to the ionic profile for liquids means also that the ionic pseudo-potentials should be uniformly distributed up to the jellium edge. Since the edge of the step profile is the plane of zero adsorption of the electrons, the system is globally neutral. The situation differs from that in the case of solids, where for reasons of electroneutrality the jellium edge is at half an interplanar spacing from the last ionic plane. Technically, most authors use the Smith scheme for describing the properties of the electron gas. Exceptions are the work of Goodisman[90] and of Halley and co-workers.[85,113,135] Concerning the type of metal, the formalism for *sp* metals is often used without great care for most metals, including, for example, mercury, gallium, or noble metals. In the case of mercury, even bulk properties cannot be described well within the simple metal formalism.[132] Gallium is a metal which exhibits peculiar properties (structure factor with a shoulder, density of the solid phase lower than the density of the liquid phase, polymorphism, etc.).[133] In order to reproduce its special behavior, one must consider the pseudo-potential as an operator in the quantum mechanical sense.[134] Finally, for noble metals, *s–d* hybridization is a characteristic feature [Section III.2(v)]. All these aspects are not only a matter of theoretical sophistication, especially when one seeks quantitative agreement between theory and experiment.

2. The Solution Side

In the first models, the solution was described in the same frame as in the classical models of the inner layer. A dielectric film of monolayer thickness was considered in Refs. 86, 98, 106, and 136. Schmickler also used a lattice model for the layer next to the electrode.[137] Besides these monolayer models, a more general description of the solution has been considered, by extending to inhomogeneous systems the theoretical methods introduced for describing the properties of ionic solutions in bulk phase. Beyond the "primitive model" in which the solvent acts as a dielectric continuum, it is now possible to consider some "civilized models" in which the solvent is treated on a microscopic scale. The first civilized model is the treatment of the solution as a mixture of hard spheres with embedded point dipoles (solvent) or charges (ions).[3-5] While this model may provide a first approximation in the case of aprotic solvents, it is known to give a poor description of bulk properties in the case of H-bonded solvents such as water. The statistical mechanics of this system can be solved in the mean spherical approximation (MSA), for which some interesting properties can be obtained analytically. At the interface, the results of the MSA are restricted to small surface charges. In this case, the solution contributes to the capacitance by a constant term (see, for example, Refs. 57 and 58). In the case of an arbitrary surface charge, Schmickler and Henderson have devised a heuristic extension of the MSA.[68]

The MSA has also been used in studying the physisorption of polar molecules on a neutral wall. The MSA predictions have been compared to Monte Carlo simulations.[138] While the structure of the interface is poorly described, the correct order of magnitude of the potential drop is obtained. Some shortcomings of the MSA have been detailed in Refs. 32 and 139. In particular, it has been shown that the MSA cannot reproduce some aspects related to the image potential.

In the bulk phase, the same model for the solution has also been treated with the hypernetted chain approximation (HNC).[140] The HNC has been also used in studying some models in which a more sophisticated description of the solvent is considered. These models may include the polarizability of the molecules or some quadrupolar effects (see, for example, Ref. 141). No HNC results exist for a planar charged interface. In recent work, attempts to mimic this interface by considering a solution in contact with a very large charged sphere have been described.[142] Recently, it has been shown that the HNC suffers from the same shortcomings concerning the image potential as the MSA. Moreover, in the case of a pure polar fluid, the HNC does not give the exact behavior of the profile far from the wall.[143]

3. The Coupling at the Interface

In all the models proposed so far, each side of the interface responds only to changes in the average distribution of the particles on the other side (mean-field coupling). As discussed in Ref. 144, this is equivalent to a treatment of the electron–solution interaction by first-order perturbation theory. Such an approximation will be valid only in the case of weak interfacial coupling.

Since these models are primarily concerned with interfaces without specific ionic adsorption, they do not consider short-range interactions between the metal and the ions in the solution. The metal–ion coupling is thus purely electrostatic and adds nothing to what has been discussed in the case of a charged surface [Section III.3(ii)]. Since the electron density has very small values at the position of the ions, it is sufficient for many purposes to replace the excess charge in the solution by a charged plane.

The models of the coupling are then mainly models for the metal–solvent interaction. In Section IV.3(ii), the molecule–metal interaction in the gas phase was examined. The same ingredients are retained in the models for the metal in contact with a dense phase. The models are usually developed following the example of the physisorption potential, which involves a repulsive part and an attractive one determined separately.

(i) Repulsive Contribution

A very common feature of the models for the coupling at the interface is the incorporation in the electronic energy functional [Eq. (14)] of a repulsion of the metal electrons by the electronic cloud of the solvent molecule. Various forms of this repulsion are considered on semiphenomenological grounds. These potentials or "pseudo-potentials" are characterized by a given shape and strength. The unknown parameters are either determined by fitting to experimental data or left unknown, with their effect on the results being investigated. The simplest simulation of the presence of the solvent is a repulsive potential, for example, a square barrier. The metal electrons are placed in a potential given by $V^{sol}(z) = V_0\theta(z - Z_{sol})$, where the parameter V_0 controls the intensity of the repulsion and where Z_{sol} is the edge of the solvent. This form was used by Schmickler and Henderson[87] and by Halley *et al.*[113] Another model is the barrier of constant slope, $V^{sol}(z) = V_0(z - Z_{sol})$, used also by Schmickler and Henderson.[145] In this case, Z_{sol} is the jellium edge. For a summary of the work of Schmickler *et al.*, see Refs. 146 and 147.

The discreteness of the solvent was taken into account by the introduction of electron-molecule "pseudo-potentials." Badiali *et al.*[86,98,148] and Feldman *et al.*[112] used the Harrison point potential[69] $V^{ps}(\mathbf{r}, \mathbf{R}) = \lambda\delta(\mathbf{r} - \mathbf{R})$ between an electron at $\mathbf{r}$ and a molecule at $\mathbf{R}$, where λ is a parameter controlling the intensity of the repulsion. Halley and Price[85,135] constructed from first principles a neon pseudo-potential in order to simulate the interaction between the closed-shell water molecule and the metal. It contains, in principle, no adjustable parameters. Their results are qualitatively similar to those presented in Section IV.3(ii) for metal-rare gas interaction. In our recent work,[60,114,144] we used a rectangular potential defined by $V^{ps}(\mathbf{R}, \mathbf{r}) = V_0$ for $|\mathbf{R} - \mathbf{r}| \leq R_0$ and equal to 0 otherwise, where R_0 is the radius of the solvent molecule. This form can be viewed as the average of the Harrison point potential over the volume of the molecule.

(ii) *Attractive Contribution*

Attractive contributions are not treated explicitly in the work of Schmickler and Henderson. Attractive contributions to the interaction were considered qualitatively by Kornyshev and Vorotyntsev.[106] Their quantitative effect was included in a simple way in the work of Feldman *et al.*[112] They are also implicit in the recent work of Goodisman[90] based on the effective medium theory. In the work of Halley *et al.*, the attractive forces include dipole-image dipole interactions[113] and constant force,[85,135] or they are built into the pseudo-potential.[85] By using some adjustable parameters, Halley *et al.* fit an estimated binding energy of water on some metals of about 20 kT. In our work, we used the parameter-free van der Waals potential discussed in the previous section. For any reasonable value of the repulsion parameter V_0, the depth of the potential is much smaller. This choice should be appropriate to less tightly bound water (presumably water on silver).

(iii) *Equilibrium Metal–Solvent Distance*

An important point is that the equilibrium configuration of the metal-solvent system results from their mutual interaction; that is, if one considers the effect of the solvent on the metal, one should also consider the reverse effect. A very important point must be borne in mind: the metal-solvent interaction must be determined in a self-consistent way. On the one hand, a solvent molecule in the fluid feels from the metal side an average potential $V_{ms}(Z)$, which is a generalization of the metal-molecule interaction $V(Z)$ defined in Eq. (25) and pertaining to an isolated molecule.[144] On the other hand, the electron gas responds to the presence of the solvent, so $V_{ms}(Z)$ must be a self-consistent potential.

Although $V_{ms}(Z)$ includes some effects of the other solvent molecules and is possibly charge-dependent, it exhibits the same kind of variation with Z as $V(Z)$. In other words, $V_{ms}(Z)$ does not behave like a pure hard wall potential; on the contrary, it acts as a "soft wall" just like $V(Z)$. Nevertheless, by using some appropriate criterion, we can associate with the soft wall $V_{ms}(Z)$ an equivalent hard wall from which it is possible to define a minimum metal-solvent distance as discussed in Section IV.2. In our work the rigid wall is introduced just as an intermediate quantity in the calculations. However, since the rigid wall concept is commonly used, hereafter our discussion will be in terms of the position of the rigid wall d, or, equivalently, in terms of the metal-solvent distance of closest approach.

Within this representation of the interaction, it is clear that the position of the rigid wall cannot be arbitrary, given the requirement of force balance at the interface. This can be visualized by the following thought experiment. Suppose that the interaction is tuned gradually by varying the strength of the repulsion from 0 to its actual value, say, V_0. In a model where the wall is fixed, as a result of increasing repulsion from the molecules, the electronic profile in the metal will adjust until it reaches its final value $n_{V_0}(z)$. In the real physical situation, increasing V_0 will also increase the repulsion felt by the solvent molecules. Thus, they will move away from the metal surface, with a subsequent decrease of the actual repulsion between the metal and the solvent. At equilibrium, the final value of the electronic profile will certainly be different from $n_{V_0}(z)$ for the previous case. This means that considering the position of the wall as fixed with respect to the metal surface will lead to the interface being overconstrained.

Of course, attractive forces also play an important role in the determination of the equilibrium (together with the pressure of the fluid). In the case of chemisorption, the solvent molecules are tightly bound to the metal by chemical bonds, and it is reasonable to assume that the metal–solvent distance is fixed. For the interfaces considered in this chapter, however, some experimental evidence favors an interaction more characteristic of physisorption. Then the average metal–solvent distance must be considered as a quantity which must be calculated self-consistently with the other properties of the interface. Taking the boundary for the solvent as fixed to the jellium edge, as Schmickler and Henderson do, cannot be justified *a priori*. This assumption will have serious consequences at a charged interface: in this case, the variation of the metal–solvent distance with charge will eventually lead to an analysis of capacitance curves which will be qualitatively different from that for the case of a fixed metal–solvent distance.

This is the important point on which the models differ. In most of them, the particles in the solution are forced to occupy only one half-space limited by a rigid wall simulating the metal surface. A central question then is the precise position of this wall at each surface charge. If the position of the rigid wall is fixed, the only remaining problem is to determine the quantity X_0 in the presence of the solution. X_0 contains then *all the effects of the metal* (except a much smaller influence of the electron tail on the polarization). In the other case, the position of the wall and X_0 should be determined within the same model of the interface.

In practice, the coupling introduces additional terms in the metal energy functional [see Eq. (14)], including terms due to the ions (electrostatic) and to the solvent (effect of the solvent polarization and short-range repulsion). The effect of the electron spillover in the solution is important with regard to the polarization,[87,148,149] in contrast to the reverse effect. This point, which has been checked explicitly, may be understood in terms of the energies involved. Most of the electrons in the interfacial region experience an average effective potential per electron of the order of an electron volt ($\sim 40\ kT$). This is much larger than the effect of the electric field due to the dipoles. In contrast, the coupling of one dipole with the tail of the electronic profile introduces an additional energy that is comparable to kT and is hence a significant part of its mean kinetic energy.

Two different cases depending on whether the metal–solvent distance is maintained fixed or not will now be discussed.

VI. MODELS WITH FIXED METAL–SOLVENT DISTANCE

Shortly after the work of Badiali *et al.* appeared,[148] Schmickler and Henderson (SH) proposed a similar model[87] coupling the jellium description of the metal to the ion–dipole model for the solution. In addition to some technical details (discussed, for example, in Refs. 87, 67, and 150), the major difference between the two approaches lies in the position of the hard wall for the solution. While this quantity appeared as a crucial parameter in the work of Badiali *et al.*,[148] SH always set the position "d" of the hard wall at the jellium edge, independently of charge. It has been mentioned previously that unless the solvent molecules are tightly bound to the ionic skeleton by chemisorption, there is no physical reason which justifies this assumption. In addition, the jellium edge is located at one half interplanar spacing from the last ionic plane in the metal. In the choice of SH, it is implicit that the distance d is determined by the interaction of the solvent with the metal ions. The value of d should then be linked with some characteristic property of the ion rather than with the jellium edge. Some correlations between the interplanar spacing and the diameter of the screened ion indeed exist, but they are not strictly equal.

In Ref. 145, Schmickler and Henderson considered a possible variation of d with charge. By assuming that the wall position is such that the electron density $n(z = d)$ (which depends on charge) at the position d takes the value $n(z = 0)$ at the jellium edge for the uncharged surface, they claimed from the results of this model that such a variation is a second-order effect. It is not clear whether or not SH repeated the whole (self-consistent) calculation with this new assumption for d. The hypothesis that $n(z = d)$ should be equal to the density at the jellium edge independently of charge cannot be justified. As detailed below, any reasonable model for the metal-solvent interaction leads to a variation of d with charge much more pronounced than that estimated in the work of SH. More recently, in a comment[151] of the analysis made by Kim *et al.*[152] of the SH approach, Schmickler and Henderson estimated that the maximum variation of d should be about 0.2 Å. In fact, this is in line with our results. Such a seemingly small variation can have a very significant effect on the charge-capacitance curve and on the influence of the nature of the electrode.

Since the metal-solvent distance is fixed in the SH approach, the metal contributes to the capacitance by an amount $1/C_m = -4\pi X_0$. Then the important parameter is the bulk electronic density of the metal. SH found that at the pzc, large inner-layer capacitances correlate with large electron densities.[145,146] While the effect of the electron density via X_0 is indeed the expected one (see the discussion in Section III.4), the apparently satisfactory results of SH for some *sp* metals would mean that the distance d at the pzc is almost the same for all these metals. There is no obvious explanation for this behavior [in the work of Badiali *et al.*[86] the sequence $C_i(\text{Hg}) < C_i(\text{In}) < C_i(\text{Ga})$ was obtained largely from the differences in the ionic radii]. Given the large theoretical uncertainties in the description of liquid Hg and Ga, a definitive answer to this question cannot be given now. A discussion on a qualitative level will be made in the next section.

The MSA treatment of the hard-sphere ion-dipole mixture gives a constant contribution of the solution to the capacitance. Then the sole variation of X_0 is far from being sufficient for predicting the correct behavior with charge. This point was recognized by SH, who devised an extension of the MSA result mainly through a Langevin distribution for the polarization.[145] While the curves exhibit now a pronounced hump due to the solvent, agreement with experiment is still not satisfactory. In our opinion, this is largely due to the neglect of the additional charge dependence related to the variation of the wall position with charge.

Besides studying the capacitance, Schmickler and co-workers used the jellium-ion-dipole model to investigate other interfacial properties. They found that this model, and particularly the improved description of the electrode, proved to be useful for understanding some experimental observations (see, for example, the review by Schmickler and Henderson[68] and Ref. 153).

VII. MODELS WITH VARIABLE METAL–SOLVENT DISTANCE

1. Determination of the Distance of Closest Approach

The importance of the variation of d with charge has been recognized by several authors. This point was discussed on a formal level by Kornyshev *et al.*[154] A very simple estimate of this effect based on the assumption of a linear variation was made by Badiali *et al.* in Ref. 86. This problem was discussed again qualitatively by Kornyshev and Vorotyntsev,[106] and quantitative estimates based on their ideas were made in the work of Feldman *et al.*[112] These studies were later continued by Halley and co-workers.[113,85,135] More recently, our own work has been mainly devoted to the study of this effect[114,144] and its consequences for the analysis of charge-capacitance curves for various metals[60] and solvents.[155] These studies will not be detailed here. Only some aspects will be discussed using the example of our own work. Among others, the following two important conclusions were reached in these studies:

1. The metal-solvent distance of closest approach, d, varies with the charge σ. This conclusion follows from the force balance condition at the interface as discussed in Section V.3(iii). A good discussion of the qualitative aspects of this effect has been presented by Kornyshev and Vorotyntsev.[106] A more systematic analysis was later made in the work of Halley *et al.* and in our work. The interfacial electric field affects the force balance in two different ways. On the one hand, the metal-solvent molecule potential will change with charge as a consequence of the variation of the electron density profile of the metal: at positive charging, the electrons are pulled back into the metal. Then a solvent molecule at a given position feels less repulsion from the electron cloud. The converse

effect occurs at negative charging: a larger spillover of the electrons from the surface means that a stronger repulsion is felt by the solvent molecules. On the other hand, coulomb forces between the net charge on the metal surface and the opposite charge in the solution will result in an electrostatic pressure P^{el} at the interface, similar to that existing between the plates of a classical condenser. At the interface, the motion of the solvent molecules is correlated to that of the ions; one may picture this by considering an ion and its solvation shell. Accordingly, the molecules will also be subject to this pressure, and thus a compression of the interfacial region may occur. Note that this effect is made possible because the metal behaves like a soft wall for the molecules. Thus this effect may exist even if the molecules are considered as rigid. Without invoking a solvation process, a quantitative account of this intuitive view of the interface based on a statistical mechanical derivation of the electrostatic pressure of an assembly of polar molecules subject to an applied field will be given below [see Eq. (28) or (30)].

2. The overall shape of the $d(\sigma)$ curve does not depend on details of the metal-solvent interaction. This second important conclusion, made by several authors, is valid provided one is dealing with physisorption. This point results from the fact that the force balance is determined by basic physical properties of the interface. At large enough magnitude of the charge, a decrease in the metal-solvent distance is expected from the dominant effect of the electrostatic pressure P^{el}, which varies as σ^2. This behavior is similar to the contraction of a classical condenser upon charging. At moderate charges, a very important feature is the asymmetric behavior of $d(\sigma)$ with respect to the sign of the charge. At positive charging of the metal, a decreasing repulsion of the solvent molecules by the electron cloud enhances even further the contraction effect due to the electrostatic pressure. The value of d will always decrease from its value $d(\sigma = 0)$ at the pzc. At moderate negative charging, the electron spillover increases and so will the value of d. This behavior is the continuation of the trend on the positive side of the pzc. However, this increase of d competes now with the decrease due to P^{el}. One thus expects the value of d to first reach a maximum when the two opposite effects exactly cancel, and later to decrease again, due to the ultimate trend imposed by P^{el} at high enough negative charges.

This behavior has been reported by several authors. Kornyshev and co-workers as well as Halley *et al.* define the metal-solvent distance d as the value at which the surface energy, including the interaction of the metal with the solution, is minimum. Such a procedure means that the total metal-solution interaction energy, defined as the difference between the energy of the metal in the presence of the solution and that in its absence, is minimum with respect to d. In our own work, a slightly different approach was followed. By using a statistical mechanical method based on the use of the Gibbs and Bogolioubov inequality,[156,157] we defined a distance of closest approach of the solvent molecules to the metal surface precisely from a force balance condition[144,114]:

$$P^{\text{ref}} + 2\pi\sigma^2 + \int_d^\infty dZ\,\rho(Z-d)\,\partial V_{\text{ms}}(Z)/\partial Z = 0 \tag{28}$$

where P^{ref} is the bulk pressure of the fluid, $\rho(Z-d)$ is the density profile for solvent molecules at the charged interface, and $V_{\text{ms}}(Z)$ is the metal-solvent potential. In deriving Eq. (28), we made use of the so-called contact theorem.[158] This theorem tells us that

$$\sum_\alpha kT\rho_\alpha(z = d_\alpha) = P^{\text{ref}} + 2\pi\sigma^2 \tag{29}$$

where T is the temperature, and ρ_α is the value of the density profiles ($\alpha \equiv$ ion or dipole) at the wall position d_α. The left-hand side of Eq. (29) is simply the momentum transfer to the wall. Since $V_{\text{ms}}(Z)$ varies sharply compared to $\rho(Z-d)$, Eq. (28) suggests that we need essentially values of $\rho(Z-d)$ near $Z = d$. Then from Eq. (29) we may replace $\rho(Z-d)$ by the corresponding profile at zero charge, $\rho_0(Z-d)$, plus a correction term which is of the order σ^2.[114] Hence, we can write Eq. (28) as

$$P^{\text{ref}} + 2\pi\sigma^2/\varepsilon_{\text{eff}} + \int_d^\infty dZ\,\rho_0(Z-d)\,\partial V_{\text{ms}}(Z)/\partial Z = 0 \tag{30}$$

where the parameter ε_{eff} takes into account this first deviation of $\rho(Z-d)$ from the value at zero charge $\rho_0(Z-d)$.

Equation (28) or (30) shows that the position d of the wall is such that the total force per unit area exerted by the potential $V_{\text{ms}}(Z)$ on the fluid is balanced by the total pressure of the fluid ($P^{\text{ref}} + P^{\text{el}}$), where $P^{\text{el}} = 2\pi\sigma^2$ if Eq. (28) is used or $P^{\text{el}} = 2\pi\sigma^2/\varepsilon_{\text{eff}}$

otherwise. These equations justify the intuitive introduction of P^{el} discussed above.

Our definition of the distance of closest approach has a simpler physical interpretation than that defined from the minimum of the total surface energy. The distance d is now obtained from a force balance equation directly related to the metal-solvent molecule potential $V_{ms}(Z)$. The physical meaning of d can be better visualized if we note that the value of d is such that $V_{ms}(d)$ is roughly equal to kT [if the repulsive part of $V_{ms}(Z)$ is sufficiently steep, the precise value of the pressure term is not crucial for the value of d]. $V_{ms}(d)$ is an estimate of the energy barrier that a particle of the fluid may climb, and d then has the meaning of an average turning point for the solvent molecules.

As in Eq. (25), $V_{ms}(Z)$ is the sum of the dispersion term U_{vdW} and a repulsive term given by an expression similar to Eq. (26) for V_R[144]:

$$V_{ms}(Z) = U_{\text{vdW}}(Z) + \int n(z) V^{ps}(\mathbf{r}, \mathbf{R})\, d\mathbf{r} \tag{31}$$

However, $n(z)$ is now calculated for the metal in the presence of the solution. This implies that the calculation of d and $n(z)$ must be performed self-consistently[144,114] since the electronic profile is now determined from a functional of the energy $E[n(z)]$ which contains a coupling term such as

$$U^{sol}[n(z)] = \int n(z) V^{ps}(\mathbf{r}, \mathbf{R}) \rho(Z - d)\, d\mathbf{r}\, dZ \tag{32}$$

At each surface charge, we can calculate $d(\sigma)$ and $X_0(\sigma)$ and later evaluate the contribution of the metal to the capacitance. Some results for the metal-solvent distance will now be presented.

The variation of d with charge has been studied quantitatively by Feldman *et al.*,[112] by Halley and co-workers,[113,135,85] and by ourselves.[114] Typical $d(\sigma)$ curves are given in Fig. 11. A comparison of the curves on a quantitative level is useless since they do not correspond exactly to the same physical quantity. They have also been obtained for different metals and with different models for the metal and for the coupling. However, they have very similar shapes. In the calculation of Halley and Price, the final decrease

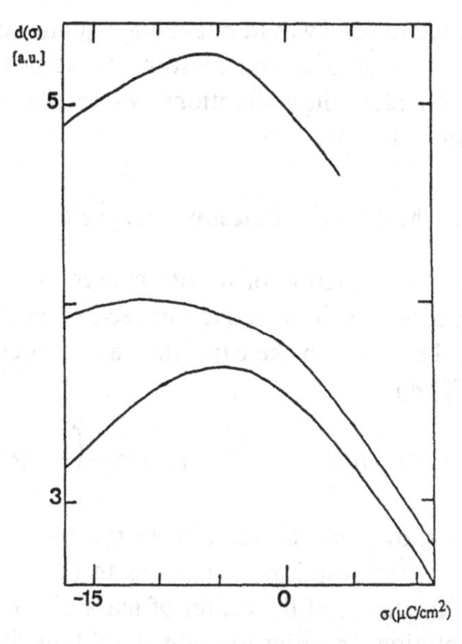

Figure 11. Results for the variation of the distance of closest approach with charge. From top to bottom: Results of Feldman *et al.*[112] (r_s = 2.7), of Halley and Price[85] (r_s = 2.59), and from our work on silver[114] (r_s = 2.64). A solvent radius (R_0 = 1.5 Å) has been added to the results of Feldman *et al.* in order to have the same definition of d for the three curves.

of d on the negative side of the pzc seems to be shifted to more negative charges. A possible explanation of this feature is that their metal-solvent potential has a very steep variation with distance at negative charges. In our work, the magnitude of d depends on the strength V_0 of the metal-solvent repulsion. However, we have checked that the variation of d with charge is virtually independent of V_0.[114]

Beyond these remarks, the observation that within different models the minimum metal-solvent distance exhibits a common behavior with charge is strong support for the previous analysis of the underlying physical mechanism. This observation is important in view of the unavoidable approximations involved in the models of the metal-solvent interaction. Most of these models may appear to be crude approximations of reality. However, it appears that more sophisticated models would not change qualitatively the conclusions on the variation of the metal-solvent distance with charge, provided one considers only situations where the metal-solvent interaction is indeed weak.

2. Effect on the Calculated Capacitance

The relevance of the variation of d with charge in the calculation of the capacitance will be discussed in this section. For this purpose, we recall Eq. (11), in which the capacitance is expressed in terms of $X_0(\sigma)$ and $X_s(\sigma)$:

$$1/C = 4\pi[X_s(\sigma) - X_0(\sigma)] - 4\pi\, \partial/\partial\sigma \int_{-\infty}^{+\infty} P(z)\, dz \tag{11}$$

This expression shows that in addition to the contribution of the polarization $P(z)$, the solution contributes to the capacitance via the variation with charge of the center of mass of $\rho_s(z)$, the excess charge in the solution. In order to understand how the distance of closest approach d for the *solvent* indeed plays a role in the above expression, let us write $\rho_s(z)$ as

$$\rho_s(z) = -\sigma\delta[z - (d + R_0)] + \tilde{\rho}_s(z) \tag{33}$$

In this expression, a planar charge distribution peaked on the plane $z = (d + R_0)$, where R_0 is the molecular radius, has been isolated. The term $\tilde{\rho}_s(z)$ is the remaining part of $\rho_s(z)$, so that the net charge corresponding to $\tilde{\rho}_s(z)$ vanishes. Then, by writing Eqs. (6) and (10) for the solution side and using the formal splitting in Eq. (33), we get

$$X_s(\sigma) = d + R_0 + \sigma\, \partial d/\partial\sigma + \tilde{X}_s(\sigma) \tag{34}$$

where $\tilde{X}_s(\sigma)$ corresponds to $\tilde{\rho}_s(z)$. We can now rewrite Eq. (11) as follows:

$$\begin{aligned} 1/C = {} & 4\pi[d - R_0 + \sigma\, \partial d/\partial\sigma - X_0(\sigma)] \\ & + 4\pi\left[2R_0 + \tilde{X}_s(\sigma) - \partial/\partial\sigma \int_{-\infty}^{+\infty} P(z)\, dz\right] \end{aligned} \tag{35}$$

At this level, this expression results only from the introduction of a particular splitting of $\rho_s(z)$ in the exact original expression, Eq.

(11). Its structure can be understood by comparing it to the corresponding expression in classical models. In these models, the inner-layer capacitance C_i corresponds to a monolayer of solvent between two charged planes. The first one is the metal surface, so $d = R_0$. An ideal behavior of the metal being further assumed, $X_0(\sigma) = 0$. The first term in brackets then vanishes in the traditional approach. We then *define* the metal contribution precisely as this term[114,60]:

$$1/C_m = 4\pi[d - R_0 + \sigma\, \partial d/\partial\sigma - X_0(\sigma)] \tag{36}$$

and the remaining part of $1/C$ as $1/C_s$:

$$1/C_s = 4\pi\left[2R_0 + \tilde{X}_s(\sigma) - \partial/\partial\sigma \int_{-\infty}^{+\infty} P(z)\, dz\right] \tag{37}$$

At the present stage these expressions are simple definitions, since only the total capacitance is a measurable quantity. C_m defined in this way has no analogue in the classical analysis of the capacitance. Its definition means also that C_m is not a purely metallic contribution since it contains the effect of the metal-solution coupling through d and its variation with charge.

Now we come to the interpretation of C_s. In the classical models and in the absence of specific ionic adsorption, the excess charge in the solution at high enough concentration is assumed to be distributed on the outer Helmholtz plane, located at $(d + R_0)$. Then it follows from the definition of $\tilde{\rho}_s(z)$ in Eq. (33) that $\tilde{X}_s(\sigma) = 0$. If one further assumes that $P(z)$ is limited to a monolayer between $d - R_0$ and $d + R_0$ and splits it as usual into orientational and distortional contributions[1,35,38]:

$$P(z) = P^{\text{pol}}(z) + P^{\text{or}}(z) \tag{38}$$

and makes the usual assumption $P^{\text{pol}}(z) = (\varepsilon - 1)4\pi\sigma/\varepsilon$, where ε is the dielectric constant at fixed orientation of the dipoles, a simple calculation leads to the following result:

$$1/C_s = 4\pi\left[2R_0 - \partial/\partial\sigma \int P(z)\, dz\right] = 4\pi e/\varepsilon - 1/C_{\text{dip}} \tag{39}$$

where $e = 2R_0$ is the thickness of the monolayer, and C_{dip} is associated with the permanent dipoles. Then C_s defined in this way is exactly equivalent to the inner-layer capacitance considered in the traditional approach, as discussed in Section II.3.

An important consequence of Eqs. (35)-(37) and (39) concerns the analysis of the capacitance at high field. From the experimental curves (Fig. 1), we have seen that $C_i(\sigma)$ exhibits a plateau at sufficiently negative charge. The corresponding capacitance is traditionally denoted $K_{\rm ion}$. It is usually accepted that $K_{\rm ion}$ corresponds to saturated orientational polarization. Its value is given by the term $4\pi e/\varepsilon$ in Eq. (39). It follows from Eqs. (36) and (37) that this limit is actually

$$\lim_{\sigma \ll 0}(1/C_i) = 4\pi[d(\sigma \ll 0) - R_0 - X_0(\sigma \ll 0)] + 4\pi e/\varepsilon \quad (40)$$

[in contrast to the metal–vacuum case, the value $X_0(\sigma \ll 0)$ indeed saturates due to the effect of the solvent through the decrease of d]. From Eq. (40), it appears that there exists in the high-field limit an extra contribution to $K_{\rm ion}$ which depends on the nature of the metal, at least through X_0. One may say that this extra term is somewhat incorporated in the inner-layer thickness e. We may consider that Trasatti[30] and Valette[22] have made a step in this direction in their interpretation of $K_{\rm ion}$ for solids although their discussion is only based on steric effects. We should say now that both d and X_0 may depend on the orientation of the surface. The presence of the first term on the right-hand side of Eq. (40) will have an important consequence on the estimation of the ratio e/ε. Instead of equating $4\pi e/\varepsilon$ to the reciprocal value of the capacitance at the plateau, this experimental value should be used now for the whole expression in Eq. (40). Since there is no reason why the extra contribution should be negligible, this will have serious consequences on the value of e/ε and hence on the estimation of the dielectric constant ε.

Coming back to the splitting of $1/C$ given in Eqs. (35)-(37), the first task for the theory is to calculate C_m by a model for the metal and for the coupling. Without going into the details of these models, we present in Fig. 12 three determinations of this quantity. They include our work[114] on Ag and the "3D jellium" calculation of Halley and Price[85] with r_s corresponding to the value for Cd. We have also plotted a curve from the data of Feldman *et al.*[112] with r_s corresponding to the value for Hg. Since the original curve

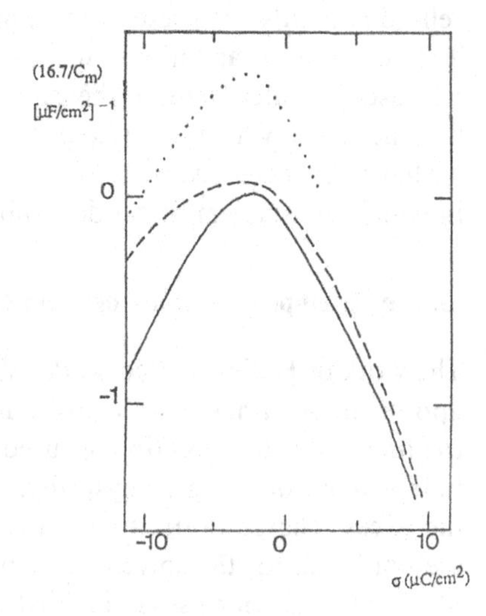

Figure 12. Examples of $(1/C_m)$ curves. ——, Ag[114]; - - -, Cd[85]; · · ·, Hg, estimated from the data given in Fig. 3 of Ref. 112. The number 16.7 is a conversion factor from atomic units to $(\mu F\ cm^{-2})^{-1}$.

of Halley and Price contains the contribution of the solution given by the MSA, we have made the appropriate correction so that the plotted curve corresponds to our definition of $1/C_m$. It differs from the original one only by a constant term. These curves are in general inverted parabolas, with a maximum at a negative charge. Their overall shape results mainly from the variation of $d(\sigma)$, slightly reduced by the monotonic variation of $X_0(\sigma)$. One may also observe that the curve corresponding to the results of Feldman *et al.* is very similar to our curve for Ag. Thus, an agreement on the shape of C_m exists in the literature. If d is maintained fixed, the metal contribution is, of course, totally different, since it is restricted to the variation of X_0 with σ. In that case, it exhibits a monotonic variation with charge.[114]

Some important points may be retained from this discussion: (1) there exists in the capacitance a contribution C_m given in Eq. (36) with no analogue in the classical models; (2) this contribution is largely determined by the variation with charge of the metal-solvent distance; (3) the overall behavior of this distance can be inferred from simple physical considerations in the case of weak metal-solution coupling; and (4) a new analysis of K_{ion} is needed.

Once C_m is known, the next step is to obtain $P(z)$ and $\tilde{X}_s(\sigma)$. *A priori*, this is an intricate problem since the two sides of the interface are coupled, and all these quantities must be calculated self-consistently. At present, two approaches have been proposed. The first one is ambitious, the challenge being to make a fully microscopic calculation of the capacitance as was attempted in the first models in which d was kept fixed.[148,87] The second one, which we have advocated, consists in a new analysis of experimental data, in which the semiempirical determination of C_s is considered.

3. An Attempt toward a Complete Calculation of the Capacitance

The work of Halley and co-workers[85,113,135] corresponds to the first approach. A variable wall position is considered, and an MSA treatment for the solution is used. Then, $\tilde{X}_s(\sigma)$ is a constant, independent of charge. Physically, this means that the centroid of the excess charge in the solution is assumed to move rigidly with the hard wall for the solvent. This behavior is the direct extension of the classical view, assuming that the ions in the outer Helmholtz plane are always separated from the metal by a layer of solvent.

After their work using a three-dimensional calculation, which was the fourth reported in a series of papers on the capacitance, Halley and Price[85] reached the rather pessimistic conclusion that the models are far too simple and that efforts in many directions are still required. While this is certainly true for quantitative purposes, we may argue against such a conclusion in several respects. Firstly, the use of the MSA immediately prevents predicting the behavior of the capacitance away from the pzc. Secondly, the law of inverse addition of capacities connected in series is the reason for the inherent difficulty of a direct calculation of the total capacitance as in Eq. (35). Since these capacities (say, C_m and C_s) are of opposite sign and comparable magnitude, the equivalent capacity (say, C_i) can be extremely sensitive to relatively minor variations of the components in the series. It is somewhat regrettable that this limitation, which applies also to the classical models, is seldom discussed in the literature. For example, in the classical models, reasonable values of C_i result very often from a precarious balance between K_{ion} and C_{dip}.[32] Just as an example, in the model proposed by Parsons,[38] a slight change in the ratio of the dipole moment of the clusters to that of the monomers affects so dramatically the balance between K_{ion} and C_{dip} that the capacitance may diverge or become negative. This discussion is also connected with the so-called "Cooper–Harrison catastrophe."[159,160] This means, in general, that the direct calculation of C may require a high accuracy on each term in Eqs. (35)-(37). Such an accuracy is dictated by the mathematical form of the equation for C and does not correspond to a physical requirement. This also means that it would be better to deal with inverse capacitances. From our point of view, this difficulty makes direct calculation of C not possible at present.

4. Basis for a Semiempirical Approach

The alternative route indicated in Section II.5 avoids the problem that we have just discussed. As noted in Section V.3(iii), $P(z)$ has a rather small effect on the calculation of the electronic profile although the converse is not true. Thus, when focusing on the calculation of $X_0(\sigma)$ and $d(\sigma)$, the effect of $P(z)$ can be neglected. More generally, we have checked that C_m given by Eq. (36) can be calculated independently of the state of polarization of the solution. Then, with a given metal-molecule pseudo-potential $V^{ps}(\mathbf{r}, \mathbf{R})$ and the expression for the van der Waals potential given in the appendix, we calculate the total metal-solvent potential $V_{ms}(Z)$ from Eq. (31). At the same time the coupling term given by Eq. (32) is introduced in the density functional $E[n(z)]$ [see Eq. (14)], and Eq. (30) is solved with a given approximation of $\rho_0(Z-d)$. This procedure, which requires a self-consistent treatment of d and $n(z)$, gives C_m. From C_m and C_i deduced from experiment after correction for the Gouy-Chapman contribution (or at high enough concentration), we obtain a semiempirical estimate of C_s from the relation

$$1/C_s = 1/C_i - 1/C_m \tag{41}$$

This relation is the basis of the new analysis of the capacitance that we have recently proposed.[60,155] The behavior of C_s obtained in this way can also be compared to the result of model calculations based on Eq. (37). This procedure may finally be used as a check of hypotheses concerning $\tilde{X}_s(\sigma)$. A route similar in concept to the one we propose has been presented by Molina *et al.* in Ref. 161. In this work, they estimated C_m from the knowledge of C_i and a model for C_s.

VIII. A SEMIEMPIRICAL APPROACH

In Refs. 60 and 155, we proposed an analysis of the capacitance based on the ideas presented above. The case of silver will be discussed first. For this metal, the present model of the electronic structure allows a reliable calculation of C_m. In addition, well-established experimental results for monocrystalline silver electrodes are available.

1. The Case of Silver

In Ref. 114, we estimated C_s by using Eq. (41) and the experimental C_i for Ag/NaF[21,23] and Ag/KPF_6.[22] The C_s curves are given in Fig. 13. We first observe a striking similarity of the curves for the two interfaces, more pronounced than in the case of C_i curves. If $C_i(\sigma)$ and $C_s(\sigma)$ differ in magnitude both are simple bell-shaped curves, at least in the range that we considered, $-15\ \mu\text{C cm}^{-2} < \sigma < +15\ \mu\text{C cm}^{-2}$. However, an important difference is that while the maximum in $C_i(\sigma)$ is at a slightly positive charge (NaF) or almost

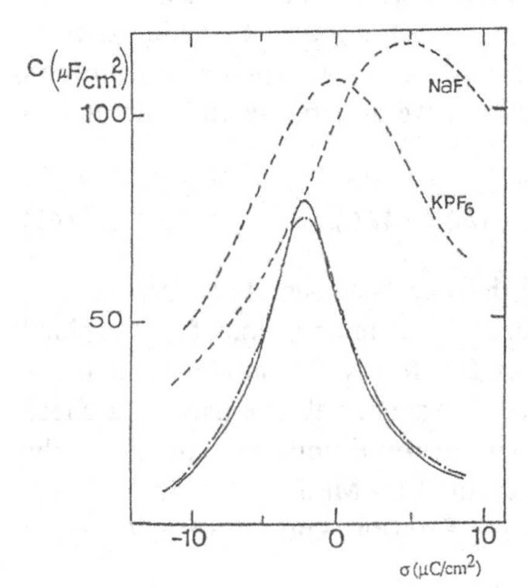

Figure 13. $C_s(\sigma)$ curves for Ag(111)/NaF (—·—) and Ag(110)/KPF_6 (——). The experimental C_i (---) are given for comparison.

at the pzc (KPF_6), the maximum in the corresponding C_s curves is at a negative charge σ_s in both cases.

In the case of silver, for which C_m was calculated, these C_s curves are a direct output of the method. They involve no approximations other than those made in the calculation of C_m. Since they have been obtained independently of any assumption about solvent polarization, they may be considered as the first semiempirical determination of the contribution of the solvent to the capacitance. Since C_s determined in this way cannot be compared directly to experiment, the interesting question is whether or not it helps in rationalizing the role of the solvent. Indeed, we can consider C_s as a quantity from which metal effects have been separated to a large extent. Before discussing $C_s(\sigma)$, it is interesting to consider what may be learned from other interfaces.

2. Other Interfaces

For other interfaces, we should in principle calculate C_m and use again Eq. (41) with the respective experimental C_i curves. For the case of gold, mercury, and gallium that we will discuss here, we adopted a different approach for two reasons. First, there are theoretical difficulties encountered with liquid metals (Hg, Ga), as we have mentioned several times. In addition, a good description of the electronic properties of gold is not available even in the bulk phase. The second and main reason is that we ultimately found out that it is even unnecessary to recalculate C_m for each metal, at least at the present stage of the theory, as we detail it now.

If one compares C_i curves for the same solvent but with different metals, still having in mind the splitting into C_m and C_s, the most natural hypothesis is to consider that C_s is the same in all cases, at least as a first approximation. This assumed "universal" behavior[60] is the natural consequence of the way C_s was *defined*, since the effect of the coupling via the distance d and its variation with charge is included in C_m. We stress that the model we used for the coupling does not contain all possible sorts of coupling. For example, it does not include, as it should in principle, an interaction which favors at the pzc a preferential orientation of the water dipole with the negative end toward the metal. We did not include in the calculation of C_m such a "residual interaction,"[1,35] since it adds truly negligible terms in the electronic energy functional. For the same reason as discussed for the effect of solvent polarization in Sections V.3(iii) and VII.4, such an additional interaction involves energies that are negligible compared to the typical energy experienced by the electrons. On the other hand, since this energy is comparable to that experienced by a dipole, this effect will certainly be required in a model for C_s. It is then known from the early models[1,35,159] that the main effect of such a residual interaction is to shift the maximum of the $C_s(\sigma)$ curves, depending on the assumed preferential orientation at the pzc. Then, except for the shift due to this residual interaction, C_s should reflect mainly solvent properties. In this case, since the difference between interfaces is due to C_m, one may estimate C_m for the interface between a metal M′ and a solution S′ if it is known for another interface between a metal M and a solution S by using

$$1/C_m(\mathrm{M'/S'}) = 1/C_m(\mathrm{M/S}) + [1/C_i(\mathrm{M'/S'}) - 1/C_i(\mathrm{M/S})] \quad (42)$$

which is obtained from Eq. (41) by taking C_s as the same for the two interfaces. Knowing C_m from the calculation for Ag, we obtained the respective curves for Hg and Ga.[60] Figure 14 also includes results for Au. We see that while the associated C_i curves in Fig. 1 were dissimilar, all the C_m curves have the shape of an inverted parabola, asymmetric with respect to zero charge, which resulted from the calculations for silver.

A tentative explanation of this behavior of C_m from metal to metal may be advanced from the following physical arguments. We first recall Eq. (36): $1/C_m = 4\pi[d - R_0 + \sigma\, \partial d/\partial\sigma - X_0(\sigma)]$. Since the curves in Fig. 14 have a similar shape, it is sufficient to restrict the discussion to the pzc. Then different values of $1/C_m$ are associated with differences in $d - X_0$. From Fig. 14 we find that the maximum variation of C_m at the pzc corresponds roughly to a difference of about 0.2 Å in the quantity $(d - X_0)$. In our model the values of d and X_0 are coupled. However, for not too different values of d, X_0 depends mainly on the density of conduction electrons. Differences in d result now from differences in the metal-molecule interaction potential $V_{ms}(Z)$, which in turn may be

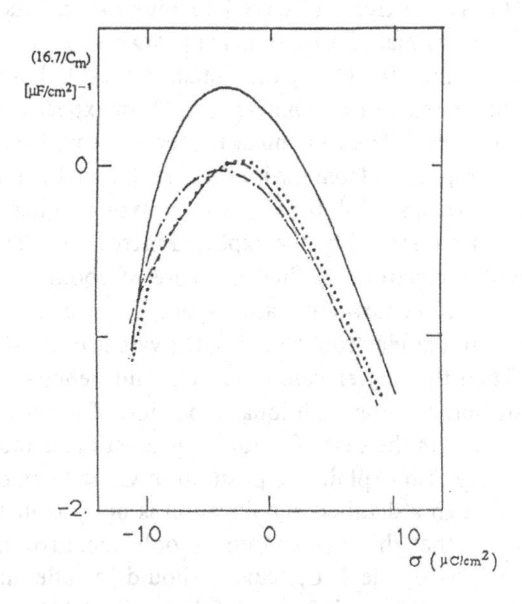

Figure 14. $1/C_m$ curves for different metals: ——, Hg; —·—, Ga; · · ·, Au(210); - - -, Ag(111). The number 16.7 is a conversion factor from atomic units to $(\mu F\ cm^{-2})^{-1}$.

different if we are considering a solid versus a liquid. In our model, $V_{ms}(Z)$ results [Eq. (31)] from the repulsive term V_R and the attractive contribution U_{VdW}. As discussed in Section III.2(iv), V_R for solids stems mainly from the free electrons while it should also include the effect of the ionic cores in the case of liquids. Insofar as U_{VdW} is concerned, its value in the asymptotic region depends weakly on the nature of the metal, but its value increases with the electron density of the metal at short distances.[107,108] Since Au and Ag are two solids with very similar electron densities, they are expected to have very close values of X_0 and d. Then similar $1/C_m$ for Ag and Au are expected, just as observed in Fig. 14. If we compare now two liquids, Hg ($r_s = 2.66$) and Ga ($r_s = 2.18$), the discussion is a little more complicated. From the work of Lang and Kohn,[33] we expect a larger value of X_0 by about 0.1 Å for Ga since its electron density is much higher than that of Hg. Then if the values of d were similar, $(d - X_0)$ should be smaller for Ga than for Hg. This is the trend observed in Fig. 14. In order now to reproduce the estimated variation from Hg to Ga, d should be smaller for Ga than for Hg by the remaining 0.1 Å. Do we expect such a slight decrease of d from Hg to Ga? An expected larger van der Waals attraction due to a much higher electron density for Ga and a weaker repulsion from the ions due to the smaller ionic radius (0.6 Å for Ga versus 1.1 Å for Hg) are in favor of a decrease in d. However, this is opposed by the expected increase of the repulsion due to the free electrons. A final decrease of about 0.1 Å is then not surprising. If we now compare Au or Ag to Hg, since all these metals have similar electron densities, the values of X_0 should also be close. Then the higher value of $1/C_m$ and hence of d for Hg may be attributed to the additional repulsion of the molecules by the metallic ions in the case of a liquid, as mentioned above. These arguments may also explain the position of Ga with respect to Au and Ag, although a detailed classification is not possible.

We stress that this explanation is only tentative and that a detailed analysis of the $1/C_m$ curves should include all terms in Eq. (36). For example, an influence of the term $\sigma\, \partial d/\partial\sigma$ is expected at nonzero charge. In addition, the discussion of the influence of the ionic radii should also take into account some effect due to screening. In this discussion we mainly intend to show that since our definition of C_m incorporates the metal-solvent coupling and the relaxation of the surface dipole, the natural hypothesis attributing the difference in experimental capacities to the effect of the metal can indeed be largely substantiated by the variation of the associated physical properties.

In our opinion, a more important point is the fact that the estimated variation of the physical quantities which determine C_m, say, a variation of $(d - X_0)$ by about 0.2 Å, indicates that very little change in these quantities is needed in order to explain the observed variation of C_m from metal to metal. In addition to the implications of this observation for the interpretation of the specificity of the capacitance curves for various metals, it also means that a quantitative account of this behavior can be very demanding. Bearing in mind that $\Delta(d - X_0) \cong 0.2$ Å, if one recalls that the water molecule is often modeled by a sphere with a radius of about 1.5 Å, it is clear how accurate the calculations must be in order to account for such a slim difference between metals.

3. Effect of Temperature

A very interesting consequence of the previous analysis of the splitting of the capacitance into C_m and C_s is found in examining the temperature dependence of the solvent contribution. From the classical data of Grahame for Hg/NaF (see Fig. 2), the effect of temperature on C_s was estimated by using Eq. (41) and C_m^{-1} for Hg given in Fig. 14, assumed to be temperature independent.† Therefore, C_s curves (Fig. 15) at different temperatures are just another representation of the original C_i curves. However, the difference between the two sets of curves is striking. In contrast to the complicated appearance of the original curves, the behavior of C_s is rather trivial. At a given temperature, $C_s(\sigma)$ is a simple

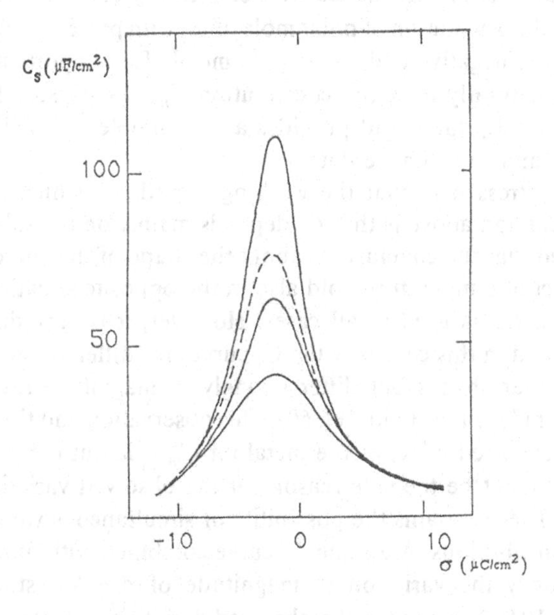

Figure 15. C_s curves for Hg/NaF at different temperatures. From top to bottom: $T = 0$, 25, 45, and 85°C. The dashed curve corresponds to C_s(Ag/NaF).

† This is the usual assumption. Arguments for this can be given. Here, we only mention that since in our approach the behavior of C_m is dominated by that of d, which depends little on temperature, the assumption that C_m also does not depend on temperature is quite reasonable.

bell-shaped curve. A maximum in the C_s curve due to the maximum randomness of dipole orientations is naturally observed at all temperatures. The interplay of the variations of C_m and C_s results in the occurrence of a hump in C_i at a slightly positive charge and only at low temperature. At a given charge now, increasing T reduces the magnitude of C_s. The so-called temperature-invariant point is also much less marked in the $C_s(T, \sigma)$ curves in Fig. 15 than in the corresponding C_i curves. The rapid drop of C_s observed away from the maximum may simply reflect saturation of the polarization.

For the aqueous solutions considered above, another important point is the systematic occurrence of the maximum of C_s at a negative charge. As substantiated by entropy curves,[27-29] this observation is in accordance with elementary considerations concerning the orientation of polar molecules, with preferential orientation of the negative end toward the metal. This important feature follows naturally once the contribution C_m is separated from the inner-layer C_i curve and provides a very simple reconciliation of entropy and capacitance data.

We stress here that the working hypothesis which underlies the discussion above is that C_s depends mainly on the solvent. We observed that the conclusions about the shape of the curve C_s and the effect of temperature hold also in the opposite situation where the same C_m is used in all cases. However, it is very difficult to understand in this case why the C_s curves for different metals have very similar shapes but differ strongly in magnitude around the maximum (see Fig. 3b in Ref. 60). This observation and the remarks on the expected effect of the metal on C_m rule out the possibility that C_s might be the sole reason for the observed variation of C_i curves. There remains the possibility of simultaneous variations in both contributions. A unique C_s curve combined with different C_m (via mostly the variation in magnitude of $d - X_0$ estimated in Section VIII.2) results in the observed variation of C_i for different metals. Thus, a possible variation of C_s combined with a change in $d - X_0$ smaller than that estimated above should be at most of the same importance, which is relatively small. Of course, there is also a shift from the pzc of the position σ_m of the maximum of C_s, which may depend on the nature of the metal. However, a metal-dependent value of σ_m should be in a very narrow range. Indeed, a large change in the maximum of $1/C_m$ is required in order to have a significant change of the position of the maximum of the C_s curves deduced from the C_i curves. Such a change would hardly be compatible with the mechanism which determines $1/C_m$, via $d - X_0$, as a function of charge.

Of course, one cannot strictly exclude the case where both C_s and C_m change considerably with the nature of the metal in some unpredictable way. However, given the physical meaning of C_s and C_m as we defined them, this hypothesis is very unlikely. It seems very reasonable that, in reality, both C_s and C_m change with the nature of the metal, but only slightly. What should also be made clear is that while dissimilar C_i curves may result from minor changes in C_s or C_m or both, the splitting of a given C_i curve into C_s and C_m results always in stable curves. This ultimately means that the sensitivity of our analysis to small variations in C_s and C_m, as a consequence of the law of inverse addition of capacitances connected in series, is an advantage and not a drawback as it may appear at first sight.

4. Role of the Solvent

As a final illustration of the previous analysis, we consider now the effect of changing the nature of the solvent. Details can be found in our recent paper.[155] In this paper, we developed arguments showing that, for qualitative purposes, one may first use the same C_m in analyzing C_i data for nonaqueous solvents at the mercury electrode. From inner-layer capacitance curves $C_i(\sigma)_{\mathrm{Hg/sol}}$ for different solvents, we defined the solvent contribution as

$$C_s^{-1}(\sigma)_{\mathrm{sol}} = C_i^{-1}(\sigma)_{\mathrm{Hg/sol}} - C_m^{-1}(\sigma)_{\mathrm{Hg}} \tag{43}$$

where $C_m^{-1}(\sigma)$ is the curve for Hg given in Fig. 14. $C_s(\sigma)$ obtained by using Eq. (43) is obviously a very indirect estimate of the solvent contribution so that details of the $C_s(\sigma)$ curves may not be reliable. However, the curves obtained in this way and shown in Fig. 16 present some features which are very unlikely to be artifacts of the method.

While their overall shape is the same, the $C_s(\sigma)$ curves depend, as expected, on the nature of the solvent in terms of their relative magnitudes and the charge σ_m corresponding to their maximum.

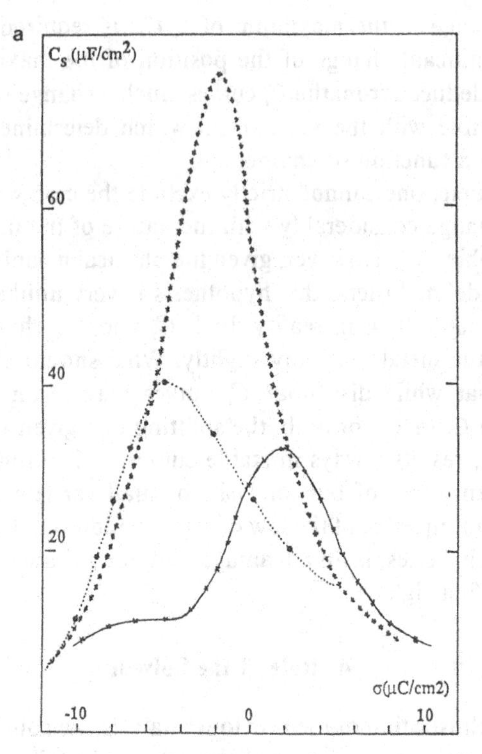

Figure 16. $C_s(\sigma)$ curves: (a) water (× × ×), methanol (— × —), and formamide (• · · · •); (b) dimethylformamide (— · —), dimethyl sulfoxide (- - -), acetonitrile (+ + +), propylene carbonate (——), ethylene carbonate (—•—), and acetone (· · ·).

For formamide, water, propylene carbonate, ethylene carbonate, and dimethyl sulfoxide, σ_m is negative. It is almost zero for dimethylformamide and acetone and positive for methanol and acetonitrile.

We can now discuss the kind of information one may derive from σ_m. As is well known, models based on a simple mechanism for the solvent orientation as a function of charge predict that the maximum of $C_i(\sigma)$ corresponds to the maximum randomness of the orientation of the dipoles. In this class of models, the sign of the charge at which the theoretical $C_i(\sigma)$ is maximum is directly related to the assumed preferred orientation of the dipoles at the

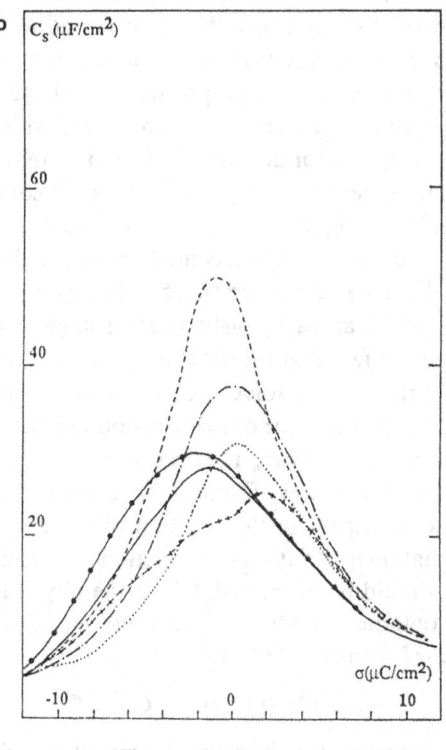

Figure 16. (*Continued*)

pzc. The maximum of the capacitance is also at the same charge as the maximum of the entropy. The latter is thought of as a more direct probe of solvent orientation, although the analysis of entropy data also requires a model. However, if one considers a more complex mechanism such as, for example, the presence of water clusters, the capacitance hump in the theoretical C_i does not correlate with the maximum of the entropy.

In the class of models attributing the hump to the orientation of the solvent dipoles, this absence of correlation between capacitance and entropy data led to the conclusion that maxima or humps in the capacitance curves give no information on solvent orientation at the pzc. Since no single criterion allows a definitive conclusion, the preferred orientation has to be assessed by comparing results from different experiments. On the basis of such a coherent set of observations, most of the solvents in Fig. 16 are usually considered as being oriented at the pzc with the negative end of the molecular dipole toward the electrode (see, for example, Refs. 162 and 163). It can be seen in Fig. 16 that the position σ_m of the maximum of $C_s(\sigma)$ is negative or very close to zero. Thus, with the possible exception of the curves for acetonitrile and methanol, these curves show a correlation between the position of the maximum of $C_s(\sigma)$ and the generally accepted preferred orientation at the pzc. The precise value of σ_m depends in this method on details of the slopes of $C_i(\sigma)$ and $C_m(\sigma)$ at moderate charges. In any case, the fact that $C_s(\sigma)$ curves exhibit an easily distinguished single maximum illustrates very well the fact that information present at the level of the "real" contribution of the solvent $C_s(\sigma)$ is simply masked at the level of $C_i(\sigma)$ by the presence of the contribution $C_m(\sigma)$. However, a reliable determination of σ_m requires accurate values of $C_m(\sigma)$.

Since $C_s(\sigma)$ curves were found to have similar shapes, almost symmetrical with respect to the position of the single maximum, this suggests that their variation as a function of the effective charge $\sigma^* = \sigma - \sigma_m$ should be examined. Given the physical significance of $C_s(\sigma)$, we obtained the dipolar contribution C_{dip} to C_s in terms of the charge σ^* from Eq. (39) as

$$C_{\text{dip}}^{-1}(\sigma^*) = 4\pi e/\varepsilon - C_s^{-1}(\sigma^*) \tag{44}$$

where C_{dip} is associated with the orientational polarization of the solvent, while the distortional polarization is related to $4\pi e/\varepsilon$ as usual. For the reason discussed below, we took the value $\varepsilon = 2$ for all solvents, and e was taken as equal to the molecular diameter determined in such a way as to have a fixed packing fraction η in the bulk for all solvents: $\eta = \pi\rho e^3/6 = 0.45$, where ρ is the bulk density. $C_{\text{dip}}(\sigma^*)$ curves obtained in this way are plotted in Fig. 17, from which the following observations can be made.

Around $\sigma^* = 0$, the variation of $C_{\text{dip}}(\sigma^*)$ for different solvents is considerably reduced, when compared to that of $C_s(\sigma)$ in Fig. 16. This confirms the dominant role of molecular size advocated many times (see, for example, the review by Payne[164]). At moderate σ^* the curves for dimethyl sulfoxide, dimethylformamide, propylene carbonate, ethylene carbonate, and acetone merge almost into a single curve, while those for H_2O, methanol, and formamide are clearly separated. Note that the curve for acetonitrile has an inter-

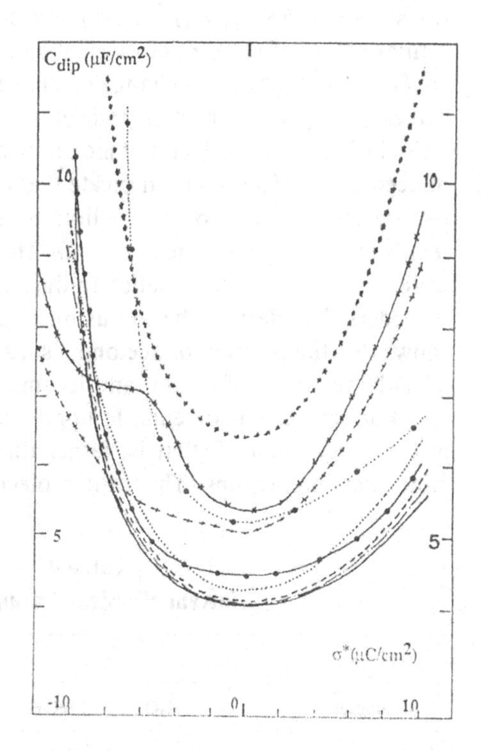

Figure 17. The contribution C_{dip} of the permanent dipoles as a function of the effective charge σ^*. For definition of symbols, see caption to Fig. 16.

mediate position. At higher magnitudes of σ^*, the saturation of the polarization depends on the solvent.

The behavior of $C_{dip}(\sigma^*)$ for nonzero σ^* reflects the role of intermolecular interactions. A detailed discussion has been presented in our recent paper.[155] The sequence near $\sigma^* = 0$ can be understood qualitatively, from the competition between the orienting effect imposed by the external field and the effect of intermolecular interactions which oppose it. With regard to the field, simple considerations show that $C_{dip}(\sigma^*)$ is proportional to $(e/\mu)^2/\partial\langle\cos\theta\rangle/\partial\xi$, where ξ is a reduced quantity defined by $\xi = \mu E^*/kT$. $E^* = E - E^r$ is the effective field, which reflects the competition between the applied field E and the residual field E^r, μ is the dipole moment, and θ is the angle that the dipole makes with the normal to the electrode surface.

For aprotic solvents, intermolecular interactions are mainly dipolar. An approximate measure of the strength of dipolar interactions is the ratio $\upsilon_{dd} = (\mu^2/kTe^3)$. For these aprotic solvents with similar values of the dipole moment and of the radius, υ_{dd} and the coefficient $(e/\mu)^2$ do not change much (see Table 4), so that similar values of C_{dip} are expected. Indeed, the values of C_{dip} in Fig. 17 differ little for these solvents at moderate charges with the exception of acetonitrile. This would indicate that deviations from the simple picture given above (polarizability, nonspherical shape, etc.) are relatively unimportant near $\sigma^* = 0$. The position of acetonitrile, which has a relatively smaller radius, can be understood if one considers the effect of the radius on C_m.[155] Inspection of Table 4 shows that the position of acetone results from the combined effect of its larger value of $(e/\mu)^2$ and its smaller value of υ_{dd}.

For associated solvents, H bonds, corresponding to energies in the range 5-20 kT, that is, larger than $\upsilon_{dd}kT$, dominate intermolecular interactions. Thus, the molecules of these solvents are

Table 4
Solvent Physical Parameters[a]

Parameter	Solvent				
	H_2O	MeOH	FA	AC	EC
e (Å)	3	3.86	3.83	4.71	4.55
μ (D)[b]	1.84	1.66	3.66	2.88	4.87
$(e/\mu)^2$ (Å/D)^2	2.65	5.40	1.09	2.67	0.87
ν_{dd}[c]	3.02	1.15	5.75	1.91	6.07
$\varepsilon/4\pi e$ (μF cm^{-2})[d]	5.89	4.58	4.61	3.75	3.88

Parameter	Solvent			
	PC	DMSO	DMF	ACN
e (Å)	4.94	4.66	4.78	4.21
μ (D)[b]	4.94	3.96	3.82	3.92
$(e/\mu)^2$ (Å/D)^2	1	1.38	1.56	1.15
ν_{dd}[c]	4.88	3.73	3.22	4.96
$\varepsilon/4\pi e$ (μF cm^{-2})[d]	3.57	3.79	3.70	4.20

[a] Abbreviations: MeOH, methanol; FA, formamide; AC, acetone; EC, ethylene carbonate; PC, propylene carbonate; DMSO, dimethyl sulfoxide; DMF, dimethylformamide; ACN, acetonitrile.
[b] Values taken from the literature.
[c] $kT\nu_{dd}$ is half the maximum dipole-dipole interaction energy in vacuum.
[d] Calculated from the tabulated values of the solvent diameter, e, and with $\varepsilon = 2$.

more difficult to orient than those of unassociated solvents and hence should have smaller values of $\partial\langle\cos\theta\rangle/\partial\xi$. Except for formamide, they also have larger values of the ratio $(e/\mu)^2$. As observed in Fig. 17, $C_{\rm dip}(\sigma^*)$ should be larger for water, formamide, and methanol than for aprotic solvents. At higher charges, the competition between the applied field and intermolecular interactions makes the rise of $C_{\rm dip}(\sigma^*)$ specific to each solvent. A detailed discussion is, however, difficult since the precise value of $C_{\rm dip}(\sigma^*)$ at these charges is very sensitive to uncertainties in e/ε and $C_s(\sigma^*)$.

In conclusion, we may say that when $C_m(\sigma)$ is extracted from $C_i(\sigma)$, a clear picture of the behavior of different solvents emerges. It can be rationalized largely in terms of their physical parameters (dipole moment, molecular size) and the nature of intermolecular interactions. This is to be compared with the often adopted classification of solvents into different groups according to the shape of $C_i(\sigma)$. In addition, the preferred orientation of the solvent dipoles at the pzc can be deduced from the position of the maximum of C_s.

5. A Simple Model for C_s

The features discussed above and the overall behavior of C_s with charge and temperature suggested that a simple model for dipole orientation in an electric field be tried. We indeed found that a model of the Langevin type for the polarization gives an excellent account (see Fig. 18) of the semiempirical C_s for water. In this model, the polarization is given by

$$P = N\langle\mu\cos\theta\rangle = N\mu[\coth(x) - 1/x]; \qquad x = \mu E_m/kT \quad (45)$$

where N is the density of dipoles, and μ is their dipole moment ($6.1\ 10^{-30}$ C m, or 1.84 D, for water). The mean field orienting the dipoles was taken as

$$E_m = 4\pi\sigma^*/\varepsilon \quad (46)$$

where ε is an adjustable parameter, and $\sigma^* = \sigma - \sigma_m$ takes into account any preferential orientation of the dipoles at the pzc.[35,165] The resulting capacitance is

$$1/C_s = 4\pi e - (4\pi N_s\mu^2/kT)[1 - \coth^2(x) + 1/x^2] \quad (47)$$

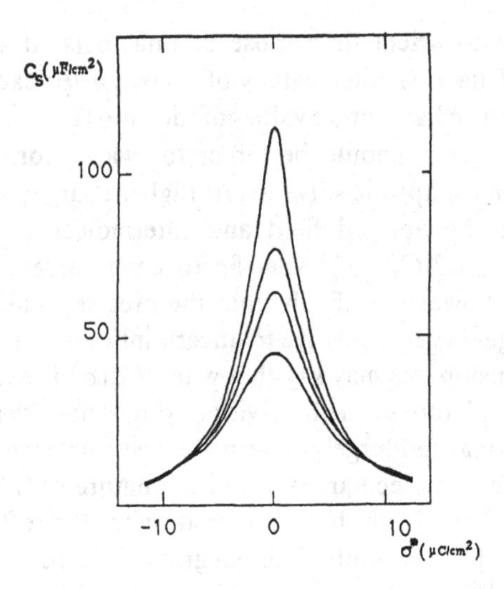

Figure 18. Theoretical $C_s(\sigma^*)$ curves with a one-parameter model; σ^* is the effective charge on the electrode. From top to bottom: $T = 0$, 25, 45, and 85°C.

where e is the thickness of the solvent layer, and N_s is the number of dipoles per unit area. Considering a monolayer of thickness $e = 2R_0 = 3$ Å and hexagonal packing, one has $N_s \cong 1.2 \times 10^{19}$ m^{-2}.[38] The values of C_s at the maximum were then fitted at different temperatures with the help of ε, giving:

T(°C)	0	25	45	85
ε	14.35	13.31	12.60	11.40

These values of ε are in the range of the dielectric constant at weak field[35,166] and decrease with temperature as expected. Note that ε may incorporate various effects negelected in the model, such as lateral interaction, image forces, or H bonds. See also the discussion of Marshall and Conway[160] for a more refined treatment of this kind of model.

As detailed in our previous work,[60] a better fit of the saturated value of C_s in Fig. 18 is obtained by introducing into the model the polarizability of the molecule. In contrast to Eq. (47), the saturated value of $1/C_s$ is now $4\pi e/\varepsilon_\infty$. We emphasize that the parameters in this "improved" model were given in Ref. 60 for the purpose of illustration and are not to be taken literally, especially in the expression of the dielectric constant $\varepsilon(\sigma)$. The overall treatment of the model is also not self-consistent. However, the important point is that the saturated value $C_s(|\sigma| \gg 0)$ given in the model by the term $4\pi e/\varepsilon_\infty$ (usually denoted by K_{ion}) is obtained with ε_∞ close to $\varepsilon_\infty = 2$. Incidentally, this value of ε_∞ happens to correspond to the value obtained from optical measurements. Such a small value of ε_∞ is obtained because the contribution of the metal was extracted from the experimental C_i, so that $C_s(|\sigma| \gg 0)$ is smaller than the classical value of about 17 μF cm^{-2} (recall the discussion in Section VII.2). While the approximations in the theory make it difficult to state that this value should be taken as definitive, this point strongly suggests that the accepted value of the "dielectric constant of the inner layer" should be examined.

Finally, our analysis of C_s remains with the traditional view: we deal with capacitances corrected for the Gouy-Chapman contribution, consider a monolayer model for C_s and a local dielectric constant, and so on. This may appear rather strange if we recall the important developments in statistical mechanics which provide, in principle, the tools for a more rigorous treatment of the model. These developments also cast some doubt on the validity of conclusions drawn from molecular models which do not incorporate an accurate description of the short-range interfacial structure (see the recent work of Torrie *et al.*[167] and references therein). We are aware of the fact that the simple calculation given above does not constitute rigorous treatment of the model. Moreover, the introduction of unknown parameters is related to physical effects, such as dipole-dipole or ion-dipole interactions, that can be dealt with properly only by an elaborate statistical mechanical treatment. The practical reason for our choice of a less rigorous and hence much simpler treatment was the observation that the shape of C_s is simple. One may guess that a rigorous treatment of the model will possibly remove the need for adjustable parameters or clarify their physical significance and precise values. Thus, it will be interesting to investigate in the future the possibility that an elaborate treatment of the solution ultimately gives the same results as a less sophisticated one (say, a Langevin function for the polarization) even if only on a formal level.

IX. CONCLUSION

The theory of the capacitance of an ideally polarized electrode has been examined in some studies of this last decade. In this chapter, we have discussed the works in which the contribution of the metal is reanalyzed. From an extension of the ideas of the physics of metal surfaces, these studies have confirmed that the metal does not behave at the interface like an "ideal" conductor which otherwise acts as a rigid wall for the particles in the solution.

This nonideality of the metal can be introduced formally in the expression for the capacitance, which appears as the result of two capacities C_m and C_s connected in series. C_m arises from the metal and its coupling with the solution, while C_s, associated with the solvent polarization, has the same physical meaning as in the traditional view.

A reliable calculation of the contribution C_m is now possible for solids, especially for silver. However, a completely *ab initio* calculation of the capacitance which also incorporates the best available description of the solution is still a very difficult task. A more efficient route consists in analyzing the semiempirical remaining part C_s obtained by separating the calculated C_m from experimental data. In the case of a silver/aqueous solution interface, the contribution C_s was characterized by a very simple variation with the electrode charge, with a single maximum at a negative charge. Such behavior is consistent with the physical meaning embodied in the definition of C_s and so was expected to persist under more general conditions.

Such an analysis has shown that the capacitance curves of other metal/aqueous solutions could be obtained from relatively small variations of the contributions C_m and C_s determined for silver. The estimated variation of C_m with the nature of the metal is in agreement with well-accepted physical properties of these metals. The variation of C_s with the electrode charge, as inferred from the data for water, remains simple when the temperature is changed. The C_s versus charge curves could be reproduced easily in a model where the value of the saturated dielectric constant is consistent with that determined at optical frequencies.

More generally, the analysis of capacitance data for the mercury electrode in contact with protic and aprotic solvents confirmed this view. The C_s curves maintain the same form, although their magnitude and the position of their maximum depend as expected on the nature of the solvent. The position of the maximum showed a good correlation with the preferred orientation at the pzc. In addition, the solvent specificity can be analyzed in a very simple way in terms of the contribution C_{dip} due to the orientation of the permanent dipoles in an electric field.

Of course, since we now have a satisfactory understanding of the dominant phenomena which determine C_m and C_s, simple models with a minimum number of adjustable parameters can be devised quite easily in order to fit the experimental curves. It does not seem that such a procedure will add more clarity to our present view of the interface.

The only measurable quantity is indeed the total capacitance. Thus, its separation into several components is not unique. Besides the fact that the partial contributions C_m and C_s have well-defined physical meanings, the ultimate test of the adequacy of this route was its ability to provide a coherent explanation of an appreciable number of experimental facts. Indeed, unless the importance of the contribution C_m was considerably overestimated in our work and in the work of many others, all the features discussed above form a very coherent set of observations in favor of the correctness of the analysis we have proposed.

More generally, an important conclusion of this analysis is that inner-layer capacitance curves $C_i(\sigma)$ corresponding to a wide variety of ideally polarizable interfaces can be discussed in a common physical framework. The idea that the qualitative behavior of apparently different curves might be determined to a large extent by relatively unspecific phenomena may appear quite at odds with ordinary thinking in electrochemistry. In fact, this idea suggests only that this specific information should be traced back to a less global quantity than the Helmholtz capacitance $C_i(\sigma)$. This means that one should try to isolate, if possible, the contributions which have a dominant physical nature, in the same way as one extracts from the total capacitance the unspecific contribution due to the diffuse layer. On the other hand, this route will be efficient only if a truly robust and reliable method for calculating this unspecific contribution is available. Improvements of the theory in many directions are then required for a more quantitative analysis or if more complicated interfaces are to be investigated.

This route remains indeed an "analysis" in the sense that we proposed an interpretation of known experimental facts that is only an alternative to the traditional one. Although this interpretation is very coherent, the predictive power of this analysis is at present limited, since none of the main quantities it considers is amenable to direct experimental test. However, we expect that some aspects of the description of the metal and its coupling with the solution discussed in this work may be relevant in areas other than capacitance studies, as is suggested by the role of the metal–solvent distance in the interpretation of optical data.

ACKNOWLEDGMENTS

We express our gratitude to F. Forstmann, M. L. Rosinberg, V. Russier, and M. A. Vorotyntsev for very stimulating discussions. One of us (S.A.) acknowledges the C.N.R.S. for support during the completion of this work.

APPENDIX. EVALUATION OF U_{vdW}

Details of the derivation of the dispersion force are given in the paper of Annett and Echenique[124] and references therein. Here we mention that the evaluation of $U_{\text{vdW}}(Z)$ is based on:

1. The approximation of the imaginary part of the response function by that corresponding to a single pole, the surface plasmon dispersion relation being taken as

$$\omega_s(q) = \omega_s^2 + aq + bq^2 + q^4/4$$

where ω_s is the surface plasmon frequency, and a is related to the static image plane position by $a = -2\omega_s^2 X_0$. X_0 for the metal in vacuum is calculated as described in Section III.3(ii). Following the procedure described in Ref. 168, b is determined from a by requiring that the surface plasmon line join the single-particle continuum at the same point as the bulk line does.

2. The expression of the atomic polarizability in terms of discrete dipolar transitions characterized by oscillator strengths f_{n0} and frequencies ω_{n0}.

Then,

$$U_{\mathrm{VdW}}(Z) = -\frac{1}{2}\sum_n \frac{f_{n0}}{\omega_{n0}} \int dq \frac{q^2 e^{-2qZ}\omega_s^2}{\omega_s(q)[\omega_s(q)+\omega_{n0}]}$$

The input parameters for the metal are r_s, X_0, and b. In our own work, we calculated X_0 and determined b as indicated above. For the atom, f_{n0} and ω_{n0} can be taken from tabulated values. For water, we used the values of f_{n0} and ω_{n0} given by Margoliash and Meath.[169]

REFERENCES

[1] R. M. Reeves, in *Modern Aspects of Electrochemistry*, No. 9, Ed. by B. E. Conway and J. O'M. Bockris, Plenum Press, New York, 1974, p. 239.
[2] M. A. Habib, in *Modern Aspects of Electrochemistry*, No. 12, Ed. by B. E. Conway and J.O'M. Bockris, Plenum Press, New York, 1977, p. 131.
[3] S. L. Carnie and G. M. Torrie, *Adv. Chem. Phys.* **56** (1984) 141; S. L. Carnie, *Ber. Bunsenges. Phys. Chem.* **91** (1987) 262.
[4] D. Henderson, in *Trends in Interfacial Electrochemistry*, Ed. by A. Fernando Silva, Reidel, Dordrecht, 1986, p. 473.
[5] J. Goodisman, in *Modern Aspects of Electrochemistry*, No. 20, Ed. By J. O'M. Bockris, R. E. White, and B. E. Conway, Plenum Press, New York, 1989, p. 1.
[6] S. Trasatti, in *Modern Aspects of Electrochemistry*, No. 13, Ed. by B. E. Conway and J. O'M. Bockris, Plenum Press, New York, 1979, p. 81.
[7] S. Trasatti, in *Trends in Interfacial Electrochemistry*, Ed. by A. Fernando Silva, Reidel, Dordrecht, 1986, p. 1.
[8] S. Trasatti, *J. Electroanal. Chem.* **138** (1982) 449.
[9] G. Valette, *J. Electroanal. Chem.* **139** (1982) 285.
[10] J. O'M. Bockris and A. K. N. Reddy, in *Modern Electrochemistry*, Vol. 2, Plenum Press, New York, 1973.
[11] P. Delahay, *Double Layer and Electrode Kinetics*, Interscience, New York, 1965.
[12] A. Frumkin, B. Damaskin, N. Grigoryev, and I. Bagotskaya, *Electrochim. Acta* **19** (1973) 69.
[13] A. Frumkin, I. Bagotskaya, and N. Grigoryev, *Denki Kagaku* **1** (1975) 43.
[14] D. C. Grahame, *Chem. Rev.* **41** (1947) 441.
[15] G. Gouy, *J. Phys.* **9** (1910) 457.
[16] D. L. Chapman, *Phil. Mag.* **25** (1913) 475.
[17] R. Parsons and F. G. R. Zobel, *J. Electroanal. Chem.* **9** (1965) 333.
[18] G. Valette, *J. Electroanal. Chem.* **269** (1989) 191; **260** (1989) 425; **224** (1987) 285.
[19] D. C. Grahame, *J. Am. Chem. Soc.* **76** (1954) 4819.
[20] A. Hamelin, Z. Borkowska, and J. Stafiej, *J. Electroanal Chem.* **189** (1985) 85.
[21] G. Valette and A. Hamelin, *J. Electroanal. Chem.* **45** (1973) 301; G. Valette, Thesis, Paris, 1983.
[22] G. Valette, *J. Electroanal. Chem.* **122** (1981) 285.
[23] A. Hamelin, T. Vitanov, E. Sevastyanov, and A. Popov, *J. Electroanal. Chem.* **145** (1983) 225.

[24] D. C. Grahame, *J. Am. Chem. Soc.* **79** (1957) 2093.
[25] A. Hamelin and L. Stoicoviciu, *J. Electroanal. Chem.* **236** (1987) 283.
[26] A. Hamelin, L. Doubova, L. Stoicoviciu, and S. Trasatti, *J. Electroanal. Chem.* **244** (1988) 133.
[27] A. Fernando Silva, in *Trends in Interfacial Electrochemistry*, Ed. by A. Fernando Silva, Reidel. Dordrecht, 1986, p. 49.
[28] J. A. Harrison, J. E. B. Randles, and D. J. Schiffrin, *J. Electroanal. Chem.* **48** (1973) 359.
[29] G. J. Hills and S. Hsieh, *J. Electroanal. Chem.* **58** (1975) 289.
[30] S. Trasatti, *J. Electroanal. Chem.* **150** (1983) 1.
[31] W. Schmickler, *J. Electroanal. Chem.* **176** (1984) 383.
[32] J. P. Badiali, *Ber. Bunsenges. Phys. Chem.* **91** (1987) 270.
[33] N. D. Lang and W. Kohn, *Phys. Rev. B* **8** (1973) 6010.
[34] P. Gies and R. R. Gerhardt, *Phys. Rev. B* **31** (1985) 6843; **33** (1986) 982.
[35] J. R. MacDonald and C. A. Barlow, *J. Chem. Phys.* **36** (1962) 3062.
[36] R. Guidelli, *J. Electroanal. Chem.* **197** (1986) 77.
[37] G. Aloisi and R. Guidelli, *J. Electroanal. Chem.* **260** (1989) 259.
[38] R. Parsons, *J. Electroanal. Chem.* **59** (1975) 229.
[39] R. Parsons, *J. Electroanal. Chem.* **150** (1983) 51.
[40] R. J. Watts-Tobin, *Phil. Mag.* **6** (1961) 133; N. F. Mott and R. J. Watts-Tobin, *Electrochim. Acta* **4** (1961) 79.
[41] J. O'M. Bockris, M. A. V. Devanathan, and K. Müller, *Proc. Roy. Soc. A* **274** (1963) 55.
[42] J. O'M. Bockris and M. A. Habib, *J. Electroanal. Chem.* **68** (1976) 367.
[43] S. Levine, J. Bell, and A. Smith, *J. Phys. Chem.* **73** (1969) 35.
[44] W. R. Fawcett, *J. Phys. Chem.* **82** (1978) 1385.
[45] W. R. Fawcett, S. Levine, R. M. de Nobriga, and A. C. MacDonald, *J. Electroanal. Chem.* **111** (1980) 163.
[46] B. B. Damaskin and N. A. Frumkin, *Electrochim. Acta* **19** (1974) 173.
[47] N. D. Lang and W. Kohn, *Phys. Rev. B* **1** (1970) 4555.
[48] N. D. Lang and W. Kohn, *Phys. Rev. B* **3** (1971) 1215.
[49] O. K. Rice, *Phys. Rev.* **31** (1928) 1051.
[50] F. Schreier and F. Rebentrost, *J. Phys. C: Solid State Phys.* **20** (1987) 2609.
[51] D. M. Newns, *Phys. Rev. B* **8** (1969) 3304.
[52] J. Heinrichs, *Phys. Rev. B* **8** (1973) 1346.
[53] A. A. Kornyshev, B. Rubinshtein, and M. A. Vorotyntsev, *Phys. Status Solidi B* **84** (1977) 125.
[54] A. G. Eguiluz, D. A. Campbell, A. A. Maradudin, and R. F. Wallis, *Phys. Rev. B* **28** (1984) 1667.
[55] K. Heinzinger and E. Spohr, *Electrochim. Acta* **34** (1989) 1849.
[56] E. Spohr and K. Heinzinger, *Electrochim. Acta* **33** (1988) 1811.
[57] S. L. Carnie and D. Y. C. Chan, *J. Chem. Phys.* **73** (1980) 2949.
[58] L. Blum and D. Henderson, *J. Chem. Phys.* **74** (1981) 1902.
[59] F. Vericat, L. Blum, and D. Henderson, *J. Electroanal. Chem.* **150** (1983) 315.
[60] S. Amokrane and J. P. Badiali, *J. Electroanal. Chem.* **266** (1989) 21.
[61] A. N. Frumkin and A. Gorodetzkaya, *Z. Phys. Chem.* **136** (1928) 451.
[62] J. E. Inglesfield, *Rep. Prog. Phys.* **45** (1982) 2223.
[63] M. Hasegawa, *J. Phys. F: Met. Phys.* **18** (1988) 1449.
[64] G. Allan, *Handbook of Surfaces and Interfaces*, Ed. by L. Dobrynsky, Garland STPN, New York, 1978.
[65] N. D. Lang, *Solid State Phys.* **28** (1973) 225.
[66] R. O. Jones and O. Gunnarsson, *Rev. Mod. Phys.* **61** (1989) 689.
[67] J. P. Badiali, *Electrochim. Acta* **31** (1986) 149.

[68] W. Schmickler and D. Henderson, *Prog. Surf. Sci.* **22** (1986) 323.
[69] W. Harrison, *Pseudo-Potential in the Theory of Metals*, Benjamin, New York, 1966.
[70] V. Heine, M. L. Cohen, and D. Weaire, *Solid State Phys.* **24** (1970) 249.
[71] N. W. Ashcroft, *Phys. Lett.* **23** (1966) 48.
[72] P. Hohenberg and W. Kohn, *Phys. Rev. B* **136** (1964) 864.
[73] W. Kohn and L. J. Sham, *Phys. Rev. A* **140** (1965) 1133.
[74] J. R. Smith, *Phys. Rev.* **181** (1969) 522.
[75] C. Q. Ma and V. Sahni, *Phys. Rev. B* **16** (1977) 4249.
[76] J. P. Perdew and Y. Wang, *Phys. Rev. B* **38** (1988) 12228.
[77] R. Monnier, J. P. Perdew, D. C. Langreth, and J. W. Wilkins, *Phys. Rev. B* **18** (1978) 656.
[78] V. Russier and M. L. Rosinberg, *J. Phys. C.* **21** (1988) L333.
[79] H. F. Budd and J. Vannimenus, *Phys. Rev. Lett.* **31** (1973) 1218.
[80] J. P. Perdew and R. Monnier, *Phys. Rev. Lett.* **37** (1976) 1286; R. Monnier and J. P. Perdew, *Phys. Rev. B* **17** (1978) 2595.
[81] G. C. Aers and J. E. Inglesfield, *Surf. Sci.* **217** (1989) 367.
[82] J. H. Rose and J. F. Dobson, *Solid State Commun.* **37** (1981) 91.
[83] W. H. Mulder, J. H. Sluyters, and J. H. van Lenthe, *J. Electroanal. Chem.* **261** (1989) 273.
[84] A. L. G. van den Eeden, J. H. Sluyters, and J. H. van Lenthe, *J. Electroanal. Chem.* **208** (1986) 243.
[85] J. W. Halley and D. Price, *Phys. Rev. B* **38** (1988) 9357.
[86] J. P. Badiali, M. L. Rosinberg, and J. Goodisman, *J. Electroanal. Chem.* **150** (1983) 25.
[87] W. Schmickler and D. Henderson, *J. Chem. Phys.* **80** (1984) 3381.
[88] J. G. Harris, J. Gryko, and S. A. Rice, *J. Chem. Phys.* **86** (1987) 1067.
[89] J. G. Harris, J. Gryko, and S. A. Rice, *J. Chem. Phys.* **87** (1987) 3069.
[90] J. Goodisman, *J. Chem. Phys.* **90** (1989) 5756.
[91] J. A. Moriarty, *Phys. Rev. B* **38** (1988) 3199.
[92] J. W. Wills and W. A. Harrison, *Phys. Rev. B* **28** (1983) 4363.
[93] V. Russier and J. P. Badiali, *Phys. Rev. B* **39** (1989) 13,193.
[94] M. Chelvayohan and C. H. B. Mee, *J. Phys. C* **15** (1982) 2305.
[95] J. K. Grepsted, P. O. Gartland, and B. J. Slogsvold, *Surf. Sci.* **57** (1976) 348.
[96] R. M. Eastment and C. H. B. Mee, *J. Phys. F* **3** (1973) 1738.
[97] F. J. Arlinghaus, J. G. Gay, and J. R. Smith, *Phys. Rev. B* **23** (1981) 5152.
[98] J. P. Badiali, M. L. Rosinberg, and J. Goodisman, *J. Electroanal. Chem.* **130** (1981) 31.
[99] H. B. Michaelson, *J. Appl. Phys.* **48** (1977) 4729.
[100] AIP Physics Vade Mecum, American Institute of Physics, New York, 1981.
[101] G. Valette, *J. Electroanal. Chem.* **178** (1984) 179.
[102] A. K. Theophilou and A. Modinos, *Phys. Rev. B* **6** (1972) 801.
[103] W. Schmickler and D. Henderson, *Phys. Rev. B* **30** (1984) 2081.
[104] M. B. Partenskii and V. I. Feldman, *Elektrokhimiya* **24** (1988) 369.
[105] A. A. Kornyshev, *Electrochim. Acta* **34** (1989) 1829.
[106] A. A. Kornyshev and M. A. Vorotyntsev, *J. Electroanal. Chem.* **167** (1984) 1.
[107] E. Zaremba and W. Kohn, *Phys. Rev. B* **13** (1976) 2270.
[108] E. Zaremba and W. Kohn, *Phys. Rev. B* **15** (1977) 1769.
[109] E. M. Stuve, K. Bange, and J. K. Sass, in *Trends in Interfacial Electrochemistry*, Ed. by A. Fernando Silva, Reidel, Dordrecht, 1986, p. 255.
[110] P. A. Thiel and T. E. Madey, *Surf. Sci. Rep.* **7** (1987) 211.
[111] M. Klaua and T. E. Madey, *Surf. Sci.* **136** (1984) L42.
[112] V. I. Feldman, A. A. Kornyshev, and M. B. Partenskii, *Solid State Commun.* **53** (1985) 157.

[113] J. W. Halley, B. Johnson, D. Price, and M. Schwalm, *Phys. Rev. B* **31** (1985) 7695.
[114] S. Amokrane, V. Russier, and J. P. Badiali, *Surf. Sci.* **217** (1989) 425.
[115] S. Amokrane, Thesis, Paris, 1989.
[116] S. K. Lyo and R. Gomer, in *Topics in Applied Physics*, Vol. 4, Ed. by R. Gomer, Springer-Verlag, Berlin, 1975, p. 41.
[117] T. N. Rhodin and G. Ertl, Eds., *The Nature of the Surface Chemical Bond*, North-Holland, Amsterdam, 1979.
[118] J. E. Müller, *Surf. Sci.* **178** (1986) 589.
[119] S. Holloway and K. H. Bennemann, *Surf. Sci.* **101** (1980) 327.
[120] J. E. Müller and J. Harris, *Phys. Rev. Lett.* **53** (1984) 2496.
[121] M. W. Ribarsky, W. D. Luedtke, and Uzi Landman, *Phys. Rev. B* **32** (1985) 1430.
[122] A. M. Kuznetsov, R. R. Nazmutdinov, and M. S. Shapnik, *Electrochim. Acta* **34** (1989) 1821.
[123] J. Harris and A. Liebsch, *J. Phys. C* **15** (1982) 2275.
[124] J. F. Annett and P. M. Echenique, *Phys. Rev. B* **34** (1986) 6853.
[125] J. K. Norskov and N. D. Lang, *Phys. Rev. B* **21** (1980) 2131.
[126] N. D. Lang and J. K. Norskov, *Phys. Rev. B* **27** (1983) 4612.
[127] N. D. Lang and A. R. Williams, *Phys. Rev. B* **18** (1978) 616.
[128] H. Hjelmberg, *Phys. Scr.* **18** (1978) 481.
[129] D. Sluis and S. A. Rice, *J. Chem. Phys.* **79** (1983) 5658.
[130] S. A. Rice, in *Fluid Interfacial Phenomena*, Ed. by C. A. Croxton, Wiley, New York, 1986.
[131] L. Bosio and M. Oumezine, *J. Chem. Phys.* **80** (1984) 459.
[132] S. A. Rice and M. P. D'Evelyn, *J. Chem. Phys.* **78** (1983) 5081.
[133] J. L. Bretonnet and C. Regnaut, *Phys. Rev. B* **31** (1985) 5071.
[134] R. W. Shaw, *Phys. Rev.* **174** (1968) 769.
[135] J. W. Halley and D. Price, *Phys. Rev. B* **35** (1987) 9095.
[136] J. P. Badiali, M. L. Rosinberg, and J. Goodisman, *J. Electroanal. Chem.* **143** (1983) 73.
[137] W. Schmickler, *J. Electroanal. Chem.* **150** (1983) 19.
[138] V. Russier, M. L. Rosinberg, J. P. Badiali, D. Levesque, and J. J. Weis, *J. Chem. Phys.* **87** (1987) 5012.
[139] J. P. Badiali and M. E. Boudh-Hir, *Mol. Phys.* **53** (1984) 1399.
[140] P. H. Fries and G. N. Patey, *J. Chem. Phys.* **82** (1985) 429.
[141] P. G. Kusalik and G. N. Patey, *J. Chem. Phys.* **88** (1988) 7715.
[142] G. M. Torrie, P. G. Kusalik, and G. N. Patey, *J. Chem. Phys.* **88** (1988) 7826.
[143] J. P. Badiali, *J. Chem. Phys.* **90** (1989) 4441.
[144] S. Amokrane and J. P. Badiali, *Electrochim. Acta* **34** (1989) 39.
[145] W. Schmickler and D. Henderson, *J. Chem. Phys.* **85** (1986) 1650.
[146] D. Henderson, E. Leiva, and W. Schmickler, *Ber. Bunsenges, Phys. Chem.* **91** (1987) 280.
[147] W. Schmickler, *Ber. Bunsenges. Phys. Chem.* **92** (1988) 1203.
[148] J. P. Badiali, M. L. Rosinberg, F. Vericat, and L. Blum, *J. Electroanal. Chem.* **158** (1983) 253.
[149] M. L. Rosinberg, Thesis, Paris, 1983.
[150] W. Schmickler, in *Trends in Interfacial Electrochemistry*, Ed. by A. Fernando Silva, Reidel, Dordrecht, 1986, p. 453.
[151] W. Schmickler, *J. Electroanal. Chem.* **265** (1989) 11.
[152] Z. B. Kim, A. A. Kornyshev, and M. B. Partenskii, *J. Electroanal. Chem.* **265** (1989) 1.
[153] W. Schmickler, *J. Electroanal. Chem.* **249** (1988) 25.
[154] A. A. Kornyshev, W. Schmickler, and M. A. Vorotyntsev, *Phys. Rev. B* **25** (1982) 5244.

[155] S. Amokrane and J. P. Badiali, *J. Electroanal. Chem.* **297** (1991) 377.
[156] J. P. Hansen and I. R. McDonald, *Theory of Simple Liquids*, 2nd ed., Academic Press, London, 1986.
[157] J. P. Badiali and M. E. Boudh-Hir, *Mol. Phys.* **53** (1984) 1399.
[158] L. Blum and D. Henderson, *J. Chem. Phys.* **74** (1981) 1902.
[159] I. L. Cooper and J. A. Harrison, *J. Electroanal. Chem.* **86** (1978) 425.
[160] S. L. Marshall and B. E. Conway, *J. Chem. Phys.* **81** (1984) 923.
[161] F. V. Molina, G. J. Gordillo, and D. Posadas, *J. Electroanal. Chem.* **239** (1988) 405.
[162] S. Trasatti, *Electrochim. Acta* **32** (1987) 843.
[163] S. Borkowska, *J. Electroanal. Chem.* **244** (1988) 1.
[164] R. Payne, in *Advances in Electrochemistry and Electrochemical Engineering*, Vol. 7, Ed. by P. Delahay and C. W. Tobias, Interscience, New York, 1970, p. 1.
[165] I. L. Cooper and J. A. Harrison, *J. Electroanal. Chem.* **66** (1975) 85.
[166] J. R. MacDonald and S. W. Kenkel, *J. Chem. Phys.* **80** (1984) 2168.
[167] G. M. Torrie, P. G. Kusalik, and G. M. Patey, *J. Chem. Phys.* **91** (1989) 6367.
[168] P. M. Echenique, R. H. Ritchie, N. Barberan, and J. C. Inkson, *Phys. Rev. B* **23** (1981) 6486.
[169] D. J. Margoliash and W. J. Meath, *J. Chem. Phys.* **68** (1978) 1426.
[170] A. A. Kornyshev and M. A. Vorotyntsev, *Surf. Sci.* **101** (1980) 23.

2

Electrocatalytic Oxidation of Oxygenated Aliphatic Organic Compounds at Noble Metal Electrodes

Bernard Beden, Jean-Michel Léger, and Claude Lamy

Laboratoire de Chimie 1, Electrochimie et Interactions, URA CNRS No. 350, Université de Poitiers, 86022 Poitiers Cedex, France

I. INTRODUCTION

The investigation of electrocatalytic processes involved in the electrooxidation of organic compounds became a subject of growing interest at the beginning of the 1960s. A tremendous effort was undertaken to develop fuel cells, which theoretically have the ability to directly convert the chemical energy of hydrocarbons into electrical energy without the limitations due to Carnot's theorem (i.e., with a high energy efficiency), an advantage which is particularly attractive for autonomous power sources.[1-3] In such applications the rate of conversion of the organic compound by oxidation into carbon dioxide has to be as high as possible, in order to obtain the maximum energy available from the fuel and thus the maximum efficiency. However, the incomplete oxidation of organic compounds is of great interest as well, if one considers the potential applications for organic electrosynthesis, so that many attempts have been made in this field as well. Rigorous control of the experimental conditions may allow such reactions to become highly

Modern Aspects of Electrochemistry, Number 22, edited by John O'M Bockris *et al.*
Plenum Press, New York, 1992.

selective, leading to the development of industrial processes based on electrochemical reactions for the production of chemicals.[4,5]

These two aspects, namely, energy conversion and chemical production through the electrocatalytic oxidation of oxygenated organic compounds, will be discussed in this chapter. Emphasis will be given to detailed kinetic studies of electrode reactions. Conversely, the large field of organic polarography and of electroanalytical methods in organic chemistry, which is mainly oriented toward electrochemical reactions whose rates are mass-transfer controlled, will not be examined here.

Oxygenated aliphatic molecules with one to six carbon atoms in length, including monosaccharides, with single or multifunctional groups (alcohols, aldehydes, ketones, carboxylic acids) will be considered here. Most of these molecules are highly soluble in water, so that their electrochemical transformation can be studied in aqueous electrolytes. Organic and nonaqueous solvents will not be discussed here, nor will hydrocarbons or aromatic compounds, which are beyond the scope of this chapter. Some data on the oxidation of hydrocarbons may be found in other review papers.[1,4,6,7]

Electrocatalysis, which plays a key role in the mechanism of electrochemical reactions at solid electrodes, can be defined as the heterogeneous catalysis of electrochemical reactions by the electrode material.[8] Thus, the role of the nature and structure of the electrode is emphasized.[9,10] For example, depending on the nature of the electrocatalytic metal or alloy, the rate of oxidation of methanol to carbon dioxide, expressed as the exchange current density, varies over several orders of magnitude, from 10^{-10} A cm^{-2} for iridium to 10^{-4} A cm^{-2} for platinum-iron alloys.[8]

The structure of a given electrocatalytic metal, such as platinum, also greatly influences the reaction rate, as clearly seen using electrodes of well-defined and controlled structure (such as low-index single crystals). Since the pioneering work of Clavilier *et al.*,[11] who were able to control the preparation of platinum single crystals and to characterize the electrode surface by electrochemical techniques, several papers have been published that definitively show that electrocatalytic reactions such as the oxidation of methanol, formaldehyde, formic acid, and carbon monoxide are structure-sensitive.[12]

Other ways of controlling the catalytic properties of the electrode surface, such as metal alloying and underpotential deposition (upd) of foreign metal adatoms on noble metal electrodes, will be considered. In these two cases, enhanced catalytic effects have been observed, insofar as the electrooxidation of small organic molecules on platinum electrodes is concerned.[13]

Most of the electrode materials considered in this chapter are pure noble metals (Pt, Rh, Pd, Au, Ir, Ru, etc.), either smooth, such as single crystals or preferentially oriented surfaces, or rough, such as polycrystalline metals. They can also be alloyed with other noble or non-noble metals or modified by upd of foreign metal adatoms (Ag, Bi, Cd, Cu, Pb, Re, Ru, Sn, Tl, etc.).

In electrocatalysis, as in heterogeneous chemical catalysis, the substrate affects the reaction kinetics mainly through the adsorption of reactants, reaction intermediates, and products. Therefore, special emphasis will be put on the adsorption of organic molecules at noble metal electrodes. Since the last reviews on the subject, by Damaskin *et al.*[14] and by Breiter (in this series),[15] tremendous progress has been made in the understanding of adsorption processes at solid electrodes, particularly due to both the use of well-controlled surfaces and the development of *in situ* spectroscopic techniques, such as infrared (IR) reflectance spectroscopy, which can provide information on the structure of adsorbed species at the molecular level.[16,17]

Weakly bonded species, acting very often as reactive intermediates, and strongly bonded species, which are irreversibly adsorbed at the electrode surface and which block the electrode active sites (poisoning species), have both been identified by IR reflectance spectroscopy. It is now recognized from spectroscopic measurements that the electrocatalytic poisons formed during the electrooxidation processes of many small organic molecules are adsorbed CO (both linearly and bridge-bonded to the surface),[18-20] and not COH as previously believed.[21,22] It is now understood that relatively long spectral accumulation times favor the formation of strongly bonded intermediates, which accumulate at the electrode surface and displace practically all the weakly bonded species. Therefore, as previously pointed out,[15] whereas the poisoning species are rather easily detected by spectroscopic methods, it is much more difficult to obtain information on weakly bonded species and on reactive intermediates. However, new improvements in the measurement techniques, particularly in the case of mass spectroscopy and real-time IR reflectance spectroscopy, have begun to provide some solutions to this difficult problem. Mass spectroscopy, for instance, has been used to analyze "on-line" the volatile products generated at a porous platinum catalyst supported on a Teflon membrane.[23,24] As only desorbed species are observed, this implies that they come from weakly adsorbed species.[12] In the case of methanol adsorption, the structure of the adsorbed species was suggested to be CHO[25] or COH.[26] With recent IR spectroscopic techniques, very high sensitivity can be achieved (10^{-4} to 10^{-6} in relative changes of absorbance), and shorter adsorption times (on the order of a few seconds) can now be monitored. It has therefore become possible to investigate the early stages of organic adsorption and to observe the time evolution of the IR spectra of the adsorbed species.[27] Similarly, by combining Fourier transform infrared reflection-absorption spectroscopy (FTIRRAS) and electrochemical thermal desorption mass spectroscopy (ECTDMS), it has been possible to detect both reactive intermediates and poisoning intermediates (of the CO type) during ethanol adsorption and oxidation at platinum electrodes.[28]

In this chapter, as mentioned above, detailed reaction mechanisms will be discussed, particularly those concerning the electrocatalytic oxidation of C_1 molecules (CO, HCOOH, HCHO, CH_3OH), for which the extensive work of the last decades has led to the unambiguous identification of the reactive intermediates and poisoning species. Purely electrochemical methods (voltammetry, chronopotentiometry, chronoamperometry, coulometry, etc.) will be discussed first, as they are the basis of any electrochemical study of adsorption and electrooxidation reactions. Then, the newly developed spectroscopic and microscopic techniques will be particularly emphasized, as they have become unique tools, not only for studying *in situ* the structure of the electrode/electrolyte interface at the molecular level, but also for investigating the nature and the structure of adsorbed intermediates. The quantitative analysis of the reaction products by modern analytical techniques (gas and liquid chromatography, mass spectroscopy, nuclear magnetic resonance, Fourier transform infrared spectroscopy, etc.) will also be discussed. These methods provide information on the mass balances and the faradaic yields, which are essential for correctly writing the reaction mechanisms.

In the last part of the chapter, selected examples, all of them dealing with the electrocatalytic oxidation of small aliphatic organic molecules, will be presented. The electrochemical reactivity of different functional groups (alcohols, aldehydes, ketones, carboxylic acids) at various catalytic anodes and the influence of chain length in a homologous series of molecules (e.g., the aliphatic monoalcohols from C_1 to C_6) and of the position of the functional group in structural isomers (e.g., the 4-butanol isomers) will be successively examined. The electrooxidation of multifunctional molecules, particularly dialcohols, dialdehydes, and diacids, will be considered as well. Finally, the oxidation of polyols (ethylene glycol, glycerol, sorbitol, etc.) and monosaccharides (glucose, fructose, etc.) will be discussed, particularly in terms of reaction mechanisms and selectivity of electrode reactions.

II. EXPERIMENTAL METHODS

The investigation of the kinetics of electrocatalytic reactions, such as the electrooxidation of organic compounds, requires the use of several complementary methods, both at the macroscopic level and at the molecular level. The reactivity of the organic molecules, the activity of the electrode materials, and the overall reaction mechanisms are conveniently studied by electrochemical methods, among which voltammetry is particularly suitable.

Adsorbed intermediates, which play a key role in electrocatalytic reactions, may be investigated *in situ* either by electrochemical methods (e.g., pulse voltammetry, coulometry, impedance measurements), by spectroscopic methods (e.g., UV-visible, IR, Raman, mass spectroscopy), or by other physicochemical methods (e.g., radiometric techniques, quartz microbalance, scanning tunneling microscopy).

The nature and structure of the electrode, which determine the electrocatalytic properties of the surface, have to be controlled and may be analyzed by convenient *ex situ* methods, such as electron microscopies (e.g., SEM, TEM), electron diffraction techniques (e.g., LEED, RHEED), and electron spectroscopies (e.g., XPS, ESCA, AES).

To determine the mass balance and the faradaic yields, the reaction products and by-products must also be quantitatively analyzed by analytical methods, such as chromatographic techniques (e.g., GC, HPLC) and other quantitative analytical methods (e.g., FTIR, NMR, MS).

The different experimental techniques now available to investigate electrocatalytic reactions will be outlined briefly in this section, focusing on the type of information obtained by each technique. For more details on the techniques themselves, the reader will be referred to specialized books or reviews.

1. Investigation of the Overall Reaction Kinetics by Electrochemical Methods[29]

The usual kinetic parameters of an electrochemical reaction, such as the exchange current density i_0 (or the standard rate constant k^0), the reaction orders p_i, the stoichiometric number ν, the transfer coefficient α (or symmetry factor), the standard heat of activation ΔH_0^*, and the rate-determining step, are usually determined by steady-state methods (direct potentiostatic or direct galvanostatic methods).

However, for most electrocatalytic reactions involving small organic aliphatic compounds, self-poisoning phenomena arise from the dissociative chemisorption of the molecule. This leads to a blockage of the electrode active sites, so that the current intensity (which is proportional to the electrode area, that is, to the number of active sites), at a given constant applied potential, falls rapidly to small values. Therefore, only by using transient methods on a time scale far shorter than the usual time scale of adsorption processes (i.e., usually a few seconds or a few tens of seconds) and/or by allowing the electrode surface active sites to be regenerated periodically by oxidizing the blocking species at higher potentials, kinetic information on such electrocatalytic reactions can be obtained.

Among transient methods, voltammetry (which is a potentiodynamic method) is one of the most popular electrochemical methods for the investigation of the overall kinetics of electrocatalytic reactions, since it combines the advantages of short time scales (at high sweep rates up to a few thousand volts per second) and large potential variations (able to oxidize the adsorbed poisons). The method consists in applying to the electrode under study a linear potential sweep between two potential limits, E_a and E_c, and recording the current flowing through the electrode as a function of the applied potential, which is equivalent to recording current versus time curves (Fig. 1). A single potential sweep is applied in linear potential sweep voltammetry (LPSV), whereas repetitive triangular waveforms are used in cyclic voltammetry (CV). This method was first used by Will and Knorr to study electrocatalytic processes at noble metal electrodes, namely, the adsorption of hydrogen and oxygen at platinum.[30]

The analysis of adsorption processes in the case of reversible and irreversible charge transfer reactions was discussed by Srinivasan and Gileadi[31] and later by Angerstein-Kozlowska *et al.*[32] and Alquié-Redon *et al.*[33,34] The case of electron transfer reactions coupled to diffusion processes was also considered and treated theoretically,[35-39] but is not directly relevant to electrocatalytic reactions, since, in the latter cases, electrode processes are mostly controlled by adsorption rather than by diffusion.

The voltammetric response $I(E)$, that is, the current intensity (I) versus potential (E) curve, which is called a "voltammogram," displays a current peak I_p at a given potential E_p. This peak results from a depletion of the electrode surface in the electroactive species as a consequence of a limited rate of either mass transfer or the adsorption process (Fig. 1). The analysis of the peak characteristics (E_p, I_p) as a function of the potential sweep rate $v = dE/dt$ allows, in the simplified case of a single kinetic control (either by diffusion or by adsorption), diagnostic criteria to be established for the reaction mechanisms (Table 1). However, for most cases where mixed kinetic control occurs, the analysis is much more complicated, so that the elucidation of the reaction mechanism is more difficult.

Voltammetry is very sensitive to surface processes (electron transfer, adsorption steps, charge of the double layer), so that it allows nonfaradaic adsorption–desorption currents, capacitive currents, or faradaic currents to be recognized immediately. It is also

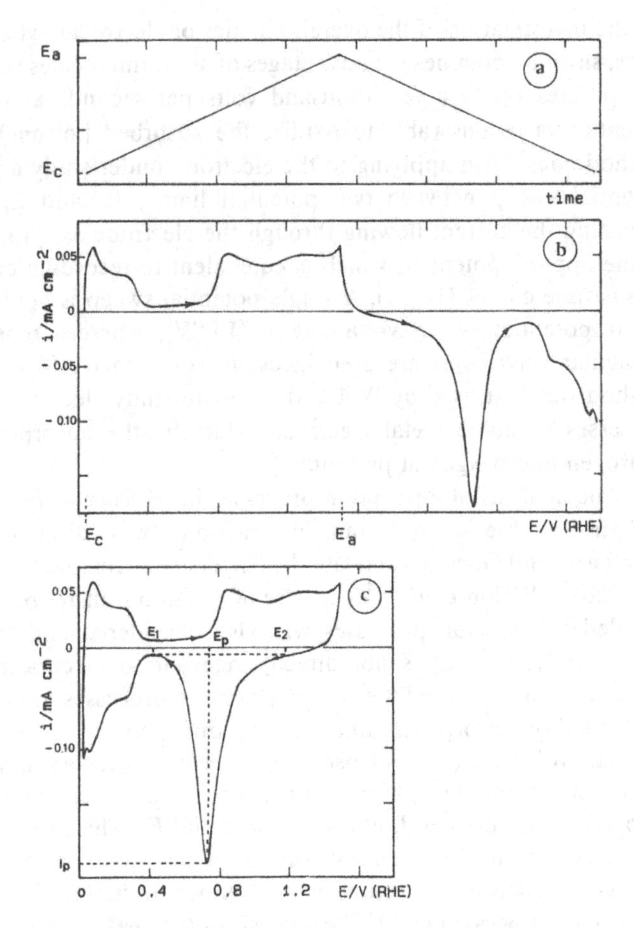

Figure 1. Schematic principle of cyclic voltammetry, illustrated by voltammogram of a Pt electrode in 0.1 *M* $HClO_4$ (25°C, $v = 50\ mV\ s^{-1}$). (a) Cyclic potential sweep; (b) cyclic current response versus time; (c) cyclic voltammogram.

important to emphasize that at high sweep rates ($v > 1\ V\ s^{-1}$) the capacitive current density ($i_c = C_d v$) for charging the double-layer capacity C_d becomes predominant and may obscure the non-faradaic and faradaic currents. Ohmic drop for resistive solutions, and/or at high currents, may also be a problem, particularly in reducing the time scale of analysis.

Table 1
Peak Characteristics of Voltammograms in the Case of a Simple First-Order Transfer Reaction Controlled Either by Adsorption or by Semi-Infinite Linear Diffusion

	Diffusion control	Adsorption control
	Peak potential E_p	
Reversible transfer	$E^0 + \frac{0.0285}{n}$	E^0
Irreversible transfer	$E^0 + \frac{RT}{\alpha n_a F}\left[\ln\left(\frac{\alpha n_a F}{RT} D_i \frac{v}{k_s^2}\right)^{1/2} + 0.78\right]$	$E^0 + \frac{RT}{\alpha n F}\ln\left(\frac{\alpha n F v}{RT k_s}\right)$
	Peak current I_p	
Reversible transfer	$0.446 nF\left(\frac{nFD_i}{RT}\right)^{1/2} C_i^0 \sqrt{v}$	$0.25 \frac{n^2F^2}{RT}[N_{ads}]v$
Irreversible transfer	$0.496 nF\left(\frac{\alpha n_a F D_i}{RT}\right)^{1/2} C_i^0 \sqrt{v}$	$0.368 \frac{\alpha n^2 F^2}{RT}[N_{ads}]v$

With voltammetry, it is possible to check, in one single triangular scan, the reactivity of the organic molecule and the activity of the catalytic electrode. For example, the electroreactivity of butanol isomers at a platinum electrode in alkaline medium is given in Fig. 2, showing that the primary isomers *n*-butanol and isobutanol behave similarly and that the reactivity of *t*-butanol is quite low.[40] Likewise, the catalytic activity of different noble metal electrodes (e.g., Pt, Rh, Pd, Au) for the oxidation of formic acid in neutral medium is shown in Fig. 3. Both the shape of the voltammograms and the maximum current densities obtained are quite different for each metal electrode.[41]

Voltammetry can be carried out at slow potential sweeps ($v \leq 10\ \mathrm{mV\,s^{-1}}$) to obtain quasi-steady-state current-potential curves $I(E)$, which allow kinetic parameters to be determined.[42,43] Practical criteria for the quasi-steady-state character of the polarization curves are that the voltammogram does not vary with the sweep

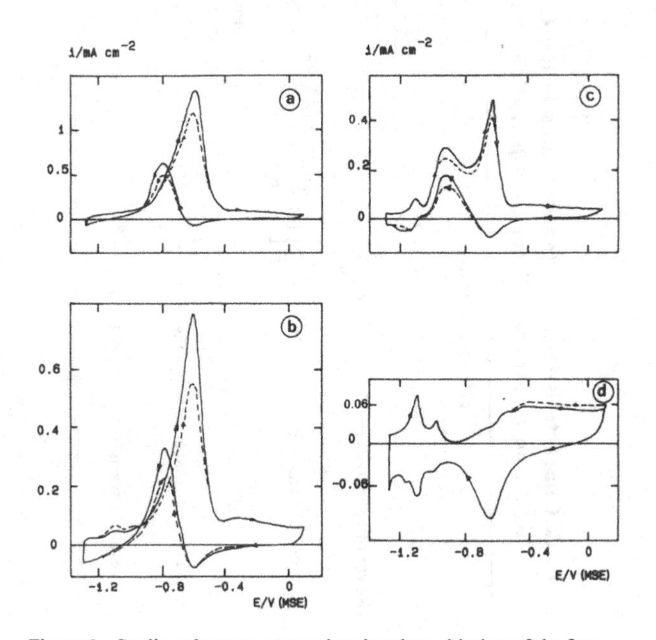

Figure 2. Cyclic voltammograms showing the oxidation of the four butanol isomers at a Pt electrode in alkaline solution (0.1 M NaOH + 0.1 M butanol; 25°C; 50 mV s^{-1}). (a) *n*-Butanol; (b) isobutanol; (c) *s*-butanol; (d) *t*-butanol.

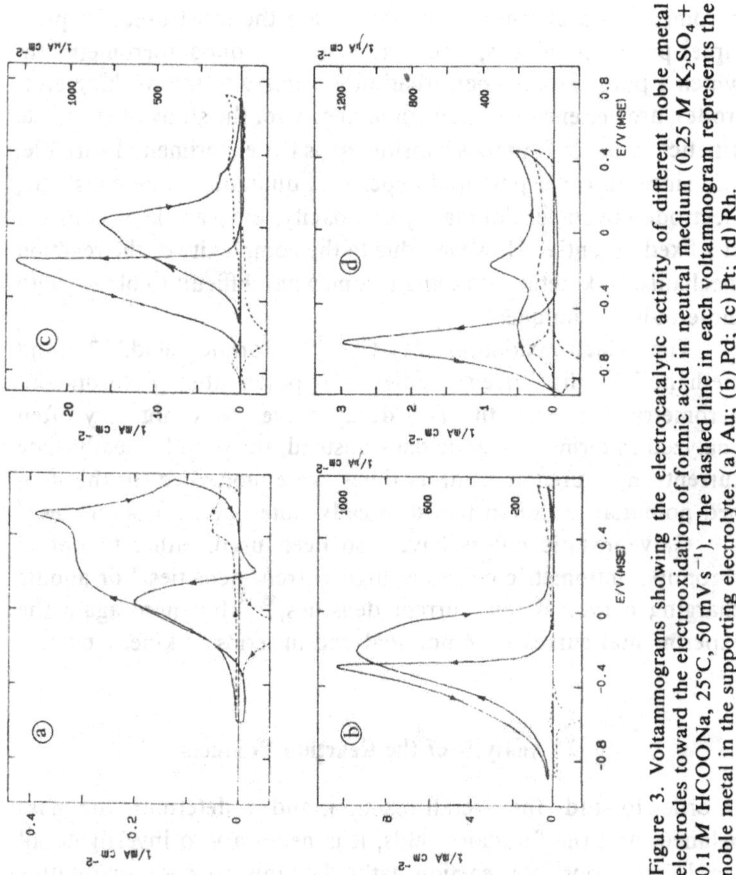

Figure 3. Voltammograms showing the electrocatalytic activity of different noble metal electrodes toward the electrooxidation of formic acid in neutral medium (0.25 *M* K_2SO_4 + 0.1 *M* HCOONa, 25°C, 50 mV s^{-1}). The dashed line in each voltammogram represents the noble metal in the supporting electrolyte. (a) Au; (b) Pd; (c) Pt; (d) Rh.

rate v and that forward and backward sweeps are nearly superimposed. This means that practically no poisoning phenomena are occurring in the course of the reaction.[42]

Of the other transient electrochemical techniques, only a few have been considered for the study of the kinetics of electrocatalytic reactions. Among them, pulse techniques, such as chronoamperometry and chronopotentiometry, are the most used. In principle, pulse techniques, and particularly chronoamperometry (in which a potential-step perturbation is applied to the working electrode), are better suited than voltammetry for the study of electrode kinetics, since only time is considered as the experimental variable, and since all other potential-dependent quantities (rate constants, electrode coverage, double-layer capacity, etc.) are kept constant at a fixed potential. However, due to the complexity of the reaction mechanisms, kinetic data remain somewhat difficult to obtain with these pulse techniques.

The electrooxidation of CO,[44,45] formic acid,[46-48] and methanol[49-53] has been studied by potential-step chronoamperometry. However, the $I(t)$ decay curves were not very often analyzed in terms of kinetic data. Instead, the pseudo-steady-state currents, measured at arbitrary times, were plotted versus the stepped potential to obtain pseudo-steady-state polarization curves.[52]

Galvanostatic pulses have also been used, either to obtain chronopotentiometric curves at high current densities[54] or anodic charging curves at low current densities,[55,56] but here again the experimental curves were not analyzed in terms of kinetic data.

2. Analysis of the Reaction Products

In order to study the overall reaction, and to determine the mass balances and the faradaic yields, it is necessary to investigate all the different possible reaction paths. For this purpose, qualitative and quantitative analysis of the primary, and also the secondary, reaction products must be carried out with suitable analytical techniques.

On the other hand, the quantity of electricity needed to transform the electroreactive compound must be determined in order to calculate the faradaic yield or current efficiency. The faradaic yield may be defined as the theoretical quantity of electricity required for producing a given compound divided by the total experimental quantity of electricity consumed in the process. If the electrolysis is carried out at constant current (as is usually done for large-scale preparations), the quantity of electricity is easily calculated, knowing the duration of the electrolysis. However, in this case, the electrochemical reaction is not controlled, since the electrode potential shifts continuously during the electrolysis, as the consequence of the disappearance of the electrolyzed compound, and since undesired electrochemical reactions may occur. Therefore, it is better to carry out the electrolysis at constant potential, particularly on the laboratory scale, in order to make the electrochemical process more selective. In that case, the quantity of electricity consumed during the electrolysis reaction is better determined using an analog or digital coulometer.[4,7]

For most electrocatalytic reactions involving organic species with more than one carbon atom, the number of reaction products is relatively high. Moreover, some of the secondary products appear only in trace amounts (with concentrations smaller than 10^{-4} M). Furthermore, the different reaction products are very often similar molecules, so that some of their physicochemical properties are not very different, which complicates their detection and separation. Thus, the electroanalytical techniques used must be able to analyze a mixture of similar organic compounds at low concentrations (concentrations ranging between 10^{-6} M and a few molar) dissolved in an aqueous electrolytic solution containing a concentrated supporting electrolyte.

Among quantitative electroanalytical techniques, polarography is one of the most widely used.[57] The current height of the polarographic wave is directly proportional to the concentration of the electroactive species. However, the half-wave potentials of the waves, which are characteristic of the electrochemical system, are generally not very well separated for similar organic compounds, so that polarography may not always be adequate for the analysis of the products and by-products resulting from the electrocatalytic oxidation of organic compounds.

Analytical methods employed in organic chemistry are much more adaptable to the quantitative analysis of electrolysis reaction products. Both chromatographic and spectroscopic methods will be discussed in this section, particularly those allowing "on-line" analysis.

(i) *Chromatographic Techniques*[58]

Chromatography is an analytical method in which the components of a mixture in a mobile phase, gaseous or liquid, are separated on an immobile phase, called the stationary phase. The stationary phase may be a solid, or a liquid supported on a solid, or a gel, either packed in a column or spread on a supporting plate as a layer or distributed as a film. The solvent of a liquid mobile phase is usually called the "eluent," whereas the inert gas of a gaseous mobile phase is called "carrier gas."

After separation of the mixture, either on a plate or through a column depending on the particular technique used, the different components are located as spots on the plate or as concentration peaks at the output of a detector. The detector output signal, plotted against time of analysis, or against the volume of the mobile phase, is called a "chromatogram" (Fig. 4).

A concentration peak in a chromatogram is characterized by its position (relative retention time, t_r, or relative retention volume, V_r), referenced to the position of a given compound which does

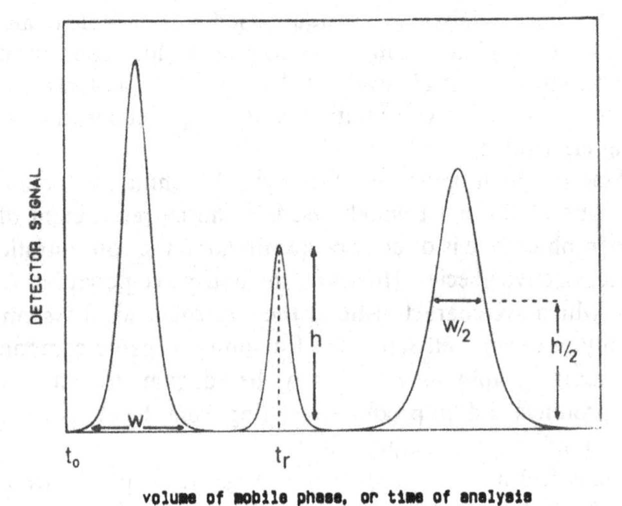

Figure 4. Theoretical chromatogram.

not interact with the stationary phase, and its intensity (peak height h and width at half-height $w/2$). The retention time depends on many factors, such as the distribution coefficient of the components between the stationary phase and the mobile phase, the geometric characteristics of the column, the temperature, the pressure, and the flow rate of the eluent or of the carrier gas. Reference compounds, whose chromatograms are recorded under exactly the same experimental conditions, are used to assign a chromatogram peak to a given substance. The amount (mass or concentration) of the analyzed products is determined by comparing the integrated intensity of the chromatogram peaks (the surface area under the peak, determined by integration of the detector signal) to those of the reference samples at different known concentrations.

Many different types of chromatographic techniques are available, differing in the method of separation and/or the nature of the phases involved. The most widely used are described below.

(*a*) *Thin-layer chromatography* (*TLC*)

In thin-layer chromatography a solution of the sample in a volatile solvent is deposited at the bottom of a uniform layer of an inert adsorbent, such as silica or alumina, spread over a supporting plate made of glass or plastic. Then the plate is placed vertically in a container, the bottom of which is filled with the mobile phase. The solvent rises by capillarity and separates the sample mixture into discrete spots. At the end of the experiment, the solvent is allowed to evaporate, and the separated spots are identified either by physical methods or by chemical reactions. A reference compound is usually used for comparison.

For example, the detection of carboxylic acids, or of ketones and aldehydes, is achieved using a chromogenic reagent, called a "locating agent," which colors the separated spots (the different spots are thus directly visualized, or more easily detected by fluorescence, after illumination by a UV lamp at 254 nm). In the case of carboxylic acids, the locating agent, such as Bromophenol Blue, is added to the eluent, a solvent mixture of ethyl acetate, acetic acid, and water, and the separation is carried out on a cellulose plate. For detection of ketones and aldehydes, a derivatizing agent, such as 2,4-dinitrophenyl hydrazine, is used, in order to transform these compounds into hydrazones, which are easily located on a silica gel plate as yellow spots, after elution by a mixture of chloroform, methanol, and acetic acid.

(b) Gas chromatography (GC)

In gas chromatography, the separation of the sample mixture, transported by an inert carrier gas (e.g., N_2, H_2, He), is achieved in a column containing the stationary phase. In gas–solid chromatography (GSC or GC), the stationary phase is a solid adsorbent, which retains selectively the different molecules of the sample (adsorption chromatography), whereas in gas–liquid chromatography (GLC or GC) the stationary phase is a liquid, which dissolves selectively the different compounds contained in the sample (partition chromatography). Separated products are characterized using various detectors, such as the flame ionization detector (FID), the thermal conductivity detector (TCD), the electron capture detector (ECD), and the nitrogen phosphorus detector (NPD); among these, the FID is the most widely used because of its reliability, high sensitivity, and good linearity.

A schematic diagram of a gas chromatograph is given in Fig. 5a. Gas chromatography has been used to identify gaseous reaction products during the electrooxidation of methanol,[59,60] methylal,[60] and glyoxal.[61]

(c) Liquid chromatography (LC)

In liquid chromatography, the mobile phase is a moving liquid, called the "eluent." A packed column is used as the stationary phase. This column contains either a solid adsorbent (adsorption chromatography), an ion-exchange resin (ion-exchange chromatography), a porous gel matrix (gel permeation chromatography), or a support for a liquid stationary phase (partition chromatography). High-pressure liquid chromatography, or high-performance liquid chromatography (HPLC), makes use of microparticulate materials in the column packing, which necessitates the use of high pressure pumps to overcome the large pressure drops across the column. Various detectors are now commonly used, such as the refractive index detector (RD), which detects the passage of the solute by recording the corresponding refractive index change of the eluent,

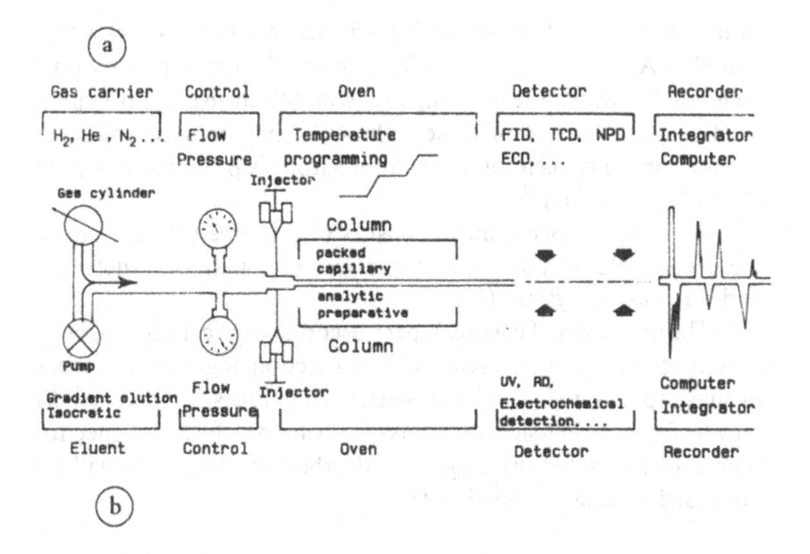

Figure 5. Schematic diagram of a gas chromatograph (a) and a liquid chromatograph (b).

or a UV spectrophotometer (working either at a fixed wavelength or at variable wavelengths), which monitors the change in absorbance of the eluent as the solute passes through the detector flow cell.

A schematic diagram of a liquid chromatograph, which is similar to that of a gas chromatograph, is shown in Fig. 5b.

The development of automated liquid chromatographs has made this technique a very useful analytical tool for the qualitative and quantitative analysis of mixtures of organic compounds (even at concentrations as low as 10^{-6} M) in electrolyzed aqueous solution.

Liquid chromatography has been successfully applied to the quantitative analysis of liquid reaction products of many electrochemical reactions, for example, reduction of glucose[62] and oxidation of methanol,[63,64] glyoxal,[61] and 1,2-propanediol.[65]

(ii) *Spectroscopic Techniques*

Spectroscopic techniques are currently used in organic chemistry for qualitative and quantitative analysis of reaction mixtures. Many good textbooks covering spectroscopic techniques are available, among them the series *Physical Methods of Chemistry*, edited by A. Weissberger and B. W. Rossiter.[66] Except spectroscopic techniques that require ultra-high vacuum (electron spectroscopies, such as ESCA), most of these techniques may be used "on-line" to monitor the concentration of the electrolysis products inside the electrolytic medium.[67]

Basically, a spectrometer consists of a source (of light, mass, etc.), a sample-inlet system, an analyzer or separator, a detector, and a recording system (Fig. 6).

The principles of classical spectroscopies (optical and magnetic spectroscopies) will be discussed in this section, together with those of mass spectrometry and radiometric techniques. Further details may be found in classical textbooks,[68,69] whereas the use of spectroscopic methods for investigating adsorbed intermediates will be discussed in detail in Section II.3(iv).

(*a*) *Optical spectroscopies*

Optical spectroscopies, that is, spectroscopic techniques involving the absorption by molecules of light in the UV-visible and infrared ranges, are particularly useful for the qualitative and quantitative analysis of reaction products. They are based on the dipolar interaction between the electric field vector $\mathbf{E}(\nu)$ of the incident light (electromagnetic waves characterized by their electric and magnetic field vectors, oscillating at frequency ν, perpendicular to each other and to the direction of propagation) and the electrical dipole moment $\mathbf{p}$ of the organic molecule:

$$W_{\text{dip}} = -\mathbf{p} \cdot \mathbf{E}(\nu) \tag{1a}$$

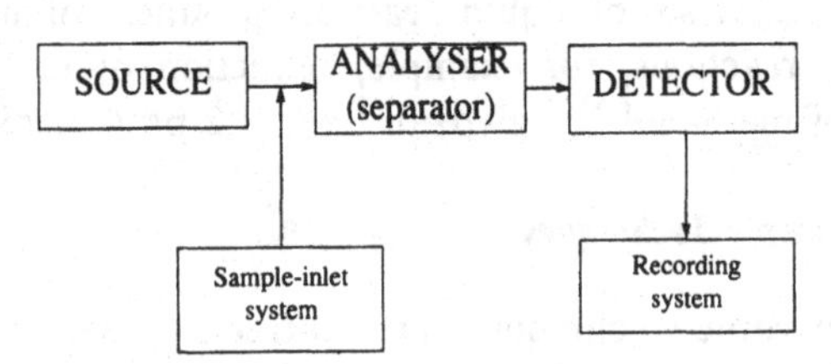

Figure 6. Schematic diagram of a spectroscopic experiment.

The electrical dipole moment $\mathbf{p} = q\mathbf{r}$ (where q is the electric charge, and $\mathbf{r}$ is the vector linking the two charges of opposite signs) results either from the electron motion inside the molecule (electronic spectroscopy in the UV-visible range) or from the asymmetric distribution of the electronic charge in the chemical bonds (vibrational spectroscopy in the infrared range and Raman spectroscopy).

Absorption spectra are recorded as the light intensity at the output of the spectrophotometer versus the frequency ν (or versus the wavenumber $\bar{\nu} = 1/\lambda = \nu/c$, or versus the wavelength $\lambda = c/\nu$, where c is the speed of light in vacuum) of the incident light. The spectra obtained exhibit absorption bands at frequencies ν_0 characteristic of the sample analyzed and related to the difference in energy between the ground state E_1 and the excited state E_2 of the molecule (Planck's relation):

$$h\nu_0 = E_2 - E_1 \tag{2}$$

The intensity of the spectrum is proportional to the transition probability, P_{12}, between these two quantum states under the effect of the incident light:

$$P_{12} = \frac{1}{\hbar^2} |W_{12}|^2 \rho(\nu) \tag{3}$$

where $\hbar = h/2\pi$, $W_{12} = \int \varphi_1 W_{\text{dip}} \varphi_2 \, d\tau$ is the matrix element of the dipolar interaction W_{dip} taken over the wave functions φ_1 and φ_2 of the quantum states, $d\tau$ is a small volume element of the corresponding coordinates, and $\rho(\nu)$ is the energy density of the electromagnetic radiation. The quantum states (E_i, φ_i) are associated either with the motion of electrons in the molecule (UV-visible spectroscopy), with the vibrations of the atoms in the molecule (infrared and Raman spectroscopies), or with the rotation of the molecule as a whole (rotational spectroscopy, which is not considered here because the rotational levels are smeared out in condensed media).

Qualitative analysis is based on the assignment of the characteristic frequencies to some functional groups of the molecules (fingerprints; see Tables 2 and 3). Quantitative analysis makes use of the Beer-Lambert law:

$$I = I^0 e^{-\alpha(\nu)Cl} \tag{4}$$

where I^0 and I are the intensity of incident and transmitted light,

Table 2
Characteristic Frequencies of Electronic Transitions

Group or compound	λ_{max} (Å)	ε_{max} (l mole^{-1} cm^{-1})
C—C	1350	
C=C	1900	10,000
C=O	1900	1,000
	2800	20
O—H	1850	200
C_6H_5 (phenyl)	1950	8,000
	2500	200
H_2O	1667	1,480
MeOH	1835	150
Me_2O	1838	2,520
Benzene	2550	200

respectively, l is the thickness of the absorbing layer, C is the molar concentration of the absorbing species, and $\alpha(\nu)$ is the linear absorption coefficient at a given frequency. This law is usually written as a semilogarithmic law, which allows us to define the absorbance A of the sample as

$$A = \log_{10}(I^0/I) = \varepsilon Cl \tag{5}$$

where $\varepsilon = \alpha/2.3$ is the extinction coefficient.

A plot of absorbance versus the frequency ν of the incident light, or the wavelength $\lambda = c/\nu$, is an absorbance spectrum.

The unknown concentration C of a given absorbing species can be obtained directly from the measurement of the absorbance if the extinction coefficient ε is known at the frequency of the measurement. Otherwise, a calibration curve is first plotted from the absorbance data for different known concentrations of the species in the sample cell.

Absorbance spectroscopy in the UV-visible range is used much more for quantitative analysis than absorbance spectroscopy in the infrared range. However the latter technique is very useful for qualitative analysis, because infrared bands are narrower and are thus much more easily assigned to particular functional groups (fingerprinting) than UV-visible bands.

Table 3
Characteristic Wavenumbers ($\bar{\nu}$) for Vibrational Transitions of Diatomic Molecules and Wavenumber Ranges for Vibrational Modes

	$\bar{\nu}$ (cm^{-1})
Diatomic molecules	
HF	3958
HCl	2885
HBr	2559
HI	2230
HD	3817
CO	2143
NO	1876
Vibrational modes	
C—H stretching	2650-2960
C—H bending	1340-1465
C—C stretching	700-1250
C=C stretching	1620-1680
C≡C stretching	2100-2260
O—H stretching	3590-3650
C=O stretching	1640-1780
CO_3^{2-}	1410-1450
H—bonds	3200-3570
C—F stretching	1000-1400
C—Cl stretching	600-800
C—Br stretching	500-600
C—I stretching	500

There are only a few papers on the analysis of electrolysis reaction products by UV-visible spectroscopy. One may cite the study of aldehyde oxidation at glassy carbon, mercury, copper, silver, gold, and nickel anodes[70] and the study of the reduction of substituted glyoxal at a mercury electrode.[71]

Similarly, infrared spectroscopy has been used to characterize the reaction products resulting from the oxidation of alcohols at a nickel anode in alkaline *t*-butanol/water mixtures.[72]

The development of Fourier transform infrared (FTIR) spectroscopy, associated with large computational facilities, makes infrared spectroscopy a very powerful technique for qualitative analysis of reaction products inside the electrolysis cell.[18] The availability of large computer memories leads to a refinement of the analysis whereby a stored reference spectrum is subtracted from the sample spectrum recorded either at a given potential (SNIFTIRS)[73] or during a potential sweep (SPAIRS)[20]; see Section II.3(iv.e). This has allowed qualitative and quantitative analysis, not only of bulk species in the reaction layer adjacent to the electrode surface, but also of adsorbed species at the electrode surface.[74]

(*b*) *Magnetic spectroscopies*

Magnetic spectroscopies are based on the dipolar interaction between the magnetic field vector $\mathbf{B}(\nu)$ of the incident electromagnetic wave at a frequency ν and the magnetic dipole moment $\boldsymbol{\mu}$ of the molecule:

$$W_{\mathrm{mag}} = -\boldsymbol{\mu} \cdot \mathbf{B}(\nu) \tag{1b}$$

The magnetic dipole moment $\boldsymbol{\mu}$ is usually proportional to the spin angular momentum $\mathbf{I}$:

$$\boldsymbol{\mu} = \gamma \mathbf{I}$$

with γ the gyromagnetic ratio ($\gamma = q/2m$). $\mathbf{I}$ is quantized; that is, its projection I_z on the z axis takes $2I + 1$ values, where I is the spin quantum number.

If the dipole moment of the molecule arises from the nuclear spin $\mathbf{I}$ of the atoms inside the molecule, the method is called nuclear magnetic resonance (NMR). If the dipole moment is associated with the electronic motion, particularly with the electronic spin angular momentum $\mathbf{S}$ (free radicals), the method is called electron spin resonance (ESR) or electron paramagnetic resonance (EPR).

Because the nuclear (or electron) spins are randomly distributed inside the sample, there is no net magnetization vector $\mathbf{M}$ in the absence of a magnetic field. Moreover, the magnetic energy levels are degenerate, with a degeneracy factor equal to $2I + 1$ (or $2S + 1$ for electron spins), where the spin quantum number I (or S for electrons) is integer or half-integer.

To observe spectral transitions, it is thus necessary to apply a static magnetic field $\mathbf{B}_0$, which will induce a net magnetization vector $\mathbf{M}_0 = \chi \mathbf{B}_0$ (where $\chi > 0$ is the paramagnetic susceptibility), because the energy levels are now separated into $2I + 1$ (or $2S + 1$) magnetic levels, whose population follows the Boltzmann distribution law.

For the transitions between the magnetic energy levels, Planck's relation [Eq. (2)] still applies, leading to the resonance equation:

$$h\nu_0 = E_2 - E_1 = g\beta B_0 \tag{6}$$

with β the Bohr magneton ($\beta_e = e_0\hbar/2m_e$ for the electron, or $\beta_N = e_0\hbar/2M_p$ for the proton), and g the "g factor" (the nuclear "g factor," g_N, is related to the gyromagnetic ratio γ_N by $\gamma_N\hbar = g_N\beta_N$).

It follows that magnetic resonance spectra are usually recorded as the absorption intensity versus the strength of the applied static magnetic field B at constant frequency ν_0 of the electromagnetic wave. In the different types of magnetic spectroscopy, the frequency ranges considered are quite different. For NMR spectroscopy, radio frequency waves of a few tens to a few hundreds of megahertz (corresponding to wavelengths of a few meters) are used depending on the nuclear spin investigated (60 MHz for protons, 15 MHz for ^{13}C nuclei at a magnetic field of about 15 kT), whereas for ESR spectroscopy, microwaves of 10 GHz, or 3-cm wavelength, are used (for an X-band ESR spectrometer).

The absorption intensity is proportional to the number of spins per unit volume, thus allowing a quantitative determination of the unknown concentration of a sample. Absolute concentration measurements are usually very difficult, so that a reference sample is often used [e.g., α,α'-diphenyl-β-picrylhydrazyl (DPPH) in ESR spectroscopy].

Apart from the main interaction with the applied magnetic fields [static field $\mathbf{B}_0$ and electromagnetic field $\mathbf{B}(\nu)$], there are small dipolar interactions inside the molecule, leading to small variations in the positions of the energy levels. This leads to small displacements of the positions of the absorption bands ("chemical shift" in NMR spectroscopy, ligand field interaction in ESR spectroscopy) and to the occurrence of several lines in the spectrum (the so-called fine structure in NMR due to dipolar coupling between different nuclear spins and the hyperfine splitting in ESR, which arises from dipolar coupling between nuclear and electronic spins). The fine and hyperfine structures are characteristic of the molecules and, in most cases, allow their identification (in a similar way to fingerprinting in optical spectroscopies).

NMR spectroscopy has been successfully used for qualitative and quantitative analysis of electrolyzed solutions, but very few results are given in the literature.[70-72] One may also cite the analysis of reaction products during prolonged electrolysis of glycerol.[75]

ESR spectroscopy, which is a very sensitive method (as low as 10^{12} spins, i.e., about 10^{-12} mol, can be detected), has been mainly applied to the investigation of radical intermediates, which occur in the reaction mechanisms of many electroorganic reactions.[76] Specially designed electrochemical flow cells, small enough to fit the resonant cavity, have been used so far.[77,78] Most applications have dealt with organic radicals and are beyond the scope of this chapter. In the case of the oxidation of methanol and formaldehyde at a platinum electrode, simultaneous electrochemical electron spin resonance spectroscopy (SEERS) has allowed $\cdot$CHO radicals to be identified as reaction intermediates,[79] but because of the way in which the experiment was done, using spin traps, its relevance to electrocatalysis under normal experimental conditions is questionable.[80]

ESR spectroscopy can also help in the understanding of electron exchange inside an electrode material, such as intercalation compounds.[77]

(*c*) *Other spectroscopic and related techniques*

Mass spectroscopy.[81,82] In classical mass spectroscopy, the organic molecule is ionized by electron bombardment, and then the molecular ion beam is accelerated by an electric field and focused on the detector, by a uniform magnetic field (mass analyzer) with a deflection depending on the ionic mass. The principle of a mass analyzer based on the deflection of the ion beam in a magnetic field is the following. When an ionic species (mass m, charge q) is accelerated by a potential difference V, it acquires a kinetic energy

$$\tfrac{1}{2}mv^2 = qV$$

that is, a velocity

$$v = \left(2V\frac{q}{m}\right)^{1/2}$$

When the ions enter a uniform magnetic field of strength B_0, acting perpendicular to the initial direction of motion, they experience the Laplace force $\mathbf{f} = q\mathbf{v} \times \mathbf{B}$, so that their motion becomes semicircular. The radius of the circle R is obtained by equating the magnetic force strength, $f = qvB$, acting centripetally, to the centrifugal force $f = mv^2/R$, so that

$$R = \frac{m}{q}\frac{v}{B} = \left(\frac{m}{q}\right)^{1/2}\left(\frac{2V}{B^2}\right)^{1/2}$$

or

$$\frac{m}{q} = \frac{B^2R^2}{2V} \tag{7}$$

For a given magnetic geometry (R fixed), the mass-to-charge ratio is thus related to the magnetic field strength B and to the accelerating potential V, so that the entire mass spectrum can be obtained, in principle, by scanning either B or V. Practically, V cannot be varied by more than a factor of 10 (e.g., from 4000 to 400 V), so that magnetic field scanning is much more advantageous for obtaining the entire mass spectrum (for $V = 4000$ V and $R = 50$ cm, a mass spectrum from 20 to 1000 m/q units is obtained by scanning the magnetic field from 8.2×10^{-2} T to 0.407 T). Other types of mass analyzer are also used, such as the quadrupole mass filter, the ion trap detector, and the time-of-flight analyzer.

Several types of detectors have been considered, such as the Faraday cup, the electron multiplier, the electro-optical ion detector, and a photographic plate. The electron multiplier detector is the most widely used, because of its excellent sensitivity (allowing the analysis of a few nanograms) and its good response time.

The introduction of the sample molecules may be either direct or through a gas-chromatographic inlet. The advantage of the latter is that the sample mixture is separated before its introduction into the ionization chamber of the mass spectrometer. The gas chromatography/mass spectroscopy (GC–MS) combination is a very powerful technique for the analysis of mixtures, which, without chromatographic separation, would lead to extremely complicated mass spectra (with many peaks for each component) that would be impossible to interpret correctly.

The ionization of the molecule is usually achieved by interaction with accelerated electrons (typically having an energy of 70 eV) emitted from a hot filament. Electron bombardment of a molecule AB leads to several ions, including parent ions (AB^+, AB^{2+}, or AB^-) produced by ionization of the molecule, fragment ions (A^+, B^+, A^-, B^-, etc.) produced by dissociation of the molecule, and rearrangement ions produced by redistribution of atoms or groups of atoms. Thus, a given molecule will give rise to a large number of peaks in the mass spectrum (fingerprint analysis).

Other modes of ionization, such as field ionization, which employs an electric field of 10^7 to 10^8 V cm^{-1}, transfer a much smaller energy to the molecule, so that predominantly parent ions are produced, making the spectra easier to interpret.

On-line mass spectrometry was first used in electrochemistry by Bruckenstein and Comeau,[83] to monitor the gas evolved during electrochemical reactions. The method was further developed by Heitbaum, Vielstich, and co-workers to investigate reaction mechanisms of the electrocatalytic oxidation of small organic molecules (methanol, formic acid, carbon monoxide). In these experiments, the cyclic voltammogram and the mass intensity versus potential curves of the gaseous and volatile species involved were recorded simultaneously.[84]

The nature of the primary reaction products of the electrocatalytic oxidation of glucose in phosphate buffer solutions has been determined by "on-line" mass spectroscopy.[85,86] Adsorbed intermediates produced during the chemisorption of small organic molecules (e.g., HCOOH, CH_3OH, C_2H_5OH) have also been investigated by differential electrochemical mass spectroscopy (DEMS),[87] by electrochemical thermal desorption mass spectroscopy (ECTDMS),[88] and by secondary ion mass spectroscopy (SIMS).[89] More details will be given in Section II.3(v).

Tracer methods.[90,91] Tracer methods consist in introducing into the electrochemical system trace amounts of an isotope of a particular element (tracer) in the same chemical form as the investigated compound and following its fate during the electrochemical reaction. The detection of the isotopically labeled compound is achieved either by mass spectroscopy (tracer method) for stable isotopes or by radiometry (radioactive tracer method) for radioactive isotopes.

Although tracer methods are available for qualitative analysis of reaction products, particularly after their separation by chromatographic techniques,[58] they have been mainly used in electrochemistry for adsorption studies, as discussed in Section II.3(vi).

3. Study of the Adsorbed Species

(*i*) *Introduction*

In electrocatalytic reactions, as in heterogeneous catalysis, the surface species, particularly the adsorbed species, play a key role in the activity of the catalytic electrode (as reactive intermediates, poisoning species, etc.) and in the selectivity of the reaction (as primary determinant of the elementary steps and the pathways taken in complex reactions).

Let us consider an electrocatalytic reaction involving an oxygenated organic compound, $C_xH_yO_z$. Its complete oxidation to carbon dioxide may require the donation of oxygen atoms, which come from water molecules, or water ions, in aqueous medium:

$$C_xH_yO_z + pH_2O \rightarrow xCO_2 + nH^+ + ne^-$$

with $p = 2x - z$ and $n = 4x + y - 2z$.

In fact, even for the oxidation of a small molecule, such as methanol, the overall reaction is relatively complex and involves several electron transfers; for example, in acid medium,

$$CH_3OH + H_2O \rightarrow CO_2 + 6H^+ + 6e^-$$

or, in alkaline medium,

$$CH_3OH + 8OH^- \rightarrow CO_3^{2-} + 6H_2O + 6e^-$$

Such complex reactions proceed through multiple elementary steps and pathways, each of them involving no more than one or two electrons. The interaction of the catalytic surface with the electroreactive species leads to the formation of different adsorbed intermediates, some of them acting as reactive intermediates (RI), others as poisoning intermediates (PI); see Section III.1.

This is illustrated by the oxidation of formic acid on platinum electrodes, which is sometimes considered as one of the most simple electrocatalytic reactions. At potentials greater than 0.45 V versus

RHE, where adsorbed hydrogen is ionized,[21,92] this reaction can be written as follows:

$$\mathrm{HCOOH} \begin{cases} \nearrow -\mathrm{COOH_{ads}(RI)} + \mathrm{H_{aq}^{+}} + e^{-} \\ \searrow -\mathrm{CO_{ads}(PI)} + \mathrm{H_2O} \end{cases}$$

$$-\mathrm{COOH_{ads}} \rightarrow \mathrm{CO_2} + \mathrm{H_{aq}^{+}} + e^{-}$$

$$\mathrm{H_2O} \rightarrow -\mathrm{OH_{ads}} + \mathrm{H_{aq}^{+}} + e^{-}$$

$$\mathrm{CO_{ads}} + \mathrm{OH_{ads}} \rightarrow \mathrm{CO_2} + \mathrm{H_{aq}^{+}} + e^{-}$$

The poisoning intermediates are strongly bonded to the electrode surface, thus blocking the electrode active sites and causing the electrode activity to decrease with time. The role of poisoning intermediates in the electrooxidation of most organic molecules has been recognized for a long time. They were called O-type species by Breiter,[15] whereas the weakly bonded desorbable species were called C—H-type species. The strongly bonded species only oxidize at potentials greater than 0.6 V versus RHE, potential at which the platinum surface begins to oxidize, and behave similarly to chemisorbed carbon monoxide [see Section III.1(iii)].

(ii) Adsorption Isotherms and Adsorption Kinetics

The reaction kinetics depend strongly on the amount of adsorbed species, that is, on the degree of coverage θ_i of the electrode surface by adsorbed species *i*.

The electrode surface coverage θ_i may be defined as the ratio of the number of adsorbed species N_{ads} per unit area S (surface concentration $\Gamma = N_{\mathrm{ads}}/S$) to the maximum surface concentration Γ^m:

$$\theta = \Gamma/\Gamma^m \qquad 0 \leq \theta \leq 1 \tag{8a}$$

It can also be defined as the ratio of the number of adsorbed species, N_{ads}, to the number of adsorption sites on the substrate, N_S:

$$\theta = N_{\mathrm{ads}}/N_S \tag{8b}$$

However, in the latter definition, there remains some ambiguity, because the number of adsorption sites on the substrate may depend on the nature of the adsorbed molecule and on the heterogeneity of the surface. Moreover, multilayer adsorption may occur so that

θ may be greater than unity. So, another alternative is to define θ as the fraction of the surface area occupied by the adsorbed species ($0 \leq \theta \leq 1$). This implies that the total surface area which can be blocked by adsorbed organic species is known.

The determination of the real active surface area S is a very important point, since the current density $i = I/S$, where I is the measured current intensity, has to be calculated in order to compare the electrocatalytic activity of different electrodes. As discussed in Section II.4(ii.a), S can be easily determined for catalytic electrodes (e.g., Pt, Rh) from the quantity of electricity Q_H^o required to deposit or to oxidize a full monolayer of adsorbed hydrogen, provided that the theoretical quantity of electricity, Q_H^{theor}, required per unit surface area (1 cm^2 usually) is known.

If Q_H is the charge required to deposit hydrogen on the adsorption sites free of organic adsorption (free surface), θ can be written as

$$\theta = 1 - \frac{Q_H}{Q_H^o} = \frac{Q_H^o - Q_H}{Q_H^o} \tag{9}$$

Thus, this definition implicitly assumes that the adsorption sites are the same for hydrogen and for the organic molecule.

One may determine the number of adsorbed organic species from the knowledge of the quantity of electricity Q_{org} necessary to oxidize them completely to CO_2; that is,

$$Q_{\text{org}} = ne_0N_{\text{ads}}$$

with n the number of electrons involved (number of electrons per molecule or N_{epm}), and e_0 the elementary charge.

Thus,

$$\theta = \frac{Q_{\text{org}}}{Q_{\text{org}}^m} \tag{8c}$$

where Q_{org}^m is the maximum value of Q_{org} achievable at saturation coverage of the surface by the organic species. The number of electrons per molecule can be calculated as

$$N_{\text{epm}} = \frac{Q_{\text{org}}}{e_0N_{\text{ads}}} \tag{10}$$

provided that N_{ads} can be determined independently (e.g., by radio-tracer methods). Otherwise, some hypothesis regarding the value of N_{epm} (which must be an integer) can be made.

The number of electrons per site (N_{eps}) is then defined as the number n of electrons required to oxidize the adsorbed species divided by the number x of hydrogen sites that the adsorbed species occupies[15,93]:

$$N_{eps} = \frac{n}{x} = \frac{Q_{org}}{e_0 N_{ads}} \Big/ \left(\frac{Q_H^o - Q_H}{e_0 N_{ads}} \right)$$

Therefore,

$$N_{eps} = \frac{Q_{org}}{Q_H^o - Q_H} \tag{11}$$

The number of electrons per site is easily measurable since one needs only to determine the quantity of electricity required either to completely oxidize the adsorbed residue (Q_{org}) or to deposit or remove adsorbed hydrogen (Q_H^o and Q_H).

The determination of N_{eps}, which is a ratio of two integers, n and x, may help in the elucidation of the nature of the chemisorbed species χ_{ads} arising from the adsorption of C_1 molecules,[93] as shown in Table 4. However, the assumptions are not at all unambiguous, as discussed in Section III.1, since the same N_{eps} (e.g., 1) holds for different adsorbed species, and since a mixture of different adsorbed species may lead to an averaged intermediate value of N_{eps}.

In the electrocatalytic oxidation of small organic molecules, the chemisorption process is usually dissociative, that is, quite

Table 4
N_{eps} for Adsorbed C_1 Molecules

χ_{ads}	N_{eps}
—CHO	3
—CO	2
\\CO/ (bridged, two bonds)	1
\\—COH/ (three bonds)	1
—COOH	1
\\—CO/ (three bonds)	0.66

irreversible. However, it is usual to relate θ_i to the concentration C_i of a reactive species i in solution and to describe adsorption isotherms in the same way as at the gas/solid interface.[7,14,94] Details on the derivation of isotherms and on the different types of isotherms can be found in review papers.[7,94] The adsorption isotherms most often encountered at the electrode/electrolyte interface are the Langmuir isotherm:

$$\frac{\theta_i}{1-\theta} = K_i C_i \qquad \text{with } \theta = \sum_i \theta_i \tag{12a}$$

and the Temkin-Frumkin isotherm:

$$\frac{\theta_i}{1-\theta} = K_i C_i \, e^{-g\theta_i} \tag{12b}$$

where K_i is the adsorption equilibrium constant (related to the adsorption Gibbs free energy, $\Delta G_i^a = -RT \ln K_i$), and g is an interaction factor between adsorbed species.

The adsorption of organic species at the electrode/electrolyte interface also depends strongly on the electrode potential E and is therefore called "electrosorption." The dependence of dissociative chemisorption on E is relatively complex, since E affects the interaction of both the water adsorption residues (H_2O, H_{ads}, OH_{ads}) and the adsorbed organic species with the electrode surface and changes the amount of adsorbed species by oxidation or reduction. Therefore, bell-shaped adsorption curves $\theta(E)$ are usually obtained for the electrosorption of organic compounds.[6,95]

The kinetics of adsorption is also an important limiting factor which may play a significant role in the overall reaction kinetics.[14,94] Kinetic laws may be obtained by plotting the measured concentration of adsorbed species (or a quantity, such as Q_{org}, directly related to it) as a function of adsorption time. The simplest kinetic law is obtained under Langmuir conditions. For the adsorption process of species i, it is written as

$$v_{ads} = \frac{d\theta_i}{dt} = k_a C_i (1-\theta) \tag{13}$$

and for the desorption process as

$$v_{des} = -\frac{d\theta_i}{dt} = k_d \theta_i \tag{14}$$

where k_a and k_d are the rate constants for adsorption and desorption, respectively.

The adsorption kinetics for small organic C_1 molecules is very often governed by the Roginskii–Zel'dovich equation[96]:

$$v_{\text{ads}} = \frac{d\theta_i}{dt} = k_a C_i \exp(-\beta f \theta_i) \tag{15}$$

where f is the heterogeneity factor of the surface, and β is a symmetry factor (equivalent to the transfer coefficient α in electrode transfer reactions). This kinetic law will lead to the Temkin adsorption isotherm:

$$\theta_i = a + \frac{\ln C_i}{f} \tag{16}$$

which is an approximation of the Temkin–Frumkin isotherm for $0.2 \leq \theta_i \leq 0.8$. This shows that the surface heterogeneity factor, f, is equivalent to the interaction factor, g, between adsorbed molecules.

When the reaction rate is controlled by both adsorption and diffusion, the rate expressions are much more difficult to write.[97] In simplified cases, such as at small adsorption times for which the concentration $C_i(0, t)$ of the electroactive species at the electrode surface remains small compared to the bulk concentration C_i^0 $[C_i(0, t) \leq C_i^0]$, the adsorption kinetics under diffusion control may be written as

$$\theta_i(t) = \frac{C_i^0}{\Gamma_i^m}\left(\frac{2D_i}{\pi}\right)^{1/2} t^{1/2} \tag{17}$$

where D_i is the diffusion coefficient of species i in solution, and Γ_i^m is the maximum surface concentration. This predicts a linear $\theta(t)$ versus $t^{1/2}$ law for small adsorption times.

However, for the general case, the equations are much more complicated, and the reader is invited to refer to a few specialized papers.[98-100]

The elucidation of electrooxidation reaction mechanisms, and of the role of catalytic electrodes, thus requires a detailed knowledge of the nature, structure, and amount of the different adsorbed intermediates involved.

In the next section, the experimental methods that may be employed to determine the amount of adsorbed species, that is, the degree of coverage θ_i, and the nature and structure of adsorbates will be presented. Electrochemical methods, the description of which may be found in many textbooks of electrochemistry,[6,14,15] will be briefly presented. Then, spectroscopic methods and other physical methods which allow investigation *in situ* of the adsorbed species at the electrode/electrolyte interface will be discussed in more detail.

(*iii*) *Electrochemical Methods*

(*a*) *Coulometric method*

Most of the electrochemical methods used to study adsorption processes are based on the quantitative determination of the electrical charge needed to oxidize the adsorbed organic substance, Q_{org}, or to deposit an adsorbed layer of other adsorbed species, such as Q_H for hydrogen adsorption and Q_O for oxygen adsorption. The degree of coverage and the number of electrons per site are easily determined from these quantities, as discussed in Section II.3(ii) [see Eqs. (9)-(11)].

Many electrochemical methods are currently used for determining the degree of coverage θ_i of adsorbed species i; among these, cyclic voltammetry, pulse voltammetry, and multipulse potentiodynamic methods are the most popular. All of them are based on the application of a programmed sequence of potentials to the electrode surface and on the measurement of the quantity of electricity Q involved.[101-103]

Discussions on the choice of experimental conditions (sweep rate, etc.) to obtain reliable results and on the different correcting factors (charge of the double-layer capacity, displacement of oxygen adsorption by organic adsorption, etc.) may be found in books[6,14] and in specialized papers.[101-105]

Before the development of potentiodynamic and voltammetric methods, coulometry was performed at constant current intensity

I, so that the quantity of electricity Q involved in the oxidation or in the reduction of the adsorbed organic species was easily obtained as the product of I and some transition time τ corresponding to the end of the electrochemical process. The so-called "charging curves," $E = f(Q)$, were thus obtained as the variation of electrode potential E with the electrolysis time, that is, with the quantity of electricity Q.[6,14,15] The method is equivalent to chronopotentiometry, but the variation of electrode potential is due to variation of the surface concentration of the adsorbed species, whereas in classical chronopotentiometry the variation of E with time results from a diffusion process (concentration overvoltage). The transition time τ is relatively well defined, since the potential arrest during which the organic substance is oxidized (anodic charging curve) or reduced (cathodic charging curve) is followed by a rapid potential variation resulting from oxygen or hydrogen adsorption, respectively. Fast galvanostatic charging curves allow the determination of the coverage in the presence of organic molecules dissolved in the electrolyte, whereas for slow galvanostatic charging curves, it is necessary to remove the parent molecule using a flow cell or to transfer the electrode surface under inert atmosphere to a cell containing the supporting electrolyte alone.

Voltammetric curves $I(E)$ (where the electrode potential is a linear or triangular function of time, with a sweep rate $v = \pm dE/dt$), are in fact equivalent to the derivative of the charging curves $E(Q)$, since

$$I(E) = \frac{dQ}{dt} = \frac{dQ}{dE}\frac{dE}{dt} = v\frac{dQ}{dE} = \frac{v}{(dE/dQ)}$$

Moreover, the resolution (that is, the ability to distinguish between different processes occurring at similar potentials) of derivative curves, such as voltammetric curves $I(E)$, is much better than that of integral curves, such as charging curves.

The quantities of electricity Q are thus easily calculated by integration of the part of the voltammograms corresponding to current peaks:

$$Q = \int_{t_1}^{t_2} I(t)\,dt = \int_{E_1}^{E_2} \frac{I(E)\,dE}{v} \tag{18}$$

where E_1 and E_2 are the potentials which define the beginning and the end of the voltammetric peaks (see Fig. 1).

(b) *Differential capacity and impedance methods*

The differential capacity C_d of the double layer can be measured from the potential decay curves obtained once the electrode, previously polarized at a given overvoltage η_0, has been disconnected from the external electrical current. Thus, the double-layer capacity discharges into the resistance of charge transfer R_c, and the analysis of the $\eta(t)$ curve allows C_d to be determined[14]:

$$\eta(t) = \eta_0 \, e^{-t/R_c C_d}$$

More generally, C_d can be determined from impedance measurements. In the simple case where it can be assumed that the electrical double-layer capacity is in parallel with the charge transfer resistance R_c (or more generally with the faradaic impedance Z_f), both of them being in series with the electrolytic resistance R_e, the following equivalent circuit (Randles equivalent circuit)[106,107] applies:

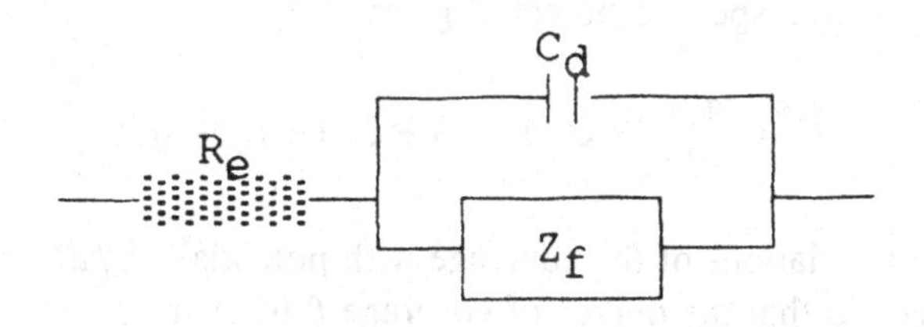

the impedance of which is given by

$$Z = \frac{dE}{dI} = R_e + \frac{Z_f}{1 + j\omega C_d Z_f} \tag{19}$$

with ω the angular frequency of the sinusoidal voltage, and $j = \sqrt{-1}$. In the case of semi-infinite linear diffusion, Z_f is expressed by

$$Z_f = R_c + Z_w \tag{20}$$

where the Warburg impedance Z_w is

$$Z_w = (1 - j)(\sigma_O + \sigma_R)/\sqrt{\omega} \tag{21}$$

with

$$\sigma_O + \sigma_R = \frac{(\delta I/\delta C_O)_{E,C_R}}{nFS\sqrt{2}\sqrt{D_O}\,(\delta I/\delta E)_{C_O,C_R}} + \frac{(\delta I/\delta C_R)_{E,C_O}}{nFS\sqrt{2}\sqrt{D_R}(\delta I/\delta E)_{C_O,C_R}} \tag{22}$$

The frequency analysis of the electrode impedance $Z(\omega)$ allows the different components of the Randles equivalent circuit to be determined, in particular, C_d.

The determination of the degree of coverage θ from C_d is based on the Frumkin equation,[108] which states that the electrical charge of the metal surface is a linear function of θ; that is,

$$q(\theta) = q_0(1 - \theta) + q_1\theta \tag{23}$$

where q_0 and q_1 are the charges on the bare electrode and on the completely covered electrode, respectively. Differentiating this equation with respect to potential gives

$$C_d(\theta) = \frac{dq(\theta)}{dE} = C_0(1 - \theta) + C_1\theta + (q_1 - q_0)\frac{d\theta}{dE}$$

For small variations of the coverage with potential, $d\theta/dE$ can be neglected, so that the degree of coverage θ is given by

$$\theta = \frac{C_d(\theta) - C_0}{C_1 - C_0} \tag{24}$$

This equation applies in the range around the potential of zero charge, where the adsorption is maximum and $d\theta/dE \approx 0$.

When the adsorption kinetics is rate-limiting, the faradaic impedance Z_f is modified and contains additional capacitance and resistance terms associated with the adsorption process. This complicates greatly the impedance analysis.[107]

(*iv*) *Spectroelectrochemical Studies of the Adsorbed Species*

Electrochemical methods are not sufficient for elucidation of the detailed nature and structure of the adsorbed species, since they are based on measurements of macroscopic quantities, proportional to the number of species involved, such as the current intensity, the quantity of electricity, or the double-layer capacity. They are not able to recognize from which molecule the electrical current, that is, the electron flow, comes, and to which species it will go. Only methods based on atomic and molecular phenomena, such as spectroscopic techniques, particularly those allowing *in situ* analysis of the electrode/electrolyte interface, will be able to give information on the exact nature and structure of the adsorbed layers.

(*a*) *In situ spectroscopic methods*

Elucidation of the nature of adsorbates and their conformation on catalytic surfaces has always been a challenge.

Recent years have seen considerable progress made in the study of the solid/gas interface with the development of *ex situ* ultra-high-vacuum (UHV) techniques, such as electron diffraction techniques (e.g., LEED, RHEED) as well as electron spectroscopies (e.g., UPS, XPS, ESCA, AES). Presently, similar information is needed for further development of interfacial electrochemistry, not only on the structure of the electrode surface itself, but also on the adsorbates and on the solution in the vicinity of the electrode surface.

Transferring the electrode to the ultra-high-vacuum chambers required by the above techniques is certainly an interesting idea, and various approaches have been tried by several groups[110,111] (for a review, see Ref. 109). Valuable information is obtained on the structure of the electrode surface, especially with single crystals, provided that the transfer is done with great care to avoid any contamination by residual gas and provided that removal of the electrode from solution is not followed by surface restructuring during transfer to high vacuum. Obviously, only very limited information on adsorbates can be obtained from these techniques (the species have to be strongly chemisorbed), and no information at all can be expected on the structure or the composition of the solution in contact with the electrode surface (i.e., in the double layer).

Therefore, there has been a need for the development of other types of techniques that are nondestructive with respect to the electrode/solution interface. Since the beginning of the 1980s, successful approaches have been made in different directions. We will consider those that can combine electrochemical techniques with *in situ* methods, such as reflection–absorption spectroscopies, diffraction techniques, and mass spectrometry (with a special emphasis on the former, due to the considerable recent development of related techniques), separately from those based on the measurement of the amount of matter (i.e., radiometry or gravimetry with quartz microbalances).

If there is one factor common to the various *in situ* techniques available to investigate the electrode/solution interface (i.e., reflection–absorption spectroscopy in the UV-visible and in the infrared range, or X-ray reflection or diffraction), it is the necessity to design a special electrochemical cell compatible with both the inescapable experimental constraints inherent to electrochemistry (particularly because of the presence of an absorbing solvent) and the geometric restrictions due to reflectance at the interface. Thus, the following criteria must be met:

(i) all the materials have to be stable in contact with the electrolytic solutions;

(ii) windows, if not the whole cell, have to be transparent to the radiation used;

(iii) if the external reflection mode is selected (Fig. 7a), the solution may drastically absorb the radiation, but the solution thickness cannot be decreased too much (typically not to less than a few microns) for the potentiostatic control of the electrode surface to operate normally; and

(iv) if the internal reflection mode is chosen (Fig. 7b), there are no restrictions on the solution thickness, but, as the transparent substrate is generally not the material that one wants to use as the electrode (and may even not be conducting), a metallic layer has to be deposited onto its surface to serve as the working electrode; the catalytic properties of this metallic layer certainly differ considerably from those of the smooth metal. Moreover, it is not easy to vary the electrode structure, particularly by using different single-crystal faces.

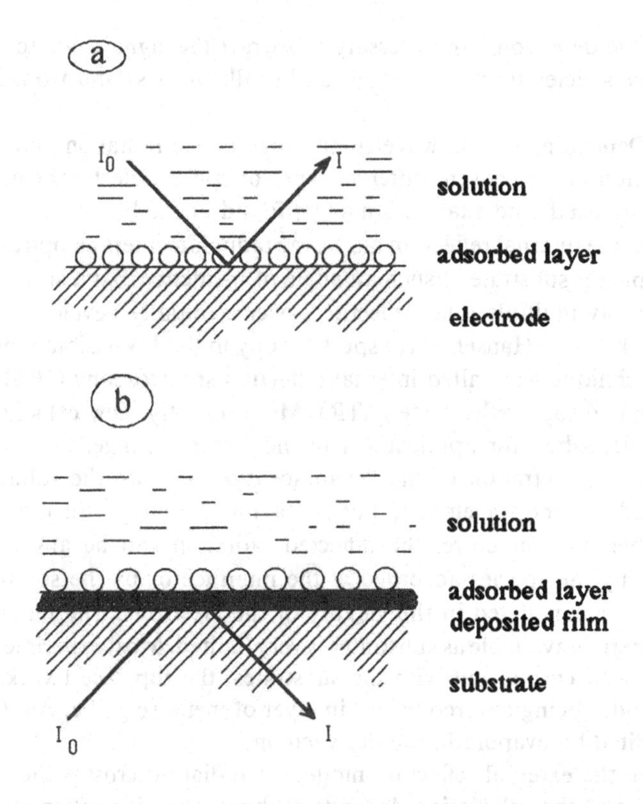

Figure 7. External (a) and internal (b) reflection modes.

Furthermore, in any case, the polarization of the incident beam is affected upon reflection. This leads to the so-called "surface selection rule," as first recognized by Francis and Ellison[112] and quantitatively established by Greenler *et al.* in the infrared range.[113,114]

The concentration of species to be detected in studies of monolayers or submonolayers of adsorbates is always low relative to that of the bulk species (the number of surface species is of the order of magnitude of the number of crystallographic sites, that is, ca. 10^{15} species per square centimeter, to be compared with 10^3 to 10^5 more species in solution), so that conventional spectroscopy is totally unable to sort out the signals due to surface species. More sophisticated techniques, using data acquisition coupled with signal processing or enhancement of the signal-to-noise ratio with phase-sensitive detection, are necessary to extract the signals due to the surface species from the background (bulk species) and from the noise.

Depending on the wavelength range of the radiation and the reflection mode chosen, different types of spectro-electrochemical cells are used, and examples may be found in the literature.

In the internal reflection mode, the radiation enters an optically transparent substrate at such an angle of incidence that a single or preferably multiple total reflections occur. Initially developed by Harrick[115] and Hansen[116] for spectroscopy in the UV-visible range, the technique was called internal reflection spectroscopy (IRS) or attenuated total reflectance (ATR). More recently, new cells have been described for applications in the infrared range.[117-120] The depth of penetration of the "evanescent wave" into the solution depends on the wavelength, but is maximum at the critical angle (see below). Therefore, the reflected radiation can be absorbed either by the species adsorbed at the interface or by the solution species accumulated in the vicinity of the electrode. The number of materials available as substrates is very limited. Most experiments have been carried out with Ge substrates, the top face (working electrode) being covered by a thin layer of metal (e.g., Fe, Au, Pt), deposited by evaporation under vacuum.

In the external reflection mode, the radiation crosses the solution, and the cell design depends on how much it is attenuated, that is, how long the path into the solution can be, the longest being the best from the electrochemistry point of view.

For spectroscopy in the UV-visible range, the solution is generally weakly absorbing, so that the simplest cell design involves sealing two quartz windows (oriented at angles close to 90°) to a Pyrex cylinder, with the flat disk working electrode (about 1 cm^2) attached at the extremity of a syringe barrel, the trunk of the syringe being also sealed to the glass cell and oriented along the bisector of the angle formed by the axes of the two windows. The grid counter electrode is positioned away from, and parallel to, the working electrode, while the tip of a Luggin capillary (linked to the reference electrode) is placed close to the working electrode. As the beam path into the solution can be as long as a few centimeters without inconvenience and as the syringe allows the position of the working electrode to be adjusted accurately, these cells are very convenient for optical alignment. They were described originally by Bewick and Tuxford[121] and Kötz and Kolb.[122]

The case of external reflectance infrared spectroscopy is more complicated, in that the solution absorbs the radiation so strongly (especially with water) that a solution layer more than a few microns thick is not acceptable. In the original design by Bewick *et al.*,[123] the infrared radiation crosses the window at an angle of incidence of 65 to 70° and then passes through a thin layer of solution before being reflected by the electrode surface and then passing again through the solution and the window. A wire ring counter electrode is placed round the hollow syringe, and the tip of the Luggin capillary is positioned on the side, as close as possible to the working electrode. Although the design is far from being perfect (from the perspective of optics, the angle of incidence is too low, and from the perspective of electrochemistry the solution layer is too thin, especially if the reaction is mass-transfer limited), but it is a good compromise and the cell has the advantage of simplicity. No real improvement has been achieved despite some attempts to design different shapes of windows (especially to diminish the energy losses by reflection at the external surface of the window) or to monitor the solution thickness for optimization of the signals,[124-127] but always at the expense of simplicity and rapidity of use. A limited number of infrared materials are stable in contact with acids or bases and can be used as windows. The list includes CaF_2, Si, SiO_2, and ZnSe to cover the whole range from 500 to 5000 cm^{-1}, but none of them can be assumed to be totally inert under the experimental conditions so that frequent polishing of the window is necessary.

A very similar cell was used by Fleischmann *et al.*[128] for *in situ* X-ray diffraction studies. In that case the window was made of Mylar, and 100-μm-thick spacers were used to control the solution thickness.

(*b*) *Physical basis for reflection-absorption spectroscopy in the UV-visible and infrared ranges*

Radiation, of initial intensity I°, transmitted through an absorbing layer of thickness x (Fig. 8a) emerges with intensity I, according to the Beer–Lambert law [see Eq. (4)].

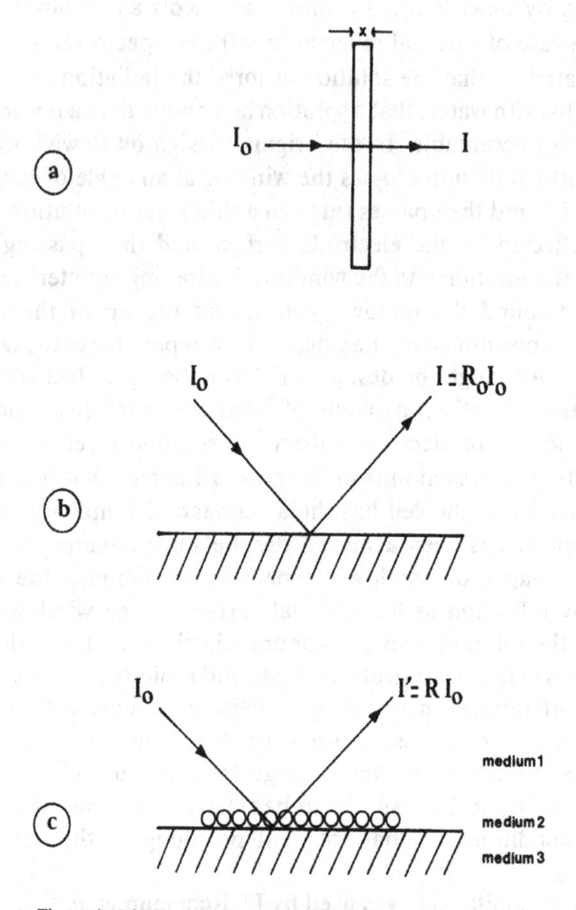

Figure 8. Different modes of light propagation: (a) transmission spectroscopy; (b) reflection at a nonabsorbing plane; (c) reflection-absorption spectroscopy.

Similarly, in the reflection mode, let us consider now the two limiting cases where the reflecting surface is (i) practically free of any absorbing species (Fig. 8b) and (ii) fully covered by a monolayer of absorbing species (Fig. 8c). Let us also assume that medium 1 is a solution, that the substrate (medium 3) is a metal (thus conductive), and that the thin film (medium 2) is isotropic. Thus, if $R°$ is the reflectivity coefficient of the surface covered by the transparent layer (ideally a two-phase model), and R the reflectivity coefficient of the surface in the presence of the absorbing layer (the three-phase model), then the resulting intensities of the reflected beams can be expressed, respectively, by:

$$I = R^\circ I^\circ \tag{25a}$$

$$I' = RI^\circ \tag{25b}$$

and the relative reflectivity change, $\delta R/R^\circ$, of the surface becomes

$$\delta R/R^\circ = (R - R^\circ)/R^\circ = (I' - I)/I \tag{26}$$

The variation of the absorbance, as defined in Eq. (5), due to the presence of the film is then

$$\delta A = \ln(I^\circ/I') - \ln(I^\circ/I) = -\ln(I'/I) \tag{27}$$

Thus, using Eq. (26):

$$\delta A = -\ln(1 + \delta R/R^\circ) \approx -\delta R/R^\circ \tag{28a}$$

for $\delta R/R^\circ \ll 1$, which is usually the case for adsorbed films or thin layers.

δA is therefore equivalent to the absorption factor originally defined by Greenler[113] as

$$\delta A = (R^\circ - R)/R^\circ \tag{28b}$$

Therefore, the absorption spectrum of an adsorbed layer can be obtained by reflectivity measurements, and the technique is called *reflection-absorption spectroscopy.*[114]

According to the theory, a linearly polarized electromagnetic plane wave becomes elliptic upon reflection, the degree of ellipticity depending on the angle of incidence, that is, on how much the two components of the electric field [the p-polarized vector parallel to the incidence plane (defined by the propagation direction of the incident beam and the normal to the surface) and the s-polarized vector perpendicular to it, and thus parallel to the interface; see Fig. 9] are affected by phase changes and amplitude changes. A technique based on the discrimination of these two vectors in the infrared range was successfully developed by Stobie *et al.*[129] and is called infrared ellipsometry.

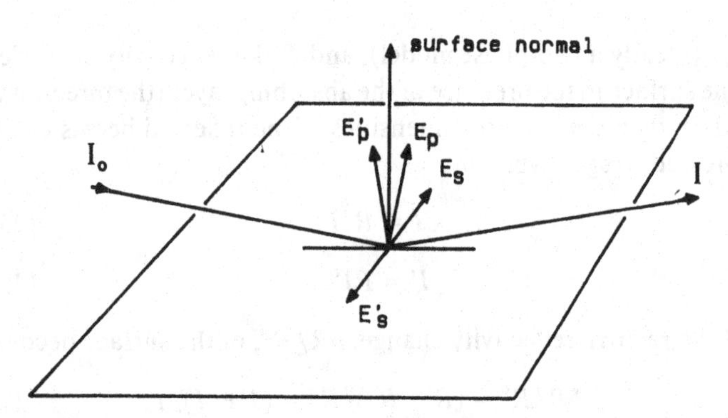

Figure 9. Changes in *s*- and *p*-polarized components of the electric field upon reflection at a high angle of incidence.

Table 5 gives the changes in phase angles upon reflection for the two *s*- and *p*-polarized electric field vectors, where φ_i and φ_r are the angles of incidence and refraction, respectively. The particular value φ_{iB} of the angle of incidence such that $(\varphi_{iB} + \varphi_r) = \pi/2$ is the critical angle, or "Brewster angle," at which the reflected light becomes totally *s*-polarized [see Eqs. (30) below]. Using Descartes' law [see Eq. (31)], φ_{iB} is calculated from the relation $\tan \varphi_{iB} = n_2/n_1$, where n_1 and n_2, the refractive indices of the two media, are taken as real (but they would become complex for highly absorbing media; see below).

Whatever the angle of incidence and the refractive indices are, the *s*-component of the electric field is reversed upon reflection (i.e., its phase angle is changed by π). Assuming that the reflectivity coefficients are not very different, the consequence is that the resultant of the incident and reflected *s*-vectors is nearly zero at the surface. The situation is different for the *p*-vectors. In the usual

Table 5
Changes in Phase Angles upon Reflection

	Changes in phase angle for:	
	$\varphi_i + \varphi_r < \pi/2$	$\varphi_i + \varphi_r > \pi/2$
p-Polarization	π	0
s-Polarization	π	π

case where n_2 is greater than n_1 (this is the case for the electrochemical interface), *the reflected p-polarized component is reversed at $\varphi_i < \varphi_{iB}$ and not reversed at $\varphi_i > \varphi_{iB}$*. In the latter case, the incident and reflected p-polarized vectors of the electric field, at the surface, coadd in such a way that the resultant vector is enhanced. Actually, its magnitude depends strongly on the angle of incidence. It is usual to denote by $E_{p\perp}$ the projection of the resultant field vector on the normal to the reflection surface, and by $E_{p\parallel}$ its projection on the surface (Fig. 10). It is clear from Fig. 11 that $E_{p\perp}$ can be nearly doubled at high angles of incidence, while $E_{p\parallel}$, as well as E_s, always remains very small.

The consequence for *reflection-absorption spectroscopy* is that the s-polarized component of the electric field is always inactive while the p-polarized component can interact strongly with those vibrational modes that correspond to dipoles oscillating perpendicularly to the surface. This is the origin of the so-called "*surface selection rule*" in infrared reflectance spectroscopy, which holds for Raman and UV-visible spectroscopies as well.[130]

Complete analytical expressions for $(\delta R/R)_p$ and $(\delta R/R)_s$ were given by McIntyre and Aspnes,[131] starting from the Fresnel equations for the reflection coefficients.

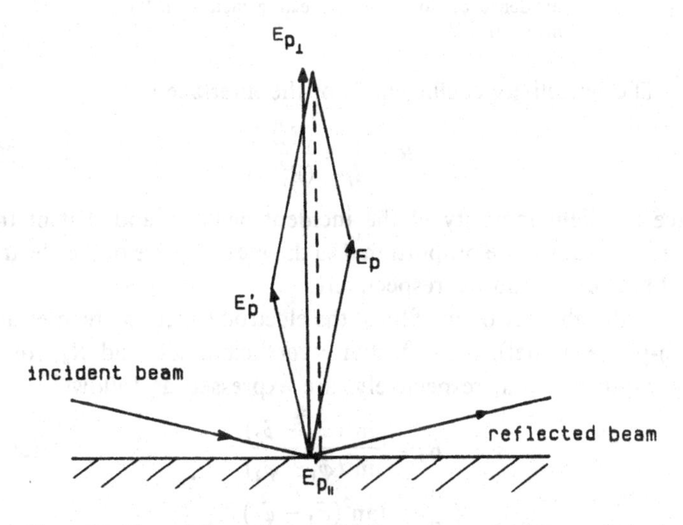

Figure 10. Normal $E_{p\perp}$ and tangential $E_{p\parallel}$ components of the p-polarized resultant electric field vector.

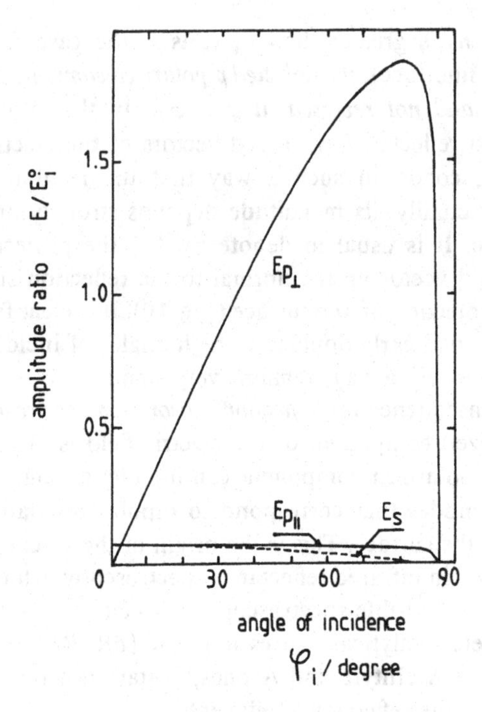

Figure 11. Dependence of E_p and E_s on angle of incidence φ_i, for highly reflecting metals, in the near infrared.

The reflectivity coefficient R of the interface is

$$R = \frac{I_r}{I_i} = \frac{\langle \mathbf{E}_r^2 \rangle}{\langle \mathbf{E}_i^2 \rangle} \tag{29}$$

since the light intensity of the incident beam I_i and that of the reflected beam I_r are proportional to the mean square of the electric field vectors $\mathbf{E}_i$ and $\mathbf{E}_r$, respectively.

In the absence of any film at the electrode/electrolyte interface (two-phase model), the reflectivity coefficients R_s and R_p, for s- and p-polarization, respectively, are expressed as follows[17]:

$$R_s = \frac{\sin^2(\tilde{\varphi}_1 - \tilde{\varphi}_2)}{\sin^2(\tilde{\varphi}_1 + \tilde{\varphi}_2)} \tag{30a}$$

$$R_p = \frac{\tan^2(\tilde{\varphi}_1 - \tilde{\varphi}_2)}{\tan^2(\tilde{\varphi}_1 + \tilde{\varphi}_2)} \tag{30b}$$

where the angle of incidence $\tilde{\varphi}_1$ and the angle of refraction $\tilde{\varphi}_2$ are complex quantities, since they are related by Descartes' law (or Snell's law):

$$\tilde{n}_1 \sin \tilde{\varphi}_1 = \tilde{n}_2 \sin \tilde{\varphi}_2 \tag{31}$$

where the imaginary part of the complex refractive index $\tilde{n} = n - ik$ takes into account the attenuation of light intensity by the absorbing medium (extinction coefficient k). The refractive index and the dielectric constant are related by Maxwell's equation $\tilde{n} = \sqrt{\tilde{\varepsilon}}$ (assuming that the magnetic permeability of the medium is unity, which is usually the case), so that the dielectric constant $\tilde{\varepsilon}$ of an absorbing medium is also complex:

$$\tilde{\varepsilon} = \varepsilon' - i\varepsilon'' = \tilde{n}^2 \leftrightarrow \begin{cases} \varepsilon' = n^2 - k^2 \\ \varepsilon'' = 2nk \end{cases} \tag{32}$$

For normal incidence ($\varphi_1 = \varphi_2 = 0$) and assuming that the first medium (i.e., the electrolyte solution) is nonabsorbing ($k_1 = 0$; i.e., $\tilde{n}_1 = n_1$), both reflectivity coefficients reduce to

$$(R)_{\varphi=0} = \frac{(n_2 - n_1)^2 + k_2^2}{(n_2 + n_1)^2 + k_2^2} \tag{33}$$

For the pseudo-Brewster angle $\tilde{\varphi}_{1B}$, defined by $\tan \tilde{\varphi}_{1B} = \tilde{n}_2/\tilde{n}_1$, so that $\tan(\tilde{\varphi}_1 + \tilde{\varphi}_2) \to \infty$, the reflectivity coefficient R_p for p-polarization as a function of the angle of incidence φ_1 passes through a minimum.[17]

In the presence of an adsorbed layer or of an absorbing film, of thickness d and complex refractive index $\tilde{n}_2 = n_2 - ik_2$, at the interface between a nonabsorbing electrolyte ($\tilde{n}_1 = n_1$) and an absorbing metal electrode ($\tilde{n}_3 = n_3 - ik_3$) (three-phase model), there is a relative reflectivity change $\delta R/R^\circ$ at the interface, normalized as follows:

$$\left(\frac{\delta R}{R}\right)_{s,p} = \left(\frac{R(d) - R(0)}{R(0)}\right)_{s,p} = \left(\frac{I(d) - I(0)}{I(0)}\right)_{s,p} \tag{34}$$

where $R(d)$ and $I(d)$ are the reflectivity coefficient and the reflected light intensity, respectively, in the presence of the film.

Assuming that the film thickness d is much smaller than the wavelength of the lights, λ (i.e., $d/\lambda \ll 1$), McIntyre and Aspnes[131] calculated the normalized reflectivity change for both s- and p-polarization:

$$\left(\frac{\delta R}{R}\right)_s = \frac{8\pi d\sqrt{\varepsilon_1}\cos\varphi_1}{\lambda}\cdot \operatorname{Im}\left(\frac{\tilde{\varepsilon}_2 - \tilde{\varepsilon}_3}{\varepsilon_1 - \tilde{\varepsilon}_3}\right) \tag{35a}$$

$$\left(\frac{\delta R}{R}\right)_p = \frac{8\pi d\sqrt{\varepsilon_1}\cos\varphi_1}{\lambda}\cdot \operatorname{Im}\left\{\left(\frac{\tilde{\varepsilon}_2 - \tilde{\varepsilon}_3}{\varepsilon_1 - \tilde{\varepsilon}_3}\right)\right.$$

$$\left.\times\left(\frac{1-(\varepsilon_1/\tilde{\varepsilon}_2\tilde{\varepsilon}_3)(\tilde{\varepsilon}_2+\tilde{\varepsilon}_3)\sin^2\tilde{\varphi}_1}{1-(1/\tilde{\varepsilon}_3)(\varepsilon_1+\tilde{\varepsilon}_3)\sin^2\tilde{\varphi}_1}\right)\right\} \tag{35b}$$

where $\operatorname{Im}(x)$ stands for the imaginary part of the complex function x.

For normal incidence ($\varphi_1 = 0$), both expressions reduce to

$$\left(\frac{\delta R}{R}\right)_{\varphi_1=0} = 8\pi\sqrt{\varepsilon_1}\frac{d}{\lambda}\left(\frac{(\varepsilon_1-\varepsilon_2')\varepsilon_3'' - (\varepsilon_1-\varepsilon_3')\varepsilon_2''}{(\varepsilon_1-\varepsilon_3')^2+\varepsilon_3''^2}\right) \tag{36}$$

This simplified expression allows the estimation of the relative reflectivity change at a wavelength $\lambda = 800$ nm due to the presence of a Pb monolayer ($\varepsilon_2' = -15$, $\varepsilon_2'' = 20$, $d = 4$ Å) formed by underpotential deposition on a silver electrode ($\varepsilon_3' = -24$, $\varepsilon_3'' = 1$) in contact with an aqueous electrolyte of $\varepsilon_1 = 1.77$. (Ref. 132). Equation (36) gives $(\delta R/R)_{\varphi_1=0} \approx 1\%$, which is in good agreement with experimental results.

For highly conductive metals ($k_3 > 10$; i.e., $\tilde{\varepsilon}_3 \gg \tilde{\varepsilon}_2 > \varepsilon_1$) and taking $\varepsilon_1 = 1$, Eqs. (35) can be further simplified:

$$\left(\frac{\delta R}{R}\right)_s \approx 0 \tag{37a}$$

$$\left(\frac{\delta R}{R}\right)_p \approx \frac{8\pi d\sin\varphi_1\tan\varphi_1}{\lambda}\cdot \operatorname{Im}(-1/\tilde{\varepsilon}_2) \tag{37b}$$

This shows that only the p-component of the electric field vector plays an active role in light absorption by a film at the surface of a conducting medium and justifies the "surface selection rule." Therefore, p-polarizers are usually used in the experimental setup, to eliminate completely any s-polarization.

If the absorbing film is not too highly absorbing, that is, for a relatively small extinction coefficient k_2, Eq. (37b) reduces to

$$\left(-\frac{\delta R}{R}\right)_p \approx \frac{16\pi k_2 d \sin\varphi_1 \tan\varphi_1}{\lambda n_2^3} \approx \frac{4\sin\varphi_1 \tan\varphi_1}{n_2^3}\cdot \alpha_2 d \qquad (38)$$

where $\alpha_2 = 4\pi k_2/\lambda$ is the absorption coefficient of the film [see the Beer-Lambert law, Eq. (4)].

From Eq. (38), it follows that the relative change δA in the absorbance A due to the presence of the film is

$$\delta A \approx -\left(\frac{\delta R}{R}\right)_p \approx \frac{4\sin\varphi_1 \tan\varphi_1}{n_2^3}\cdot A_2 \qquad (39)$$

where $A_2 = \alpha_2 d$ is the absorbance of the film. This shows that, for high angles of incidence ($\varphi_1 > 80°$), the reflectance experiment is easier to perform than the transmittance experiment, since the factor $4\sin\varphi_1 \tan\varphi_1/n_2^3$, which is greater than unity, behaves as an enhancement factor.

This last equation allows the evaluation, in the infrared range, of the relative reflectivity change of a platinum electrode at $\lambda = 5\ \mu m$ due to a monolayer of a strong IR absorber, such as adsorbed CO, to about 1%, which is easily detectable.[17]

(c) *Symmetry of the adsorbent sites and vibrational modes of adsorbates. Spectroscopic selection rules*

Since the first *in situ* infrared spectroscopic experiments have already been done on single-crystal electrodes [see the discussion in Sections II.4(i) and III.1] and since they have showed "structural effects," it is of primary importance to consider the symmetry properties of the adsorbent/adsorbate systems. The aim of this section is just to give some general ideas about the problem and its implications for interfacial electrochemistry. The reader who wants to know more about the subject is invited to refer to the abundant literature related to the gas phase.[133-137]

The recent improvement of infrared reflectance spectroscopy at the electrode/solution interface (higher sensitivity and shorter spectral accumulation times) makes it now theoretically possible to carry out a much more rigorous analysis of spectroscopic data by application of the selection rules based on symmetry properties. To gain a better knowledge of the way in which species are adsorbed onto the electrode surface is not only a challenge for academic research but is also of interest for practical applications in electrocatalysis, in that both the activity (of the surface) and the reactivity (of adsorbates) involve symmetry considerations. Obviously, a lot of new information should be available with the help of reflectance spectroscopy, at least in the near future.

Let us start by considering a species in the gas phase. Its N atoms correspond to $3N$ kinetic degrees of freedom. Classically, the translation of the center of mass accounts for three degrees of freedom, and rotation accounts for three as well, but only for nonlinear molecules (linear molecules have only two rotational degrees of freedom). Such species have low symmetry and, normally, a number of vibrational motions equal to the number of vibrational degrees of freedom, that is, $3N$-6 or $3N$-5 for nonlinear and linear molecules, respectively.

When the same species are adsorbed onto a metal, the three translational motions are not possible anymore, which results in an extra three vibrational degrees of freedom. In other words, the translational motions are converted into vibrations of the species at the surface. It follows that the degeneracy of the vibration modes due to adsorption leads to a higher symmetry ($3N-3$ or $3N-2$ vibrations, if rotation is still possible, or $3N$ if rotational motion is also converted into vibrations). In fact, the exact description of vibrational motions depends strongly on the adsorption mode, and it is easy to see that *physisorption* will affect very little the vibrational modes defined in the gas phase, while *chemisorption* may involve large structural changes of the adsorbates and therefore major changes in the vibrational frequencies.

Furthermore, at the interface, there is a reduction in the number of symmetry elements. Thus, improper rotation axes as well as centers of inversion have no meaning, and only the mirror planes perpendicular to the surface and the rotation axes normal to the surface would have to be considered.

For instance, starting from the atomic arrangements of an fcc metal, the only point symmetry groups that apply to the (100), (110), and (111) crystallographic planes are C_{6v}, C_{4v}, C_{3v}, C_{2v}, C_6, C_4, C_3, C_2, C_s, C, and C_1. In fact, the overall symmetries of unit meshes are determined by their geometry, that is, square for (100), rectangular for (110), and hexagonal for (111). However, more important for the adsorbent-adsorbate interaction is the definition of symmetry elements (or point groups) related to bare sites, such as on-top, bridge, threefold hollow, and fourfold hollow sites, according to the usual nomenclature. These sites are represented in Fig. 12, and their symmetry elements are listed in Table 6.

In considering Table 6, it is worth stressing that:

(i) because of surface crystallography, point groups C_5 and $C_{n>6}$ have no significance;

(ii) the same types of sites are not equivalent along the different unit meshes. Thus, on-top sites will have different symmetry properties determined by the point groups C_{4v}, C_{3v}, and C_{2v}, respectively, for the (100), (111), and (110) planes;

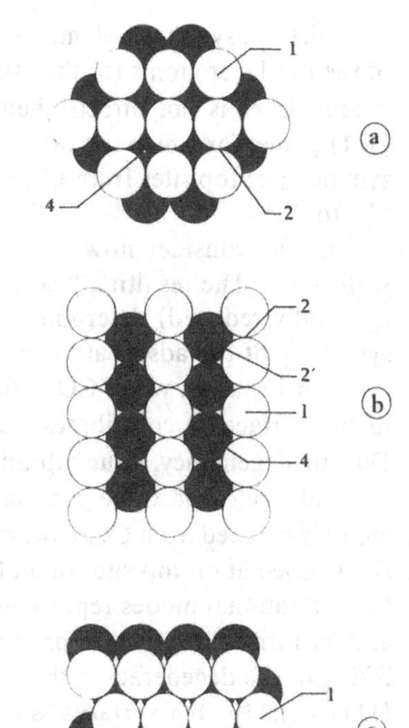

Figure 12. Site symmetries for (100) (a), (110) (b), and (111) (c) crystallographic planes. See Table 6 for identification of the sites.

Table 6
Bare Sites for fcc Surfaces and Symmetry Elements

Unit mesh	Site numbering	Type of site	Point group
(100)	1	On-top	C_{4v}
	2	Bridge	C_{2v}
	4	Fourfold hollow	C_{4v}
(110)	1	On-top	C_{2v}
	2	Bridge	C_{2v}
	2′	Bridge	C_{2v}
	4	Fourfold hollow	C_{2v}
(111)	1	On-top	C_{3v}
	2	Bridge	C_s
	3	Threefold hollow	C_{3v}
	3′	Threefold hollow	C_{3v}

(iii) the symmetry elements are determined by the arrangement of the top-layer atoms for the (100) and (110) planes (although the second layer is not directly below the first one), but not for the (111) plane, for which the presence of the second layer reduces the symmetry of top sites from C_{6v} to C_{3v} and that of bridge sites from C_{2v} to C_s.

Let us consider now an *isolated species adsorbed* on a *bare surface site.* The resulting "*surface complex*" will have a symmetry (generally reduced) determined by the compatibility between the symmetry of the adsorbate and that of the adsorbent.

For instance, when CO is linearly bonded and perpendicular to the surface, it contributes six degrees of vibrational freedom. Due to degeneracy, four vibrations are observed, one being the internal mode and three coming from translation. Thus, the symmetry is reduced from $C_{\infty v}$ (free molecule) to C_{4v} if the CO molecule is adsorbed at on-top sites of an fcc (100) unit mesh, leading to the four vibrational modes represented in Fig. 13. In C_{4v} symmetry, ν_3 and ν_4 form a degenerate pair of vibrations. However, there is a lifting of the degeneracy if the adsorption occurs at on-top sites of (111) or (110) fcc surfaces, since the symmetry groups become, respectively, C_{3v} and C_{2v} (cf. Table 6). Similarly, adsorption onto bridge sites would lead to a complex with C_{2v} symmetry on (100) and (110) planes (Fig. 13), but reduced to C_s on the (111) plane.

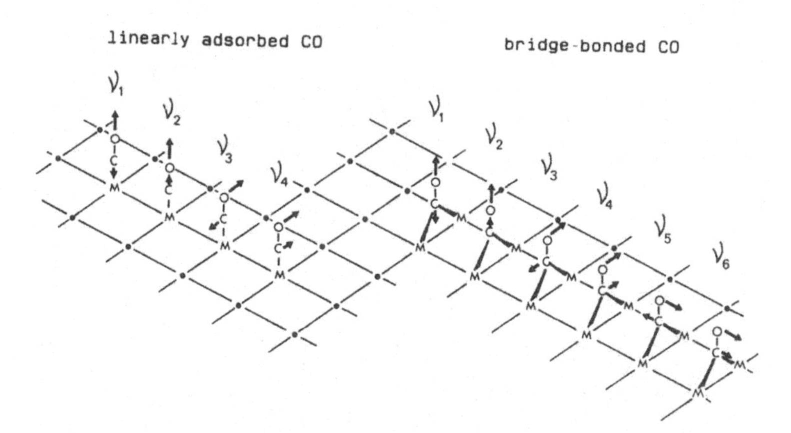

Figure 13. Vibrational modes of linearly and bridge-bonded CO adsorbed on an fcc (100) plane.

If the CO molecule is tilted, an asymmetric bridging would result, as a consequence of which the point group would become C_s on all planes.

Apart from CO, two important adsorbed species in electrocatalysis (see the results given in Section III.1) are formate and methoxy species. While the former is planar and has C_{2v} symmetry, the latter has a C_{3v} symmetry determined by the C—H bonds—with all the consequences of this symmetry with respect to adsorption on fcc metals. Figure 14 illustrates this particular situation in the case of the complex formed by a methoxy species on a (100) surface. It is clear that the symmetry of the methoxy species is reduced to C_s, due to incompatibility between the point groups C_{4v} of on-top sites and C_{3v} of the adsorbate, while the symmetry of the complex formed between formate and the surface is C_{2v} when the formate group is bonded to the surface via the two oxygens, and only C_s if it is singly bonded to the surface (Fig. 15). Of course, further reduction in symmetry to point groups C_s or C_1 would result from any orientation of the molecular axis other than normal to the surface.

All the discussion above applies for isolated adsorbates. At increased coverages, lateral interactions are known to take place. When they are strong enough, islands of adsorbates are formed, leading to new local arrangements with their own symmetry properties.

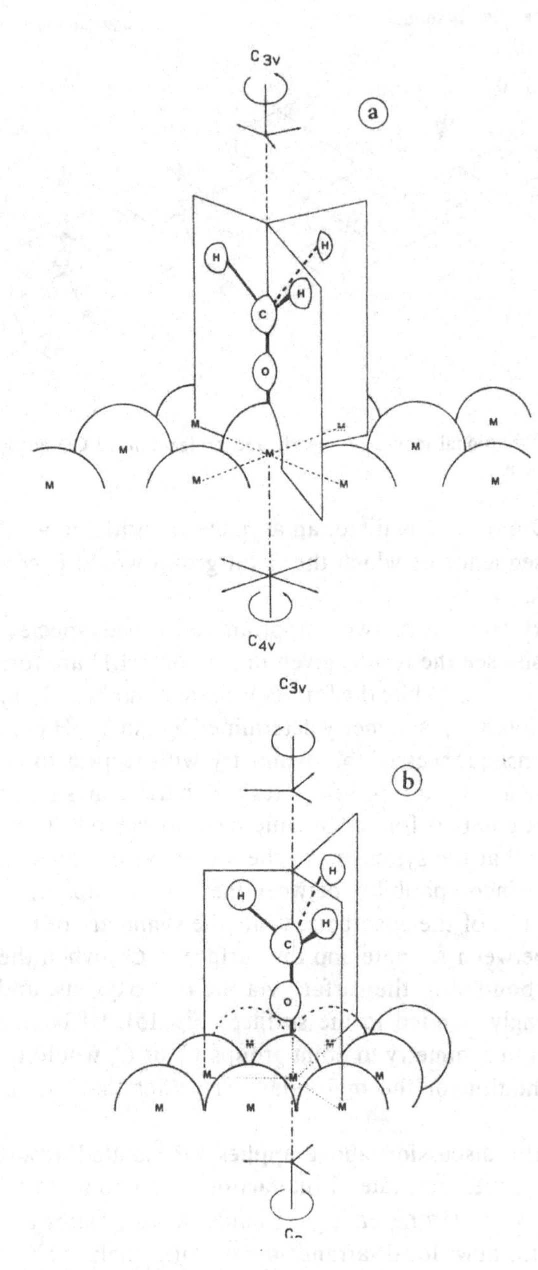

Figure 14. Methoxy adsorption at (a) C_{4v} and (b) C_{3v} sites.

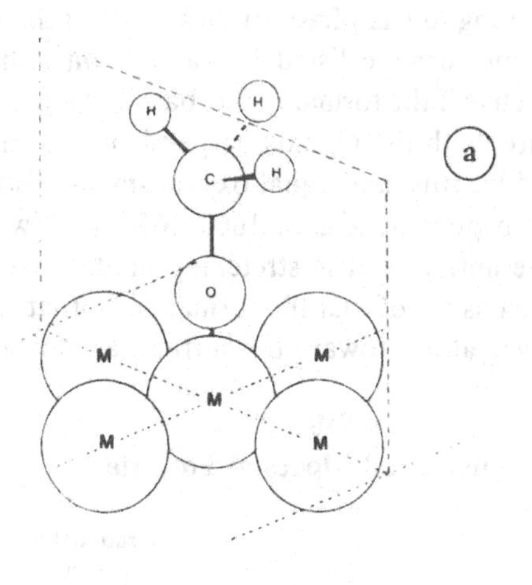

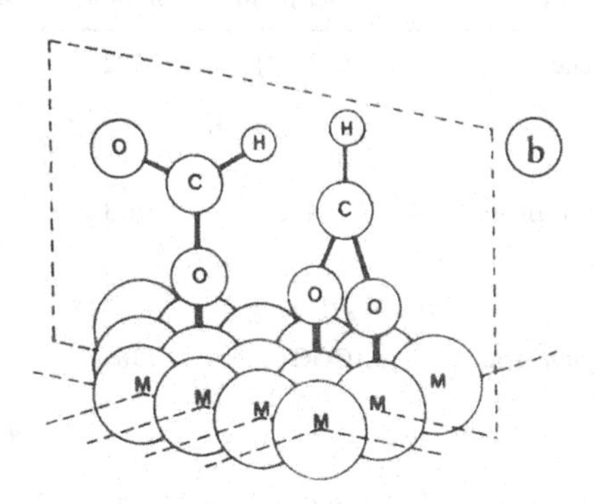

Figure 15. Adsorbed species at C_{4v} sites: (a) methoxy group; (b) monodentate and bidentate formate.

Spectroscopic selection rules. The selection rules for surface spectroscopy are not different from those for transmission spectroscopy. However, as mentioned above, an additional "surface selection rule" applies to surface species (for infrared, Raman, and UV-visible spectroscopies). The rule states that the only observable surface molecules are those having a component of their dynamic electric dipole moment perpendicular to the surface.

Hence, according to gas-phase studies,[138,139] of the six normal modes of free ionic formate listed in Table 7, only three would remain infrared active if the formate adsorbate had a C_{2v} symmetric bidentate structure, with the C_2 axis perpendicular to the surface, equal C—O bond lengths, and equal oxygen-metal distances. The absence of any absorption band at ca. 1600–1650 cm^{-1} (which would correspond to the antisymmetric stretching mode of free formate ions) may be taken as proof that the formate adsorbate is oriented with the two oxygen atoms toward the surface, as indicated above.

Table 7
Vibrational Modes of Formate

	Mode	Description	Vibrational frequency (cm^{-1}) in aqueous solution[a]	Infrared activity in the C_{2v} configuration
H | C O→ ←O	OCO bend	δ(OCO)	772	Yes
H + | C O O	CH out-of-plane bend	π(CH)	1073	No
H | C O O	OCO symmetric stretch	ν_s(COO)	1366	Yes
H → | C O→ ←O	CH in-plane bend	δ(CH)	1377	No
H | ←C O→ ←O	OCO antisymmetric stretch	ν_a(COO)	1567	No
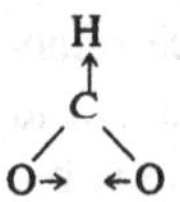 H ↑ C O→ ←O	CH stretch	ν(CH)	2841	Yes

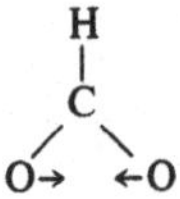

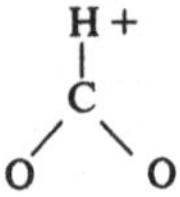

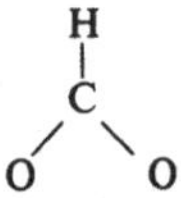

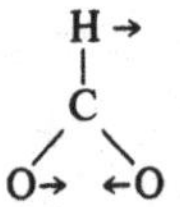

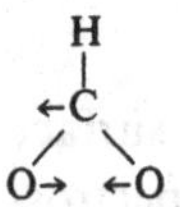

[a] From Ref. 141.

However, recent electrochemically modulated infrared reflectance spectroscopy (EMIRS) investigations at an electrolyte solution/smooth platinum electrode interface[140] led us to conclude that the rather complex band which extends from 1620 to 1700 cm^{-1}, and which includes the δ(HOH) deformation mode of water at 1625 cm^{-1}, may also contain near 1640 cm^{-1} a contribution from the antisymmetric stretch of formate [see also Section III.1(iii)]. The likely surface configuration that would result in such a mode becoming infrared active would be a formate species oriented with only one oxygen toward the surface.

Similarly, in the case of methanol adsorption, the multiplicity of C—H stretches, as observed by EMIRS[140] in the range 2800-3100 cm^{-1}, could be interpreted as due to different orientations at the surface. Only one mode, due to the C—H symmetric stretch, should be active if the C—O axis is perpendicular to the surface, whereas any tilt from this position would make the asymmetric stretching modes also active.

Thus, the potential utility of infrared reflectance spectroscopy, even though the technique is still difficult, for understanding the mechanisms of electrocatalytic reactions has been demonstrated. It should help, for instance, in understanding why the reactivity of methanol (or methoxy species) is so different from that of formic acid on various catalytic metals and on different crystallographic planes of the same metal.

(*d*) *Surface spectroscopy in the UV-visible range*

The first attempts to use reflectance spectroscopy in the UV-visible range were done, with the aim of understanding the structure of the electrode/electrolyte interface.[131,142,143] However, only limited information readily usable for electrocatalysis was obtained.

Real interest in this approach came a few years later when it was demonstrated that optical studies (i) were sensitive enough to characterize single-crystal electrodes,[144] (ii) could help in understanding the formation of passive films on non-noble metals,[145-147] (iii) could be used to investigate the formation of adsorbed hydrogen atoms on smooth Pt,[121] and, more surprisingly, could be used to study CO adsorption on catalytic metals by following the changes in the formation of their oxide layers.[148]

Several extensive reviews are available in the literature, to which the reader may refer for experimental details.[132,149-153] Since the beginning of reflectance spectroscopy, various experimental setups have been developed, either using modulation techniques with phase-sensitive detection or signal processing and averaging with the help of computers.

Briefly, potential-modulated reflectance spectroscopy (PMRS), the most commonly used technique, is an *in situ* technique in which periodic reflectivity changes (δR) in the monochromatic light reflected from the electrode surface (whose reflectivity is R) are produced by a sine or square-wave potential modulation and detected by a lock-in amplifier. By this technique, typical $\delta R/R$ values of 10^{-5} can be routinely obtained, which is generally sufficient for most studies.

However, the PMR technique itself has its own limitations. If slow processes occur on the electrode surface, the frequency of the electrochemical response may fall below the hertz range, which is usually the lower limit for normal operation of phase-sensitive amplifiers. The situation is even worse with non-noble metals, at the surface of which totally irreversible processes are known to occur, or with single crystals, which may reconstruct under potential modulation. Moreover, the interpretation of PMR spectra is often difficult. Recently, it was suggested that the maxima might correspond to the normal absorption bands of species, which, if confirmed, would make PMRS interesting as an analytical tool.[153]

An alternative technique is to use rapid-scan spectrometers coupled with high-speed signal averaging.[115,154,155] The very short time necessary for acquisition of a single spectrum allows the use of on-line data-recording systems to collect and average a large number of spectra in a few seconds or a few tens of seconds. Thus, signal-to-noise ratio improvement is obtained, and sensitivities of 10^{-3} (in $\delta R/R$) are currently achieved. In differential reflectance spectroscopy (DRS), series of a few hundred spectra are usually collected at different potentials. The subtraction of the averaged series, one serving as reference, gives differential spectra which allow the dependence of the layer on potential to be followed. Alternatively, the potential may be kept constant during the successive series. Thus, after subtraction of the reference series, the time dependence of the phenomenon is observed. Such information is particularly useful for studying the growth of metal oxide layers[156] or corrosion processes.

The new generation of rapid-scan spectrometers with diode arrays combined with a polychromator and multichannel analyzer certainly offers interesting possibilities for DRS. Some applications have been described.[157]

A very different approach to optical measurements for the study of electrocatalysis was developed by Beden *et al.*,[158] on the basis of the coupling of cyclic voltammetry with UV-visible reflectance spectroscopy (CVUVRS). The technique is based upon the recording of "reflectograms" synchronously to the voltammograms. The reflectograms represent the changes of reflectivity $\delta R/R$ versus the electrode potential E, at fixed wavelengths. Once a series of reflectograms is recorded, the reflectivity changes are plotted in a three-dimensional diagram $\delta R/R = f(E, \lambda)$ which, if cut at constant E, yields the usual reflection–absorption spectra $\delta R/R = f(\lambda)_E$. Despite the duration of the experiments, the above technique is particularly suitable for interfacial electrochemical investigations, in that cyclic voltammetry provides a permanent control of the surface state during spectral acquisition, which the other techniques cannot offer. CVUVRS has been successfully applied to the study of adsorbed organic species on catalytic electrode surfaces,[158] as well as to follow the changes that may affect unstable surfaces, such as oxide layer growth[156] or corrosion processes.[159] CVUVRS has also been very successfully employed to investigate the state and valency of oxide layers at noble metal electrodes. This is a very important point in electrocatalysis, since the extra oxygen atoms needed to completely oxidize the adsorbed oxygenated organic molecules to CO_2 can come, in aqueous solution, from adsorbed water and hydroxyl species or from oxygen and oxide layers. As a typical example, the nature of the oxide layers which are involved in the electrocatalytic oxidation of formate at rhodium (and which contribute to the oscillating behavior encountered) was elucidated by CVUVRS.[160]

(*e*) *Surface spectroscopy in the infrared range*

The case of external reflectance infrared spectroscopy is more complicated, because of the strong absorption of infrared radiation by aqueous solutions, so that the technique was long considered by spectroscopists to be infeasible.

The first *in situ* experiments, reported by Bewick and coworkers in 1980 and 1981,[161-164] demonstrated the feasibility of a technique now known as electrochemically modulated infrared reflectance spectroscopy (EMIRS) (Fig. 16a). Since this initial work, many improvements have been made, either by modifications of the technique or by enlarging considerably the variety of subjects investigated. Several extended reviews and chapters in books may be consulted by the reader interested in the technical details.[16,17,73,123,165]

The variants of infrared reflectance spectroscopy differ in the type of spectrometer employed (dispersive or Fourier transform), the reflection mode (external or internal), and, principally, the technique used for signal enhancement, without which no signals due to adsorbed species can be extracted. In Table 8 the main infrared reflectance techniques are listed.

In all infrared reflectance techniques, data processing plays an important role. The tasks of data processing are:

- to sort out the information that corresponds to surface species from that arising from bulk species (generally much more intense); and
- to improve the weak signal-to-noise ratio (S/N) inherent to this type of experiment.

Techniques coupled with dispersive instruments. Except in some pioneering experiments,[115,116,173] the most powerful technique, EMIRS, consists in modulating the electrode potential between two limits (the lower potential, E_c, at which the reflectivity is R_c, and the upper potential, E_a, at which the reflectivity is R_a), with a modulation amplitude of 50 to 500 mV and a frequency of a few hertz. By synchronous analysis of the signals, it is possible to reject the nonmodulated information (due to absorption by species in the electrolytic solution) and to amplify the signal-to-noise ratio of the absorption bands due to vibrations of adsorbed species. In fact, a frequency of about 8 to 15 Hz is a good compromise that allows the electrochemical system to respond to the change in potential and the synchronous detection to work satisfactorily. The dc output of the phase-sensitive detector, $\delta R(\bar{\nu})$, at a wavenumber $\bar{\nu}$ is then stored, averaged if necessary, and divided by the electrode reflec-

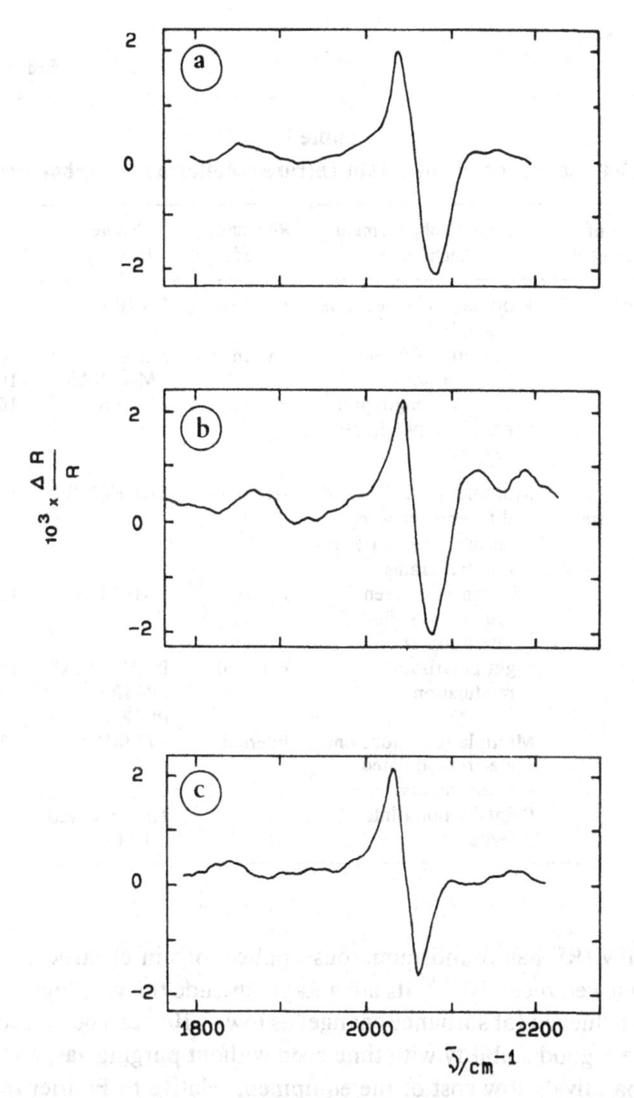

Figure 16. EMIRS spectra of CO species resulting from the adsorption on a Pt electrode (in 0.5 *M* $HClO_4$ at room temperature) of: (a) 0.25 *M* CH_3OH; (b) gaseous CO; (c) 0.25 *M* HCOOH.

tivity $R(\bar{\nu})$ in order to obtain the spectrum in its final dimensionless form $(\delta R/R, \bar{\nu})$. The technique, which was introduced by Bewick *et al.*, is fully described in Refs. 16, 17, and 123.

Table 8
Characteristics of the Main Infrared Reflectance Techniques

Type of spectrometer	Signal enhancement technique	Reflection mode	Name of techniques	Refs.
Dispersive	Modulation of electrode potential	External	EMIRS	123, 16
	Modulation of light polarization	External	IRRAS PM-IRRAS	166 167
	Fixed wavelength and repetitive potential sweeps	External	LPSIRS	168
Fourier transform	Multiple reflections and difference between accumulated series of interferograms	Internal	MIRFTIRS	117
	Difference between sequenced series of interferograms	External	SNIFTIRS	124
	Light polarization modulation	External	PMFTIRRAS SPAIRS PDIRS	169 170 171
	Multiple reflections and electromodulated interferogram	Internal	FTEMIRS	120
	Point-by-point interferogram		Time-resolved FTIR	172

EMIRS has found numerous applications in electrocatalysis, as reviewed recently.[17,19] Its advantages include the very high sensitivity achieved (absorbance changes as low as 10^{-6} can be detected), the very good stability with time even without purging gas, and the comparatively low cost of the equipment relative to Fourier transform spectrometers. Figure 17a illustrates the case of methanol chemisorption on platinum electrodes in acid medium. Different EMIRS bands are seen. They correspond to various surface species, including formate and carbon monoxide. Furthermore, with the same basic modular equipment, it is possible to set up other types of techniques such as linear potential sweep infrared spectroscopy (LPSIRS), introduced by Kunimatsu,[168] which allows the investigation, at fixed wavelengths, of the species produced at different electrode potentials during the voltammetric sweep. On the other hand, there are some inconveniences associated with EMIRS. The most important one is that the electrode potential is not fixed at a given value (because of potential modulation, whose amplitude can reach 0.5 V), so that many electrochemical processes may occur in the potential range of modulation, leading to highly complex EMIRS spectra. The second one is that the modulation technique combined with the so-called Stark effect (shift of frequency with potential; see Refs. 174–176) leads in most cases to signals having a pseudo-derivative shape, which makes it difficult to extract quantitative information.[17,177] The third one is that the commercial grating spectrometers specially designed for EMIRS, for many technical reasons (related to infrared materials, wavelength scanning rates, change of filters for sorting out the grating orders, computerized data acquisition and processing), do not allow the whole range of 400–4000 cm^{-1} to be covered in a single experiment.

Alternatively, a very informative technique is to modulate the polarization state of the infrared radiation, instead of the potential. Derived from IRRAS, originally developed for investigations at the solid/gas interface, the technique is a direct application of theory [see Section II.3(iv.b), on *s*- and *p*-polarizations] and was adapted to investigations at the electrode surface,[166] to confirm EMIRS results by quantitative measurements and to obtain conformational information with respect to the orientation of adsorbates. Recent applications have shown this technique to have very good sensitivity.[167,178]

Techniques coupled with Fourier transform spectrometers. By alternatively coadding short series of interferograms at two given electrode potentials, it is possible to take advantage of the very high luminosity of Fourier transform infrared spectrometers and to obtain, after subtraction and ratioing, spectra of adsorbed species of the form

$$\frac{\delta R}{R} = \frac{R_a - R_c}{R_c} = \frac{R_a}{R_c} - 1$$

Coupled with an external reflection spectroelectrochemical cell, the technique is called subtractively normalized interfacial Fourier transform infrared spectroscopy (SNIFTIRS) and was developed by Pons *et al.*[16,124,179] The successive storage of interferograms at E_c and E_a allows drifts, whatever their electrical or

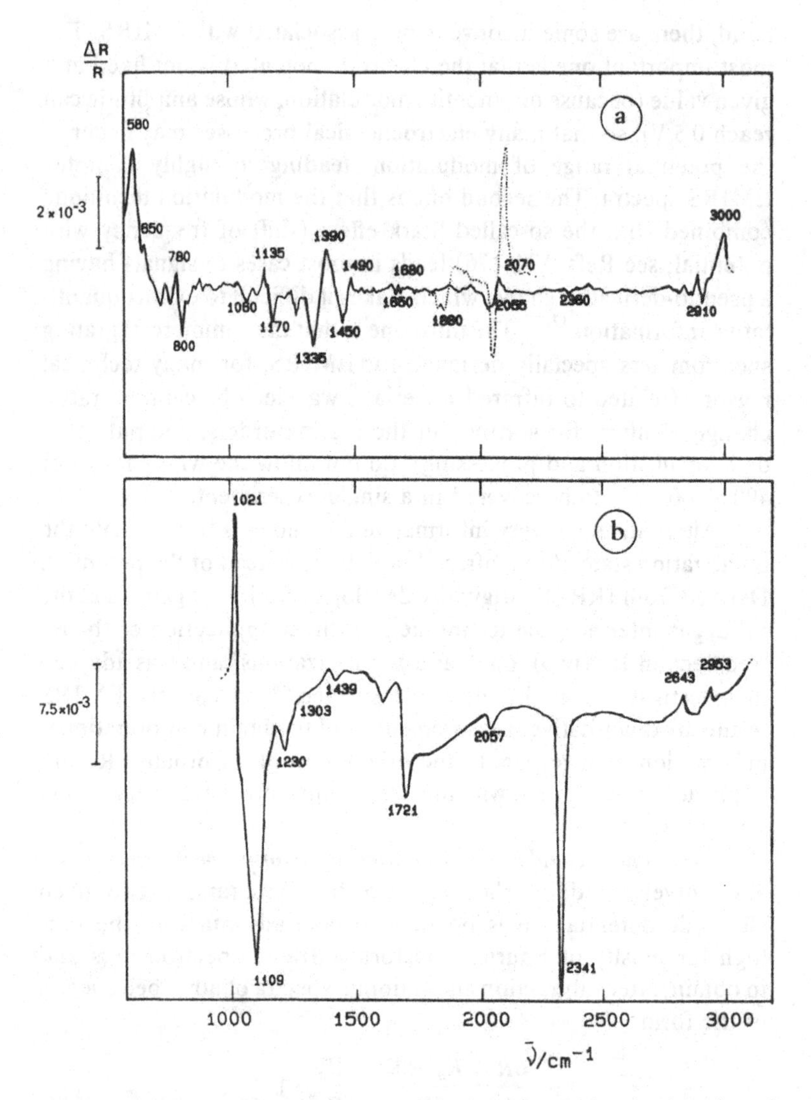

Figure 17. Infrared reflectance spectra of the layer formed by chemisorption of CH_3OH from perchloric acid medium (room temperature) onto a polycrystalline platinum electrode. (a) EMIRS spectrum of the only surface species resulting from chemisorption of 10^{-3} *M* CH_3OH (after Ref. 140). CO (band at ca. 2050 cm^{-1}) and formate (bands at ca. 1450 cm^{-1}) result from oxidation. The strong band at ca. 580 cm^{-1} corresponds to Pt—O or Pt—OH stretches. Increasing the concentration of CH_3OH in solution leads to an increase in surface CO (dotted band). (b) SNIFTIRS spectrum of both adsorbed and solution species resulting from

chemical origin, to be minimized. Thus, long-duration experiments are possible, up to several hours. Typically, 50 interferograms are stored in a few minutes at each potential limit. Once the potential is switched, data accumulation is only triggered after the short latency time necessary for the capacitive currents to go to zero. Figure 17b gives an example of a SNIFTIRS spectrum. The strongest peaks correspond to solution species (perchlorate ions, methanol, CO_2, and probably formic acid). Surface species (CO and formate) give rise to comparable intensities as in the EMIRS experiment of Fig. 17a.

A variant uses a coupling of Fourier transform spectroscopy with light polarization modulation (PMFTIRRAS).[127,169] The frequency of the photoelastic modulator (70 kHz) is much higher than that of the trigger signal used in recording the interferograms, so that, using phase-sensitive detection, it is possible to demodulate the detected signal before its Fourier transformation. Actually, the resulting spectra are differences between spectra recorded with *s*- and *p*-polarized infrared light and are supposed to contain information only on surface species.[180] However, because of anisotropy (as a result of which the baselines do not cancel totally), they may also contain contributions from the solvent and those solution species which are produced or consumed in the vicinity of the electrode surface. To sort out these different contributions, it is therefore necessary to accumulate spectra at two different potentials and to normalize them, as in the SNIFTIRS technique.[181] The PMFTIRRAS technique has a very high sensitivity, due to the use of high modulation frequencies, which allows reduction of the spectral accumulation times.

The idea of using a multiple internal reflection cell coupled with a Fourier transform spectrometer was introduced by Neugebauer *et al.*[117] and later developed by Pham *et al.*[119] under the name of multiple internal reflection Fourier transform infrared spectroscopy (MIRFTIRS). The technique seems to be particularly suitable for following the growth of layers, either of oxides[117] or

---

oxidation of 0.1 *M* CH_3OH (after Ref. 239). By comparison with panel a, it is clear that strong solution bands, such as those corresponding to perchlorate (at ca. 1100 cm^{-1}) or possibly to formic acid (C=O stretch at ca. 1720 cm^{-1}), are detected as well. However, the band intensities for surface species have approximately the same intensities as in the EMIRS experiment of panel a.

organic polymers,[182,183] on electrode surfaces. As in PMFTIRRAS, the main difficulty remains the separation of the contributions of surface species from the solvent contribution.

This particular problem may have been solved recently by Ozanam and Chazalviel,[120] who have developed a new technique combining the advantages of interferometry with those of potential modulation. In Fourier transform electromodulated infrared spectroscopy (FTEMIRS), it is the interferogram itself that is electromodulated. To respect the constraints imposed by the electrochemical system, the mobile mirror of the interferometer had been designed to move at the extremely slow speed of 6 $\mu m\ s^{-1}$. Hence, the electrode potential can be modulated at a frequency of around 100 Hz, compatible with the response of the electrochemical reaction. At the Si–aqueous electrolyte interface, using an ATR electrochemical cell with 15 reflections, signals as weak as 10^{-6} (in absorbance units) could be detected over the entire range 800–4200 cm^{-1}, with 10-cm^{-1} resolution and after only one hour of spectral accumulation. No application to electrocatalysis has been published yet.

Also of great interest, for both fundamental research and applied studies, is the time dependence of electrochemical processes. Transient species often play a kinetic role as reaction intermediates and may be involved in the rate-determining step. However, the first infrared reflectance experiments, because of too long spectral accumulation times, were far beyond the time scale required for identification of transient intermediates.

Recently, attempts have been made to use single potential alteration infrared spectroscopy (SPAIRS) with reduced spectral acquisition times of a few seconds.[184] These "real-time FTIR" experiments are a good approach to the time dependence problem and allowed the authors to follow some slow adsorption processes on electrode surfaces. The first experiments involving time-resolved spectroscopy on a millisecond scale were reported by Pons and co-workers.[172] In their experiments, the interferogram is collected as a set of discrete points, the moving mirror being displaced step by step. A final interferogram is reconstructed from the averaged intensities at each point and then Fourier transformed. Interesting applications of the technique have been made in the far infrared, where the detection of metal–metal vibration was reported.[185]

(v) *Other Spectroscopic and Diffraction Techniques*

In the past ten years, most investigations of the electrode/electrolyte interface have been done by infrared reflectance spectroscopy, using the versatile EMIRS or SNIFTIRS techniques [see Section II.3(iv.e)]. The impact of these techniques has been considerable, in terms of the importance of the results, which in many cases led to new concepts, and the variety of electrochemical systems investigated so far, ranging from electrocatalysis to electrosynthesis or corrosion.

However, insofar as applications to interfacial electrochemistry and electrocatalysis are concerned, many other techniques, involving absorption, reflection, or diffraction of electromagnetic waves or of electrons, are interesting as well. These will now be reviewed briefly.

(*a*) *Raman spectroscopy*

Raman spectroscopy was recognized as a useful tool for surface investigation after the original discovery in 1974 by Fleischmann and co-workers[186,187] that the Raman spectrum of pyridine adsorbed on a silver electrode in aqueous KCl was considerably amplified if the electrode was previously subjected to oxidation-reduction cycles (ORC) under potentiostatic control. Since then, surface-enhanced Raman scattering (SERS) has been observed with different active metals, and numerous papers have been devoted to the enhancement mechanism itself[187-189] and to the study of adsorbed molecules, or layers, on electrode surfaces.

While the Raman effect was originally predicted by Smekal,[190] the technique was developed by Raman,[191] who recorded the first spectra. Briefly, when an electromagnetic wave irradiates a molecule, the associated electric field induces a small dipole moment through its polarizability. If the source emits an intense radiation, such as that emitted by a Laser, in the visible part of the spectrum, the collision between the incident photon and the molecule is nearly elastic, and the oscillating dipole radiates light at the same frequency, but not necessarily in the same direction, leading to intense Rayleigh scattering. Additional scattering comes from the contribution of the normal modes of vibration of the molecule, which can add to or subtract from the main scattered radiation. Thus, very weak spectral bands can be detected at frequencies slightly shifted from the Rayleigh frequency, and these shifts reflect the characteristic vibrational motions of the molecule. This is the basis of normal Raman spectroscopy (NRS). However, if the frequency of the exciting monochromatic radiation is chosen so that it corresponds to an electronic absorption, amplified Raman spectra are obtained. The technique, which is called resonance Raman spectroscopy (RRS), may be useful when the NR spectra are too weak to be accurately measured. As mentioned above, a more interesting technique for electrocatalysis is SERS. Although not yet totally understood, it is now believed that the SERS effect, which can lead in some cases to enhancement factors as high as 10^6 or more, originates when several conditions are fulfilled, namely:

(i) the molecules have to be either adsorbed (preferably chemisorbed) or in the vicinity of the electrode surface;

(ii) the metal surface has to be highly reflecting at the wavelength of the exciting radiation. In the visible, Ag, Au, and Cu, respectively, are the strongest SERS active metals, but recently the first reports have appeared of SERS with other metals, such as Ni, Pd, Al or even Hg;

(iii) the surface has to be rough, the minimum microscopic roughness being one order of magnitude greater than the atomic scale. Such a surface can be obtained by different means, the simplest process probably being electrochemical cycling. It has been shown that, in some cases, a single oxidation–reduction cycle of the surface was enough to produce an intense Raman spectrum.[192,193]

Under these conditions, most adsorbed molecules can give rise to SERS, but with quite different amplification effects, as pointed out by Otto and co-workers.[194,195]

Experimentally, the basic equipment for SERS is the same as for NRS or RRS. Laser sources are used to provide enough power for weak Raman signals to be detected. High discrimination against stray light (due to the intense Rayleigh band) is necessary, which can be achieved using double or triple monochromators. Multichannel detection is now often preferred to the photomultiplier tubes employed in the original experiments.[192,193] A recent and detailed discussion of experimental problems may be found in Refs. 187

and 196–198, together with a description of spectroelectrochemical cells.

Applications of SERS to different domains of interfacial electrochemistry may be found in the literature (see the references cited above). The lack of understanding of the exact nature of the amplification effect, as well as the restriction of SERS to a few active metals, has certainly limited the applicability of the technique. Thus, much effort has been directed toward understanding the enhancement effect itself, rather than toward the use of SERS as an analytical tool.[197] However, in the past ten years, extremely valuable qualitative information has been obtained by several groups on certain systems of interest for electrocatalysis, especially those systems for which information is also available from infrared reflectance spectroscopy. Some criteria that may help in distinguishing the signals of the surface from those of the bulk have been investigated.[197] Thus, in several articles, Weaver and co-workers have compared the band frequencies, bandwidths, and selection rules for different adsorbates on gold and silver electrodes, two metals which are SERS active and catalytically interesting.[199-201] Different applications of SERS to electrochemistry may be found in Refs. 187 and 197. New developments have been reported in the field of corrosion[202-204] and in the field of polymer films.[205]

It is now true that the SERS and the infrared spectroscopic approaches are complementary for probing molecular structure of adsorbates at the electrode/solution interface. For instance, infrared spectroscopy, which does not require surface roughness, can be applied to any type of surface, including well-defined single-crystal faces of any metal. Conversely, despite recent developments, SERS is still limited to a relatively low number of active metals. However, it has great sensitivity and is particularly suitable for investigations in the highly interesting spectral range of 100–600 cm^{-1} (where most substrate-adsorbate bonds absorb IR radiation); in this spectral range, infrared spectroscopy suffers from severe limitations due to weakness of sources, low detectivity of detectors, and increasing absorbance of materials.

Why then is Raman spectroscopy not more popular? As pointed out recently by Hendra and Mould,[206] the reason is the common occurrence of stray absorbance and fluorescence interference. These authors estimated that, because of this problem, less than 20% of the samples that they have analyzed so far have provided usable spectra.

A new development of the technique might solve this problem. Recent papers have described Fourier transform Raman spectroscopy (FT-Raman) as a promising alternative to conventional Raman spectroscopy, especially if it is applied in the near infrared. According to several authors, not only is the new technique more versatile, but it would also be applicable to samples which normally exhibit fluorescence in conventional Raman work.[206–208]

(b) *Ellipsometry*

Ellipsometry is an analytical technique based on the measurement of changes in the polarization state of light caused by the interaction of the incident beam with a physical system, for example, after reflection at the interface between two different media. It has been mainly employed within the UV-visible and IR spectral regions, allowing thus the *in situ* investigation of the electrode/electrolyte interface.

The use of ellipsometry in interfacial electrochemistry is now relatively widespread, and several review papers have been published, to which the reader may refer in order to find discussions on the basic principles, on the instrumentation, and on the information obtained.[209,210]

In classical ellipsometry, the wavelength of the incident light is fixed, and the ellipsometric parameters (ψ, Δ) are measured. Because ψ and Δ are directly related to, respectively, the ratio of the reflectivity coefficients and the difference in phase angles of *p*- and *s*-polarizations, ellipsometry is highly surface selective. Moreover, its high sensitivity makes it a powerful technique for the investigation of very thin films and submonolayers at the electrode surface.

Depending on the spectral range used (UV-visible or IR), information on the electronic properties, particularly the values of the real and imaginary components of the complex dielectric constant, or on the vibrational states of the film is obtained. The thickness of the film can also be estimated from ellipsometry measurements. When measurements are taken over a wide range of wavelengths, in so-called "spectroscopic ellipsometry," information can be obtained on the film structure, because the ellipsometry response is equivalent to that obtained in reflection-absorption spectroscopy, with the advantage of a higher specificity for monitoring the surface film (which thus eliminates the need to obtain difference spectra by subtracting a reference spectrum that includes the spectral response of the species in solution).

Ellipsometry in the UV-visible range, with potential control for application to interfacial electrochemistry, was first developed by Reddy and Bockris in 1964.[211] Later on, the technique was extensively used for *in situ* measurements of the optical constants of passive oxide layers, mainly for purposes of studying corrosion[212] and characterizing electrochemically grown polymeric films, such as polyaniline.[213]

The study of chemisorbed species is theoretically possible, but so far, due to the complexity of the experiments and to the difficulty of interpretation, only investigation of H and O adsorption and of anionic adsorption in the double layer has been considered.[214] However, ellipsometry has been much more successful in the study of upd of some metallic adatoms on metallic substrates, for example, Pb/Ag and Pb/Cu.[215] Some attempts were also made to develop ellipsometry in the infrared range. Applications to the metal/solution interface were reported by Dignam and Baker[216] and also by Graf *et al.*[217]

(*c*) *Nonlinear optical spectroscopy*

Nonlinear optical techniques, such as second-harmonic generation (SHG) and sum frequency generation (SFG), are highly specific techniques for studying interfaces between two centrosymmetric media.[218] Due to its high surface sensitivity, as the result of a filtering system that rejects the bulk signal (at fundamental frequency ω), but passes the second harmonic (at frequency 2ω), SHG appears to be a powerful spectroscopic technique for investigating *in situ* the structure of the electrode/electrolyte interface, and particularly submonolayers of adsorbed molecules.[219] The first SHG studies have mainly concerned electronic transitions (due to the use of a visible laser and photomultipliers as detectors), so that they only gave information on the electronic structure of the electrode/electrolyte interface. However, the low sensitivity of infrared detectors has been overcome in the sum frequency generation technique, which uses a tunable infrared laser to probe the vibrational spectrum of the adsorbates and a visible laser to convert this spectrum to a visible spectrum by sum frequency generation. SFG can provide *in situ* vibrational spectra of adsorbed molecules at the electrode/electrolyte interface. This was effectively demonstrated recently in the case of the adsorption of carbon monoxide at platinum in contact with 0.1 *M* $HClO_4$ aqueous electrolyte.[220]

(d) X rays

Compared to spectroscopies in the UV-visible or in the infrared range, which give information on the electronic structure of the electrode surface as well as on the molecular structure and local environment of adsorbates, *in situ* X-ray techniques have the potential ability to provide details about longer range ordering. Indeed, with the advent of appropriate sources, many different techniques have been developed in the past ten years that use either scattering or absorption of X rays, in both the transmission and the reflection mode.

As X rays are scattered by all the matter in their path, they are inherently not surface sensitive. Therefore, for surface studies, it is necessary to design special experiments with improved surface selectivity. Various approaches can be used. Table 9 gives the names of the techniques, together with the type of X-ray source, the way in which surface sensitivity is achieved, and the applicability to electrochemistry.

Investigations at the electrode/solution interface are theoretically possible, since X rays at suitable wavelengths are able to penetrate a thin layer of solution. However, cell design remains difficult [see Section II.3(iv.a)]. One of the most successful techniques, initially developed by Fleischmann *et al.*,[221] utilizes a position-sensitive X-ray detector in conjunction with potential modulation, in a very similar way as in the modulation spectroscopies described previously. Its versatility, compared to other X-ray techniques, comes from the use of a conventional laboratory X-ray source instead of a synchrotron. Up to now, however, only a few studies with this technique have been related to electrocatalysis, the most relevant ones concerning hydrogen or carbon monoxide

Table 9
***In Situ* X-Ray Techniques and Their Applicability to Interfacial Electrochemistry**

Technique	X-ray source	Method of achieving surface sensitivity	Applicability to *in situ* investigations of electrochemical systems	Refs.
Scattering methods				
TRBD (total reflection Bragg diffraction)	Synchrotron	Glancing angle of incidence	Seems possible but not yet tried	228
X-ray standing waves	Synchrotron		Only *ex situ* experiments were carried out	229
In situ X-ray diffraction	Conventional laboratory X-ray source	Position-sensitive X-ray detection with electrode potential modulation	Yes, tested on several electrochemical systems	128, 221
Absorption methods				
EXAFS (extended X-ray absorption fine structure)	Synchrotron	Conventional EXAFS not surface sensitive		230
ReflEXAFS	Synchrotron	Reflection below the critical angle	Seems difficult	224, 231
SEXAFS (surface extended X-ray absorption fine structure)	Synchrotron	Measurements of electron yield	None	232
XANES (X-ray absorption near edge structure)	Synchrotron	Measurements of fluorescence yield coupled with glancing incidence geometry	Yes, several attempts have been made	233, 234

adsorption on platinized platinum in acid medium. It was shown that the electrode surface reconstructs upon adsorption of the so-called weakly adsorbed hydrogen[222] or carbon monoxide (independently of its coordination state)[223], but not with strongly adsorbed hydrogen.

The feasibility of the application of most of the other techniques listed in Table 9 to electrochemical systems has been demonstrated, generally taking corrosion of metals[224,225] or underpotential deposition of metals[226,227] as examples. In that sense, further progress can be expected in the field of electrocatalysis. However, the necessity for routine access to a synchrotron-type source will certainly be a serious limitation.

(*e*) *Mass spectroscopy*

The basic principle of mass spectroscopy has been described in Section II.2(ii). Electrochemical mass spectroscopy (EMS) was first developed by Bruckenstein and Comeau[83] and later used to investigate the behavior of adsorbed CO at a Pt electrode in acid medium.[235] A porous membrane (of small pore size and nonwetting characteristics) is set between the porous working electrode and a fritted disk which protects the inlet of the mass spectrometer. Thus, volatile products can pass freely through to the inlet to be analyzed.

Later on, the system was improved by Wolter and Heitbaum,[236] especially in terms of a considerably reduced time response, leading to the so-called DEMS technique (see Table 10). Examples of DEMS results are given in Fig. 18. It is striking how the production of CO_2 and methyl formate follows the rate of oxidation of methanol. However, the amount of the latter product is only 1% of that of the former.

Wilhelm *et al.* introduced ECTDMS, which combines electrochemical techniques with thermal desorption and mass spectrometry.[237] In this case, the electrode has to be transferred from

Table 10
Mass Spectroscopic Techniques

Technique	Necessitates transfer to ultra-high-vacuum (UHV)	Applicable to interfacial electrochemistry	Refs.
EMS (electrochemical mass spectroscopy	No	Yes	83, 235
DEMS (differential electrochemical mass spectroscopy)	No	Yes	236
ECTDMS (electrochemical thermal desorption mass spectroscopy)	Yes	Yes	237
SIMS (secondary ion mass spectroscopy)	Yes	Yes	89, 238

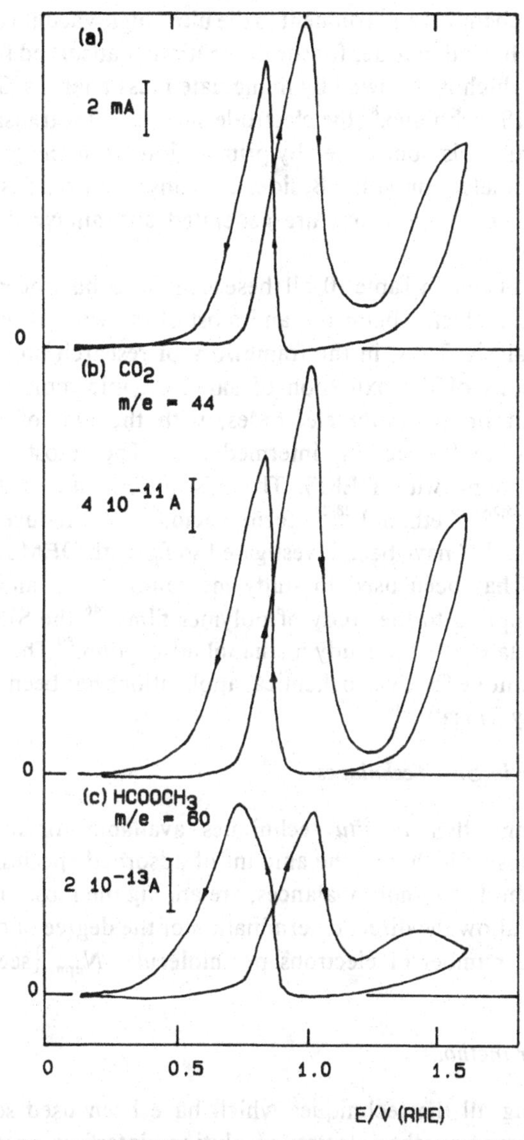

Figure 18. DEMS study of oxidation of 0.1 M CH_3OH at a porous Pt electrode (roughness factor of about 50) in perchloric acid medium. Current (a) and mass intensity (b and c) signals are shown. Formation of CO_2 ($m/e = 44$) and methyl formate ($m/e = 60$) was followed as a function of potential at a potential sweep rate of 20 mV s^{-1} (after Ref. 239).

the electrochemical environment to the ultra-high-vacuum chamber, where it is heated in order for the desorption of adsorbed fragments to occur, which is followed by immediate mass analysis. Similarly, in the SIMS technique,[89] the electrode surface, after transfer to the UHV chamber, is bombarded by primary ions with energies on the order of kiloelectron volts. Besides photons and particles, a small quantity of secondary ions are generated and analyzed by mass spectrometry.

As indicated in Table 10, all these techniques have been applied to interfacial electrochemistry, and most of the work is relevant to electrocatalysis. Thus, in the framework of research on fuel cells, investigations of the oxidation of small organic molecules were carried out on platinum electrodes, with the aim of detecting adsorbates and reaction intermediates. The most extensive work was done with DEMS. The adsorption and oxidation of methanol,[25,26,239] ethanol,[28,240] formic acid,[87,241,242] reduced CO_2,[87] and propanol[243] have been investigated so far with DEMS, whereas ECTDMS has been used to study methanol[88,237,244] and urea.[245] Initially applied to the study of polymer films,[246] the SIMS technique was later used to study methanol adsorption.[89] The potential of the technique for electrochemical applications has been reviewed recently by Trasatti.[238]

(*vi*) *Other in Situ Techniques*

Among other *in situ* techniques available for adsorption studies, those which give the amount of adsorbed species, such as tracer methods and microbalances, are among the most important, since they allow the direct determination of the degree of coverage, θ, and the number of electrons per molecule, N_{epm} [see Section II.3(ii)].

(*a*) *Tracer methods*

Among all the techniques which have been used so far for investigations at the electrode/solution interface, radiotracer methods represent certainly the oldest nonspectroscopic approaches. Actually, although they started to be developed as early as 1960, by Blomgren and Bockris,[247] their value in studies of interfacial electrochemistry is still great, and they still compare favorably with the wide range of much more sophisticated, newly developed spectroscopic techniques.

Based upon the properties of radioactive isotopes, radiotracer techniques provide a direct access to the surface concentration of adsorbed species, which is the most fundamental variable of surface science. Furthermore, in more recent applications, it has been demonstrated that their combination with other experimental techniques (voltammetry, infrared or mass spectroscopies) can be very successful for a better understanding of surface mechanisms.

Basically, tracer methods use trace amounts of an isotope of a particular element (the "tracer") in the same chemical form as the bulk electroactive species. Once the tracer is introduced into the cell, its behavior is followed as a function of time, potential, or any other parameter by means of either radiometry if the isotope is radioactive or of mass spectrometry or IR spectroscopy if the isotope is stable.

A variety of techniques have been developed (Table 11). They are described in detail in several recent extended papers by Horanyi,[91] Kazarinov and Andreev,[90] and Wieckowski,[248] in which all the technical aspects are discussed.

The direct methods, which do not require the removal of the electrode from solution and do not rely on the measurement of analyte concentration, are now preferred to indirect methods, which were not very accurate nor very sensitive.[90]

The principle of the foil method was initially proposed by Frédéric Joliot (as cited in Ref. 90), but the first electrochemical

Table 11
Tracer Methods

Method	Reference
Indirect methods	
Based upon the change of adsorbate concentration in solution and the subsequent change of radioactivity	For review, see 90
Based upon the measure of the radioactivity of the electrode surface, after adsorption and its removal from solution	For review, see 90
Direct methods	
Foil method	247
Thin-layer method	See 90
Electrode lowering method	249

application was undoubtedly due to Blomgren and Bockris.[247] In this technique, a gold foil, thin enough for radioactive radiation to cross it, is glued to the external face of a flowing-gas counter. The counter is moved vertically, slowly, toward the solution containing the radiotracer. The radioactivity, which is followed during the approach to the solution surface, increases sharply when the counter comes into contact with the solution, due to adsorption.

The thin-layer method has to be used in conjunction with flow cell techniques. It necessitates that the solution be circulated and allows a fixed position for both the electrode and the radioactive counter. Although interesting, it does not seem to have been greatly developed for electrochemical applications, probably because of the difficulty in maintaining the cleanliness of the solution.

The electrode lowering method was originally proposed by Kazarinov *et al.*[249] The radioactive counter is not in contact with the thin metal electrode, as in the foil method, but is placed below the bottom of the cell, from which it is separated by a thin-film membrane. Once adsorption has taken place, the electrode is lowered, so that only a thin layer of solution remains between the electrode and the membrane, for electrochemical control. A correct estimation of the radioactivity of the surface requires that blank experiments be carried out to measure the various radioactive contributions of the species in the bulk or of those adsorbed onto the membrane. Good sensitivity and accuracy are obtained, and the measurements can be carried out on both smooth and developed surfaces. Isotopes emitting any kind of radiation can be used.

Improvements of the original foil technique have been made by Horanyi[91] and by Sobkowski and Wieckowski,[250] by using metallized plastic foils instead of thin-foil electrodes. Other improvements came as well with the use of plastic or glass scintillators as radioactive detectors, on the surface of which metallic electrodes were vacuum deposited. In the technique developed by Wieckowski,[251] the scintillator is connected to a photomultiplier tube via a light pipe. More recently, combinations of scintillator technology with the electrode-lowering method have been proposed.[252]

The usefulness of radiotracer methods for interfacial electrochemistry is illustrated by 30 years of applications. Most of the studies were done on noble metal electrodes. Besides the adsorption of ions, which has recently been the subject of renewed interest with the coupling of FTIR and tracer methods,[253] the adsorption of a great number of organic compounds was studied. Horanyi[91] demonstrated that organics could be classified into two groups on the basis of their adsorption behavior on platinum. With the first group, the mobility of the adsorbed species is high, so that there is a fast exchange between adsorbed and nonadsorbed molecules. The saturated aliphatic carboxylic acids (apart from formic acid) belong to this group. With the second group, the adsorption is much stronger, leading in most cases to chemisorption. Nearly no exchange occurs with the species in solution. Most alcohols, unsaturated aliphatic acids, and aromatic acids belong to this second group. It should be emphasized that all the so-called "small organic molecules" which are often considered for possible use in fuel cells belong to this second group as well.

As pointed out by Wieckowski,[248] the capability of radiotracer methods to provide information for fuel cell applications has not yet been fully exploited. This is especially true now that improvements in the techniques are making them nearly applicable to well-defined surfaces and single-crystal electrodes. It is also true now that a thin-layer spectroelectrochemical cell fitted with a polished glass scintillator has been successfully designed for application to coupled infrared spectroscopic and radiotracer studies of ^{14}C organic adsorbates.[254]

(*b*) *Isotopic labeling*

Although it is not a technique by itself, recourse to the use of isotopic labeling may be of great help for the interpretation of spectra, whether mass, infrared, or Raman. The idea is to induce spectral shifts due to the different mass of the labeled compound as compared to that of the normal molecule. For example, labeled isotopes have been used in conjunction with on-line mass spectroscopy to investigate the adsorbed intermediates resulting from the oxidation of methanol,[26] ethanol,[240] propanol,[243] formic acid,[255] urea,[245] and glucose.[86]

Using infrared spectroscopy, $^{12}CO + {}^{13}CO$ mixtures of various compositions were studied with the aim of evaluating the coupling effects between adsorbates at various coverages.[256,257] Similarly, the formation of CO in the course of electrooxidation of methanol and formic acid on Pt was studied with FTIR spectroscopy on the basis of $^{12}C/^{13}C$ isotopic substitution.[258]

Depending on the type of study, H_2O, D_2O, or $H_2{}^{18}O$ may be considered as solvents. If deuterated organics are used, it must be taken into account that during the time scale of the experiment, some of the D atoms do not exchange (those, for instance, which are associated with $-CD_3$, $-CD_2H$, or $-CDH_2$ groups) while others do (such as those of the terminal OD groups of alcohols).

(c) *Quartz microbalance*

Tracking *in situ* the variations of mass which affect an electrode during electrochemical processes, as a function of potential or as a function of time, has become possible since the development of quartz microbalances sensitive enough to work in the 10^{-9} g range.[259] Briefly, these devices are based upon the properties of a piezoelectric quartz plate, whose resonance frequency varies proportionally with the quantity of a foreign mass added on its surface. Linearity is assumed, provided that the quantity deposited is small. Therefore, using evaporated metals on quartz as electrodes, it is possible to measure changes in their mass during electrochemical processes.[260] The technique has been popularized simultaneously by Bruckenstein and Shay[261] and Kaufman *et al.*[262] under the name electrochemical quartz crystal microbalance (EQCM). Other variants have been developed more recently; one is based on coupling with a rotating disk electrode, and another is based on a transient analysis of the mass-voltage relationship (Table 12).

Among the applications to electrochemical systems which have been reported so far, those relevant to electrocatalysis concern the upd of metals,[264] the absorption of hydrogen by Pd,[266] and the adsorption of ions[264] and polymer films.[267,268] Work has also been done in the field of corrosion.[265,269]

However, further development of the technique is expected in the field of electrocatalysis and adsorption studies, since its sensitivity (on the order of a few nanograms) is sufficient to monitor a full monolayer of adsorbed species (e.g., a monolayer of linearly adsorbed CO would correspond to a mass change of about 60 ng). The main inconvenience, for fundamental studies with single-crystal electrodes, is the need to use evaporated metallic films as electrodes,

Table 12
Techniques Derived from Quartz Microbalances

Technique	Characteristics	Refs.
QCM (quartz crystal microbalance)	Initially developed for use at the solid/gas interface	263
EQCM (electrochemical quartz crystal microbalance)	Designed for *in situ* use at the electrode/solution interface	261, 262
(rde)-EQCM	Combined rotating disk electrode with EQCM	264
ac-QEG (ac-quartz electrogravimetry)	Sinusoidal perturbation of current. A transient analysis of the mass-voltage relationship is possible	265

since this does not allow the nature and the structure of the electrode surface to be changed easily.

(*d*) *Electron microscopy*

Electron microscopy makes use of a thin beam of thermally excited electrons, which are focused on the sample under investigation. In transmission electron microscopy (TEM), the incident electrons penetrate through a thin film of the sample (or a replica of the sample), whereas in scanning electron microscopy (SEM), the incident electrons are scattered by the surface.[270]

Scanning electron microscopy, scanning tunneling microscopy (STM), and their variants (see Table 13) are recent techniques which have seen a large expansion in their use since their ability to examine surfaces on a subnanometer scale was demonstrated. Furthermore, STM can now operate in an electrolytic environment, which makes the technique a new and powerful tool for *in situ* examination of electrode surfaces.

Some of the methods listed in Table 13 are still under development. As yet, STM is the only one which has been applied to a wide range of electrochemical problems, leading to direct real-space information on the topography of electrode surfaces. Several extensive reviews are now available in the literature.[272,273,279]

In STM, a chemically inert fine metal tip is positioned so close to the sample surface (typically a few angstroms) that overlapping

Table 13
***In Situ* Scanning Electron Microscopy and Related Techniques**

Technique	Applicable to *in situ* studies of the electrode/solution interface	Reference(s)
SEM (scanning electron microscopy)	No: operates in UHV	
TEM (transmission electron microscopy)	No: operates in UHV	
STEM (scanning transmission electron microscopy)	No: operates in UHV	
STM (scanning tunneling microscopy)	Yes: can operate in various media including solutions	271, 272, 273, 274
SECM (scanning electrochemical microscopy)	Yes: can operate in various media including solutions	275, 276
SICM (scanning ion-conductance microscopy)	Yes: but seems more suitable for biological applications	277
AFM (atomic force microscopy)	No? (designed to work on insulating materials)	278

of the electron wave functions of the tip and the sample occurs. Thus, when a low voltage is applied between the tip and the sample, a very small current flows as a result of electron tunneling. Reference and counter electrodes have to be added for electrochemical applications, in order that the potentials of the tip and the sample are monitored independently. Of great importance are the design of the mechanical setup, especially to avoid vibrations, and the fabrication of the tip.

In the field of electrochemistry, different types of experiments have proved to be successful, since the first report by Dovek *et al.*[280] of a specially designed microscope. As predicted by Arvia,[281] topographic images provide direct evidence that electrochemical activation may produce surface roughening, leading to parallel ridges as well as domelike structures. The formation of preferentially oriented electrodes is also possible, depending on the potential programs used.

Underpotential deposition of metals has now been widely investigated by STM.[272,279] Such studies are highly relevant for electrocatalysis but, probably more interesting with respect to the potential applications, is the discovery that adsorbed molecules could be detected as well. Thus, among other examples, the coadsorption of benzene and carbon monoxide on Rh(111) was recently studied.[282] However, as pointed out by Cataldi *et al.*,[272] the role of adsorbed species in the tunneling process has not yet been clarified, so that one has to be careful in the interpretation of imaging.

Scanning electrochemical microscopy (SECM) is a different approach, developed by Bard and co-workers,[275,276] that uses an ultramicroelectrode with a tip radius of less than 10 μm. Recent applications have demonstrated the feasibility of the technique.

With an increasing number of groups working in the field of scanning microscopy, there is no doubt that new technological improvements of the technique are to be expected. Therefore, more applications to electrocatalysis, especially with respect to fast surface processes, should follow in the near future.

(*e*) *Other experimental approaches*

Some other techniques are still in their infancy with respect to their application to interfacial electrochemistry. It is still not clear whether they have any advantages over the well-established methods described above. However, it must be emphasized that their potentialities have not yet been explored. In this sense, worth mentioning are, for instance, photoacoustic spectroscopy (PAS) and Mössbauer spectroscopy. However, the list is not restrictive.

Although the photoacoustic effect is quite old, the feasibility of its application to electrochemistry has only recently been demonstrated.[283,284] Either gas-microphone or piezoelectric detectors can be used. Details on basic principles and on cell design may be found in the recent review by Vallet.[283] Up to now, most applications of PAS seem to have concerned the study of oxide films on electrodes.

Mirage spectroscopy, or photothermal deflection spectroscopy (PDS), which is about two orders of magnitude more sensitive than PAS, is able to record both the concentration gradient and the absorption spectrum of species present at the electrode/electrolyte interface, by monitoring the refractive index gradient produced in the electrolyte by an exciting light (a laser beam perpendicular to the electrode surface). A probe laser beam, parallel to the surface, is deflected by the refractive index gradient, and the amplitude of this deflection as a function of wavelength may provide the absorption spectrum.[285]

Applications of PDS have also mainly concerned the study of oxide layers at copper electrodes[285] or zinc electrodes,[286] but a recent study mentions the investigation of (Ir + Ru)/Ti mixed-oxide electrocatalysts for chlorine evolution.[287]

Mössbauer spectroscopy[288] has been developed for *in situ* electrochemical experiments and has been recently reviewed.[289] Due to the nature of the technique, applications are restricted to a few metals such as Fe, Co, and Ni. Electrochemically grown films of oxides and of hydroxides, as well as transition-metal macrocycles, used as electrocatalysts for oxygen reduction, have been studied so far.

4. Characterization of the Electrode Material and of the Catalytic Surface

Since in electrocatalysis both the electrode material and the catalytic surface play a key role in the reaction mechanism and kinetics, particularly for the oxidation of small organic molecules, it is of fundamental importance to use well-controlled electrode materials and to characterize their surface properties.

(i) *Role of the Electrode Material*

The kinetics of electrocatalytic reactions depend strongly on the chemical nature, the electronic structure, and the geometric texture of the electrode material.

Varying the chemical nature of the electrode material from noble metals to transition metals and *sp* metals changes the electrocatalytic activity greatly. For example, the oxidation of ethylene glycol in alkaline medium on four different noble metals electrodes (Au, Pd, Pt, Rh) is shown in Fig. 19. Gold gives the highest current densities, but at too anodic potentials for practical use in fuel cells. Platinum and palladium display moderate electroactivity, whereas rhodium is nearly inactive for alcohol oxidation, due to a strong

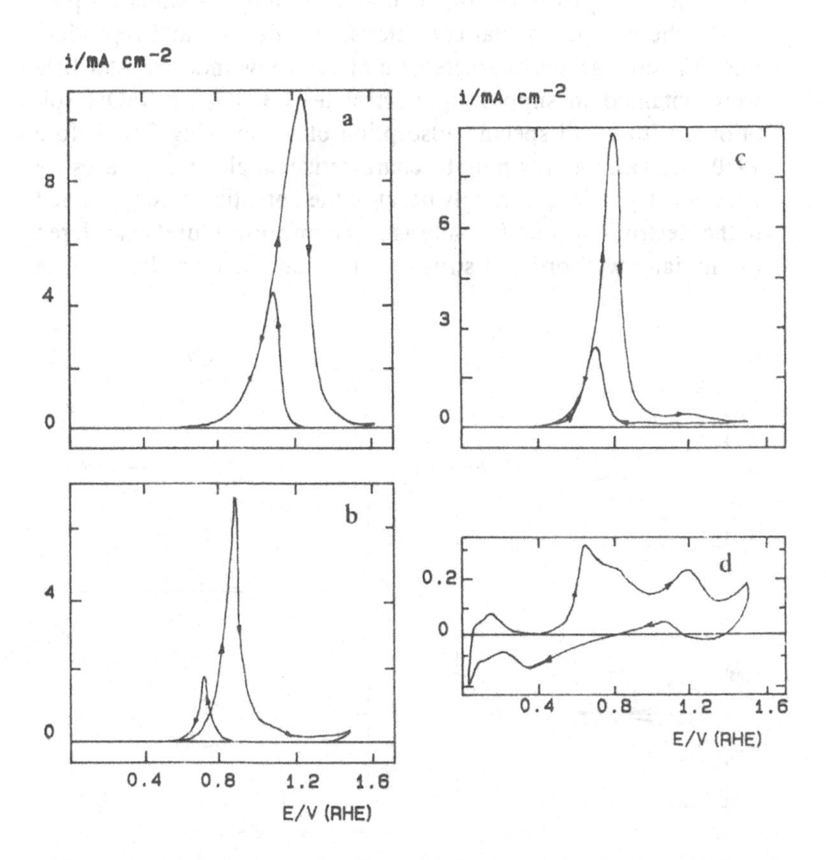

Figure 19. Voltammograms of noble metal electrodes, showing the oxidation of 0.1 *M* ethylene glycol in alkaline medium (0.1 *M* NaOH, 25°C, 50 mV s^{-1}): (a) Au; (b) Pd; (c) Pt; (d) Rh.

oxidation of the surface, which blocks the catalytic sites. In acid medium, platinum appears to be the only electrocatalyst sufficiently active for practical applications.

The geometry and the electronic structure of electrode materials is easily varied using different faces of single crystals. Well-defined and well-controlled platinum single-crystal electrodes were first developed and characterized in 1980 by Clavilier *et al.*[11] They used flame annealing of the electrode surface and then quenched the electrode structure with a droplet of ultrapure water, before transferring the electrode to the electrochemical cell, the electrode surface still being protected from contamination by the water droplet. Under these experimental conditions, well-defined and reproducible voltammograms characteristic of each low-index crystal face were obtained in supporting electrolytes ($HClO_4$ or NaOH solutions) with small specific adsorption of anions (Fig. 20). Before 1980, the various attempts to characterize single-crystal faces by voltammetry failed,[290] mainly because the potential sweeps applied to the electrode surface (usually as a cleaning procedure) disordered the initially well-ordered structure. It is now known that surface

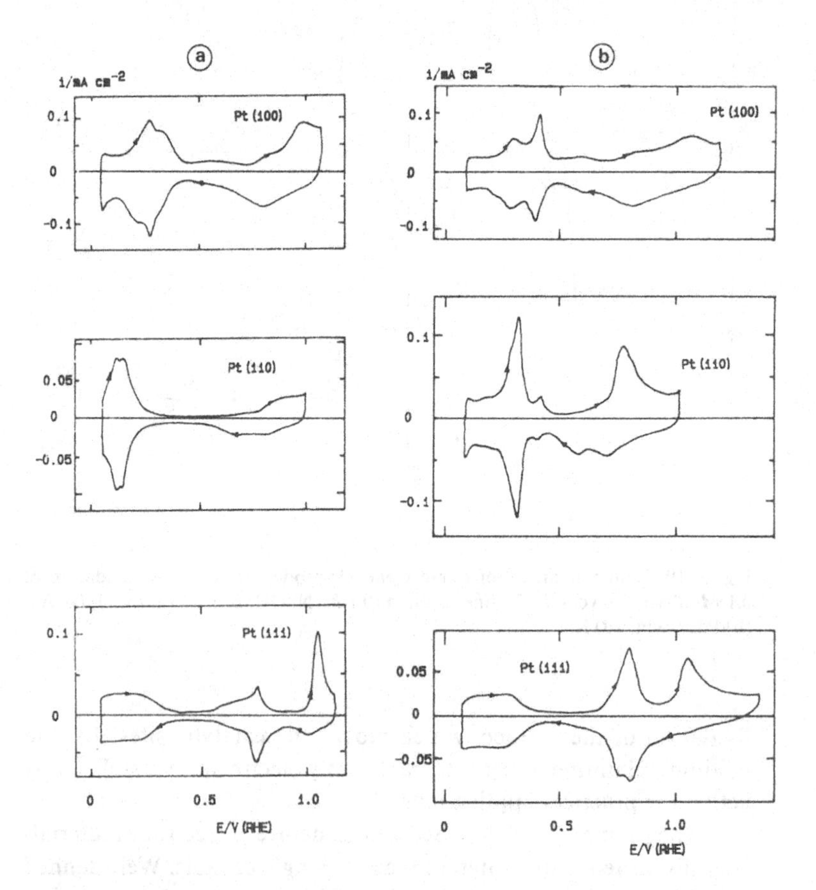

Figure 20. Voltammograms of platinum low-index single crystals in ultrapure supporting electrolytes (25°C, 50 mV s^{-1}): (a) 0.1 *M* $HClO_4$; (b) 0.15 *M* NaOH.

reconstruction is particularly due to oxygen adsorption. Surface reconstruction is also observed in the presence of other adsorbed species (e.g., anions, molecules such as CO), so that it is necessary to perform the flame treatment of the electrode surface after each experiment, to recover the initial surface structure.[291] On the other hand, fast repetitive pulsed potential programs applied to polycrystalline surface have been shown to lead to preferentially oriented surface electrodes, exhibiting similarities to certain low-index faces.[292,293] Thus, these electrodes behave more or less like single-crystal electrodes in the electrooxidation of various oxygenated species, such as CO[294] and CH_3OH.[295]

Each of the three low-index platinum single-crystal faces, Pt(100), Pt(110), and Pt(111), behaves very differently in the electrooxidation of aliphatic oxygenated molecules, such as CO,[296] HCOOH,[297] CH_3OH,[298] C_2H_5OH,[299] C_4H_9OH,[300] glucose,[301] and sorbitol,[302] as discussed in detail in Section III. In the oxidation of small molecules (e.g., CO, HCOOH, CH_3OH, C_2H_5OH), the behavior of polycrystalline surface can be easily simulated as the weighted contribution of each single-crystal plane. For very smooth polyoriented platinum surfaces prepared in a similar way to single crystals,[11] the main contribution to the electrocatalytic behavior comes from Pt(110), and the second largest contribution from Pt(100), with a small fraction (about 5%) from Pt(111). Conversely, for larger molecules (e.g., butanol, glucose), it is quite impossible to simulate the electrocatalytic behavior of polycrystalline platinum by that of low-index single-crystal planes, their activity always being much larger (by a factor of 2 to 10) than that of the polycrystalline electrode. This different behavior is probably related to the symmetry and the geometry of the single-crystal surface, which may favor, or may or may not accommodate, the particular structure of the adsorbed species, depending on its size. With small organics, the main adsorbed species is CO, as evidenced by infrared reflectance spectroscopy—see Section III.1—so that the electrocatalytic behavior of the different single-crystal faces is primary determined by CO poisoning. It is therefore observed, in acid medium, that Pt(111), although less active on the first potential sweep, is almost not poisoned at all by CO, so that after a few sweeps it becomes the most active and the most stable face. Conversely, Pt(110), which gives the highest current density on the first sweep, is rapidly poisoned, so that its activity decreases rapidly. The electrocatalytic activity of Pt(100) is more stable with time, although the electrode surface is strongly blocked until potentials of 0.7–0.75 V versus RHE, of which adsorbed CO is oxidized. With larger molecules, such as butanol and glucose, the main adsorbed species is not CO (although a small amount of CO was found by IR reflectance spectroscopy), so that the single-crystal faces are not as sensitive to poisoning, and, consequently, Pt(111) is the most active plane, especially in alkaline medium: it gives a current density of up to 18 mA cm^{-2} for the oxidation of *n*-butanol, whereas Pt(110) gives 12 mA cm^{-2}, and Pt(100) 3.2 mA cm^{-2} under the same experimental conditions (0.1 *M* butanol in 0.1 *M* NaOH, 25°C, 200 mV s^{-1}), the polycrystalline electrode giving only 2.8 mA cm^{-2} (Ref. 300).

Another way to greatly change the electrocatalytic activity of a given electrode is to modify its surface by underpotential deposition (upd) of various foreign adatoms, such as Ag, As, Bi, Cd, Cu, Ge, Pb, Re, Ru, Sn, and Tl.[13,149,303,304] Depending on their size and on their redox potential, which governs the potential range in which they desorb from the surface by oxidation, as well as on the kind of organic molecule that is oxidized, different enhancement effects or inhibiting effects are found in the literature. One can say roughly that, except for methanol, the more the electrode is sensitive to poisoning, particularly by adsorbed CO, the more important are the enhancement effects of some adatoms. Among the first experimental findings that demonstrated the effect of adatoms was the striking increase of formic acid oxidation currents at a platinum electrode modified by upd of Pb adatoms[13,303] [see Section III.1(iii.b)]. Another impressive example is the oxidation of ethylene glycol in alkaline medium on modified platinum electrodes.[305] Some of the adatoms considered (Pb, Bi, Tl) greatly increase the current density (by a factor from 6 to 15), leading to values as high as a few hundred milliamperes per square centimeter, whereas others, such as Cd, do not change the current density, but shift the electrode potential cathodically (by about 200 mV), which reduces the oxidation overvoltage [see Section III.3(i)]. In acid medium the behavior of some adatoms is quite different, since, for example, Bi and Tl greatly inhibit the electrooxidation of ethylene glycol.[306]

The interpretation of such effects is not entirely clear, since both geometric and electronic effects modify the catalytic properties of the electrode surface and its ability to adsorb the organic molecule. Many possible explanations have been discussed in the literature—see, for example, the review by Parsons and Vander-Noot[12]—such as the modification of the electronic and redox properties of the surface, the prevention of the formation of blocking species acting as poisoning intermediates, and explanations in terms of the bifunctional theory of electrocatalysis. Moreover, recent experiments using EMIRS clearly showed that, at least in the case of formic acid oxidation in acid medium, at rhodium electrodes, lead adatoms acted by replacing, one by one, bridge-bonded CO, thus reducing the poisoning of the electrode surface.[19,177] Conversely, Cd adatoms, which are not as strongly adsorbed, were not able to displace adsorbed CO[19] [see Section III.1(iii.c)].

Similarly, the electrocatalytic activity of electrode materials may be improved by using alloys of different kinds, such as binary alloys, plurimetallic alloys, and intermetallic compounds.[307,308] This approach has been particularly developed for the electrooxidation of hydrogen and the electroreduction of oxygen in fuel cells.[309] For the electrooxidation of small organic molecules, many attempts have also been made to improve the catalytic behavior of the electrode surface by metal alloying. For example, methanol electrooxidation can be enhanced by using platinum electrodes modified by a second metal[310,311] [see Section III.1(i.d)]. Homogeneous binary alloys, such as Pd/Au, Pt/Pd, and Pt/Rh, which form continuous series of solid solutions, are particularly interesting since they allow the lattice parameters and the electronic properties of the metal catalyst to be varied in a continuous manner. These metallic alloys usually lead to enhancement of the electrocatalytic oxidation of small organic molecules for a given alloy composition. Such "synergistic effects" were observed, for example, with the oxidation of formic acid at Pt/Rh,[312] Pt/Au,[313] Pt/Pd,[314] and Pd/Au[315] alloy electrodes, with the oxidation of methanol at Pt/Rh[316] and Pt/Pd[43] alloy electrodes, and with the oxidation of ethylene glycol at a Pt/Au alloy electrode.[317] Ternary alloys are even more efficient for the electrocatalytic oxidation of oxygenated molecules, as illustrated by the electrooxidation of ethylene glycol at Pt/Pd/Bi and Pt/Pd/Pb alloy electrodes[318] and at Pt/Au alloy electrodes modified by lead adatoms.[319]

The interpretation of the enhancement of activity is similar to that discussed previously for noble metal electrodes modified by upd of foreign metal adatoms, provided that the surface composition of the catalytic material is similar.[320,321] The surface composition of metallic alloys is not easy to determine in most cases, as discussed in Section II.4(ii.c). Furthermore, there is generally a preferential dissolution of one metal compound in acid solutions, such as Rh in the case of Pt/Rh alloys and Pd in the case of Pd/Au alloys.[322] Even in neutral solutions, it was shown that cyclic voltammetry in the oxygen adsorption region greatly increased the rate of dissolution of one metallic component of the alloy.[315]

In practical systems, such as fuel cells, the catalytic material must be dispersed on a convenient substrate. The first idea was to use smooth foils of the same metal for electrodeposition of the catalytic metal, which gives metallized metal electrodes or metal blacks, platinized platinum being the most popular.[323] Very often, these rough surfaces (the roughness factor of which varies from single-digit values to a few hundred) behave relatively differently from low-surface-area electrocatalysts. For example, the rate of methanol electrooxidation, which is a structure-dependent reaction, does not increase linearly with the true surface area, but, instead, tends to a limiting current intensity, mainly diffusion controlled, when the roughness factor, ρ, reaches 200.[324] This particular behavior may be related to the change, on the one hand, in the rate-determining step and, on the other hand, in the nature and in the surface distribution of the adsorbed species, as shown by EMIRS.[325]

However, metallized metals and noble metal blacks do not lead to high-surface-area catalysts, since the specific surface area only reaches a few square meters per gram. To achieve a greater dispersion of the electrocatalytic material, it is thus necessary to select high-surface-area electrically conducting substrates, among which graphite, carbon blacks, and activitated carbons are widely used.[326] Specific surface areas, as measured by the method of Brunauer, Emmett, and Teller (BET), vary from a few hundred to a few thousand square meters per gram. Depending on the origin of the carbon substrate, for a given dispersion of platinum or for a given size of the platinum particles, the electrocatalytic properties vary greatly, due to specific metal–support interactions.[326] Ultrahigh dispersions of Pt electrocatalysts, supported on carbon blacks, with specific surface areas of up to 200 m^2/g of Pt, corresponding to particle sizes of about 15 Å, display a linear relationship between the mass activity (in A/g) for methanol oxidation and the platinum surface area.[327,328] The preparation of the electrocatalyst deposit on the carbon surface may be carried out by several methods, including impregnation, ion-exchange adsorption, chemical reduction, thermal decomposition, attack of a Raney metal alloy in base electrolyte, and electrodeposition.[326]

Other supports can also be considered. There is increasing interest in solid polymer electrolytes (SPE), as originally proposed by Hamilton Standard, for the General Electric H_2/O_2 fuel cells, in the framework of the NASA Gemini Space Program, and in perfluorosulfonated membranes (Nafion of Dupont de Nemours and Proton Exchange Membrane of Dow Chemicals), with high ionic conductivity. The use of these has led to the development of extremely powerful systems (a few watts per square centimeter).[329,330] In recent attempts, the platinum electrocatalyst is directly dispersed at the surface of the SPE membrane, which plays the role of both solid electrolyte and separator. This gives highly efficient systems for water electrolysis and H_2/O_2 fuel cells.[331] These Pt–SPE composites have been also used for the electrocatalytic oxidation of methanol.[332,333] Binary Pt alloys were also dispersed on SPE membranes, leading again to enhancement effects for methanol electrooxidation, with Sn, Ru, Ir,[334] and Mo.[335–337]

Another way to achieve very high dispersion of the metal electrocatalyst is by formation *in situ* of metal microparticles inside a polymer matrix. The recent development of numerous electronic conducting organic polymers, among which polyaniline,[338,339] polypyrrole,[340] and polythiophene[341,342] are the most stable in aqueous solutions, offers the possibility of obtaining catalytic electrodes with a very high efficiency. However, apart from studies of hydrogen evolution or oxygen reduction,[343–346] very few studies concern the electrocatalytic oxidation of organic compounds. One of the first reports, on the oxidation of formic acid at modified platinum polyaniline-coated electrodes, underlined the tenfold increase of the electrocatalytic activity, as compared to that of platinized platinum electrodes, and correspondingly the net decrease in the rate of formation of poisoning species.[347] Comparable behavior was recently found for the oxidation of small organics (HCOOH, CH_3OH) at modified platinum polyaniline-, polypyrrole-, polythiophene-, and copolymer (pyrrole-dithiophene)-coated electrodes.[348,349] Excellent current densities (up to 10 mA cm^{-2}, i.e., about 100 times higher than with smooth platinum) were obtained, with a small amount of platinum (0.1 mg cm^{-2}), and in most cases the forward and backward sweeps were superimposed, showing that strongly chemisorbed poisoning species were not extensively formed on such electrodes.

(ii) *Characterization of the Electrocatalyst Surface*[350]

To establish precise correlations between the catalytic properties of the electrode material and the kinetics of the electrooxidation of organic compounds, the electrocatalyst surface must be characterized by appropriate methods. This is the only way to conceive and develop new electrocatalysts with high activity for a given reaction, such as the electrooxidation of methanol in a direct methanol fuel cell (DMFC).

The real surface area and the surface structure and texture, as well as the surface composition, must be determined, preferably by *in situ* methods. However, some *ex situ* techniques, particularly those developed so far for studying the solid/vacuum interface (electron diffraction and electron spectroscopies), will also be discussed, because of their great interest with regard to the determination of the surface structure and nature of the electrocatalysts. As they require ultra-high vacuum (UHV), their application to electrode surfaces will necessitate the transfer of the electrode from the electrochemical environment to the vacuum chamber. Therefore, the results may be subject to controversy, in that there always remain doubts about whether the double layer and the interface structure are still intact after removal from the solution. However, this is not sufficient reason for rejecting methods that have proven so successful at the solid/vacuum interface. Kolb *et al.* have recently examined this crucial point of the link between electrochemistry and UHV techniques.[351]

(a) *Determination of the real surface area*

A key physical quantity in comparing electrocatalytic reactions on electrodes of different nature and structure is the real surface area S_r. The real surface area of the electrocatalyst can be measured by adsorption processes, either physical adsorption involving van der Waals forces or chemisorption leading to the formation of a chemical bond between the catalytic surface and the adsorbate.[326,352,353]

Ex situ measurements make use of the adsorption of a gaseous molecule, such as N_2, H_2, O_2, or CO, on the catalyst surface.[354] In the well-known BET method, the volume V_m of the adsorbed gas (usually N_2) corresponding to a full monolayer is first determined, allowing the specific surface area S to be calculated as

$$S = V_m \mathcal{N}_A \sigma / V_0 \tag{40}$$

where V_0 is the molar volume of the gaseous molecule under standard conditions, σ is the cross-sectional area of the adsorbed gas ($\sigma \approx 16.2$ Å^2 for N_2 at -195°C), and $\mathcal{N}_A$ is Avogadro's number. The BET method gives relatively reliable S values for high specific surface areas (on the order of a few tens to a few hundreds of square meters per gram). A variant of the method using radioactive gases, such as ^{85}Kr-enriched krypton ($\sigma \approx 15.6$ Å^2), is more sensitive, giving surface areas on the order of $0.1\ m^2\ g^{-1}$ (Ref. 355).

Gas-phase chemisorption of H_2, CO, or O_2 has been applied to the measurement of surface areas of supported catalysts, such as Pt, Rh, Pd, and Ir, deposited on carbon substrates. The amount of adsorbed gas on the catalyst surface can be determined by static volumetric measurements or, in the continuous-flow method, by gas chromatography after desorption or after titration based on a given surface reaction.[356] For example, preadsorbed oxygen on a Pt surface can be titrated by reaction with either hydrogen or CO at room temperature:

$$Pt{-}O_{ads} + 3/2H_2(gas) \rightarrow Pt{-}H_{ads} + H_2O$$

$$Pt{-}O_{ads} + 2CO(gas) \rightarrow Pt{-}CO_{ads} + CO_2$$

These two reactions illustrate the main problem encountered in determining the surface area from the amount of chemisorbed gases, namely, the adsorption stoichiometry. For hydrogen the ratio

H/M is usually assumed to be 1 for most metals M. However, for oxygen the O/M ratio can vary between 1 and 2, depending on the catalytic material. Similarly, for chemisorbed CO, the CO/M ratio varies from 1 for linearly bonded CO to 0.5 for bridge-bonded CO, depending on the nature of the catalytic material and on the size of the metallic particles, as discussed in Section II.3(iv).

The gas-phase adsorption methods are not suitable for determining the surface area of smooth electrodes, particularly single crystals, because their sensitivity in measuring the adsorbed volume is too low. Moreover, the adsorption of gases does not necessarily give the real surface area, S_r, of the electrode catalyst, that is, the surface in contact with the electrolyte solution and the electroreactive species. *In situ* measurements of S_r by electrochemical techniques, such as galvanostatic charging curves and potential sweep techniques, are thus much more preferable. Information on both very smooth electrodes (including single crystals) and highly developed electrodes with catalytic particles deposited on high-surface-area substrates can be obtained.

For metals that adsorb hydrogen well (e.g., Pt, Rh, Ir), S_r can be easily determined from the quantity of electricity Q_H° corresponding to a full monolayer of adsorbed hydrogen ($\theta_H = 1$), assuming that each surface atom is associated with one adsorbed hydrogen atom.[357] Then the theoretical charge per unit surface area (usually per square centimeter), corresponding to deposition or ionization of adsorbed hydrogen, may be calculated provided that the geometry of the surface, that is, the distribution of the surface atoms, is known. This information is only available for single-crystal planes, particularly for the low-index faces. For example, the adsorption of a hydrogen monolayer on the low-index planes of platinum single crystals corresponds to 243 μC for Pt(111), 209 μC for Pt(100), and 195 or 149 μC for Pt(110), depending on whether the second row of atoms situated below the upper layer is counted or not. For a polycrystalline electrode, the number of surface atoms will depend on the distribution of the different crystallographic sites, so that the charge per real unit surface associated with the adsorption of hydrogen will be relatively difficult to estimate. For a smooth polycrystalline platinum electrode, a value of 210 μC cm^{-2}, based on the equally weighted contribution of the three low-index planes, is usually assumed, so that the real surface area S_r is calculated as

$$S_r\,(\text{cm}^2) = \frac{Q_H^\circ\,(\mu\text{C})}{210\,(\mu\text{C/cm}^2)} \tag{41a}$$

From the knowledge of S_r, one may evaluate the roughness factor ρ of the electrode as

$$\rho = S_r/S_g$$

with S_g the geometric area.

For metals that do not adsorb hydrogen (such as gold), the real surface area may be estimated from oxygen adsorption, provided that a full monolayer of adsorbed oxygen can be isolated. This is not so easy, because the charge Q_O associated with the surface oxidation continuously increases with increasing potential before oxygen evolution. However, if the Q_O versus E curve exhibits an inflection point corresponding to a given stoichiometry of the oxygen layer, for example, PtO for platinum, it is possible to calculate the real surface area; for example,

$$S_r\,(\text{cm}^2) = \frac{Q_O\,(\mu\text{C})}{420\,(\mu\text{C/cm}^2)} \tag{41b}$$

where 420 μC cm^{-2} is the charge associated with a full monolayer of oxygen, corresponding to PtO, which requires two electrons per site for its removal.

Instead of adsorbing hydrogen or oxygen on the electrocatalyst surface, and particularly for those metals that do not adsorb them, or that absorb either one or both of them, it is also possible to adsorb foreign metal adatoms by upd. This leads to another way of evaluating S_r, namely, from the quantity of deposited metal, provided that the stoichiometry of the layer, the degree of oxidation of the metal, and its atomic radius are known. For a full monolayer of Cu on Pt, the corresponding quantity of electricity is 417 μC cm^{-2} (Ref. 358), consistent with a Cu/Pt stoichiometry of 1 : 1 and two electrons per site [i.e., $Cu^{2+} + 2e^- \rightarrow Cu(0)$]. The method is particularly useful in the case of ruthenium electrodes, where hydrogen adsorption, because of competition with hydrogen absorption, or oxygen adsorption, because of the formation of different oxygenated species, cannot be used for surface area measurements. Underpotential deposition of Cu on Ru allows S_r to be evaluated with good accuracy, assuming a Cu/Ru ratio of 1 and a value of 502 μC cm^{-2} for the oxidation of a Cu monolayer.[359] However, the method seems to be limited to S_r measurements for Ru electrodes of small roughness factors ($\rho < 30$), because of the growth of copper multilayers at higher ρ.[360]

The real surface area, S_r, can also be determined from the measurement of the double-layer capacity C_d, assuming that a unit real surface area (1 cm^2) corresponds to a known capacity, for example, about 20 μF cm^{-2} for platinum or gold in the double layer, so that

$$S_r\,(\text{cm}^2) = \frac{C_d\,(\mu\text{F})}{20\,(\mu\text{F/cm}^2)} \tag{42}$$

Unfortunately, the reference value is not very well established (values of 16 to 80 μF cm^{-2} have been reported in the literature), and the measured capacity of a rough surface is an underestimate as a result of current line distribution in porous media.

(b) *Determination of the surface structure*

The catalyst microstructure, or porosity, of porous electrodes is an important factor, which controls both the diffusion and/or adsorption processes and the electron transfer step through the distribution of current and potential lines. The distribution of pore sizes inside the electrode material can be determined from N_2 adsorption isotherms.[361] The method is limited to pore diameters smaller than 300 Å, so that mercury porosimeters have to be used for larger pore diameters. The latter technique presents a major disadvantage, in that after the measurements there remains some Hg, which is known to be an electrocatalytic poison. Moreover, the high pressure necessary for Hg to penetrate into the porous structure may disrupt the pore system.

For some structure-sensitive reactions, the specific activity of the catalyst may depend on the mode of preparation and on the size of the particles.[362] The metallic properties, and thus the electrocatalytic properties, of high-surface-area electrodes depend strongly on the particle size, particularly in the case of small particles.[363]

From the knowledge of the specific surface area S, it is easy to estimate the average particle size, assuming spherical (diameter d) or cubic (dimension d) particles. This gives

$$d = 6/\rho S \tag{43}$$

where ρ is the density of the metallic particle.[326]

Various physicochemical methods, such as X-ray diffraction (XRD), small-angle X-ray scattering (SAXS), and electron microscopy (EM), allow the particle sizes to be determined.

The presence of small particles in polycrystalline materials causes line broadening of the XRD pattern. The average crystallite size, d, can be evaluated from Scherrer's equation[353,364]:

$$d = \frac{k\lambda}{\beta \cos \theta} \tag{44}$$

where k is a constant depending on the crystallite shape (k can vary from 0.7 to 1.7), λ is the wavelength of the X-ray beam, β is the linewidth free of broadening (the natural linewidth), and θ is the Bragg angle [from the Bragg equation $2d_{hkl} \sin \theta = n\lambda$, with d_{hkl} the interplanar spacing in the crystal lattice planes of Miller indices (h, k, l), and n an integer]. This method is useful for crystallite sizes ranging from 50- to 1000-Å average diameter. SAXS allows measurement of smaller particle sizes ranging from 20 to 500 Å.[365] In this technique corrections have to be made for the X-ray scattering intensity of the catalyst support.

The size and shape of electrocatalyst particles can be evaluated from transmission electron microscopy (TEM), by observing electron micrographs.[366] Particle sizes as low as 10 Å can be resolved by EM, but the accuracy is very poor for particles less than 20 Å in diameter.[367] Particle sizes determined from surface area measurements, X-ray diffraction, and electron microscopy are roughly in agreement.[363]

Surface morphology and surface topography are better studied by scanning electron microscopy (SEM) [see Section II.3(vi.d)].

The crystallographic structure of metallic single-crystal electrodes can be conveniently investigated by electron diffraction techniques, such as low-energy electron diffraction (LEED) and reflection high-energy electron diffraction (RHEED).[368,369] In LEED, the electrons incident on the electrode surface, having an energy typically in the range 10 to 500 eV (which corresponds to a wavelength range of 1200 to 24 Å), can be diffracted by periodic surface structure, that is, by single-crystal planes. This is equivalent to X-ray diffraction by a single-crystal lattice and provides an image of the surface lattice in reciprocal space, from which the surface structure in real space can be derived. The method can also give informations on periodic surface defects (e.g., steps, terraces, long-range disorders), in terms of density, size distribution, etc.[370]

LEED and RHEED have provided very useful information on the reconstruction of the surface of noble metal single-crystal electrodes, which results from adsorption–desorption processes of the oxygen layers in cyclic voltammetry.

LEED also provides useful information on the adsorption of different atoms and molecules on well-defined surfaces, such as single-crystal planes.

Finally, Rutherford backscattering spectroscopy (RBS) is also an *ex situ* technique which can be applied for electrode surface analysis, as reviewed by Kötz.[371] Although most of the work was done on oxides and oxide formation, the RBS technique has also been applied by Hyde *et al.* to characterize the distribution of Pt particles in a fuel cell catalyst.[372]

It is also worth mentioning that RBS can be performed *in situ* using a thin silicon membrane, on which a metallic film electrode is evaporated and through which the accelerated ionic beam passes, reaching the electrode/electrolyte interface before being backscattered. Such an arrangement allowed Kötz *et al.* to study Cu deposition on Ir electrodes.[373]

The present list of techniques is only a brief introduction to the possibilities offered by *ex situ* UHV spectroscopies. It is not intended to cover all the aspects nor all the approaches to problems connected with interfacial electrochemistry. It was just presented in order to show the profit that electrochemists may draw from the use of powerful techniques that have long been developed for surface chemical investigations of the metal/vacuum interface, as recently pointed out by Bockris and González-Martin.[374] Their main application in the field of electrocatalysis is the control and characterization of well-defined electrode surfaces, such as single-crystal electrodes and preferentially oriented surfaces.

(c) *Determination of the surface composition of multicomponent catalysts*

The surface composition of many alloys and bi- or plurimetallic electrodes is different from that of the bulk material, because of preferential dissolution of one component in an aggressive electrolytic solution (acid electrolyte) and/or the influence of adsorbed species (adsorbed oxygen resulting from potential cycling in base electrolyte or adsorbed electroreactive species such as CO).[357]

X-ray diffraction and microprobe electron microscopy (analytical TEM, which makes use of the energy of the X rays generated by electron bombardment to identify elements with atomic number $Z > 10$) can be used to determine the surface composition. XRD gives the composition of binary alloys from an accurate measurement of the lattice parameter, which is, to a first approximation, a linear combination of the lattice parameters of the pure metals. This is the case for homogeneous alloys, such as Pt/Rh alloys.[375] However, the determined composition is characteristic of a surface layer a few microns thick, which is not very significant for electrocatalysis, since only the first atomic layers are electroactive.

Electron spectroscopies,[376,377] such as ultraviolet photoelectron spectroscopy (UPS), X-ray photoelectron spectroscopy (XPS), also known as electron spectroscopy for chemical analysis (ESCA),[378,379] and Auger electron spectroscopy (AES)[380] are very powerful techniques for determining the surface composition of metallic electrodes. Since the penetration of electrons (on the order of a few tens of angströms, corresponding to two to five atomic layers) is much lower than that of X rays, electron spectroscopy gives information much more relevant to the surface composition. Electron spectroscopy can give qualitative and quantitative information about the surface from the position in energy and the intensity of the spectral line characteristic of the electron levels. Because the position of the line is strongly dependent on the environment of the atom (chemical shift), it can be used to determine the oxidation state of the element. For example, Pt(II) species have been identified by ESCA as one valence state of platinum dispersed on a carbon support that is active for the electrocatalytic oxidation of methanol.[381,382] XPS has also been used to analyze polymer films[246] and to monitor the surface concentration of anions and cations on gold emerged electrodes.[351]

AES is much more sensitive than ESCA, and surface quantities as small as 10^{-3} of a monolayer can be monitored. However, quantitative analysis by AES is much more delicate, since numerous Auger peaks may overlap and are difficult to separate. AES can be performed in the same UHV chamber as LEED, and both methods are usually simultaneously available. Therefore, structural information on the investigated surface can be obtained under the same experimental conditions.[369,377]

Similarly, AES can be coupled with electrochemical experiments, as initially shown by Ishikawa and Hubbard[383] and Revie *et al.*[384] As pointed out in recent reviews by Aberdam[380] and Durand *et al.*,[385] AES can provide detailed information on the upd of metals, which is relevant to electrocatalysis.

Combining these different techniques with cyclic voltammetry and energy loss electron spectroscopy (EELS) enabled Batina *et al.*[386] to demonstrate the utility of *ex situ* techniques for electrochemistry. These surface spectroscopies in UHV were used to study the adsorption on Pt single crystals from aqueous solutions of a wide series of organic compounds of biological or pharmacological interest.

As stated before, most of the techniques described above require the transfer of the electrode surface to the UHV chamber of the apparatus, which always raises the question of the validity of measurements in a nonelectrochemical environment (with thus no potential control of the electrode surface, possible change of the surface properties in vacuum, risk of contamination of the surface during the transfer operation, etc.). *In situ* methods are thus much more preferable.

The only method described in the literature to evaluate *in situ* the surface composition of metal alloys is the voltammetric determination of the peak potential, E_p, associated with oxygen desorption during the negative sweep of a voltammogram recorded in the pure supporting electrolyte.[357] For a homogeneous alloy (e.g., Pt/Rh,

Au/Pd), there is only one desorption peak, the position of which is located between those of the pure metals and is, to a first approximation, a linear function of the surface composition.[315] A good correlation was found between such *in situ* determinations and *ex situ* results obtained with ESCA.[387] For a heterogeneous alloy (such as Au/Pt), the number of oxygen desorption peaks gives the number of different surface phases, and the quantity of oxygen adsorbed on each phase gives its surface area, so that the surface composition can be determined.

5. Role of the Electrolytic Solution

The composition of the aqueous electrolytic solution is also of great importance, because it can influence greatly the behavior of the electrocatalytic material, and it may modify drastically the oxidation kinetics of small organic molecules, through pH effects and interaction effects between adsorbed species.

(i) *Role of the Water Adsorption Residues*

Water molecules and their adsorbed residues (H_{ads}, OH_{ads}, O_{ads}) play a key role in the oxidation mechanism, particularly in the case of the oxidation of alcohols to carboxylic acids (or to carbon dioxide), which requires one extra oxygen atom to be supplied:

$$R{-}CH_2OH + H_2O \rightarrow R{-}COOH + 4H^+ + 4e^-$$

This oxygen atom may come either from an adsorbed water molecule, $(H_2O)_{ads}$, or more probably from adsorbed hydroxyl, OH_{ads}, or adsorbed oxygen, O_{ads}, depending on the electrode potential and the pH of the solution.

By varying the potential limit of a voltammetric sweep, it is very easy to control the amount of adsorbed hydrogen or of adsorbed hydroxyl or adsorbed oxygen at the electrode surface. For example, the effect of adsorbed hydrogen on the electrooxidation of ethylene glycol at a platinum electrode in sulfuric acid can be easily seen in Fig. 21. When the lower potential limit E_c is made more negative, the current at the first oxidation peak decreases drastically. This could be due to the presence of a poisoning species formed between the adsorbed organic residue and adsorbed hydrogen, as was checked by modifying the electrocatalytic surface by

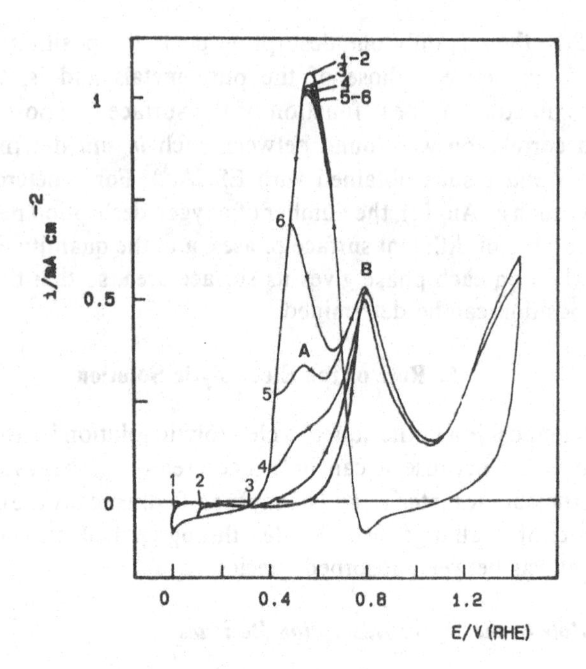

Figure 21. Effect of the lower potential limit on the voltammogram showing the oxidation of 0.1 *M* ethylene glycol in 0.5 *M* H_2SO_4 at a platinum electrode (25°C, 50 mV s^{-1}). Curves 1-6 are obtained by increasing the lower potential limit. A and B represent two different oxidation peaks during the positive sweep.

upd of cadmium adatoms.[18,306] In fact, it is now known, from voltammetric and EMIRS measurements, that cadmium adatoms are not able to displace adsorbed CO. However, they still can prevent hydrogen adsorption, and, with this modification of the surface, there was no longer an effect of the lower potential limit on the first oxidation peak A, which was as intense as the second oxidation peak B.

The effects of OH_{ads} and O_{ads} are more subtle, since, on the one hand, the oxidation reaction needs such species in order to go to completion and, on the other hand, the presence of relatively strongly adsorbed oxygenated species may block some electrode sites, preventing adsorption of the organic molecule. Therefore, the oxidation current passes through a maximum when θ_{OH} is varied, either by changing the electrode potential or the pH of the solution. For example, the oxidation current, at a given electrode potential, of methanol or *n*-butanol gives a volcano-shaped curve when plotted as a function of the bulk concentration of hydroxyl ions

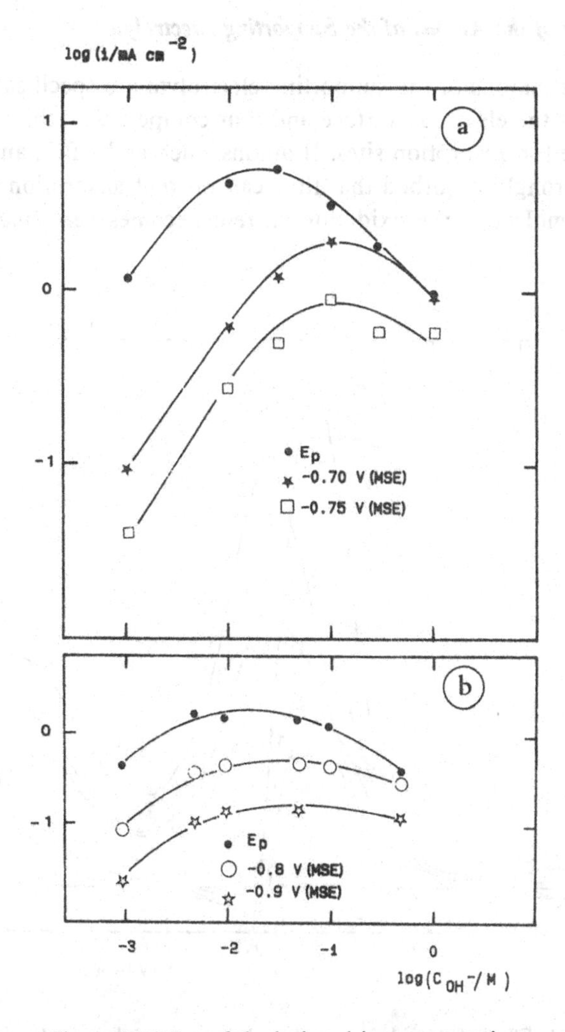

Figure 22. Effect of the hydroxyl ion concentration on the current densities of the oxidation of 0.1 *M* CH_3OH (a) and 0.1 *M* *n*-butanol (b) in 0.1 *M* NaOH at a platinum electrode (25°C, 50 mV s^{-1}).

(Fig. 22). The OH^- concentration at the maximum will depend on the nature and concentration of the alcohol. In the case of methanol oxidation, the maximum corresponds to an equimolar ratio of the alcohol and hydroxyl ions.

(ii) *Role of the Anions of the Supporting Electrolyte*

Most anions of the supporting electrolyte are specifically adsorbed at the electrode surface and thus compete with the organic molecule for adsorption sites. If anions, such as I^-, Br^-, and Cl^-, are so strongly adsorbed that they can prevent adsorption of the organic molecule, the oxidation current becomes very small and

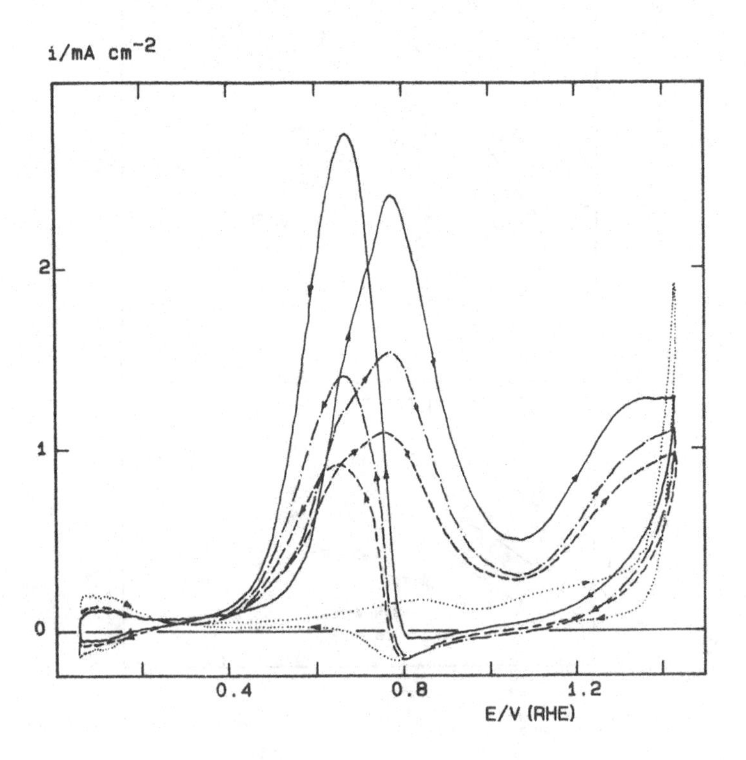

Figure 23. Effect of chloride ions on the voltammograms showing the oxidation of 0.1 *M* CH_3OH in 0.5 *M* H_2SO_4 at a platinum electrode (25°C, 50 mV s^{-1}). Concentration of Cl^-: ——, 0; —·—, 10^{-5}; - - -, 10^{-4}; · · ·, 10^{-3} *M*.

may even disappear. For example, the effect of chloride ions on the electrooxidation of methanol is illustrated in Fig. 23. Hydroxyl ions and, to a smaller extent, perchlorate ions are not specifically adsorbed at the electrode surface, while sulfate, bisulfate, and phosphate ions are much more strongly adsorbed. For example, the adsorption of sulfate and bisulfate ions on Pt(111) is so strong that it prevents adsorption of methanol, and the oxidation currents are drastically reduced by comparison to those obtained in perchloric acid medium, which are ten times higher (Fig. 24).[388]

Other acid electrolytes, such as trifluoromethanesulfonic acid, CF_3SO_3H, have been also considered as supporting electrolytes in a direct methanol fuel cell.[311] However, although less adsorbed than sulfate or bisulfate ions, $CF_3SO_3^-$ results in lower activity of platinum electrocatalysts for methanol oxidation, particularly at elevated temperatures (60 to 80°C). This is probably due to decomposition of the electrolyte, leading to adsorbed sulfur species, which poison the catalytic surface and cause the oxidation current to decrease.

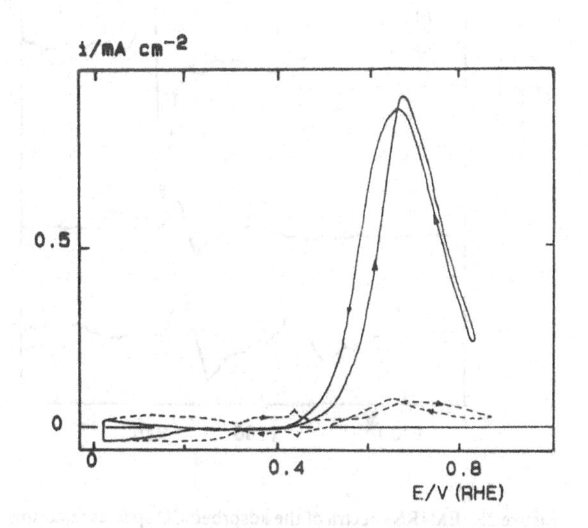

Figure 24. Voltammograms showing the oxidation of 0.1 M CH_3OH at a Pt(111) single-crystal electrode (25°C, 50 mV s^{-1}): ——, 0.1 M $HClO_4$; - - -, 0.5 M H_2SO_4.

In alkaline solutions, hydroxyl ions are not strongly adsorbed at most noble metal electrodes. Moreover, they can compete efficiently for adsorption sites with CO poisoning species, so that the amount of adsorbed CO is reduced and higher current densities for the oxidation of alcohols are obtained in alkaline medium. The smaller extent of poisoning phenomena is displayed in the voltammograms, where the forward and backward sweeps are quite well superimposed in the potential range of alcohol oxidation, without blockage of the active sites by OH_{ads} or O_{ads}. This is also evidenced in the EMIRS spectra of the CO poisoning species, whose intensity is decreased when the pH of the solution is increased; see, for example, the EMIRS spectra obtained with ethylene glycol as a function of pH (Fig. 25).

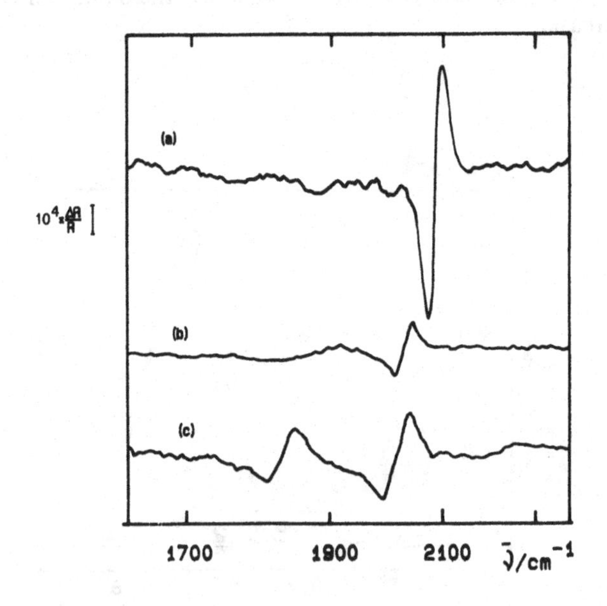

Figure 25. EMIRS spectra of the adsorbed CO species resulting from the dissociative chemisorption of ethylene glycol on a platinum electrode at room temperature as a function of pH: (a) pH ≈ 1; (b) pH ≈ 5; (c) pH ≈ 13.

III. SELECTED EXAMPLES

Most of the examples selected in the following pages will deal with the electrocatalytic oxidation of relatively small oxygenated organic molecules soluble in aqueous electrolytic solutions. They concern particularly the oxidation of alcohols and of the derivatives of their oxidation, from one to six carbon atoms in length, at noble metal electrodes (e.g., Pt, Rh, Pd, Au).

1. Oxidation of One-Carbon Molecules

Simple one-carbon-atom organic molecules such as methanol, formaldehyde, and formic acid have several advantages with respect to their use in fuel cells. Their structures are simple enough that one can hope for a complete understanding of the reaction mechanism. In addition, the high energy densities of these compounds, their ease of handling and storage, and their availability from biomass are particularly attractive.[12,18]

Unfortunately, these fuels give only low current densities except on platinum electrodes. Moreover, mainly on platinum, poisoning effects lead to a decrease of the initial performance. It was recently recognized, for example, from IR reflectance spectroscopic studies, that adsorbed CO plays a key role in the poisoning phenomenon,[164] so that the electrocatalytic oxidation of dissolved CO will also be discussed in this section. From a fundamental point of view, there have been numerous studies on the electrooxidation of these simple compounds during the past 15 years, which have led to the proposal of new ideas and hypotheses for understanding the reaction mechanisms.

(*i*) *Methanol*

Methanol is one of the best candidates as a fuel in fuel cell systems for terrestrial applications.[389]

In a recent paper, Parsons and VanderNoot[12] reviewed the electrocatalytic oxidation of small organic molecules and pointed out that more than 40 papers were devoted to the electrooxidation of methanol between 1981 and 1987. According to these authors, a distinction should be made between work done before and after

1980. Various reviews on work published before 1980 are available.[15,311,323] Before 1980, although many groups worked on this subject, the mechanism of methanol oxidation was not elucidated, so that two kinds of mechanisms were proposed in the literature, depending on the assumed nature of the adsorbed intermediates responsible for the poisoning effect. Until 1980, no definitive experimental data were available that could help in making a clear choice between the two possibilities. Since 1980, different research groups have considered the whole problem again, taking into account the surface structure of the electrode, but, above all, using more rigorous, reproducible, and well-controlled experimental conditions. The newly available *in situ* spectroscopic techniques and on-line analytical methods have allowed a detailed study of the adsorbed species (reactive and poisoning species) and of the reaction intermediates,[239] making obsolete many of the older works, as well as the reaction mechanisms based only on electrochemical measurements.

In acid medium the overall reaction of methanol electrooxidation can be written as follows:

$$CH_3OH + H_2O \rightarrow CO_2 + 6H^+ + 6e^-$$

with a standard electrode potential $E^0 = 0.02$ V versus SHE, as calculated from thermodynamic data.[390]

This reaction involves several adsorption steps, including the formation of chemisorbed residues, leading to a decrease in the catalytic activity of the electrode surface.

In fact, the general mechanism can be summarized as follows:

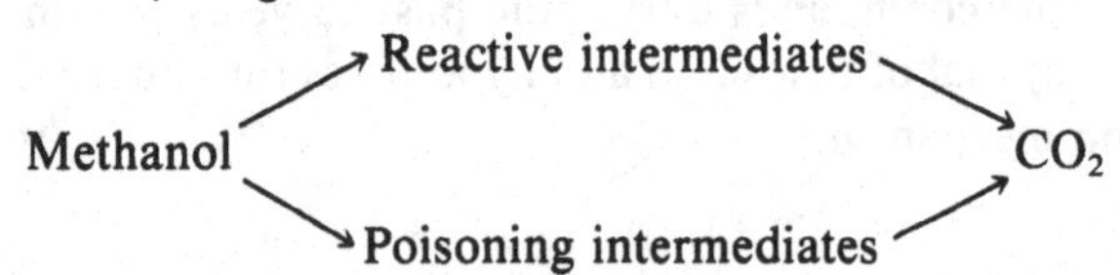

where the reactive intermediates are weakly adsorbed on the electrode surface, in contrast to the poisoning intermediates, which are strongly bonded.

(a) *Identification of the adsorbed species by spectroelectrochemical techniques*

The nature of the adsorbed intermediates responsible for electrode poisoning led, over a long period, to polemic discussions.

According to Bagotsky and Vassiliev,[391] the poisoning intermediate could be COH. This suggestion was the result of assumptions made on the basis of electrochemical measurements, such as the determination of the number of electrons involved in the oxidation of the adsorbed intermediate [N_{epm}; see Section II.3(ii)]. However, other groups proposed CO as the adsorbed intermediate, using the same type of experiments.[392,101]

This discrepancy in the interpretation of the same type of experiments could be attributed to a lack of accuracy of the measurements, especially because the electrode structure was not well controlled. Until the early eighties, no experimental methods were really suitable for the *in situ* identification of adsorbed intermediates. Only UHV techniques were available, but they provided information only after transfer of the electrode from the electrochemical cell to the vacuum chamber, which was not easy to do under controlled electrochemical conditions.

The development of *in situ* infrared spectroscopic techniques allowed significant progress to be made in the identification of the adsorbed intermediates.[16,17,123,164] In particular, electrochemically modulated infrared reflectance spectroscopy (EMIRS) led to the first unambiguous *in situ* proof of the presence of adsorbed CO on platinum electrodes during the adsorption of methanol[164] (see Fig. 16a). CO was found to be mainly linearly bonded to the electrode surface (IR absorption band at around $2060\ cm^{-1}$), but bridge-bonded species were also detected (very small band at around $1850–1900\ cm^{-1}$). These two adsorbed CO species are responsible for the poisoning phenomena observed during working of a direct methanol fuel cell.

Such results were widely confirmed by several groups, not only by EMIRS, but also by similar methods, such as SNIFTIRS and IRRAS, which are based upon the use of Fourier transform IR spectrometers[244,393,394] [see Section II.3(iv.e)].

However, these early conclusions were partially refuted by other observations made using different methods based on electrochemical measurements coupled with mass spectroscopy, such as differential electrochemical mass spectroscopy (DEMS)[23,236] and electrochemical thermal desorption mass spectroscopy (ECTDMS)[88,237] [see Section II.3(v.e)]. These techniques, which require large-area electrodes made from the catalytic material (platinum lacquer) mixed with Teflon powder, allowed the detection of adsorbed CO, but the main species identified were claimed to be either adsorbed COH[26,88,89,237] or adsorbed CHO.[23,87]

These apparently contradictory observations have to be related to the different experimental conditions used,[395] particularly the electrode structures, which were different in the two sets of experiments. Further work made clear that the chemical nature and the surface distribution of adsorbed species depend on the electrode structure and on its degree of coverage by the adsorbed species. This was confirmed recently by an EMIRS study, in which the methanol concentration was varied over a wide range.[396] It was demonstrated that different adsorbed species exist simultaneously on smooth polycrystalline platinum at low coverages, that is, at low concentrations of methanol in solution (Fig. 26). Under these conditions, it is reasonable to think that adsorbed CO and other adsorbed species such as formyl, CHO_{ads} (complex band at around 1700 cm^{-1}), exist simultaneously on the electrode surface. It follows that these results, obtained by EMIRS, could be correlated to those obtained by mass spectroscopy.

The coexistence of poisoning species (CO_{ads}) and reactive species (CHO_{ads}) on the electrode surface during the electrooxidation of methanol is the first conclusion of these *in situ* studies.

(*b*) *Correlation between electrochemical data and in situ spectroscopic measurements*

As seen above, the species responsible for poisoning is identified as adsorbed CO, mainly linearly bonded. This conclusion can be well correlated to pure electrochemical measurements. When a species is adsorbed at a constant potential during a given adsorption time, it becomes possible to calculate the electrode coverage and the number of electrons per site, N_{eps}, from the quantities of electricity associated with its oxidation and with the deposition or oxidation of adsorbed hydrogen [see Section II.3(ii)].

The technique used to obtain such results, potential programmed voltammetry (PPV), is illustrated in Fig. 27. The electrode surface is submitted to a potential program including a preparation procedure, an adsorption plateau (of duration t_{ads} at E_{ads}) followed

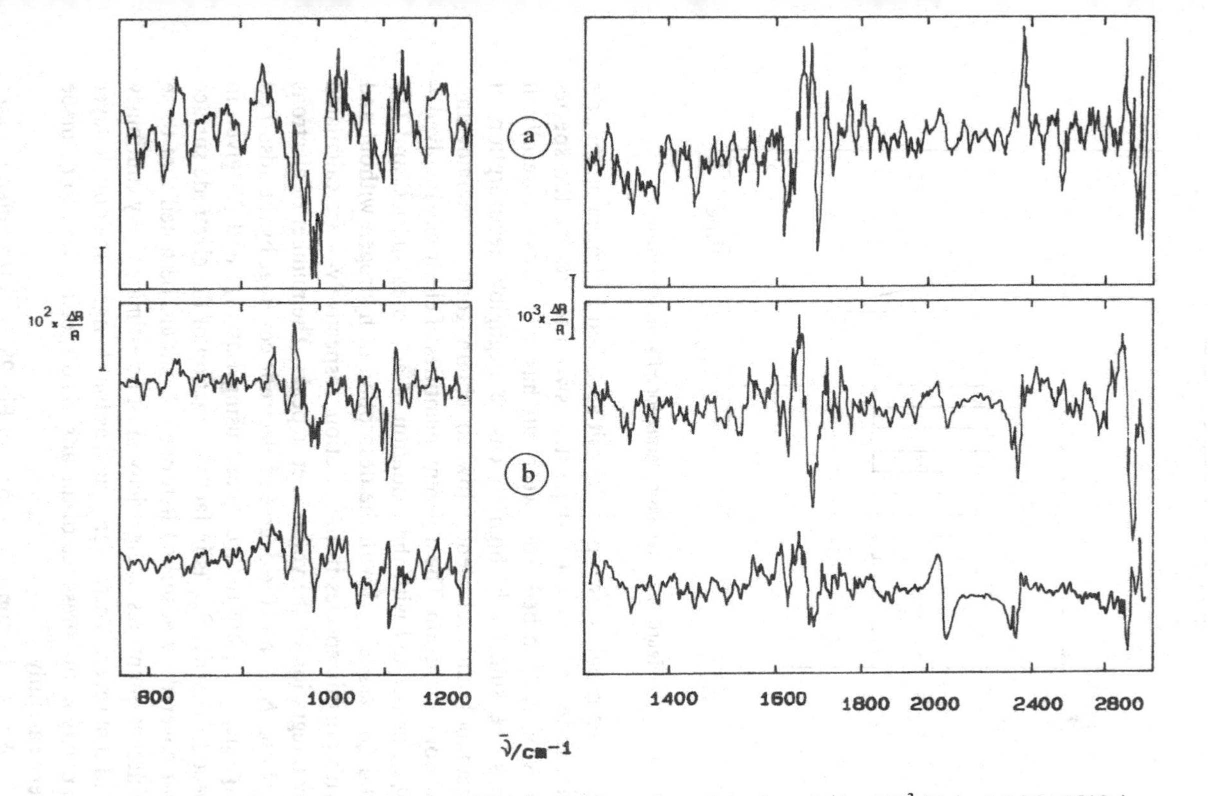

Figure 26. EMIRS spectra of the species resulting from adsorption of methanol (5×10^{-3} *M* in 0.5 *M* $HClO_4$) on a polycrystalline Pt electrode for different times of accumulation $\Delta E = 0.4$ V; $\bar{E} = 0.2$ V versus RHE; room temperature): (a) 1st scan; (b) 10th and 25th scans.

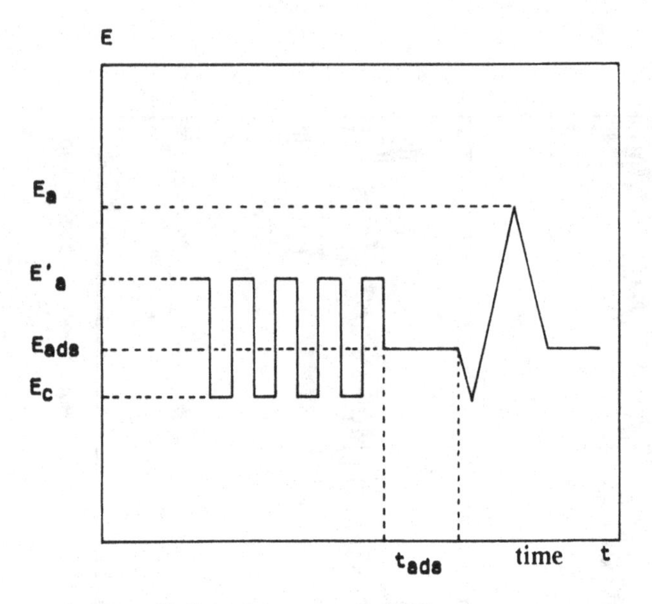

Figure 27. Potential program for PPV measurements.

by a fast negative sweep to deposit adsorbed hydrogen on the free electrode sites, and a fast positive sweep to oxidize the species previously adsorbed. The sweep rate has to be chosen carefully: it must be sufficiently high to ensure negligible readsorption of methanol and sufficiently low to obtain good voltammograms without distortion. The different quantities of electricity involved—Q_{org}, associated with the oxidation of adsorbed species, and Q_H° and Q_H, associated with the desorption of hydrogen without and with organic species in the solution, respectively— are calculated by integration of the voltammetric peaks. The number of electrons per site, N_{eps}, and the degree of electrode coverage by the adsorbed organic, Θ_{org}, are then deduced using Eqs. (11) and (9) given in Section II.3(ii). Θ_{org} is in fact the fraction of the electrode surface not covered by adsorbed hydrogen. It should be noted that these different quantities of electricity must be estimated very cautiously and that several corrections, such as the change of the double-layer capacity in the presence of the adsorbed species, need to be made very carefully.[105]

A typical example is shown in Fig. 28 for three different cases involving two structures [polycrystalline platinum and Pt(100)] and

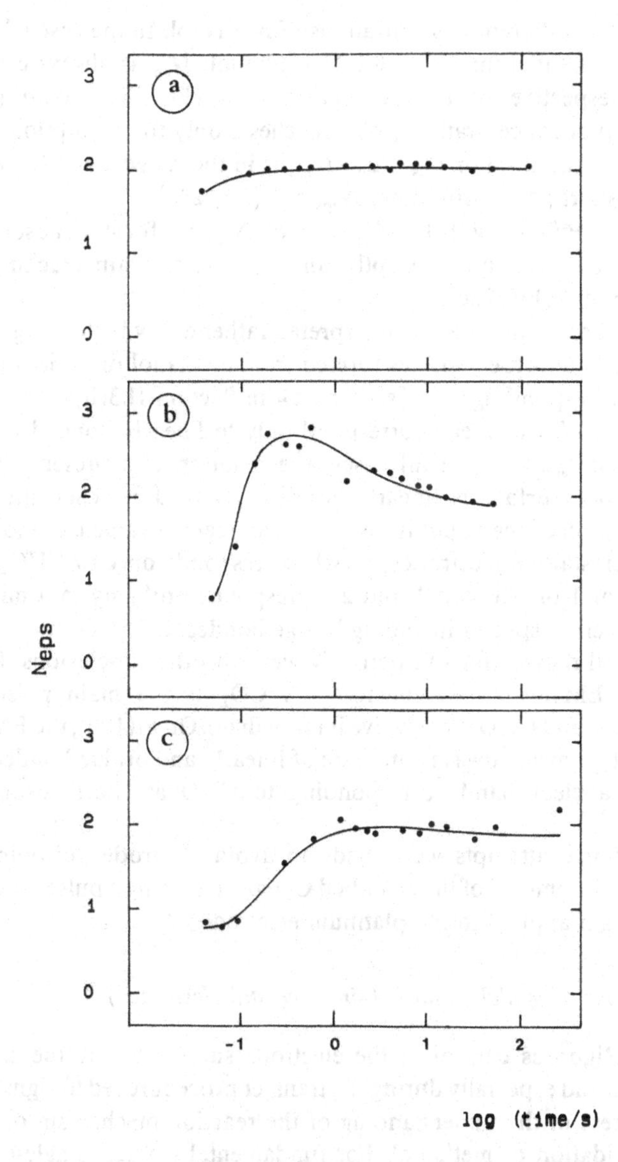

Figure 28. Number of electron per site, N_{eps}, versus the logarithm of adsorption time for: (a) 0.1 *M* CH_3OH at polycrystalline Pt; (b) 0.001 *M* CH_3OH at polycrystalline Pt; (c) 0.1 *M* CH_3OH at a Pt(100) single crystal.

with two different concentrations of methanol. In the case of polycrystalline platinum with 0.1 *M* methanol, N_{eps} is always close to 2, irrespective of the adsorption time (Fig. 28a). With lower methanol concentrations, N_{eps} reaches 2 only for adsorption times longer than 2 s. For $t_{ads} < 2$ s, N_{eps} is in the range 2.5–2.7, and for very short adsorption time, $N_{eps} \approx 1$ (Fig. 28b).

On Pt(100), with 0.1 *M* methanol, $N_{eps} \approx 1$ for short adsorption times and increases abruptly for $t_{ads} > 0.5$ s, again reaching the value of 2 (Fig. 28c).

These results can be interpreted rather easily by looking at the possible adsorbed species formed from methanol dissociation and the corresponding N_{eps} [see Table 4 in Section II.3(ii)].

A N_{eps} of 2 can correspond only to linearly bonded CO, so that, at $t_{ads} > 2$ s, in all cases, the main species present on the electrode surface is linearly bonded CO (CO_L). When the concentration of methanol is low, N_{eps} can reach a value close to 3 for polycrystalline electrodes, which corresponds only to CHO_{ads}. A N_{eps} of 1 or between 1 and 2 corresponds probably to a mixture of different species including bridge-bonded CO (CO_B).[397]

All these results fit perfectly well with the conclusions drawn from EMIRS measurements[398,399]: CO_L is the main poisoning species, and CHO the reactive intermediate. On Pt(100), the EMIRS results showed always a mixture of linearly and bridge-bonded CO with a clear band corresponding to CHO at short adsorption times.[399]

Some attempts were made to avoid electrode poisoning by periodic removal of the adsorbed CO species, using a pulse potential program applied to the platinum electrode.[400]

(c) *Use of model surfaces (single-crystal electrodes)*

Rigorous control of the electrode surface before the experiments and especially during the transfer procedure led to significant progress in the understanding of the reaction mechanism of electrooxidation of methanol. For fundamental studies, single-crystal electrodes are ideal model systems.[401] However, the first attempts to observe possible effects of superficial structure in electrocatalysis on platinum single crystals failed, probably because of the difficulty of transferring the electrode, after its preparation, into the electrolyte without contamination of its surface. Furthermore, the electrochemical pretreatments, which were supposed to "clean the surface," in fact led to drastic modifications and reconstruction of the surface structure.[290]

A new and simple method was proposed by Clavilier *et al.* in 1980, in order to perform such experiments with single-crystal platinum electrodes, well oriented and uncontaminated during transfer to the electrochemical cell.[11] It was thus possible to observe drastic structural effects on the adsorption–desorption processes of both hydrogen and oxygen at low-index single-crystal platinum electrodes in contact with the supporting electrolyte alone (see Fig. 20).

The electrocatalytic oxidation of methanol is also structure-dependent (Fig. 29a).[298,388,402,403] Pt(110) is the most active plane, but also the most sensitive to poisoning, which leads to a rapid decrease in the current densities observed. The Pt(111) plane appears to be less sensitive to poisoning phenomena, even though the current densities are rather weak. Finally, the Pt(100) plane is totally blocked over a large range of potentials, but the current increases sharply once the adsorbed blocking species are removed at higher potentials, the maximum current densities remaining very stable with time.

Confirmation of these structural effects was obtained from EMIRS experiments using platinum single-crystal electrodes (Fig. 29b).[398,399] The reactive intermediate (CHO_{ads}) was clearly observed in the EMIRS spectrum in the case of Pt(111) and Pt(100) electrodes, while adsorbed CO was observed for all three surface orientations studied. However, linearly bonded CO_{ads} was the only species detected on Pt(110), whereas two kinds of CO (linearly and bridge-bonded) were clearly present on Pt(100). The latter observation is very interesting, because it is an indication that the blocking phenomena observed with the Pt(100) plane could result from lateral interactions between these two kinds of adsorbed CO species.

(*d*) *Modification of the electrode surface composition*

The previous examples illustrated the case of methanol oxidation at platinum electrodes. In acid medium, platinum appears to be the most efficient catalyst for the electrooxidation of methanol

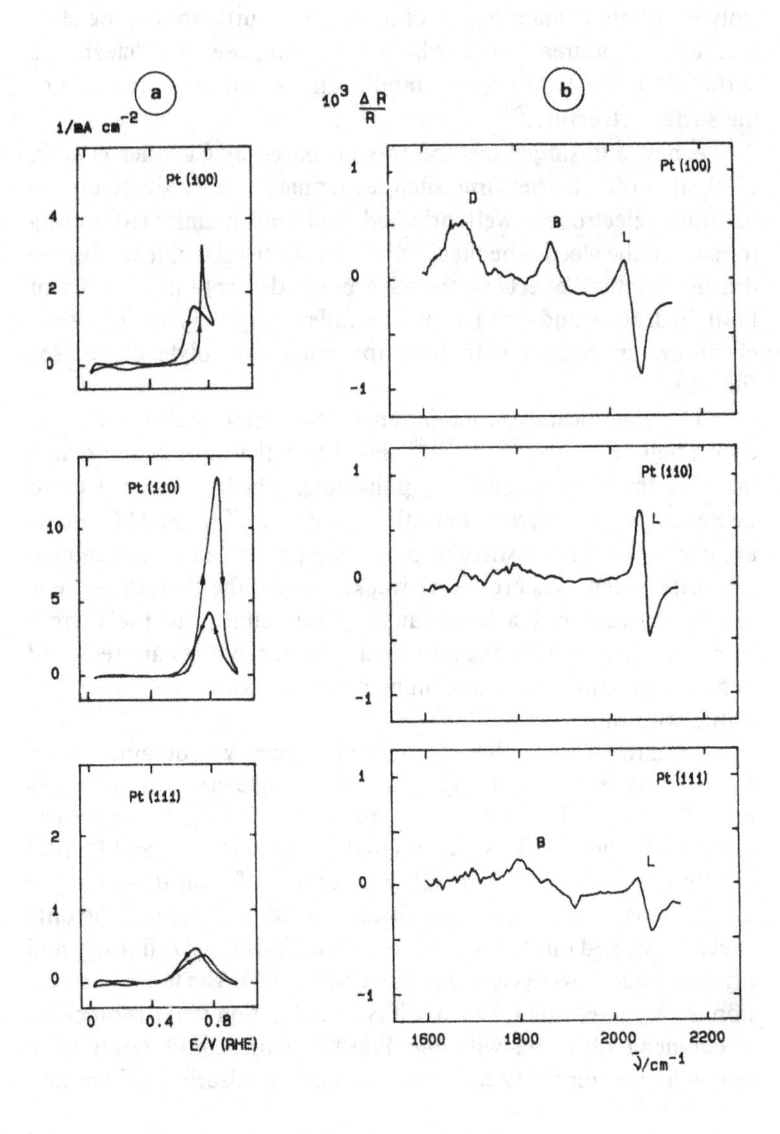

Figure 29. Adsorption and oxidation of 0.1 M CH_3OH in 0.5 M $HClO_4$ at the three low-index faces of platinum single crystals (room temperature): (a) voltammograms (50 mV s^{-1}; first sweep); (b) EMIRS spectra ($\Delta E = 0.4$ V; $\bar{E} = 0.35$ V versus RHE).

even though a few other noble metals, such as palladium and rhodium, also exhibit activity, albeit very low.[404,405] In alkaline medium, the problem is somewhat different, and while platinum appears to be the best catalyst, significant activities are obtained with palladium and rhodium electrodes and, to a smaller extent, with gold electrodes.[406]

Different attempts have been made, through the use of alloys and upd-modified electrodes, to increase the electrocatalytic activity and to decrease the extent of poisoning phenomena by modifying the nature of the surface.

In the case of alloys, different bimetallic electrocatalysts have been investigated, including Pt/Sn,[407] Pt/Rh,[316] Pt/Pd,[43] and Pt/Ru.[52,311,381,408-411] Only Pt/Ru, and perhaps Pt/Sn, electrodes display a greater activity than pure platinum. The Pt/Ru alloys appear to be the most promising,[311,381] with a negative shift of the polarization curves compared to those obtained with pure platinum and a decrease in poisoning.[381] However, surface enrichment in one metal could arise under working conditions, thus modifying the activity of the electrodes.

The second possibility is to modify the properties of the electrocatalyst by adsorption of a foreign metal on its surface by upd. Several review papers have been published on this subject.[13,149,303,304] The atoms adsorbed on the electrode surface, the so-called adatoms, modify greatly the adsorption of organics, leading generally to positive effects with significant enhancements of the electrocatalytic activity of platinum. However, methanol seems to be a particular case in which the influence of adatoms is very limited.[320] Only Pb,[321,412,413] Ru,[321] Bi,[321,412,414] Sn,[321,415] and Mo[416] adatoms increase slightly the activity of platinum in the electrooxidation of methanol, and only at low potentials.[414]

(e) *Increase of the real surface area*

For practical applications, the key point is to decrease the amount of platinum used. It is necessary to design electrodes with a large active area but with the smallest amount possible of precious metals. The electrode has generally been made using a conducting support, such as carbon, with deposition of small platinum particles, generally by reduction of a platinum salt.

Numerous studies have been carried out to examine possible effects of the particle size.[363] For example, systematic studies were made in the case of the electroreduction of oxygen and the adsorption of hydrogen[417]; no significant size effects were observed, except for O_2 reduction. Recently, two different groups investigated this problem for the case of electrooxidation of methanol in acid medium. According to Goodenough *et al.*,[418] the most efficient electrodes are those which contain both small crystallites (diameter ≈ 20 Å) and a minimum amount of ionic Pt species. Watanabe *et al.*[328] disagreed with these conclusions. They observed no effects of platinum crystallite size, even at high platinum dispersion, such as 70% (corresponding to particles as small as 14 Å). They concluded that increasing the dispersion of platinum on carbon is an interesting possibility for achieving higher catalytic activity with low amounts of platinum.

Other possible ways to disperse platinum have recently been explored, using other types of conducting matrix (conducting polymers or ionic membranes). Preliminary work[348,349] showed the possibility of obtaining significant activity with platinum loadings below 1 mg cm^{-2}, associated with an important decrease in the extent of poisoning phenomena [see also Section II.4(i)].

Similar conclusions were drawn by Aramata and Ohnishi[332] for the electrooxidation of methanol on Pt–SPE electrodes. The high activity observed is much more stable over time than that of similar platinized platinum electrodes. By modifying the electrode structure with Sn, Ru, or Ir, it is possible to enhance the activity of Pt–SPE electrodes.[334] Nakajima and Kita[333] studied the direct electrooxidation of methanol from the gas phase on Pt–SPE electrodes. The currents observed were 10^3 times larger than in the liquid phase, with a smaller extent of deactivation phenomena. When modified with molybdenum atoms, the Pt–SPE electrode exhibits a greater activity, especially at low polarizations (~0.1 V versus RHE).[335,336]

(*f*) *Analysis of the reaction products*

To elucidate the overall reaction mechanism, it is also necessary to analyze the products of the reaction. This has generally been performed during prolonged electrolysis using on-line analytical techniques (gas or liquid chromatography)[59] [see Section II.2(i)]. The main reaction product of methanol electrooxidation is carbon dioxide,[63,64] but formaldehyde and formic acid have also been detected in small amounts.[63,64,419] Methyl formate has also been mentioned as a possible by-product.[64]

(g) *Conclusions and mechanisms*

Since 1980, significant progress has been made in the understanding of the reaction mechanism for the electrooxidation of methanol both in acid and in alkaline medium. The following main points seem to be now widely accepted:

- The decrease of the current densities with time is due to poisoning effects of adsorbed residues coming from the dissociative adsorption of methanol. The species responsible for the electrode poisoning is adsorbed CO, linearly and bridge-bonded to the electrode surface (as proved by *in situ* infrared spectroscopic methods);
- Different adsorbed species exist simultaneously on the electrode surface, and their distribution depends greatly on the surface structure, on the methanol concentration, and on the electrode potential;
- EMIRS studies have allowed the identification of the reactive intermediates as probably CHO_{ads} (formyl);
- However, a complete interpretation of the enhancement of the activity of platinum seen with some alloys and some adatoms is difficult, since methanol seems to be a particular case compared to other alcohols or small organics, such as formic acid.

On the basis of all these observations and conclusions, it is now possible to propose a detailed mechanism for the oxidation of methanol at platinum electrodes in acid medium. The steps are as follows:

(1) $$Pt + H_2O \rightarrow Pt{-}(OH)_{ads} + H^+_{aq} + e^-$$

(2) $$Pt + (CH_3OH)_{sol} \rightarrow Pt{-}(CH_3OH)_{ads}$$

(3) $$Pt{-}(CH_3OH)_{ads} \rightarrow Pt{-}(CH_3O)_{ads} + H^+_{aq} + e^-$$

(4) $$Pt{-}(CH_3O)_{ads} \rightarrow Pt{-}(CH_2O)_{ads} + H^+_{aq} + e^-$$

(5) $$Pt{-}(CH_2O)_{ads} \rightarrow Pt{-}(CHO)_{ads} + H^+_{aq} + e^-$$

(6a) $$\text{Pt—(CHO)}_{\text{ads}} \rightarrow \text{Pt—(CO)}_{\text{ads}} + \text{H}^+_{\text{aq}} + e^-$$

(6b) $$\text{Pt—(CHO)}_{\text{ads}} + \text{Pt—(OH)}_{\text{ads}} \rightarrow 2\text{Pt} + \text{CO}_2 + 2\text{H}^+_{\text{aq}} + 2e^-$$

(6c) $$\text{Pt—(CHO)}_{\text{ads}} + \text{Pt—(OH)}_{\text{ads}} \rightarrow \text{Pt} + \text{Pt—(COOH)}_{\text{ads}} + \text{H}^+_{\text{aq}} + e^-$$

(7a) $$\text{Pt—(CO)}_{\text{ads}} + \text{Pt—(OH)}_{\text{ads}} \rightarrow 2\text{Pt} + \text{CO}_2 + \text{H}^+_{\text{aq}} + e^-$$

(7b) $$\text{Pt—(CO)}_{\text{ads}} + \text{Pt—(OH)}_{\text{ads}} \leftrightarrow \text{Pt} + \text{Pt—(COOH)}_{\text{ads}}$$

(8) $$\text{Pt—(COOH)}_{\text{ads}} \rightarrow \text{Pt} + \text{CO}_2 + \text{H}^+_{\text{aq}} + e^-$$

The key point is the formation of the reactive intermediate $(CHO)_{ads}$ and its further oxidation. There are two main possibilities: $(CHO)_{ads}$ is oxidized to $(CO)_{ads}$ (reaction 6a), leading thus to the poisoning species, or the reactive intermediate $(CHO)_{ads}$ is oxidized either directly to CO_2 (reaction 6b) or through $(COOH)_{ads}$ (reaction 6c), which is further oxidized to CO_2 (reaction 8). The poisoning species $(CO)_{ads}$ can also be oxidized to CO_2, either directly (reaction 7a) or through $(COOH)_{ads}$ (reaction 7b, followed by reaction 8). These different possibilities for the oxidation of $(CHO)_{ads}$ can be summarized in the following scheme, where all the species involved have been observed by EMIRS:

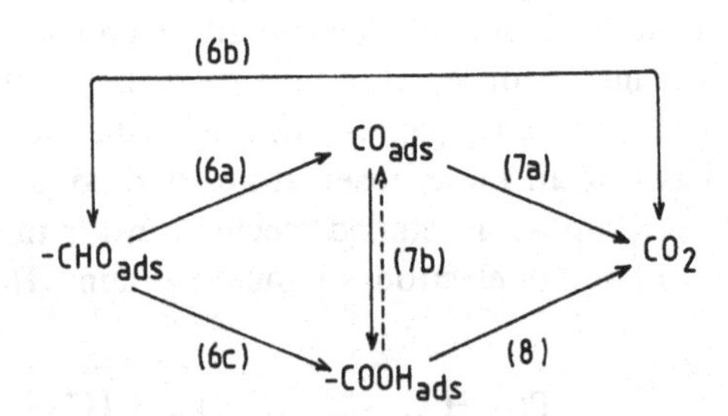

From this scheme, it is possible to understand the requirements for the use of methanol in a fuel cell in acid medium. It is necessary to avoid the formation of $(CO)_{ads}$ and to favor route (6b), leading directly to CO_2, or alternatively the indirect routes (6c) and (8) through $(COOH)_{ads}$. This is probably possible by using specific surface structures.

In alkaline medium, the mechanism is similar and could be written as follows:

(1) $$Pt + OH^- \rightarrow Pt{-}(OH)_{ads} + e^-$$

(2) $$Pt + (CH_3OH)_{sol} \rightarrow Pt{-}(CH_3OH)_{ads}$$

(3) $$Pt{-}(CH_3OH)_{ads} + OH^- \rightarrow Pt{-}(CH_3O)_{ads} + H_2O + e^-$$

(4) $$Pt{-}(CH_3O)_{ads} + OH^- \rightarrow Pt{-}(CH_2O)_{ads} + H_2O + e^-$$

(5) $$Pt{-}(CH_2O)_{ads} + OH^- \rightarrow Pt{-}(CHO)_{ads} + H_2O + e^-$$

(6a) $$Pt{-}(CHO)_{ads} + OH^- \rightarrow Pt{-}(CO)_{ads} + H_2O + e^-$$

(6b) $$\begin{array}{l} Pt{-}(CHO)_{ads} \\ + Pt{-}(OH)_{ads} + 4OH^- \end{array} \rightarrow 2Pt + CO_3^{2-} + 3H_2O + 2e^-$$

(6c) $$\begin{array}{l} Pt{-}(CHO)_{ads} \\ + Pt{-}(OH)_{ads} + OH^- \end{array} \rightarrow Pt + Pt{-}(COOH)_{ads} + H_2O + e^-$$

(7a) $$\begin{array}{l} Pt{-}(CO)_{ads} \\ + Pt{-}(OH)_{ads} + 3OH^- \end{array} \rightarrow 2Pt + CO_3^{2-} + 2H_2O + e^-$$

(7b) $$\begin{array}{l} Pt{-}(CO)_{ads} \\ + Pt{-}(OH)_{ads} \end{array} \leftrightarrow Pt + Pt{-}(COOH)_{ads}$$

(8a) $$Pt{-}(COOH)_{ads} + OH^- \rightarrow Pt{-}(OH)_{ads} + HCOO^-$$

(8b) $$\begin{array}{l} Pt{-}(COOH)_{ads} \\ + Pt{-}(OH)_{ads} + 2OH^- \end{array} \rightarrow 2Pt + CO_3^{2-} + 2H_2O$$

The main difference with respect to the mechanism in acid medium is the formation of formate and carbonate ions (steps 6b, 7a, 8a, and 8b). The amount of $(CO)_{ads}$ formed is smaller than in acid medium, leading to a greater electrocatalytic activity.

(ii) Formaldehyde

The electrooxidation of formaldehyde has been often studied as a model reaction. Only four electrons are required for complete oxidation to carbon dioxide. Formaldehyde is also an intermediate oxidation product found during the electrocatalytic oxidation of methanol.

If the reaction mechanism seems, *a priori*, rather simple with only a few possible adsorbed intermediates, there are difficulties in studying the electrooxidation of formaldehyde due to the lack of stability of this molecule. Formaldehyde is generally supplied in aqueous solution, stabilized by methanol. Obviously, the presence of methanol may present some problems for the determination of the true reactivity of the aldehyde during its oxidation, even if it is much more reactive than methanol. Formaldehyde is often available as paraformaldehyde, that is, its polymeric form. The only way to obtain the pure monomer from the polymeric form is to reflux for several hours and to rapidly use the monomer obtained.

One of the first complete studies on the electrooxidation of formaldehyde, in comparison with that of formic acid and methanol, was done by Buck and Griffith,[54] whereas Sibille *et al.*[420] and Van Effen and Evans[70] compared the behavior of several aldehydes. Several other recent works can be mentioned, such as those of Beltowska-Brzezinska and Heitbaum[421,422] on gold–platinum alloys in alkaline medium and those of Avramov-Ivić *et al.*[423] on noble metals. Other kinds of electrode material have been considered, such as copper,[424,425] copper-nickel alloys,[426] and even ternary alloys, Cu/Pd/Zr.[427] The effect of adatoms was also considered,[52,412,428-430] and possible structural effects were investigated[431] by using platinum and gold single-crystal electrodes.

(*a*) *Oxidation on pure metals*

Important differences in the electroreactivity of formaldehyde exist depending on the nature of the electrode material and on the pH of the electrolyte. On platinum in acidic medium, the voltammogram of formaldehyde electrooxidation is characterized by a significant hysteresis between the positive and the negative sweeps (Fig. 30a). This fact is due to extensive poisoning of the electrode surface, and again the species responsible for the poisoning has been clearly identified by *in situ* infrared spectroscopy as adsorbed CO.[423,432]

A mechanism similar to that given for methanol oxidation can be proposed for the electrooxidation of formaldehyde in acid medium:

(1) $$\mathrm{Pt + H_2O \rightarrow Pt{-}(OH)_{ads} + H^+_{aq} + e^-}$$

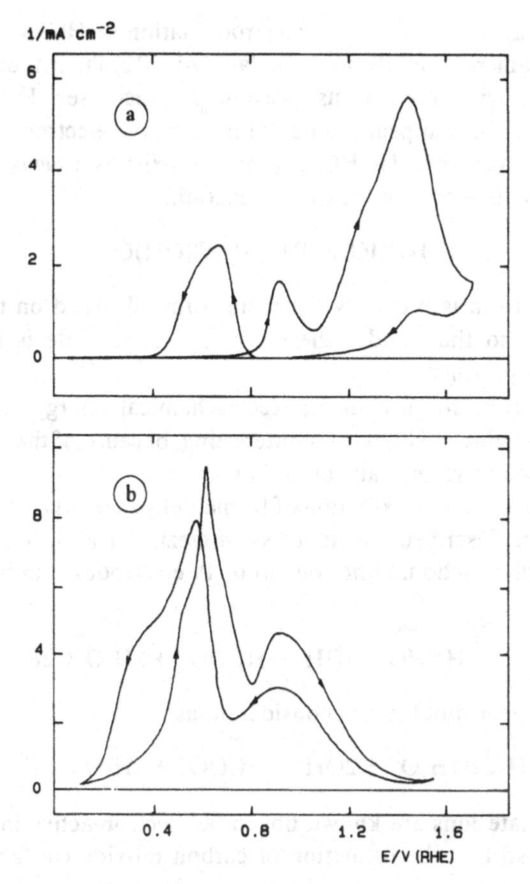

Figure 30. Voltammograms showing the oxidation of 0.1 *M* formaldehyde at a platinum electrode (25°C, 50 mV s^{-1}) in: (a) 0.1 *M* $HClO_4$; (b) 1 *M* NaOH.

(2) $$Pt + (HCHO)_{sol} \rightarrow Pt{-}(HCHO)_{ads}$$

(3) $$Pt{-}(HCHO)_{ads} \rightarrow Pt{-}(CHO)_{ads} + H^{+}_{aq} + e^{-}$$

with step (4) and subsequent steps being similar to steps (6) to (8) for methanol oxidation in acid medium.

This mechanism is derived from that proposed earlier by Spasojević *et al.*,[428] but differs in the nature of the poisoning species, which is here $(CO)_{ads}$, whose formation originates in the same process as in the case of methanol electrooxidation.

In alkaline medium, the electrooxidation of HCHO occurs at a significant rate on all noble metals (Au, Pd, Pt, Rh, but also Ir and Ag), with no serious poisoning effect (see Fig. 30b for platinum). Gold appears to be the most active electrocatalyst.

At alkaline pH, HCHO is known to exist as a *gem*-diol form, which results from the following reaction:

$$HCHO + OH^- \leftrightarrow H_2C(OH)O^-$$

This form is probably the main form adsorbed on the metal electrode, so that, under these conditions, formate is the main oxidation product.

However, for use in an electrochemical energy conversion device, alkaline medium is not interesting, because of the consumption of electrolyte by salt formation.

Mechanisms of oxidation of formaldehyde in alkaline medium have been described by Beltowska-Brzezinska and Heitbaum.[421] The overall reaction, either on Au or Pt electrodes, can be written as

$$HCHO + 3OH^- \rightarrow HCOO^- + 2H_2O + 2e^-$$

or, if the *gem*-diol form is considered, as

$$H_2C(OH)O^- + 2OH^- \rightarrow HCOO^- + 2H_2O + 2e^-$$

Formate ions are known not to be electroreactive in alkaline medium, so that the formation of carbon dioxide (in fact, CO_3^{2-}) is negligible.

The detailed mechanism proposed by Beltowska-Brzezinska and Heitbaum takes into account the possible formation of different adsorbed intermediates, including CO. The formation of such a poison was demonstrated by Avramov-Ivić *et al.*[423] in alkaline medium and by Nishimura *et al.*[432] in acid medium.

The influence of the structure of the electrode surface was investigated by Adžić *et al.*[431] The differences in electrocatalytic activity observed between the different orientations of platinum (or gold) single crystals remain small in comparison with those observed in acid medium for other molecules. This general behavior is similar to that of methanol on Pt(*hkl*) electrodes in alkaline medium.[402]

(b) Effect of adatoms

The poisoning phenomena observed, mainly in acid medium, during the electrooxidation of formaldehyde can be drastically reduced by using adatoms to modify the electrode surface.

In acid medium, the oxidation of HCHO on platinum is considerably enhanced by underpotential deposition of Pb, Bi, or Tl.[428] According to Motoo and Shibata,[430] two kinds of adatoms can be distinguished: the first type of adatoms (Cu, Ag, Tl, Hg, Pb, As, Bi) have only a geometrical effect, whereas the second kind of adatoms (Ge, Sn, Sb), which are able to adsorb oxygen at low potentials, give rise to greater enhancement factors. The authors interpreted the enhancement effects in terms of three reaction paths:

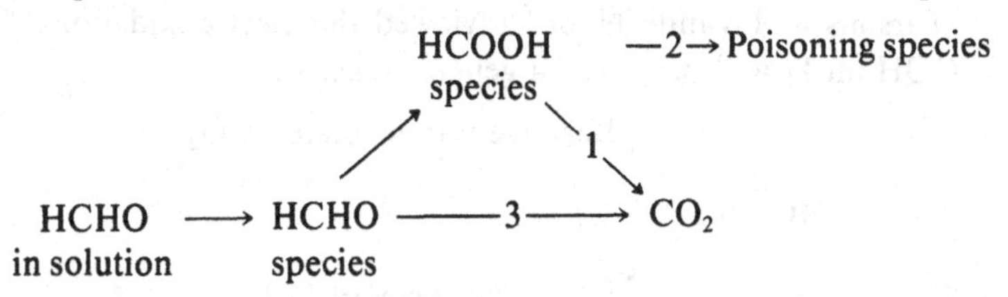

If the number of neighboring platinum sites is greater than three, the formation of poisoning species is favored. If their number is equal to three, which can be obtained by underpotential deposition of some adatoms (e.g., Bi), the formation of poisons is inhibited, and path 1 is favored. With adatoms that adsorb oxygen, the direct path 3 seems to be favored.

(iii) Formic Acid

As mentioned above, the possible use of methanol in fuel cells explains why the electrooxidation of methanol has been the most studied reaction involving a one-carbon organic molecule. However, formic acid, which is also an intermediate product of methanol oxidation, has also been considered. Thus, with the aim of elucidating the mechanism of the oxidation of methanol, various research groups have carried out parallel studies on the electrooxidation of formic acid.

Formic acid is oxidized in acid medium according to the following overall reaction:

$$HCOOH \rightarrow CO_2 + 2H^+ + 2e^-$$

Platinum is a good electrocatalyst for the oxidation of formic acid, but rhodium and especially palladium also have good activity in acid medium.[433] The reactivity of formic acid (or formate) varies dramatically with the pH of the solution. In neutral medium, at a pH corresponding to the pK_a of formic acid (p$K_a \approx 4$), gold exhibits significant activity, although it is practically completely inactive in acid medium. Moreover, platinum, palladium, and rhodium also display increased activity in neutral medium.[41]

In acid medium, platinum remains the most studied electrocatalyst for formic acid oxidation.

(*a*) *Nature of the adsorbed species*

Parsons and VanderNoot[12] reviewed the electrooxidation of HCOOH on Pt and proposed a general scheme:

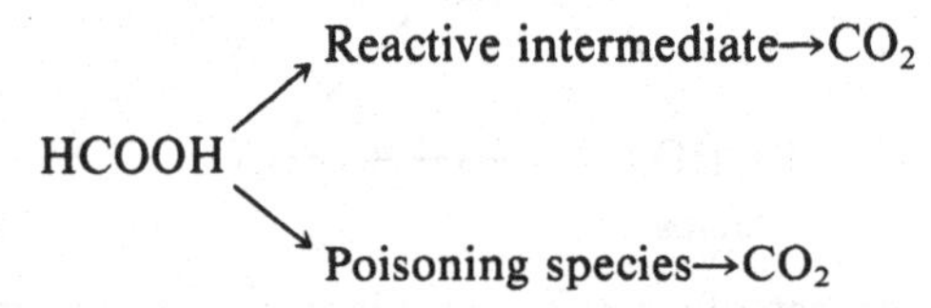

Capon and Parsons postulated —COOH as the reactive adsorbed intermediate and —COH as the poisoning species.[21,433,434] These conclusions were drawn from pure electrochemical measurements, aimed mainly at determining the quantity of electricity associated with the oxidation of the adsorbed species.

Similarly, Kazarinov *et al.*,[435] on the basis of radiometric measurements, also proposed the presence of adsorbed COH as a poisoning species. More recently, Beden *et al.*,[436] using *in situ* infrared reflectance spectroscopy (EMIRS), proved unambiguously the occurrence of adsorbed CO on Pt after adsorption of formic acid. Two IR absorption bands attributed to adsorbed CO were detected, one assigned to linearly bonded CO and the other to bridge-bonded CO (Fig. 16c). In order to confirm these assignments, the EMIRS spectrum of the adsorbed species resulting from the adsorption of gaseous CO, dissolved in the electrolytic solution, were recorded under the same experimental conditions (Fig. 16b). These experiments showed that the same poisoning species were involved during the adsorption of CO and that of both formic acid and methanol (Fig. 16).

On the other hand, Heitbaum and co-workers, using DEMS[255,437] and XPS measurements,[438] concluded that another inhibiting species, namely, $(HCO)_{ads}$, was present on the electrode surface during adsorption of HCOOH. They explained the presence of $(CO)_{ads}$, as detected by EMIRS, as the consequence of the reaction of their postulated $(HCO)_{ads}$ with residual oxygen in the electrochemical cell.

Corrigan and Weaver[20] also observed $(CO)_{ads}$ in performing infrared measurements by SPAIRS [see Section II.3(iv.e)]. They concluded from their experiments that $(CO)_{ads}$ is the species responsible for the poisoning phenomena.

(*b*) *Structural effects*

The occurrence of different kinds of adsorbed species during the oxidation of formic acid was also indirectly proved by studies carried out on electrodes with different surface structures. The extensive literature on this subject may be classified into three categories according to the type of electrode used: single crystals, alloys, and platinum electrodes modified by foreign adatoms.

Single crystals. The best way to observe structural effects in electrocatalysis is to use model electrode surfaces, such as single crystals. After the first experiments, in 1976, on gold single crystals,[439] various research groups observed important structural effects during the electrooxidation of formic acid on platinum single crystals[297,388,440-443] or on platinum single crystals modified by adatoms.[444,445] Even though the findings reported by these research groups for platinum single crystals are not all in agreement, the influence of the surface structure during the electrocatalytic oxidation of formic acid was clearly demonstrated (Fig. 31).

The current densities for formic acid oxidation show a dependence on the orientation of the Pt(*hkl*) planes, and the decrease in the current that results from electrode poisoning depends strongly on the nature of the face studied. This poisoning effect is drastic for the Pt(110) plane, with a rapid decrease of the current versus time. Conversely, it is much smaller for Pt(111), which appears to be less sensitive to poisoning species, even though the initial activity is rather low. The Pt(100) plane is blocked until the potential reaches about 0.7 V versus RHE, the current attaining appreciable values

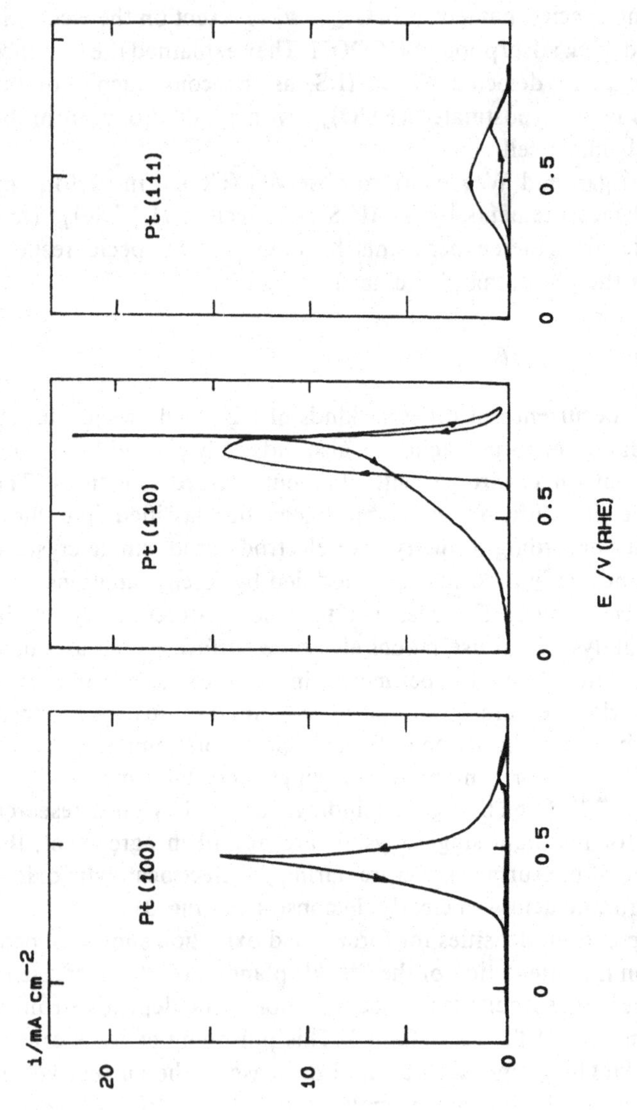

Figure 31. Voltammograms of low-index platinum single crystals showing pronounced structural effects in the electrooxidation of formic acid (0.1 *M* $HClO_4$, 0.1 *M* HCOOH, 25°C, 50 mV s^{-1}, first sweep). After Ref. 297.

only during the negative sweep. This behavior is a little bit different from that of methanol on Pt(100) [see Section III.1(i.c)].

In preliminary studies, the influence of adatoms on single-crystal electrodes was investigated in the case of platinum.[444,445] As expected, important effects were observed with lead and bismuth adatoms.

Alloys. Historically, the first attempts to modify the structure of a catalytic electrode involved the use of metallic alloys. The electrooxidation of formic acid has been investigated on different types of alloys: Pt/Au,[313,446,447] Pt/Pd,[314,448,449] Pt/Rh,[312] and Au/Pd.[315] Except for Pt/Au, these alloys are solid solutions, and it is possible to vary continuously the bulk composition of the electrode. One of the main problem remains the determination, with good accuracy, of the surface composition, which differs significantly from the bulk composition, due to surface enrichment by preferential dissolution of one metal component. Techniques to estimate the true surface composition are discussed in Section II.4(ii.c). Synergistic effects were observed in many cases: the electrode activity for some surface compositions is greater than that for the pure metals. Such effects were interpreted by taking into account the nature of the adsorbed species: strongly bonded species (poisons), for instance, are more easily formed on platinum than on gold or palladium electrodes. By modification of the platinum electrode with a second metal, the platinum sites are diluted and the formation of such poisons is less favored.

Adatoms. The electrooxidation of HCOOH is probably the most typical example of electrocatalytic reactions used to illustrate the influence of adatoms on the activity of platinum. Adžić has reviewed this subject.[13,303]

Striking catalytic effects are observed in the case of the oxidation of HCOOH on Pt, Rh, Ir, and Pd in the presence of adatoms, such as Pb, Bi, Tl, and Cd. A very significant enhancement effect on HCOOH oxidation at platinum electrodes is observed in the presence of lead adatoms. During the positive sweep in the range 0.2 to 0.8 V versus RHE, only a small peak is observed on pure platinum. In the presence of small amounts of lead salts in solution, the oxidation of HCOOH during the positive sweep leads to a large peak (of around 70 mA cm^{-2}).[13] Another important observation is the quasi-superposition of the oxidation currents during the forward and the reverse sweeps in the range 0.2-0.5 V versus RHE, a fact which confirms the dramatic decrease in the poisoning phenomena. This spectacular effect is also observed with systems such as Pt/Cd_{ads}, Pt/Tl_{ads}, and Pt/Bi_{ads}, but to varying extents. However, the effect of cadmium is somewhat different. The maximum current densities observed are not affected as much relative to those on pure platinum, but the potential at which oxidation occurs is shifted negatively.

The effect of these adatoms is also illustrated in the case of formic acid oxidation at rhodium electrodes, modified by lead or cadmium adatoms (Fig. 32). As with platinum, lead adatoms give

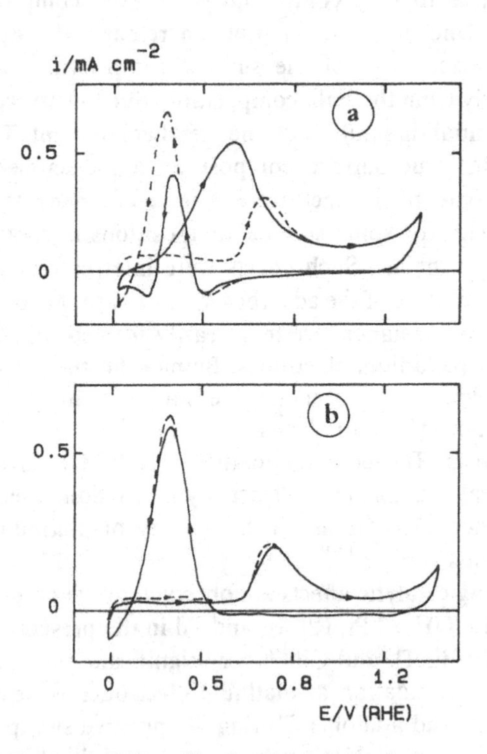

Figure 32. Effect of lead and cadmium adatoms on the voltammograms showing the oxidation of 0.1 M HCOOH in 0.5 M $HClO_4$ at Rh electrodes (25°C, 50 mV s^{-1}). The dashed curves correspond to unmodified Rh electrodes. (a) 5×10^{-4} M Pb^{2+}; (b) 10^{-3} M Cd^{2+}.

enhancement effects, but, conversely, cadmium adatoms do not change the electrocatalytic activity of rhodium.[450] An explanation of this different behavior has been given on the basis of IR spectroscopic measurements (see below).

(*c*) *Discussion of the reaction mechanism*: *Example of the rhodium electrode*

The electrooxidation of formic acid can be assumed to follow a dual-path mechanism[314]:

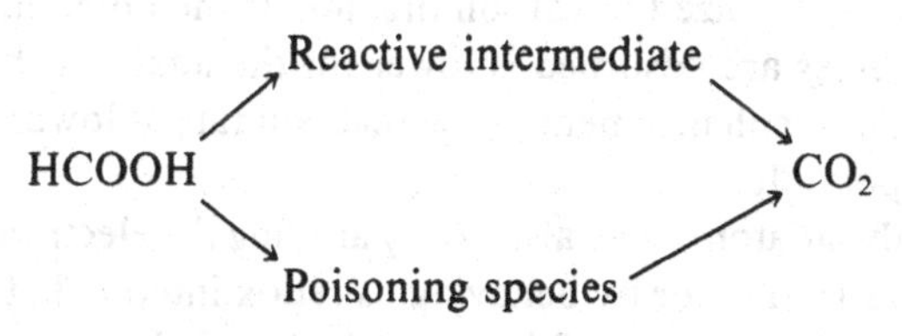

Hence, the reactive intermediate is of the $(COOH)_{ads}$ type, and, as stated above, the poisoning species is now known to be adsorbed CO. Different conclusions about the exact nature of the poisoning species led to different hypotheses being put forward to explain the effect of adatoms. According to Bagotsky and Vassiliev[391] and, later, Adžić,[13] the structure of the poisoning species is $(COH)_{ads}$, which thus needs H_{ads} to be formed. However, no clear experimental confirmation of the occurrence of this poisoning species was obtained at that time. According to the original explanation by Adžić,[303] the foreign adatoms block the surface active sites and prevent the formation of adsorbed hydrogen. In the absence of H_{ads}, according to Adžić, the formation of the postulated poisoning species, $(COH)_{ads}$, would be impossible. This hypothesis fails, as the poisoning species is now known and well accepted to be $(CO)_{ads}$, and as this latter species obviously does not need H_{ads} to be formed. Another remark is worth making: a platinum electrode can be poisoned even when it has been maintained at potentials more positive than those required for the adsorption of hydrogen.

Various other hypotheses were proposed by several authors.[48,304,412,451-454] Hartung *et al.*[451] and Kokkinidis[304] suggested that adatoms work by blocking the poison formation reaction. According to this hypothesis, a certain number of sites of a given symmetry are required to allow the formation of a poisoning species.

Generally, this number of sites is greater for the poison formation reaction than for the intermediate involved in the main reaction path.[47,48,430,452] Thus, any type of adatoms which form small islands on the electrode surface should enhance the electrocatalytic activity of the substrate, which is not the case.

Another hypothesis, proposed by Shibata and Motoo,[455] is the bifunctional theory of electrocatalysis. The principle is rather simple. A catalytic surface partially covered with foreign adatoms presents two types of active sites—firstly, the sites of the catalytic metal, namely, platinum, which are able to adsorb and dissociate the organic molecules, and, secondly, the adatom sites, which can adsorb the oxygen atoms or the hydroxyl ions necessary for adsorbed CO to be oxidized to carbon dioxide. If the potential at which oxygen species are adsorbed is lower for the adatoms than for the platinum sites, enhancement of the reaction rate at lower potentials becomes possible.

Finally, adatoms may also act by altering the electronic properties of the substrate or by behaving as redox intermediates.[456] This explanation is more satisfying for alkaline solutions, where no significant poisoning effects are observed, nor can a $(CO)_{ads}$ contribution be invoked.[413]

It is really difficult to definitively choose one of these different hypotheses. However, the following experimental facts about the electrooxidation of formic acid at noble metals are now widely accepted: (i) adsorbed CO (linearly and bridge bonded) is the poisoning species; and (ii) there is no experimental evidence for the presence of adsorbed COH species. Only some adatoms have positive effects on the electrooxidation of formic acid; among these, lead and bismuth have the most significant effects. They presumably may occupy some of the specific electrode active sites necessary for formation of the poisoning species, but at the same time they may behave according to a bifunctional mechanism as well. Modifications of the electronic properties cannot be excluded either, at least in some cases. There is, to date, no definitive experimental proof that favors one hypothesis or the other, except that the modification of the electrode coverage by strongly bonded species is firmly demonstrated. Recent results concerning the Rh/HCOOH system with lead, bismuth, or cadmium adatoms, obtained by *in situ* IR spectroscopic measurements (EMIRS), gave interesting information.[177,450] The modification of the surface distribution of adsorbed species during adsorption of an organic molecule at adatom-modified electrodes was detected for the first time. Thus, the degree of coverage by bridge-bonded CO species was found to be much more affected by lead adatoms than was that by linearly bonded CO species (Fig. 33). It was concluded that the poisoning

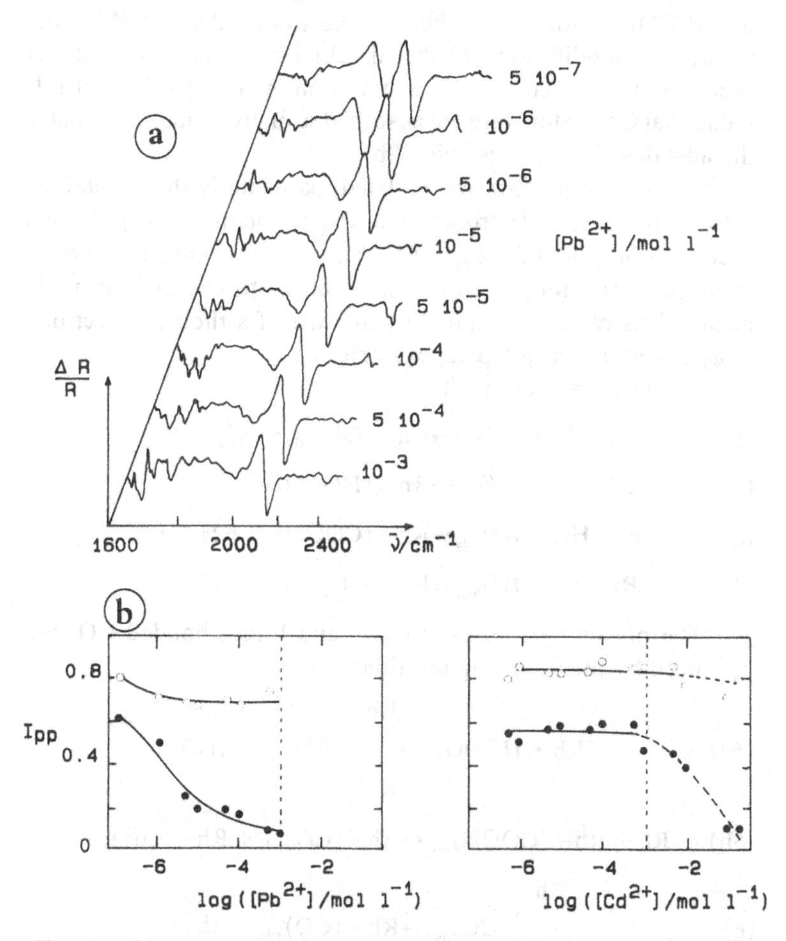

Figure 33. (a) Effect of lead adatoms on the EMIRS spectra of CO_{ads} resulting from the adsorption of 0.1 *M* HCOOH in 0.5 *M* $HClO_4$ at rhodium electrodes. (b) Plots of the peak-to-peak intensities of the EMIRS bands of CO_{ads} resulting from HCOOH adsorption at modified rhodium electrodes as a function of the bulk concentration of the precursor salt (Pb^{2+} or Cd^{2+}).

effect was due not only to the presence of both CO species (linearly and bridge-bonded), but above all to interactions between these two kinds of species. The inhibition of the formation of bridge-bonded CO is sufficient to enhance the catalytic activity of the electrodes. The same observation was made with bismuth-modified rhodium electrodes. On the other hand, the case of cadmium is completely different, in that cadmium adatoms have nearly no effect on HCOOH oxidation at rhodium electrodes. An EMIRS study showed no modification of the distribution of the adsorbed CO species in the presence of Cd adatoms on the electrode surface. It seems that Cd adatoms are not adsorbed strongly enough to displace the adsorbed CO species (Fig. 33b).

All the results now accumulated, particularly those obtained with well-defined electrodes and those obtained using *in situ* spectroscopic methods (e.g., EMIRS), allow a detailed mechanism to be proposed for the electrooxidation of formic acid at noble metals. This can be illustrated by the case of a rhodium electrode in acid medium, as a typical example.

The main reaction path is:

$$\text{Rh} + \text{H}_2\text{O} \rightarrow \text{Rh—(OH)}_{\text{ads}} + \text{H}^+_{\text{aq}} + e^- \tag{1}$$

$$\text{Rh} + (\text{HCOOH})_{\text{sol}} \rightarrow \text{Rh—(HCOOH)}_{\text{ads}} \tag{2}$$

$$\text{Rh—(HCOOH)}_{\text{ads}} \rightarrow \text{Rh—(COOH)}_{\text{ads}} + \text{H}^+_{\text{aq}} + e^- \tag{3}$$

$$\text{Rh—(COOH)}_{\text{ads}} \rightarrow \text{Rh} + \text{CO}_2 + \text{H}^+_{\text{aq}} + e^- \tag{4}$$

The poisoning species, linearly and bridge-bonded CO, are produced by the following reactions:

$$2\text{Rh} + \text{HCOOH} \rightarrow \begin{matrix}\text{Rh} \diagdown \\ \text{Rh} \diagup\end{matrix} (\text{CO})_{\text{ads}} + \text{H}_2\text{O} \tag{5a}$$

$$\text{Rh} + \text{Rh—(COOH)}_{\text{ads}} \leftrightarrow \text{Rh—(CO)}_{\text{ads}} + \text{Rh—(OH)}_{\text{ads}} \tag{5b}$$

$$\begin{matrix}\text{Rh} \diagdown \\ \text{Rh} \diagup\end{matrix} (\text{CO})_{\text{ads}} \leftrightarrow \text{Rh—(CO)}_{\text{ads}} + \text{Rh} \tag{6}$$

In the presence of lead adatoms, the formation of bridge-bonded CO is inhibited, which leads to an enhancement of the electrocatalytic activity.

The oxidation of the poisoning species can be summarized by the following reactions:

(7a) $\mathrm{Rh{-}(CO)_{ads} + Rh{-}(OH)_{ads} \rightarrow 2Rh + CO_2 + H^+_{aq} + e^-}$

(7b) $\mathrm{\begin{matrix} Rh \\ Rh \end{matrix} \rangle (CO)_{ads} + Rh{-}(OH)_{ads} \rightarrow 3Rh + CO_2 + H^+_{aq} + e^-}$

(*iv*) *Carbon Monoxide and Reduced* CO_2

Although, strictly speaking, CO is not an organic molecule, it is worth considering here, due to its virtual omnipresence as an intermediate or a product of most adsorption and electrooxidation processes involving small oxygenated organic molecules. Exceptions to this may arise in the case of long-chain aliphatic acids, as demonstrated by Horanyi,[91] using radiotracer techniques, and, more recently, by Leung and Weaver[258] on the basis of an infrared and Raman spectroscopic survey. In most cases, as for the oxidation of methanol [see Section III.1(i)], adsorbed CO acts as a poison on catalytic surfaces, drastically decreasing their activity.[12,15,18] This conclusion is now widely accepted, even though some research groups have postulated, from pure electrochemical measurements, that differences exist between species arising from the adsorption of gaseous CO and species arising from the dissociative adsorption of small oxygenated organic molecules.[457] On the other hand, the ease with which CO is formed and adsorbed on almost all catalytic surfaces led Beden *et al.* to emphasize recently the role that CO may play as a probe molecule to test the electrocatalytic activity of metallic electrodes.[458]

Since the work of Breiter,[2,15,55,459] it has been known that the electrooxidation of CO on polycrystalline noble metals is a complex reaction, leading to multiple voltammetric peaks in a potential range depending to a large extent on the adsorption potential and on the concentration of dissolved CO in the electrolytic solution.[457,460–462] Usually, with a solution saturated in dissolved CO, and for adsorption potentials greater than 0.4 V versus RHE (but below 0.6–0.7 V), only one narrow oxidation peak is detected at 0.9 V. If the adsorption potential is chosen below 0.4 V, multiple peaks are obtained.[457,462] Two main peaks are usually observed: the more cathodic one at around 0.45 V, and the second one at around

0.8–0.9 V.[462] More recently, following a study by Kita and Nakajima,[463] it has been recognized by Caram and Gutiérrez[464] that the behavior of CO depends strongly on its admission potential, that is, on the potential at which it has been introduced into the previously deareated cell (the value of 0.22 V versus RHE being especially critical).

The formation of an adsorbed CO layer results from the contributions of at least three distinguishable types of adsorbates, of different configurations, on the surface, depending on the crystallographic sites. Thus, linearly bonded CO_L is adsorbed "on top" of sites, while bridge-bonded CO_B lies between two sites, and multibonded CO_m occurs at higher coordination sites [see Section II.3(iv.c)]. However, depending on potential and/or on coverage, other configurations may have to be considered as well. For instance, Ikezawa *et al.*,[465] reconsidering the infrared band intensities, postulated that part of the surface might be covered by CO species adsorbed in a flat position on two adjacent Pt atoms. Such species would be IR inactive, which would explain their nondetection.

As stated above, the electrooxidation process of CO is complex. On Pt, or on Rh, it is known that the electrochemical behavior of CO depends on whether it was previously adsorbed in the double-layer region or in the hydrogen region. Kunimatsu *et al.*,[176,466] for instance, have shown some interesting differences in the CO infrared bands when adsorption on polycrystalline Pt was carried out at two separate potentials, one close to the hydrogen evolution potential, and the other in the middle of the double-layer region. This suggests that some of the CO species may react with hydrogen.

Recent work on Pt single crystals[388,452,455,467] demonstrated the sensitivity of CO adsorption to surface structure. Moreover, extensive spectroscopic investigations of CO adsorption on well-defined single crystals of Pt or Rh by Weaver and co-workers[468–470] revealed its dependence on surface reconstruction, through a comparison of (hkl) ordered and disordered crystal faces.

Of interest for practical applications of electrocatalysis is the fact that surface roughness acts to diminish the adsorption of CO, particularly when it results from chemisorption. This was known from fuel cell studies: developed electrodes are less poisoned than model smooth electrodes. It has also been recently demonstrated using EMIRS[295,325] with preferentially oriented Pt electrodes, as well as with electrochemically developed Pt surfaces, obtained by using high-frequency pulse potential programs.[293]

Direct oxidation of gaseous CO can be realized at metallized electrodes whose opposite side is in contact with an electrolytic solution. In an earlier work, Gibbs *et al.* used a PTFE membrane onto which a gold or a platinum layer was sputtered.[471] The electrochemical behavior of these electrodes was found to be similar to that of conventional electrodes. In a more recent work, Kita and Nakajima obtained much higher currents at Au–SPE electrodes than at smooth gold, while less poisoning was observed.[463]

In a similar way to the oxidation processes involving small organic molecules, the electrochemical oxidation of CO may be enhanced either by using adatoms or by using alloys of noble metals. Thus, in a long series of papers, Motoo and co-workers have examined many possibilities involving Pt, Rh, Au, and Ir electrodes, and their modification by Ru, As, and Sn adatoms (see, for instance, Refs. 472–476).

As mentioned above, adsorbed CO is the species responsible for electrode poisoning during the electrooxidation of numerous small oxygenated organic molecules. However, the case of the electrooxidation of the adsorbed species involved in the electroreduction of carbon dioxide has led to some controversy concerning the exact nature of the species formed in the hydrogen adsorption region. According to various authors, these species are somewhat different from those formed from dissolved gaseous CO,[477–482] and a clathrate-type structure has been proposed by Marcos *et al.*[480–482] However, confirming the early suggestion of Breiter,[483] made on the basis of pure electrochemical measurements, Beden *et al.*[484] found the same IR absorption bands for reduced CO_2 species and for adsorbed CO coming from dissolved CO, CH_3OH, HCOOH, and HCHO. The confirms that adsorbed CO is probably also formed during the electroreduction of carbon dioxide.

(v) *Summary of the Reaction Mechanisms*

To summarize the preceding discussion about the adsorbates formed from C_1 organic compounds on noble metals, Fig. 34 illustrates the more likely pathways between bulk species and adsorbates, from methanol to CO_2. In the redox reference scale used,

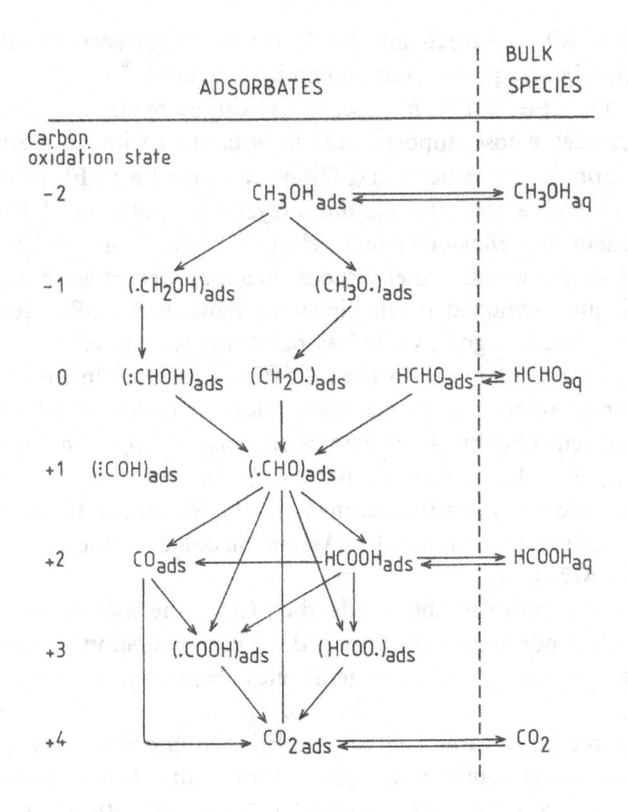

Figure 34. Pathways between bulk species and adsorbates for C_1 organic compounds with different oxidation state of carbon.

the carbon in methane (which is not represented here) would have an oxidation number of −4, while the carbon in carbon dioxide has an oxidation number of +4.

2. Oxidation of Monofunctional Molecules

Numerous oxygenated molecules with more than one carbon atom have been considered both as fuels in fuel cells and as raw materials for electrosynthesis. In this section, the oxidation of such molecules will be discussed, with selected examples concerning C_2 to C_6 molecules. Different aspects will be considered—for example, the nature of the functional group (alcohol, carboxylic acid, aldehyde, or ketone), the length of the carbon chain, and the type of isomer.

A preliminary remark concerns the possible use of such molecules in fuel cells. An increase in the number of carbon atoms leads to an increase in the number of possible intermediates and reaction products in the course of their electrooxidation, and, consequently, to more complex reaction mechanisms. Furthermore, they are much more difficult to oxidize completely to carbon dioxide, so that their practical energy density is much lower than expected (e.g., complete oxidation of glucose would liberate 4.4 kW-h kg^{-1}, whereas partial oxidation to gluconic acid gives only $\frac{1}{12}$ of this energy, i.e., 0.4 kW-h kg^{-1}). Finally, the prices of these fuels (e.g., ethanol, ethylene glycol) are much higher than that of methanol, and this may limit their practical use.

However, from a fundamental point of view, studies of the electrocatalytic oxidation of such products are quite useful for understanding the general reaction mechanisms. Moreover, mainly for multifunctional molecules, their application in the field of electrosynthesis is a further motivation for such studies.[485]

(i) *Reactivity of Alcohol Functions*

Due to the good solubility of alcohols in water, their electrooxidation has been widely studied in aqueous electrolytes. Two main topics have been investigated: the effect of the length of the carbon chain, and the influence of the position of the functional group in the carbon skeleton.

Beginning with the primary alcohols (characterized by the—CH_2OH functional group), the electrocatalytic oxidation of the first member of the series, ethanol, has been the subject of numerous studies,[27,74,240,395,486-493] but that of propanol[494-504] and of butanol[40,300,505,506] have also been considered. Several reviews of work on the electrooxidation of several alcohols on platinum and platinum-gold alloys have appeared.[507-509] Gold was also considered as a good electrode catalyst.[510-512]

Submonolayers of various adatoms were also considered to modify the electrocatalytic behavior of platinum in the oxidation of several alcohols.[413,491]

From all these studies, although performed under different experimental conditions, making comparison of the results difficult, some general conclusions can be drawn:

- The rate of alcohol electrooxidation depends strongly on the pH of the solution. Platinum is the best electrocatalyst in acid medium, whereas gold is quite inactive. Conversely, in alkaline medium, gold is usually a very active catalyst for the electrooxidation of alcohols at rather high potentials.
- The reactivity of alcohols depends on the position of the functional group. On platinum, primary alcohols are more reactive than secondary alcohols,[40,507,509] and tertiary alcohols are practically unreactive at room temperature. However, in this latter case, a small reactivity is observed at higher temperatures in a potential region corresponding to the adsorption of oxygen. This reactivity, even though small, is very interesting and unexpected. For primary and secondary alcohols, the removal of a hydrogen in the α position is the first step in dissociative adsorption.[509] Since tertiary alcohols do not have any H atom bonded to the carbon in the α position, the mechanisms of their electrooxidation must be completely different from those for primary or secondary alcohols.
- Gold is a very active catalyst for the electrooxidation of alcohols in alkaline medium.[512] The reactivity of primary alcohols is usually lower than that of secondary alcohols.[510] This difference is probably related to the inductive effect of the adjacent methyl groups present in secondary alcohols. Conversely to what happens at platinum electrodes, the reactivity of alcohols at gold electrodes increases with the number of carbon atoms, at least in a series up to six carbons.
- Values of the apparent activation energies are rather similar for the electrooxidation of all alcohols between one and four carbons in length and are in line with those expected for adsorption processes.[498]
- The first step in the electrooxidation on platinum is an adsorption process with, in most cases, dissociation of the molecule, mainly at low potentials in the double-layer region. This dissociative process is obvious on platinum, but does not seem to occur on gold electrodes.[505,510,511]
- As for the oxidation of methanol, poisoning effects are observed for all alcohols during their electrooxidation on platinum.

The nature of the adsorbed residue produced during the dissociative chemisorption of alcohol molecules has been studied by *in situ* infrared reflectance spectroscopy.[240,395,489,490] According to experiments using EMIRS[395] and SNIFTIRS,[489] the nature of the poisoning species is definitively $(CO)_{ads}$ formed during the dissociation of the alcohol molecule. This is true for all alcohols and even for aldehyde compounds.[258] Differences exist only in the extent of coverage of the surface by adsorbed CO, which is higher for the smallest molecules, reaching 0.9 for methanol[395,513-515]

The general reaction mechanism of the oxidation of aliphatic alcohols on platinum can be summarized with the following simple scheme:

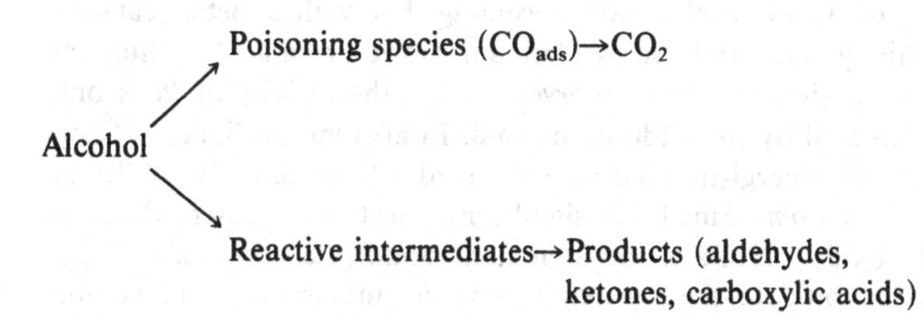

The nature of the reactive intermediates is more difficult to identify clearly. In the case of ethanol, an aldehyde-like intermediate $(CH_3{-}\overset{|}{C}{=}O)$ is postulated.[240,395]

In general, the main product formed during oxidation of primary and secondary alcohols is the corresponding aldehyde[495] and ketone, respectively:

$$R{-}CH_2OH \rightarrow R{-}CHO + 2H^+_{aq} + 2e^-$$

$$R{-}CHOH{-}R' \rightarrow R{-}CO{-}R' + 2H^+_{aq} + 2e^-$$

Further oxidation of aldehydes is always possible, leading usually to carboxylic acids:

$$R{-}CHO + H_2O \rightarrow R{-}COOH + 2H^+_{aq} + 2e^-$$

This will depend on the potential applied to the electrode. The structure of the electrode support may also influence the nature of the products formed.[516]

As a typical example, the electrocatalytic oxidation of ethanol at a platinum electrode gives mainly acetaldehyde and acetic acid, as proved by chromatographic analysis.[491] A very small amount of CO_2 is detected in the gas phase, coming probably from the oxidation of the CO poisoning species.[517]

Role of the electrode material

Alloys. As mentioned before, the electrooxidation mechanisms on gold and platinum seem to be different: no poisoning effect is observed on gold, but oxidation occurs at higher potentials than on Pt. The use of platinum-gold alloys thus could offer the possibility of avoiding electrode poisoning, but with a better catalytic activity than exhibited by platinum. However, in acid medium, no synergistic effect was observed,[497] and the activity of Pt is only decreased by the addition of gold. In alkaline medium,[508,509] significant synergistic effects are observed with the alloy 20:80 Pt/Au.

Adatoms Small, but significant, effects were observed during the oxidation of alcohols at adatom-modified electrodes.[413,491] The oxidation potential is generally lowered, and, in the most favorable cases, namely, in the presence of adatoms of lead, germanium, thallium, or bismuth, the maximum activity is increased by a factor of 2 to 3.[413] A more detailed study of ethanol oxidation[491] showed that the two steps (oxidation to acetaldehyde and then to acetic acid) should be considered separately. The first step is inhibited by the presence of some adatoms (Ag, Hg, Se, Te, and Bi), but the second one is activated by some of these same adatoms (Se, Te, and Bi).

Single crystals. In the electrooxidation of molecules with one carbon atom, important structural effects[401] were demonstrated (see Section III.1). Only a very few studies have been carried out to investigate such effects during the oxidation of alcohols heavier than methanol. The electrooxidation of ethanol on Pt(*hkl*) single crystals[492] shows structural effects similar to those encountered in the case of methanol in acid medium, but with smaller currents. Moreover, in alkaline medium, the three low-index planes of platinum appear to be much more active than polycrystalline platinum, conversely to the case of methanol. The same observations were also made in the case of the oxidation of butanol at Pt(*hkl*) electrodes in alkaline medium.[300] Thus, it has not been possible to simulate the electrocatalytic behavior of polycrystalline platinum by that of the three low-index single crystal planes.

(ii) *Reactivity of the Aldehyde Function*

The electrocatalytic oxidation of aldehydes has been much less studied than that of alcohols, except that of formaldehyde [see Section III.1(ii)]. Nevertheless, some studies, mainly devoted to the electrooxidation of acetaldehyde, can be cited.[27,28,518-522] Effects of the nature of the catalytic electrode[70,420] or the length of the carbon chain[258,523,524] were considered in some studies.

The oxidation of acetaldehyde or other aliphatic aldehydes on platinum electrodes occurs at relatively high potentials compared to those at which the oxidation of alcohols occurs. There is no oxidation in the double-layer region, while two oxidation peaks generally appear in the oxygen region. This simple observation means that the mechanism of the electrooxidation of aldehydes should be different from that of alcohols. In the latter case, oxidation of the electrode surface, leading to the formation of an adsorbed oxygen layer, inhibits the electrooxidation process, whereas such adsorbed oxygenated species seem to be necessary in the case of aldehyde oxidation. The main products of electrooxidation of aldehydes are the corresponding carboxylic acids.[524]

On other metals,[70] the electrooxidation of aldehydes also forms mainly carboxylic acids, with by-products resulting from the Cannizzaro reaction. On metals such as nickel or copper, oxide layers are necessary for the electrooxidation to proceed but on gold or silver, the oxidation takes place on the oxide-free surface.[70]

Adsorption of acetaldehyde leads to a partial dissociation of the molecule, as proved recently by spectroscopic measurements.[27,28,525] Production of carbon dioxide, detected by DEMS and ECTDMS,[525] and the presence of linearly adsorbed CO, detected by EMIRS,[27] both proved without doubt the breaking of the C—C bond. Adsorbed CO again plays the role of a catalytic poison. These observations are corroborated by experiments on platinum single crystals.[522] However, when the concentration of acetaldehyde is not too low, the main product formed remains acetic acid.

The electrooxidation of acetaldehyde is greatly enhanced by the presence of foreign adatoms on platinum.[430,456,526] Oxygen-adsorbing adatoms (Ru, Ge, Sn, As, and Sb) strikingly increase the current densities during the electrooxidation of aliphatic aldehydes (formaldehyde, acetaldehyde, and propionaldehyde). Non-oxygen-adsorbing adatoms (Bi mainly) also have significant effects. In the former case, the enhancement of platinum activity can be explained mainly by the bifunctional theory of electrocatalysis,[430,456,526] but in the latter case inhibition of the formation of poisoning species can certainly explain the observed increase in current.

(iii) *Reactivity of the Ketone Function*

A ketone function usually represents the highest possible oxidation level, and in a compound where a ketone is the only functional group present, no mild electrooxidation can occur, so that degradation of the molecule will take place, associated with the breaking of C—C bonds. However, ketones are very stable in aqueous medium, and electrochemical degradation requires high positive potentials as well as long adsorption times. For instance, the oxidation of adsorbed acetone is only possible after a rather long adsorption time, around 30 minutes.[527] Under these conditions, CO_2 was detected by DEMS during the positive sweep, in the oxygen region of the platinum electrode.[527]

Oxidation of products containing ketone groups can occur, as will be discussed in Section III.3, which deals with multifunctional compounds.

Electroreduction of ketones is possible, but this topic is not directly relevant to the subject of this chapter.

(iv) *Reactivity of the Carboxylic Acid Function*

The adsorption of carboxylic acids on platinum is a reversible phenomenon,[528-530] and significant dissociation of the acid molecule does not occur.[258] This fundamental observation means that the electrooxidation of carboxylic acids is difficult in aqueous medium, even though the degradation of such molecules is possible. As a typical example, the oxidation of oxalic acid, either at platinum[6,531-533] or at gold[512] electrodes, does occur in acid medium, but only at high positive potentials, with breaking of the C—C bond, leading to the formation of carbon dioxide.

Reduction processes have been reported to occur in the hydrogen adsorption region at platinum electrodes.[529]

3. Oxidation of Multifunctional Molecules

In the previous section, the reactivity of different functional groups was discussed, mainly from the point of view of their oxidation at catalytic electrodes. However, the environment of the functional group plays a very important role in its reactivity. The influence of the position of the reactive group in the carbon chain is clearly evident for the oxidation of aliphatic monoalcohols, such as the butanol isomers.[40,506] The length of the carbon skeleton is also an important factor.[509] Nevertheless, the presence of (at least) one other functional group in the molecule may modify greatly its electrooxidation. While there is a very large variety of such molecules, some typical examples will be given below, dealing with molecules having practical interest, particularly in fuel cells and in electroorganic synthesis.

(i) *Oxidation of Diols and Polyols*

Due to the good solubility of polyols in water, and the relatively good reactivity of alcohol groups, the number of studies concerning the electrooxidation of polyols in aqueous medium is rather large.

The first member of the series, ethylene glycol, is probably the most promising fuel among those polyols with more than one carbon atom, even though many problems still remain unsolved for its use in an electrochemical power source. The electrooxidation of ethylene glycol has been widely studied on platinum and gold[319,534–538] and on adatom-modified electrodes.[305,306,413,539–542] The influence of the electrode surface structure has also been observed with platinum single crystals.[543,544]

Ethylene glycol is easily oxidized on platinum electrodes in acid or alkaline medium. The current densities in the latter case are rather high (reaching around 7 mA cm^{-2}). With gold, in alkaline medium, the electrocatalytic activity observed is also very significant

(~12 mA cm^{-2}). The poisoning phenomena previously observed in the oxidation of monoalcohols are also seen with ethylene glycol.

As an example, in acid medium the oxidation of ethylene glycol at platinum electrodes starts with a shoulder at around 0.5 V versus RHE and with a main peak at 0.8 V. However, if the lower limit of the potential is shifted positively, the shoulder becomes a real peak showing weak poisoning phenomena. The meaning of such an observation is clear: in the hydrogen region, a poisoning species is formed by dissociative adsorption of ethylene glycol, partially blocking the electrode active sites. Spectroscopic measurements by EMIRS showed the presence of adsorbed CO,[545] responsible for the poisoning of the platinum surface. Thus, in acid medium, CO is mainly linearly bonded to the electrode surface, while in alkaline medium bridge-bonded CO is dominant.

The presence of adatoms on platinum surfaces leads to a very large enhancement effect on the electroreactivity of ethylene glycol, especially in alkaline medium.[305] Oxidation current densities as high as 100 mA cm^{-2} were observed with lead or bismuth adatoms. In acid medium, the effect is much weaker, and this can be related to the adsorption on the catalytic surface. As seen for formic acid adsorption on rhodium,[177] lead adatoms occupy two adsorption sites and consequently modify strongly the coverage of the electrode by bridge-bonded CO. It was concluded[177] that bridge-bonded CO is probably the species responsible for electrode poisoning. The electrooxidation of ethylene glycol at platinum in alkaline medium probably involves the same phenomenon, bridge-bonded CO being the main poisoning species present on the surface in this case.

On gold electrodes, the effect of adatoms is very small,[542] which confirms that poisoning phenomena are quite absent in this case.

The reaction products encountered in the electrooxidation of ethylene glycol are mainly C_2 compounds, oxalic acid being the last stage of oxidation without breaking of the C—C bond. However, some products resulting from C—C bond breaking (HCHO, HCOOH, and CO_2) have also been detected by liquid and gas chromatographies.[485,542]

Ethylene glycol electrooxidation is very sensitive to the structure of the electrode surface.[543,544] As in the case of the oxidation of monoalcohols, the Pt(110) orientation appears to be the most active plane, but also the most poisonable one.

Despite the problems involved, ethylene glycol represents a possible fuel for an electrochemical power source, and practical systems have been studied using plurimetallic anodes.[318,546,547] On the other hand, selective oxidation of ethylene glycol to glycolaldehyde was recently obtained at platinum electrodes by carefully choosing the oxidation potential and by modifying the electrode activity by introduction of foreign metal adatoms,[548] an interesting result for electroorganic synthesis.

Studies on the electrooxidation of other diols are much more scarce. The two isomers of propanediol have however been considered by several research groups.[65,549-553] On platinum, the electrooxidation of the two propanediol isomers follows different routes. 1,2-Propanediol behaves very similarly to ethylene glycol, but with smaller current densities, whereas 1,3-propanediol has a general behavior similar to that of a monoalcohol (propanol or butanol).

Numerous reaction products are formed during the prolonged electrolysis of such compounds. For example, during the oxidation of 1,2-propanediol at a platinum electrode, two main products have been identified by HPLC: lactic acid ($CH_3CHOHCOOH$) and hydroxyacetone (CH_3COCH_2OH). In addition, pyruvic acid ($CH_3COCOOH$) was also found, but in a smaller quantity. Lighter products, such as acetic acid and even formic acid, formed as the result of the breaking of the C—C bond, were also detected.[65,485]

The electrooxidation of other polyols has also been investigated, such as that of glycerol[554-558] and sorbitol.[558-561] Such molecules have been often considered as interesting raw materials for electrosynthesis. Some oxidation products of diols may be of practical interest. As a typical example, the transformation of 2,3-butanediol to 2-butanone can be achieved[562] by oxidation of butanediol to acetoin followed by reduction to 2-butanone.

(ii) *Oxidation of Other Multifunctional Compounds*

Multifunctional compounds are diverse and display a variety of molecular structures. Numerous studies have been devoted to the electrooxidation of molecules containing some combination of alcohol, aldehyde, acid, and ketone functional groups.

Among these molecules, oxalic acid,[531-533] glyoxal,[61,538,563-565] glyoxylic acid,[455,566] lactic and pyruvic acids,[567,568] and mesoxalic acid[569-571] should be mentioned. In fact, these compounds are of interest for several very different reasons.

Some of them, such a diacids, are only oxidizable by breaking of C—C bonds, leading generally to carbon dioxide. Typical examples are oxalic and mesoxalic acids, which are easily oxidized to CO_2 in acid medium on platinum, but only at relatively high potentials corresponding to the oxidation of the platinum surface.

In the case of other compounds, such as glyoxal and lactic acid, reaction products corresponding to selective oxidation are interesting for electrosynthesis purposes. To perform such nondestructive mild oxidation necessitates finding convenient experimental conditions (in terms of potential and electrode structure) that avoid breaking of C—C bonds.[61]

(*iii*) *Oxidation of Monosaccharides*

The electrooxidation of glucose has been extensively studied, because of interest in the use of this reaction in biosensors and implanted fuel cells for pacemakers.[301,572-586] Glucose is a typical product of the biomass, and practical electrosyntheses, using glucose as a raw material, have been proposed, such as production of gluconic acid by electrooxidation or of sorbitol by electroreduction. The simultaneous production of both gluconic acid and sorbitol in the same undivided cell was also carried out in order to evaluate paired synthesis.[587]

The main reaction product formed during the electrooxidation of glucose is gluconolactone, either at platinum electrodes in acid medium or at gold electrodes in alkaline medium. Gluconolactone is further transformed by hydrolysis into gluconic acid. The first step in the reaction is the dehydrogenation of the C_1 carbon, which occurs at rather low potentials at a platinum electrode. A recent study by *in situ* FTIR spectroscopy[585] showed that the platinum electrode surface is covered, at low potentials, by linearly adsorbed carbon monoxide coming from the partial dissociation of the glucose molecule. This adsorbed CO species is probably responsible for the electrode poisoning.

A significant effect of the electrode structure was observed with platinum in acid medium,[301] and also with gold in alkaline medium.[586] The three low-index single-crystal planes of platinum exhibit higher activity than polycrystalline platinum, and the Pt(111) orientation appears to be less sensitive to poison formation over the whole pH range.

The influence of foreign metal adatoms on the electrooxidation of glucose at platinum has been studied.[301,579,580,582] Adatoms such as Tl, Pb, and Bi enhance greatly the activity of platinum in acid, as well as in alkaline, medium. These effects are generally interpreted in terms of a decrease of electrode poisoning.

The other monosaccharides are much less studied, but seem also less reactive. As an example, fructose is only slightly oxidized at platinum electrodes in acid medium, but more significantly at gold electrodes in alkaline medium.[588]

IV. OUTLOOK FOR THE OXIDATION OF SMALL ORGANIC MOLECULES AT CATALYTIC ELECTRODES

If one considers the tremendous development of techniques relevant to interfacial electrochemistry and the efforts which have been made in the past ten years to track the adsorbates on electrodes, it would not be an exaggeration to say that some new concepts have been definitively established.

Thus, insofar as the oxidation of small aliphatic organic molecules is concerned, the following points can be put forward:

(i) Adsorbates belong to two groups:

• *reactive species*, which are characterized by a weak adsorption to the surface and a rather short lifetime. Kinetic intermediates are supposed to belong to this group, but none has been detected yet;

• *poisoning species*, which are characterized by a strong adsorption to the electrode surface and, therefore, high stability. CO_{ads} is the most common poison. It can be formed by chemisorption of all small organic molecules in acid medium, but only by some of them in alkaline medium.

(ii) Different adsorbates are present simultaneously at the electrode surface. Furthermore, the active surface, that is, the catalytic surface under reaction, is not static. It should be recognized that *the populations of adsorbates vary with time* even if the experimental parameters remain fixed. Thus, *poisoning phenomena*, are explained in terms of *competitive adsorption*. The more strongly adsorbed species compete and displace the weak adsorbates, until total deactivation of the surface occurs.

(iii) A limited number of adsorbates are implicated in the chemisorption of small organics. For, instance, Fig. 34 helps in understanding the correlation between the adsorbates formed by chemisorption of C_1 compounds.

(iv) *Surface arrangements* have to be considered to take into account the crystallography of active sites and the local number of adsorption sites necessary per adsorbate, as a result of which *electrocatalytic reactions are surface sensitive*. Except when the size of adsorbates increases to the point that steric effects become dominant, the behavior of a polycrystalline surface can be modeled by the weighted contributions of its basal planes. However, at the surface, further rearrangements and *disproportionation reactions* may occur as well, at least under some conditions.

(v) There is as yet *no general theory to explain the role of adatoms*. In some experiments, adatoms act by bringing together the oxygenated species necessary to trigger the electrochemical process. In other experiments, they simply act by selectively displacing the poisons or by modifying the way in which the adsorbates are bonded to the surface. Steric effects can also be invoked.

(vi) Developed surfaces and catalysts dispersed in conducting matrices are markedly less poisoned than smooth surfaces. They offer interesting possibilities for practical applications.

(vii) In the case of *multifunctional molecules*, it can be roughly said that each functional group has its own reactivity at a given surface in a given medium. Additive behavior may be expected, provided that the geometries of the adsorbates and of the surface allow the simultaneous reactions to occur.

This short summary of progress in the understanding of electrocatalytical reactions also shows the direction for future research. In particular, more work is needed on controlled model surfaces, as they are the basis of any theory.

Experimentally, now that sufficient sensitivity has been achieved for adsorbate detection, there is no doubt that further progress will come from advanced techniques relying on detection with much shorter time constants. Such techniques are required in order to study the formation of kinetic intermediates, which is the key to elucidating mechanisms.

Similarly, now that the reactive and poisoning species of catalytic surfaces have begun to be identified, further progress should be made in the development of highly stable practical catalytic electrodes. Applications will concern their industrial use, not only for electrochemical power sources burning organic fuels directly, but also for electrosynthesis processes.

ACKNOWLEDGMENTS

We are very grateful to Mrs. Danielle Laurent for typing the manuscript, to Dr. El Mustapha Belgsir for drawing the figures, and to different scientists, postdoctoral fellows, and Ph.D. students, who worked on this subject in the laboratory.

Most of these studies were supported by various research agencies, such as CNRS, DRET, EDF, AFME, and CEC-DG XII, to which we are greatly indebted.

REFERENCES

[1] J. O'M. Bockris and S. Srinivasan, *Fuel Cells. Their Electrochemistry*, McGraw-Hill, New York, 1969.
[2] M. W. Breiter, *Electrochemical Processes in Fuel Cells*, Springer-Verlag, New York, 1969.
[3] W. Vielstich, *Fuel Cells—Modern Processes of the Electrochemical Production of Energy*, John Wiley & Sons, New York, 1970.
[4] M. M. Baizer, *Organic Electrochemistry*, Marcel Dekker, New York, 1973.
[5] R. Roberts, R. P. Ouellette, and P. N. Cheremisinoff, *Industrial Applications of Electroorganic Synthesis*, Ann Arbor Science, Ann Arbor, Michigan, 1982.
[6] B. J. Piersma and E. Gileadi, in *Modern Aspects of Electrochemistry*, No. 4, Ed. by J. O'M. Bockris, Plenum Press, New York, 1966, p. 47.
[7] E. J. Rudd and B. E. Conway in *Comprehensive Treatise of Electrochemistry*, Vol. 7, Ed. by B. E. Conway, J. O'M. Bockris, E. Yeager, S. U. M. Khan, and R. E. White, Plenum Press, New York, 1983, p. 641.
[8] J. O'M, Bockris and A. K. N. Reddy, *Modern Electrochemistry*, Vol. 2, Plenum Press, New York, 1972, p. 1141.
[9] A. J. Appleby, in *Modern Aspects of Electrochemistry*, No. 9, Ed. by B. E. Conway and J. O'M. Bockris, Plenum Press, New York, 1974, p. 369.
[10] G. P. Sakellaropoulos, in *Advances in Catalysis*, Vol. 30, Ed. by D. D. Eley, H. Pines, and P. B. Weisz, Academic Press, New York, 1981, p. 218.

[11] J. Clavilier, R. Faure, G. Guinet, and R. Durand, *J. Electroanal. Chem.* **107** (1980) 205.
[12] R. Parsons and T. VanderNoot, *J. Electroanal. Chem.* **257** (1988) 9.
[13] R. R. Adžić in *Advances in Electrochemistry and Electrochemical Engineering*, Vol. 13, Ed. by H. Gerischer, John Wiley & Sons, New York, 1984, p. 159.
[14] B. B. Damaskin, O. A. Petrii, and V. N. Batrakov, *Adsorption of Organic Compounds on Electrodes*, Plenum Press, New York, 1971.
[15] M. W. Breiter, in *Modern Aspects of Electrochemistry*, No. 10, Ed. by J. O'M. Bockris and B. E. Conway, Plenum Press, New York, 1975, p. 161.
[16] A. Bewick and B. S. Pons, in *Advances in Infrared and Raman Spectroscopy*, Vol. 12, Ed. by R. J. H. Clark and R. E. Hester, Wiley Heyden, London, 1985, p. 1.
[17] B. Beden and C. Lamy, in *Spectroelectrochemistry—Theory and Practice*, Ed. by R. J. Gale, Plenum Press, New York, 1988, p. 189.
[18] C. Lamy, *Electrochim. Acta* **29** (1984) 1581.
[19] B. Beden, in *Spectroscopic and Diffraction Techniques in Interfacial Electrochemistry*, NATO ASI Series, Vol. 320, Ed. by C. Gutiérrez and C. Melendres, Kluwer Academic Publ., Dordrecht, 1990, p. 103.
[20] D. S. Corrigan and M. J. Weaver, *J. Electroanal. Chem.* **241** (1988) 143.
[21] A. Capon and R. Parsons, *J. Electroanal. Chem.* **44** (1973) 1.
[22] V. S. Bagotsky, Yu. B. Vassiliev, and O. A. Khazova, *J. Electroanal. Chem.* **81** (1977) 229.
[23] J. Willsau and J. Heitbaum, *J. Electroanal. Chem.* **161** (1984) 93.
[24] S. Wilhelm, T. Iwasita, and W. Vielstich, *J. Electroanal. Chem.* **194** (1985) 27.
[25] J. Willsau and J. Heitbaum, *J. Electroanal. Chem.* **185** (1985) 181.
[26] T. Iwasita, W. Vielstich, and E. Santos, *J. Electroanal. Chem.* **229** (1987) 367.
[27] J. M. Perez, B. Beden, F. Hahn, A. Aldaz, and C. Lamy, *J. Electroanal. Chem.* **262** (1989) 251.
[28] B. Bittins-Cattaneo, S. Wilhelm, E. Cattaneo, H. W. Buschmann, and W. Vielstich, *Ber. Bunsenges. Phys. Chem.* **92** (1988) 1210.
[29] E. Yeager, J. O'M. Bockris, B. E. Conway, and S. Sarangapani (eds.), *Comprehensive Treatise of Electrochemistry*, Vol. 9, Electrodics: Experimental Techniques, Plenum Press, New York, 1984.
[30] F. G. Will and C. A. Knorr, *Z. Electrochem.* **64** (1960) 258, 270.
[31] S. Srinivasan and E. Gileadi, *Electrochim. Acta* **11** (1966) 321.
[32] H. Angerstein-Kozlowska, J. Klinger, and B. E. Conway, *J. Electroanal. Chem.* **75** (1977) 45, 61.
[33] A. M. Alquié-Redon, A. Aldaz, and C. Lamy, *J. Electroanal. Chem.* **52** (1974) 11.
[34] A. M. Alquié-Redon, J. Vigneron, and C. Lamy, *J. Electroanal. Chem.* **92** (1978) 147.
[35] J. E. B. Randles, *Trans. Faraday Soc.* **44** (1948) 327.
[36] A. Sevcik, *Collect. Czech. Chem. Commun.* **13** (1948) 349.
[37] P. Delahay, *New Instrumental Methods in Electrochemistry*, Wiley Interscience, New York, 1954.
[38] R. S. Nicholson and I. Shain, *Anal. Chem.* **36** (1964) 706.
[39] J. M. Savéant and E. Vianello, *Electrochim. Acta* **12** (1967) 629.
[40] D. Takky, B. Beden, J. M. Léger, and C. Lamy, *J. Electroanal. Chem.* **145** (1983) 461.
[41] B. Beden, C. Lamy, and J. M. Léger, *J. Electroanal. Chem.* **101** (1979) 127.
[42] G. Crépy, C. Lamy, and S. Maximovitch, *J. Electroanal. Chem.* **54** (1974) 161.
[43] F. Kadirgan, B. Beden, J. M. Léger, and C. Lamy, *J. Electroanal. Chem.* **125** (1981) 89.
[44] C. McCallum and D. Pletcher, *J. Electroanal. Chem.* **70** (1976) 277.

[45] E. Santos, E. P. M. Leiva, W. Vielstich, and U. Linke, *J. Electroanal. Chem.* **227** (1987) 199.
[46] D. Pletcher and V. Solis, *J. Electroanal. Chem.* **131** (1982) 309.
[47] I. Fonseca, J. Lin-Cai, and D. Pletcher, *J. Electrochem. Soc.* **130** (1983) 2187.
[48] M. Shibata and S. Motoo, *J. Electroanal. Chem.* **188** (1985) 111.
[49] N. A. Hampson and M. J. Willars, *J. Chem. Soc. Faraday Trans. 1* **75** (1979) 2535.
[50] D. Pletcher and V. Solis, *Electrochim. Acta* **27** (1982) 775.
[51] A. M. de Ficquelmont and M. M. de Ficquelmont-Loizos, *J. Electrochem. Soc.* **131** (1984) 2880.
[52] M. Watanabe, Y. Furuuchi, and S. Motoo, *J. Electroanal. Chem.* **191** (1985) 367.
[53] D. B. Hibbert and F. Y. Y. Yon-Hin, *J. Electrochem. Soc.* **132** (1985) 1387.
[54] R. P. Buck and L. R. Griffith, *J. Electrochem. Soc.* **109** (1962) 1005.
[55] M. W. Breiter, *J. Electroanal. Chem.* **101** (1979) 329.
[56] M. W. Breiter, *J. Electroanal. Chem.* **157** (1983) 327.
[57] J. Kuta, in *Comprehensive Treatise of Electrochemistry*, Vol. 8, Ed. by R. E. White, J. O'M. Bockris, B. E. Conway, and E. Yeager, Plenum Press, New York, 1984, p. 249.
[58] A. Braithwaite and F. J. Smith, *Chromatographic Methods*, 4th ed., Chapman and Hall, London, 1985.
[59] K. I. Ota, Y. Nakagawa, and M. Takahashi, *J. Electroanal. Chem.* **179** (1984) 179.
[60] I. Yamanaka and K. Otsuka, *Electrochim. Acta* **34** (1989) 211.
[61] E. M. Belgsir, H. Huser, C. Lamy, and J. M. Léger, *J. Electroanal. Chem.* **270** (1989) 151.
[62] T. C. Chou, W. J. Chen, H. J. Tien, J. J. Jow, and T. Nonaka, *Electrochim. Acta* **30** (1985) 1665.
[63] E. M. Belgsir, H. Huser, J. M. Léger, and C. Lamy, *J. Electroanal. Chem.* **225** (1987) 281.
[64] M. Shibata and S. Motoo, *J. Electroanal. Chem.* **209** (1986) 151.
[65] H. Huser, J. M. Léger, and C. Lamy, *Electrochim. Acta* **33** (1988) 1359.
[66] A. Weissberger and B. W. Rossiter, *Techniques of Chemistry*, Vol. 1, *Physical Methods of Chemistry, Part III, Optical Spectroscopic and Radioactive Methods*, John Wiley & Sons, New York, 1971.
[67] G. Cauquis and V. D. Parker, in *Organic Electrochemistry*, Ed. by M. M. Baizer, Marcel Dekker, New York, 1973, p. 93.
[68] E. F. H. Brittain, W. O. George, and C. H. J. Wells, *Introduction to Molecular Spectroscopy—Theory and Experiment*, Academic Press, London, 1970.
[69] R. S. Drago, *Physical Methods in Chemistry*, W. B. Saunders, Philadelphia, 1977.
[70] R. M. Van Effen and D. H. Evans, *J. Electroanal. Chem.* **103** (1979) 383.
[71] S. Letellier, J. C. Dufresne, and M. B. Fleury, *Electrochim. Acta* **25** (1980) 1043.
[72] M. Amjad, D. Pletcher, and C. Smith, *J. Electrochem. Soc.* **124** (1977) 203.
[73] S. Pons, J. K. Foley, J. Russell, and M. Seversen, in *Modern Aspects of Electrochemistry*, No. 17, Ed. by J. O'M. Bockris, B. E. Conway, and R. E. White, Plenum Press, New York, 1986, p. 223.
[74] L. W. H. Leung and M. J. Weaver, *J. Phys. Chem.* **92** (1988) 4019.
[75] A. Kahyaoglu, Ph.D. thesis, University of Poitiers, 1981.
[76] B. Kastening, in *Comprehensive Treatise of Electrochemistry*, Vol. 8, Ed. by R. E. White, J. O'M. Bockris, B. E. Conway, and E. Yeager, Plenum Press, New York, 1984, p. 433.
[77] C. Lamy and P. Crouigneau, in *Electronic and Molecular Structure of Electrode-Electrolyte Interfaces*, Ed. by W. N. Hansen, D. M. Kolb, and D. W. Lynch, *J. Electroanal. Chem.* **150** (1983) 545.

[78] R. G. Compton and A. M. Waller, in *Spectroelectrochemistry, Theory and Practice*, Ed. by R. J. Gale, Plenum Press, New York, 1988, p. 349.
[79] A. Heinzel, R. Holze, C. H. Hamann, and J. K. Blum, *Electrochim. Acta* **34** (1989) 657.
[80] C. Lamy, B. Beden, and J. M. Léger, *Electrochim. Acta* **35** (1990) 679.
[81] J. Throck Watson, *Introduction to Mass Spectrometry*, 2nd ed., Raven Press, New York, 1985.
[82] I. Howe, D. H. Williams, and R. D. Bowen, *Mass Spectroscopy—Principles and Applications*, McGraw-Hill, New York, 1981.
[83] S. Bruckenstein and J. Comeau, *Faraday Disc. Chem. Soc.* **56** (1974) 285.
[84] O. Wolter, M. C. Giordano, J. Heitbaum, and W. Vielstich, in *Proceedings of the Symposium on Electrocatalysis*, Ed. by W. E. O'Grady, P. N. Ross, and F. G. Will, Vol. 82-2, The Electrochemical Society Pennington, New Jersey, 1982, p. 235.
[85] S. Ernst, J. Heitbaum, and C. H. Hamann, *Ber. Bunsenges. Phys. Chem.* **84** (1980) 50.
[86] A. E. Bolzan, T. Iwasita, and W. Vielstich, *J. Electrochem. Soc.* **134** (1987) 3052.
[87] J. Willsau and J. Heitbaum, *Electrochim. Acta* **31** (1986) 943.
[88] S. Wilhelm, T. Iwasita, and W. Vielstich, *J. Electroanal. Chem.* **238** (1987) 383.
[89] H. W. Buschmann, S. Wilhelm, and W. Vielstich, *Electrochim. Acta* **31** (1986) 939.
[90] V. E. Kazarinov and V. N. Andreev, in *Comprehensive Treatise of Electrochemistry*, Vol. 9, Ed. by E. Yeager, J. O'M. Bockris, B. E. Conway, and S. Sarangapani, Plenum Press, New York, 1984, p. 393.
[91] G. Horanyi, *Electrochim. Acta* **25** (1980) 43.
[92] A. Wieckowski and J. Sobkowski, *J. Electroanal. Chem.* **63** (1975) 365.
[93] M. W. Breiter, *Z. Phys. Chem. Neue Folge* **98** (1974) 23.
[94] E. Gileadi, *Electrosorption*, Plenum Press, New York, 1967, p. 1.
[95] B. J. Piersma, in *Electrosorption*, Ed. by E. Gileadi, Plenum Press, New York, 1967, p. 19.
[96] V. S. Bagotsky and Yu. B. Vassiliev, *Electrochim. Acta* **11** (1966) 1439.
[97] A. K. N. Reddy, in *Electrosorption*, Ed. by E. Gileadi, Plenum Press, New York, 1967, p. 53.
[98] P. Delahay and I. Trachtenberg, *J. Am. Chem. Soc.* **79** (1957) 2355.
[99] W. H. Reinmuth, *Anal. Chem.* **65** (1961) 473.
[100] P. Delahay and D. M. Mohilner, *J. Am. Chem. Soc.* **84** (1962) 4247.
[101] M. W. Breiter and S. Gilman, *J. Electrochem. Soc.* **109** (1962) 622, 1099.
[102] S. Gilman, *J. Phys. Chem.* **67** (1963) 78.
[103] S. Gilman, *Trans. Faraday Soc.* **62** (1966) 466.
[104] O. A. Khazova, Yu. B. Vassiliev, and V. S. Bagotsky, *Elektrokhimiya* **1** (1965) 82.
[105] A. Papoutsis, J. M. Léger, and C. Lamy, *J. Electroanal. Chem.* **234** (1987) 315.
[106] D. D. McDonald and M. C. H. McKubre, in *Modern Aspects of Electrochemistry*, No. 14, Ed. by J. O'M. Bockris, B. E. Conway, and R. E. White, Plenum Press, New York, 1982, p. 61.
[107] M. Sluyters-Rehbach and J. H. Sluyters, in *Comprehensive Treatise of Electrochemistry*, Vol. 9, Ed. by E. Yeager, J. O'M. Bockris, B. E. Conway, and S. Sarangapani, Plenum Press, New York, 1984, p. 177.
[108] A. N. Frumkin, *Z. Physik.* **35** (1926) 792.
[109] P. N. Ross and F. T. Wagner, *Adv. Electrochem. Electrochem. Eng.*, Vol. 13, Ed. by H. Gerischer, John Wiley & Sons, New York, 1984, p. 69.
[110] E. Yeager, W. E. O'Grady, M. Y. C. Woo, and P. Hagans, *J. Electrochem. Soc.* **125** (1978) 348.
[111] D. Aberdam, R. Durand, R. Faure, and F. El Omar, *Surf. Sci.* **171** (1986) 303.
[112] S. A. Francis and A. H. Ellison, *J. Opt. Soc. Am.* **49** (1959) 131.

[113] R. G. Greenler, *J. Chem. Phys.* **44** (1966) 310; *Surf. Sci.* **59** (1976) 205.
[114] R. G. Greenler, R. R. Rahn, and J. P. Schwartz, *J. Catal.* **23** (1971) 42.
[115] N. J. Harrick, *Internal Reflection Spectroscopy*, Interscience, New York, 1967.
[116] W. N. Hansen, *Symp. Faraday Soc.* **4** (1970) 27.
[117] H. Neugebauer, G. Nauer, N. Brinda-Konopik, and G. Gidali, *J. Electroanal. Chem.* **122** (1981) 381.
[118] H. Neff, P. Lange, D. K. Koe, and J. K. Sass, *J. Electroanal. Chem.* **150** (1983) 513.
[119] M. C. Pham, F. Adami, P. C. Lacaze, J. P. Doucet, and J. E. Dubois, *J. Electroanal. Chem.* **201** (1986) 413.
[120] F. Ozanam and J. N. Chazalviel, *J. Electron. Spectrosc. Relat. Phenom.* **45** (1987) 323.
[121] A. Bewick and A. M. Tuxford, *Symp. Faraday Soc.* **4** (1970) 114; *J. Electroanal. Chem.* **47** (1973) 255.
[122] R. Kötz and D. M. Kolb, *Surf. Sci.* **12** (1982) 287.
[123] A. Bewick, K. Kunimatsu, B. S. Pons, and J. W. Russell, *J. Electroanal. Chem.* **160** (1984) 47.
[124] S. Pons, T. Davidson, and A. Bewick, *J. Electroanal. Chem.* **160** (1984) 63.
[125] H. Seki, K. Kunimatsu, and W. G. Golden, *Appl. Spectrosc.* **39** (1985) 437.
[126] D. K. Roe, J. K. Sass, D. S. Bethune, and A. C. Luntz, *J. Electroanal. Chem.* **216** (1987) 293.
[127] M. A. Habib and J. O'M. Bockris, *J. Electrochem. Soc.*, **132** (1985) 108.
[128] M. Fleischmann, A. Oliver, and J. Robinson, *Electrochim. Acta* **31** (1986) 899.
[129] R. W. Stobie, B. Rao, and M. J. Dignam, *Surf. Sci.* **56** (1976) 334.
[130] R. G. Greenler and T. L. Slager, *Spectrochim. Acta* **29A** (1973) 193.
[131] J. D. E. McIntyre and D. E. Aspnes, *Surf. Sci.* **24** (1971) 417.
[132] D. M. Kolb, in *Spectroelectrochemistry, Theory and Practice*, Ed. by R. J. Gale, Plenum Press, New York, 1988, p. 87.
[133] L. H. Little, *Infrared Spectra of Adsorbed Species*, Academic Press, New York, 1966.
[134] R. F. Willis, *Vibrational Spectroscopy of Adsorbates*, Springer-Verlag, Berlin, 1980.
[135] A. T. Bell and M. L. Hair, *Vibrational Spectroscopies for Adsorbed Species*, American Chemical Society, Washington, D.C., 1980.
[136] J. Pritchard, in *Chemical Physics of Solids and Their Surface*, Vol. 7, Specialist Periodical Reports, The Chemical Society, London, 1978, p. 157.
[137] J. T. Yates, Jr., and T. E. Madey, *Vibrational Spectroscopy of Molecules on Surfaces*, Plenum Press, New York, 1987.
[138] B. A. Sexton, *Surf. Sci.* **88** (1979) 319; *Appl. Phys.* **A26** (1981) 1.
[139] B. E. Hayden, K. Prince, D. Woodruff, and A. M. Bradshaw, *Surf. Sci.* **133** (1983) 589.
[140] M. I. Lopes, B. Beden, F. Hahn, J. M. Léger, and C. Lamy, *J. Electroanal. Chem.* **258** (1989) 463; *J. Electroanal. Chem.* **313** (1991) 323.
[141] K. Nakamoto, *Infrared Spectra of Inorganic and Coordination Compounds*, 2nd ed., Wiley Interscience, New York, 1970.
[142] T. Takamura, K. Katamura, W. Nippe, and E. Yeager, *J. Electrochem. Soc.* **117** (1970) 626.
[143] D. M. Kolb and J. D. E. McIntyre, *Surf. Sci.* **28** (1971) 361.
[144] T. E. Furtak and D. W. Lynch, *Phys. Rev. Lett.* **35** (1975) 960.
[145] N. Hara and K. Sugimoto, *Trans. Jpn. Inst. Met.* **24** (1983) 236.
[146] C. Gutiérrez and A. M. Martinez, *J. Electrochem. Soc.* **133** (1986) 1873.
[147] F. Hahn, B. Beden, M. J. Croissant, and C. Lamy, *Electrochim. Acta* **31** (1986) 335.
[148] N. Collas, B. Beden, J. M. Léger, and C. Lamy, *J. Electroanal. Chem.* **186** (1985) 287.

[149] D. M. Kolb, in *Advances in Electrochemistry and Electrochemical Engineering*, Vol. 11, Ed. by H. Gerischer and C. W. Tobias, John Wiley & Sons, New York, 1978, p. 125.
[150] J. D. E. McIntyre, in *Advances in Electrochemistry and Electrochemical Engineering*, Vol. 9, Ed. by R. H. Müller, John Wiley & Sons, New York, 1973, p. 61.
[151] D. M. Kolb, G. Lehmpfuhl, and M. S. Zei, in *Spectroscopic and Diffraction Techniques in Interfacial Electrochemistry*, Ed. by C. Gutiérrez and C. Melendres, NATO ASI Series, Series C, Vol. 320 Kluwer Acad. Publ., Dordrecht, 1990, p. 361.
[152] W. Plieth, in *Spectroscopic and Diffraction Techniques in Interfacial Electrochemistry*, Ed. by C. Gutiérrez and C. Melendres, NATO ASI Series, Series C, Vol. 320, Kluwer Acad. Publ., Dordrecht, 1990, p. 223.
[153] C. Gutiérrez, in *Spectroscopic and Diffraction Techniques in Interfacial Electrochemistry*, Ed. by C. Gutiérrez and C. Melendres, NATO ASI Series, Series C, Vol. 320, Kluwer Acad. Publ., Dordrecht, 1990, p. 261.
[154] D. M. Kolb and R. Kötz, *Surf. Sci.* **64** (1977) 698.
[155] J. D. E. McIntyre, in *Optical Properties of Solids—New Developments*, Ed. by B. O. Seraphin, North-Holland, Amsterdam, 1976, p. 555.
[156] F. Hahn, D. Floner, B. Beden, and C. Lamy, *Electrochim. Acta* **32** (1987) 1631.
[157] A. Bewick, A. C. Lowe, and C. W. Wederell, *Electrochim. Acta* **28** (1983) 1899.
[158] B. Beden, O. Enea, F. Hahn, and C. Lamy, *J. Electroanal. Chem.* **170** (1984) 357.
[159] C. Gutiérrez and B. Beden, *J. Electroanal. Chem.* **293** (1990) 253.
[160] M. Hachkar, M. Choy de Martinez, A. Rakotondrainibe, B. Beden, and C. Lamy, *J. Electroanal. Chem.* **302** (1991) 173.
[161] A. Bewick, K. Kunimatsu, and B. S. Pons, *Electrochim. Acta* **25** (1980) 465.
[162] A. Bewick and K. Kunimatsu, *Surf. Sci.* **101** (1981) 131.
[163] B. Beden, A. Bewick, K. Kunimatsu, and C. Lamy, 32nd ISE Meeting, Dubrovnik-Cavtat, 1981, Extended Abstract A 28, Vol. I, p. 92.
[164] B. Beden, C. Lamy, A. Bewick, and K. Kunimatsu, *J. Electroanal. Chem.* **121** (1981) 343.
[165] P. A. Christensen and A. Hamnett, in *Comprehensive Chemical Kinetics*, Vol. 29, Ed. by R. G. Compton and A. Hamnett, Elsevier, Amsterdam, 1989, p. 1.
[166] J. W. Russell, J. Overend, K. Scanlon, M. W. Seversen, and A. Bewick, *J. Phys. Chem.* **86** (1982) 205.
[167] Y. Ikezawa, H. Sato, H. Matsubayashi, and G. Toda, *J. Electroanal. Chem.* **252** (1988) 395.
[168] K. Kunimatsu, *J. Electroanal. Chem.* **140** (1982) 205.
[169] W. G. Golden, K. Kunimatsu, and H. Seki, *J. Phys. Chem.* **88** (1984) 1275.
[170] D. S. Corrigan, L. W. H. Leung, and M. J. Weaver, *Anal. Chem.* **59** (1987) 2252.
[171] D. S. Corrigan and M. J. Weaver, *J. Phys. Chem.* **90** (1986) 5300.
[172] J. L. Daschbach, D. Heisler, and S. Pons, *Appl. Spectrosc.* **40** (1986) 489.
[173] H. B. Mark, Jr. and S. Pons, *Anal. Chem.* **38** (1966) 119.
[174] D. K. Lambert, *Solid State Commun.* **51** (1984) 197.
[175] C. Korzeniewski, S. Pons, P. P. Schmidt, and M. Seversen, *J. Phys. Chem.* **85** (1986) 4153.
[176] K. Kunimatsu, W. G. Golden, H. Seki, and M. R. Philpott, *Langmuir* **1** (1985) 245.
[177] M. Choy de Martinez, B. Beden, F. Hahn, and C. Lamy, *J. Electroanal. Chem.* **249** (1988) 265.
[178] Y. Ikezawa, H. Sato, M. Yamazaki, and G. Toda, *J. Electroanal. Chem.* **245** (1988) 245.
[179] S. Pons, *J. Electroanal. Chem.* **160** (1984) 369; K. Ashley and S. Pons, *Chem. Rev.* **88** (1988) 673.

[180] B. R. Scharifker, K. Chandrasekaran, M. E. Gamboa-Aldeco, P. Zelenay, and J. O'M. Bockris, *Electrochim. Acta* **33** (1988) 159.
[181] J. K. Foley and S. Pons, *Anal. Chem.* **57** (1985) 945A.
[182] H. Neugebauer, G. Nauer, A. Neckel, G. Tourillon, F. Garnier, and P. Lang, *J. Phys. Chem.* **88** (1984) 653.
[183] H. Neugebauer, A. Neckel, and N. Brinda-Konopik, *Springer Ser. Solid-State Sci.* **63** (1985) 227.
[184] L. W. H. Leung and M. J. Weaver, *J. Electroanal. Chem.* **240** (1988) 341.
[185] J. Li, J. L. Daschbach, J. J. Smith, M. D. Morse, and S. Pons, *J. Electroanal. Chem.* **209** (1986) 387.
[186] M. Fleischmann, P. J. Hendra, and A. J. McQuillan, *Chem. Phys. Lett.* **26** (1974) 163.
[187] M. Fleischmann and I. R. Hill, in *Comprehensive Treatise of Electrochemistry*, Vol. 8, Ed. by R. E. White, J. O'M. Bockris, B. E. Conway and E. Yeager, Plenum Press, New York, 1984, p. 373.
[188] R. K. Chang and T. E. Furtak, eds., *Surface Enhanced Raman Scattering*, Plenum Press, New York, 1982.
[189] A. Otto, *J. Electron Spectrosc. Relat. Phenom.* **29** (1983) 329.
[190] A. Smekal, *Naturwissenschaften* **11** (1923) 873.
[191] C. V. Raman, *Nature (London)* **121** (1928) 619; C. V. Raman and K. S. Krishnan, *Nature (London)* **121** (1928) 501.
[192] A. Bewick, M. Fleischmann, and J. Robinson, *DECHEMA-Monogr.* **90** (1981) 87.
[193] M. A. Tadayyoni, S. Farquharson, T. T. T. Li, and M. J. Weaver, *J. Phys. Chem.* **88** (1984) 4701.
[194] A. Otto, in *Light Scattering in Solids IV, Topics in Applied Physics*, Vol. 54, Ed. by M. Cardona and G. Güntherodt, Springer-Verlag, Berlin, 1984, p. 289.
[195] C. Pettenkoffer, J. Mrozek, T. Bornemann, and A. Otto, *Surf. Sci.* **188** (1987) 519.
[196] R. L. Birke and J. R. Lombardi, in *Spectroelectrochemistry, Theory and Practice*, Ed. by R. J. Gale, Plenum Press, New York, 1988, p. 263.
[197] S. Efrima, in *Modern Aspects of Electrochemistry*, No. 16, Ed. by B. E. Conway, R. E. White, and J. O'M. Bockris, Plenum Press, New York, 1985, p. 253.
[198] R. K. Chang, in *Spectroscopic and Diffraction Techniques in Interfacial Electrochemistry*, Ed. by C. Gutiérrez and C. Melendres, NATO ASI Series, Series C, Vol. 320, Kluwer Acad. Publ., Dordrecht, 1990, p. 155.
[199] D. S. Corrigan, P. Gao, L. W. H. Leung, and M. J. Weaver, *Langmuir* **2** (1986) 744.
[200] P. Gao and M. J. Weaver, *J. Phys. Chem.* **89** (1985) 5040; **93** (1989) 6205.
[201] D. S. Corrigan, J. K. Foley, P. Gao, S. Pons, and M. J. Weaver, *Langmuir* **1** (1985) 616.
[202] C. A. Melendres, J. J. McMahon, and W. Ruther, *J. Electroanal. Chem.* **208** (1986) 175; C. A. Melendres, N. Camillone III, and T. Tipton, *Electrochim. Acta* **34** (1989) 281.
[203] J. Dünnwald and A. Otto, *Fresenius' Z. Anal. Chem.* **319** (1984) 738.
[204] J. C. Rubim, *J. Chem. Soc., Faraday Trans. 1* **85** (1989) 4247.
[205] J. Bukowska and K. Jackowska, *Electrochim. Acta* **35** (1990) 315.
[206] P. J. Hendra and H. Mould, *Int. Lab.* **18**(7) (1988) 34.
[207] B. Chase, *Anal. Chem.* **59** (1987) 881A.
[208] T. Hirschfeld and B. Chase, *Appl. Spectrosc.* **40** (1986) 133.
[209] R. Greef, in *Comprehensive Treatise of Electrochemistry*, Vol. 8, Ed. by R. E. White, J. O'M. Bockris, B. E. Conway, and E. Yeager, Plenum Press, New York, 1984, p. 339.

[210] S. Gottesfeld, in *Electroanalytical Chemistry*, Vol. 15, Ed. by A. J. Bard, Marcel Dekker, New York, 1989, p. 144.
[211] A. K. N. Reddy and J. O'M. Bockris, Proceedings of the Symposium on the Ellipsometer and Its Use in the Measurements of Surfaces and Thin Films, Bureau of Standards, Washington D.C., 1964, p. 229.
[212] Z. Q. Huang and J. L. Ord, *J. Electrochem. Soc.* **132** (1985) 24.
[213] S. Gottesfeld, A. Redondo, and S. W. Feldberg, *J. Electrochem. Soc.* **134** (1987) 271.
[214] R. H. Muller and J. C. Farmer, *Surf. Sci.* **135** (1983) 521.
[215] P. J. Hyde, C. J. Maggiore, A. Redondo, S. Srinivasan, and S. Gottesfeld, *J. Electroanal. Chem.* **186** (1985) 267.
[216] M. J. Dignam and M. D. Baker, *J. Vac. Sci. Technol.* **21** (1982) 80.
[217] R. T. Graf, J. L. Koenig, and H. Ishida, *Anal. Chem.* **58** (1986) 64.
[218] Y. R. Shen, *The principles of Non-Linear Optics*, Wiley, New York, 1984, p. 479.
[219] Y. R. Shen, in *Spectroscopic and Diffraction Techniques in Interfacial Electrochemistry*, Ed. by C. Gutiérrez and C. Melendres, NATO ASI Series, Series C, Vol. 320, Kluwer Acad. Publ., Dordrecht, 1990, p. 281.
[220] P. Guyot-Sionnest and A. Tadjeddine, *Chem. Phys. Lett.* **172** (1990) 341.
[221] M. Fleischmann, P. J. Hendra, and J. M. Robinson, *Nature* (*London*) **288** (1980) 152.
[222] M. Fleischmann and B. W. Mao, *J. Electroanal. Chem.* **247** (1988) 311.
[223] M. Fleischmann and B. W. Mao, *J. Electroanal. Chem.* **229** (1987) 125.
[224] R. Cortes, M. Froment, A. Hugot-Le-Goff, and S. Joiret, Corros. Sci. **31** (1990) 121.
[225] J. McBreen, W. E. O'Grady, G. Tourillon, E. Dartyge, A. Fontaine, and K. I. Pandya, *J. Phys. Chem.* **93** (1989) 6308.
[226] J. H. White and H. D. Abruna, *J. Electroanal. Chem.* **274** (1989) 185.
[227] G. Tourillon, D. Guay, and A. Tadjeddine, *J. Electroanal. Chem.* **289** (1990) 263.
[228] W. C. Marra, P. Eisenberger, and A. Y. Cho, *J. Appl. Phys.* **50** (1979) 6927.
[229] P. L. Cowan, J. A. Golovchenko, and M. F. Robbins, *Phys. Rev. Lett.* **44** (1980) 1680.
[230] P. A. Lee and J. B. Pentry, *Phys. Rev.* **B11** (1975) 2795.
[231] L. Bosio, R. Cortes, and M. Froment, 3rd International EXAFS Conference, Stanford, California, 1984.
[232] E. A. Stern, *J. Vac. Sci. Technol.* **14** (1977) 461.
[233] G. G. Lang, J. Krüger, D. R. Black, and M. Kuriyama, *J. Electrochem. Soc.* **130** (1983) 240.
[234] K. I. Pandya, R. W. Hoffman, J. McBreen, and W. E. O'Grady, *J. Electrochem. Soc.* **137** (1990) 383.
[235] L. Crambow and S. Bruckenstein, *Electrochim. Acta* **22** (1977) 377.
[236] O. Wolter and J. Heitbaum, *Ber. Bunsenges. Phys. Chem.* **88** (1984) 2.
[237] S. Wilhelm, W. Vielstich, H. W. Bushmann, and T. Iwasita, *J. Electroanal. Chem.* **229** (1987) 377.
[238] S. Trasatti, *Mater. Chem. Phys.* **24** (1990) 327.
[239] T. Iwasita-Vielstich, in *Advances in Electrochemical Science and Engineering*, Vol. 1, Ed. by H. Gerischer and C. W. Tobias, VCH Verlagsgesell, Weinheim, 1990, p. 127.
[240] J. Willsau and J. Heitbaum, *J. Electroanal. Chem.* **194** (1985) 27
[241] V. Solis, T. Iwasita, A. Pavese, and W. Vielstich, *J. Electroanal. Chem.* **255** (1988) 155.
[242] N. A. Anastasijevic, H. Baltruschat, and J. Heitbaum, *J. Electroanal. Chem.* **272** (1989) 89.

[243] T. Hartung and J. Heitbaum, 170th Fall Meeting of the Electrochemical Society, San Diego, California, Extended Abstract 86-2, 1986, p. 235.
[244] W. Vielstich, P. A. Christensen, S. A. Weeks, and A. Hamnett, *J. Electroanal. Chem.* **242** (1988) 327.
[245] A. E. Bolzan and T. Iwasita, *Electrochim. Acta* **33** (1988) 109.
[246] G. Tourillon, P. C. Lacaze, and J. E. Dubois, *J. Electroanal. Chem.* **100** (1979) 247.
[247] E. Blomgren and J. O'M. Bockris, *Nature* (*London*) **186** (1960) 305.
[248] A. Wieckowski, in *Modern Aspects of Electrochemistry*, No. 21, Ed. by R. E. White, J. O'M. Bockris, and B. E. Conway, Plenum Press, New York, 1990, p. 65.
[249] V. E. Kazarinov, G. Ya. Tysyachnaya, and V. N. Andreev, *J. Electroanal. Chem.* **65** (1975) 391.
[250] J. Sobkowski and A. Wieckowski, *J. Electroanal. Chem.* **34** (1972) 185.
[251] A. Wieckowski, *J. Electrochem. Soc.* **122** (1975) 252; *J. Electroanal. Chem.* **78** (1977) 229.
[252] E. K. Krauskopf, K. Chan, and A. Wieckowski, *J. Phys. Chem.* **91** (1987) 2327.
[253] P. Zelenay, M. A. Habib, and J. O'M. Bockris, *Langmuir* **2** (1986) 393.
[254] D. S. Corrigan, E. K. Krauskopf, L. M. Rice, A. Wieckowski, and M. J. Weaver, *J. Phys. Chem.* **92** (1988) 1596.
[255] O. Wolter, J. Willsau, and J. Heitbaum, *J. Electrochem. Soc.* **132** (1985) 1635.
[256] A. Bewick, C. Gibilaro, M. Razaq, and J. W. Russell, *J. Electron Spectrosc. Relat. Phenom.* **30** (1983) 191.
[257] A. Bewick, M. Razaq, and J. W. Russell, *J. Electroanal. Chem.* **256** (1988) 165.
[258] L. W. H. Leung and M. J. Weaver, *Langmuir* **6** (1990) 323.
[259] C. Lu and A. W. Czanderna, *Applications of Piezoelectric Quartz Crystal Microbalances, Methods and Phenomena*, Vol. 7, Elsevier, Amsterdam, 1984.
[260] M. R. Deakin and D. A. Buttry, *Anal. Chem.* **61** (1989) 1147A.
[261] S. Bruckenstein and M. Shay, *J. Electroanal. Chem.* **188** (1985) 131.
[262] J. H. Kaufman, K. K. Kanazawa, and G. B. Street, *Phys. Rev. Lett.* **53** (1984) 2461.
[263] G. Sauerbrey, *Z. Phys.* **155** (1955) 206.
[264] M. Hepel and S. Bruckenstein, *Electrochim. Acta* **34** (1989) 1499.
[265] S. Bourkane, C. Gabrielli, and M. Keddam, *Electrochim. Acta* **34** (1989) 1081.
[266] G. T. Cheek and W. E. O'Grady, *J. Electroanal. Chem.* **277** (1990) 171.
[267] G. Inzelt, *J. Electroanal. Chem.* **287** (1990) 171.
[268] P. T. Varinears and D. A. Buttry, *J. Phys. Chem.* **91** (1987) 1292.
[269] R. Schumacher, J. G. Gordon, and O. Melroy, *J. Electroanal. Chem.* **216** (1987) 127.
[270] D. J. Kampe, in *Comprehensive Treatise of Electrochemistry*, Vol. 8, Ed. by R. E. White, J. O'M. Bockris, B. E. Conway, and E. Yeager, Plenum Press, New York, 1984, p. 475.
[271] G. Binnig and H. Röhrer, *Helv. Phys. Acta* **55** (1982) 726; *Surf. Sci.* **126** (1983) 235.
[272] T. R. I. Cataldi, I. G. Blackham, G. A. D. Briggs, J. B. Pethica, and H. A. O. Hill, *J. Electroanal. Chem.* **290** (1990) 1.
[273] L. Vázquez, J. Gómez, A. M. Baró, N. Garcia, M. L. Marcos, J. González-Velasco, J. M. Vara, A. J. Arvia, J. Presa, A. Garcia, and M. Aguilar, *J. Am. Chem. Soc.* **109** (1987) 1730.
[274] O. E. Husser, D. H. Craston, and A. J. Bard, *J. Electrochem. Soc.* **136** (1989) 3222.
[275] J. Kwak and A. J. Bard, *Anal. Chem.* **61** (1989) 1794.
[276] J. Kwak, C. Lee, and A. J. Bard, *J. Electrochem. Soc.* **137** (1990) 1481.
[277] P. K. Hansma, B. Drake, O. Marti, S. A. C. Gould, and C. B. Prater, *Science* **243** (1989) 641.
[278] G. Binnig, C. F. Quate, and Ch. Gerber, *Phys. Rev. Lett.* **56** (1986) 930.

[279] P. Lustenberger, H. Röhrer, R. Christoph, and H. Siegenthaler, *J. Electroanal. Chem.* **243** (1988) 225.
[280] M. M. Dovek, M. J. Heben, N. S. Lewis, R. M. Penner, and C. F. Quate, *Electrochemical Surface Science: Molecular Phenomena at Electrode Surfaces*, ACS Symp. Ser., No. 378, American Chemical Society, Washington, D.C., 1988, p. 174.
[281] A. J. Arvia, *Surf. Sci.* **181** (1987) 78.
[282] S. Chiang, R. J. Wilson, C. F. Mate, and H. Othani, *J. Microsc.* **152** (1988) 567.
[283] C. E. Vallet, in *Spectroscopic and Diffraction Techniques in Interfacial Electrochemistry*, NATO ASI Series, Series C, Vol. 320, Ed. by C. Gutiérrez and C. Melendres, Kluwer Acad. Publ., Dordrecht, 1990, p. 133.
[284] R. E. Malpas and A. J. Bard, *Anal. Chem.* **52** (1980) 109.
[285] R. E. Russo, F. R. McLarnon, J. D. Spear, and E. J. Cairns, *J. Electrochem. Soc.* **134** (1987) 2783.
[286] M. J. Weaver, F. R. McLarnon, and E. J. Cairns, 40th ISE Meeting, Kyoto, Japan, September 1989, Extended Abstracts, Vol. I, 19-05, p. 472.
[287] C. E. Vallet, D. E. Heatherly, and P. W. White, *J. Electrochem. Soc.* **137** (1990) 579.
[288] R. L. Mössbauer, *Z. Phys.* **124** (1958) 124.
[289] D. A. Scherson, in *Spectroelectrochemistry—Theory and Practice*, Ed. by R. J. Gale, Plenum Press, New York, 1988, p. 399.
[290] F. G. Will, *J. Electrochem. Soc.* **112** (1965) 451.
[291] J. M. Léger, B. Beden, C. Lamy, and S. Bilmes, *J. Electroanal. Chem.* **170** (1984) 305.
[292] W. E. Triaca, T. Kessler, J. C. Canullo, and A. J. Arvia, *J. Electrochem. Soc.* **134** (1987) 1165.
[293] R. M. Cerviño, W. E. Triaca, and A. J. Arvia, *Electrochim. Acta* **30** (1985) 1323.
[294] E. M. Leiva, E. Santos, M. C. Giordano, R. M. Cerviño, and A. J. Arvia, *J. Electrochem. Soc.* **133** (1986) 1660.
[295] B. Beden, F. Hahn, C. Lamy, J.-M. Léger, N. R. de Tacconi, R. O. Lezna, and A. J. Arvia, *J. Electroanal. Chem.* **261** (1989) 401.
[296] B. Beden, S. Bilmes, C. Lamy, and J.-M. Léger, *J. Electroanal. Chem.* **149** (1983) 395.
[297] J. Clavilier, R. Parsons, R. Durand, C. Lamy, and J.-M. Léger, *J. Electroanal. Chem.* **124** (1981) 321; 32nd ISE Meeting, Dubrovnik-Cautat, 1981, Extended Abstract III-06, Vol. II, p. 624.
[298] J. Clavilier, C. Lamy, and J.-M. Léger, *J. Electroanal. Chem.* **125** (1981) 249.
[299] M.-C. Morin, C. Lamy, J.-M. Léger, J.-L. Vasquez, and A. Aldaz, *J. Electroanal. Chem.* **283** (1990) 287.
[300] D. Takky, B. Beden, J.-M. Léger, and C. Lamy, *J. Electroanal. Chem.* **256** (1988) 127.
[301] G. Kokkinidis, J.-M. Léger, and C. Lamy, *J. Electroanal. Chem.* **242** (1988) 221.
[302] J.-M. Léger, I. Fonseca, F. Bento, and I. Lopes, *J. Electroanal. Chem.* **285** (1990) 125.
[303] R. R. Adžić, *Isr. J. Chem.* **18** (1979) 166.
[304] G. Kokkinidis, *J. Electroanal. Chem.* **201** (1986) 217.
[305] F. Kadirgan, B. Beden, and C. Lamy, *J. Electroanal. Chem.* **143** (1983) 135.
[306] F. Kadirgan, B. Beden, and C. Lamy, *J. Electroanal. Chem.* **136** (1982) 119.
[307] P. N. Ross, Oxygen Reduction on Supported Pt Alloys and Intermetallic Compounds in Phosphoric Acid, EPRI Report EM-1553, September, 1980.
[308] P. Stonehart, in *Proceedings Electrocatalysis on Non-Metallic Surfaces*, National Bureau of Standards, Special Publication No. 445, Washington, D.C., 1976, p. 167.
[309] A. J. Appleby and R. Foulkes, *Fuel Cell Handbook*, Van Nostrand Reinhold, New York, 1989.

[310] M. Watanabe and S. Motoo, *J. Electroanal. Chem.* **60** (1975) 267.
[311] B. D. McNicol, in *Power Sources for Electric Vehicles*, Vol. 11, Ed. by B. D. McNicol and D. A. J. Rand, Elsevier, Amsterdam, 1984, p. 807.
[312] N. R. de Tacconi, J.-M. Léger, B. Beden, and C. Lamy, *J. Electroanal. Chem.* **134** (1982) 117.
[313] B. E. Conway, H. Angerstein-Kozlowska, and G. Czartoryska, *Z. Phys. Chem.* **112** (1978) 195.
[314] A. Capon and R. Parsons, *J. Electroanal. Chem.* **65** (1975) 285.
[315] B. Beden, C. Lamy, and J. M. Léger, *Electrochim. Acta* **24** (1979) 1157.
[316] D. F. A. Koch, D. A J. Rand, and R. Woods, *J. Electroanal. Chem.* **70** (1976) 73.
[317] W. Hauffe, J. Heitbaum, and W. Vielstich, in *Electrode Materials and Processes for Energy Conversion and Storage*, Ed. by J. D. E. McIntyre, S. Srinivasan, and F. G. Will, The Electrochemical Society, Pennington, New Jersey, PV 77-6, 1977, p. 308.
[318] H. Cnobloch, D. Gröppel, H. Kohlmüller, D. Kühl, and G. Siemsen, *Power Sources Symp. Proc.* **24**(7) 1979, p. 389.
[319] B. Beden, F. Kadirgan, A. Kahyaoglu, and C. Lamy, *J. Electroanal. Chem.* **135** (1982) 329.
[320] M. M. P. Janssen and J. Moolhuysen, *Electrochim. Acta* **21** (1976) 869.
[321] B. Beden, F. Kadirgan, C. Lamy, and J. M. Léger, *J. Electroanal. Chem.* **127** (1981) 75.
[322] D. A. J. Rand and R. Woods, *J. Electroanal. Chem.* **36** (1972) 57.
[323] M. R. Andrew and R. W. Glazebrook, in *An Introduction to Fuel Cells*, Ed. by K. R. Williams, Elsevier, Amsterdam, 1966, p. 109.
[324] L. Mehalaine, J. M. Léger, and C. Lamy, *Electrochim. Acta* **36** (1991) 519.
[325] B. Beden, F. Hahn, J. M. Léger, C. Lamy, C. L. Perdriel, N. R. de Tacconi, R. O. Lezna, and A. J. Arvia, *J. Electroanal. Chem.* **301** (1991) 129.
[326] K. Kinoshita and P. Stonehart, in *Modern Aspects of Electrochemistry*, No. 12, Ed. by J. O'M. Bockris and B. E. Conway, Plenum Press, New York, 1977, p. 183.
[327] M. Watanabe and P. Stonehart, 40th ISE Meeting, Kyoto, Japan, Sept. 1989, Vol. II, Extended Abstract No. 19-14-6-P, p. 1250.
[328] M. Watanabe, S. Saegusa, and P. Stonehart, *J. Electroanal. Chem.* 271 (1989) 213.
[329] S. Srinivasan, D. J. Manko, M. Koch, M. A. Enayetullah, and A. J. Appleby, *J. Power Sources* **29** (1990) 367.
[330] K. Prater, *J. Power Sources* **29** (1990) 239.
[331] [illegible] Adelbert, F. Novel-Cattin, M. Pineri, P. Millet, C. Doumain, and R. Durand, *Solid State Ionics* **35** (1989) 3.
[332] A. Aramata and R. Ohnishi, *J. Electroanal. Chem.* **162** (1984) 153.
[333] H. Nakajima and H. Kita, *Electrochim. Acta* **33** (1988) 521.
[334] A. Aramata, T. Kodera, and M. Masuda, *J. Appl. Electrochem.* **18** (1988) 577.
[335] J. Wang, H. Nakajima, and H. Kita, *J. Electroanal. Chem.* **250** (1988) 213.
[336] J. Wang, H. Nakajima, and H. Kita, *Electrochim. Acta* **35** (1990) 323.
[337] H. Nakajima and H. Kita, *Electrochim. Acta* **35** (1990) 849.
[338] J. C. Chiang and A. G. McDiarmid, *Synth. Met.* **13** (1986) 193.
[339] E. Genies and C. Tsintavis, *J. Electroanal. Chem.* **195** (1985) 109.
[340] A. F. Diaz and K. K. Kanazawa, in *Extended Linear Chain Compounds*, Vol. 3, Ed. by G. S. Miller, Plenum Press, New York, 1983, p. 417.
[341] G. Tourillon and F. Garnier, *J. Electroanal. Chem.* **135** (1982) 173.
[342] G. Tourillon and F. Garnier, *J. Phys. Chem.* **88** (1984) 5281.
[343] W. H. Kao and T. Kuwana, *J. Am. Chem. Soc.* **106** (1984) 473.
[344] K. M. Kost, D. E. Bartak, B. Kazee, and T. Kuwana, *Anal. Chem.* **62** (1990) 151.
[345] S. Holdcroft and B. L. Funt, *J. Electroanal. Chem.* **240** (1988) 89.

[346] F. T. A. Vork and E. Barendrecht, *Electrochim. Acta* **35** (1990) 135.
[347] M. Gholamian, J. Sundaram, and A. Q. Contractor, *Langmuir* **3** (1987) 741.
[348] P. Ocon-Esteban, J. M. Léger, C. Lamy, and E. Genies, *J. Appl. Electrochem.* **19** (1989) 462.
[349] H. Laborde, J. M. Léger, C. Lamy, F. Garnier, and A. Yassar, *J. Appl. Electrochem.* **20** (1990) 524.
[350] F. Delannay, Ed., *Characterization of Heterogeneous Catalysts, Chemical Industries*, Vol. 15, Marcel Dekker, New York, 1984.
[351] D. M. Kolb, D. L. Rath, R. Wille, and W. N. Hansen, *Ber. Bunsenges. Phys. Chem.* **87** (1983) 1108.
[352] W. B. Innes, in *Experimental Methods in Catalytic Research*, Ed. by R. B. Anderson, Academic Press, New York, 1968, p. 44.
[353] J. L. Lemaitre, P. Govind Menon, and F. Delannay, in *Characterization of Heterogeneous Catalysts*, Ed. by F. Delannay, *Chemical Industries*, Vol. 15, Marcel Dekker, New York, 1984, p. 299.
[354] J. M. Thomas and W. J. Thomas, *Introduction to the Principles of Heterogeneous Catalysis*, Academic Press, New York, 1967, p. 180.
[355] G. Beurton and P. Bussière, *J. Pure Appl. Chem.* (Supplement), *Surface Area Determination*, Butterworths, London, 1970, p. 217.
[356] M. A. Vannice, J. E. Benson, and M. Boudart, *J. Catal.* **16** (1970) 348.
[357] R. Woods, in *Electroanalytical Chemistry*, Vol. 9, Ed. by A. J. Bard, Marcel Dekker, New York, 1976, p. 1.
[358] C. L. Scortichini and C. N. Reilley, *J. Catal.* **79** (1983) 138.
[359] M. A. Quiroz, Y. Méas, E. Lamy-Pitara, and J. Barbier, *J. Electroanal. Chem.* **157** (1983) 165.
[360] M. A. Quiroz, I. Gonzalez, H. Vargas, Y. Méas, E. Lamy-Pitara, and J. Barbier, *Electrochim. Acta* **31** (1986) 503.
[361] E. W. Lard and S. M. Brown, *J. Catal.* **25** (1972) 451.
[362] M. Boudart, *Adv. Catal.* **20** (1969) 153.
[363] K. Kinoshita, in *Modern Aspects of Electrochemistry*, No. 14, Ed. by J. O'M. Bockris, B. E. Conway, and R. E. White, Plenum Press, New York, 1982, p. 557.
[364] H. Klug and L. Alexander, *X-Ray Diffraction Procedures*, Wiley, New York, 1962, p. 491.
[365] A. Guinier, in *X-Ray Crystallography Technology*, Ed. by K. Lonsdale, Hilger & Watts, London, 1952, p. 268.
[366] G. R. Wilson and W. K. Hall, *J. Catal.* **24** (1972) 306.
[367] P. C. Flynn, S. E. Wanke, and P. S. Turner, *J. Catal.* **33** (1974) 233.
[368] M. A. Van Hove, W. H. Weinberg, and C. Chan, *Low Energy Electron Diffraction*, Springer, Berlin, 1986.
[369] D. M. Kolb, G. Lehmpfuhl, and M. S. Zei, in *Spectroscopic and Diffraction Techniques in Interfacial Electrochemistry*, Ed. by C. Gutiérrez and C. Melendres, NATO ASI Series, Series C, Vol. 320, Kluwer Academic Publ., Dordrecht, 1990, p. 309.
[370] F. T. Wagner and P. N. Ross, *J. Electroanal. Chem.* **150** (1983) 141.
[371] R. Kötz, in *Spectroscopic and Diffraction Techniques in Interfacial Electrochemistdry*, Ed. by C. Gutiérrez and C. Melendres, NATO ASI Series, Series C, Vol. 320, Kluwer Academic Publ., Dordrecht, 1990, p. 439.
[372] P. J. Hyde, C. J. Maggiore, and S. Srinivasan, *J. Electroanal. Chem.* **168** (1984) 383.
[373] R. Kötz, J. Gobrecht, S. Stucki, and R. Pixley, *Electrochim. Acta* **31** (1986) 169.
[374] J. O'M. Bockris and A. González-Martin, in *Spectroscopic and Diffraction Techniques in Interfacial Electrochemistry*, Ed. by C. Gutiérrez and C. Melendres, NATO ASI Series, Series C, Vol. 320, Kluwer Acad. Publ., Dordrecht, 1990, p. 1.

[375] P. N. Ross, K. Kinoshita, A. J. Scarpellino, and P. Stonehart, *J. Electroanal. Chem.* **59** (1975) 177.
[376] B. G. Baker, in *Modern Aspects of Electrochemistry*, No. 10, Ed. by J. O'M. Bockris and B. E. Conway, Plenum Press, New York, 1975, p. 93.
[377] J. Augustinsky and L. Balsenc, in *Modern Aspects of Electrochemistry*, No. 13, Ed. by B. E. Conway and J. O'M. Bockris, Plenum Press, New York, 1979, p. 251.
[378] J. S. Hammond and W. Winograd, in *Comprehensive Treatise of Electrochemistry*, Vol. 8, Ed. by R. E. White, J. O'M. Bockris, B. E. Conway, and E. Yeager, Plenum Press, New York, 1984, p. 445.
[379] R. Kötz, in *Spectroscopic and Diffraction Techniques in Interfacial Electrochemistry*, Ed. by C. Gutiérrez and C. Melendres, NATO ASI Series, Series C, Vol. 320, Kluwer Academic Publ. Dordrecht, 1990, p. 409.
[380] D. J. Aberdam, in *Spectroscopic and Diffraction Techniques in Interfacial Electrochemistry*, Ed. by C. Gutiérrez and C. Melendres, NATO ASI Series, Series C, Vol. 320, Kluwer Academic Publ., Dordrecht, 1990, p. 383.
[381] J. B. Goodenough, A. Hamnett, B. J. Kennedy, R. Manoharan, and S. A. Weeks, *J. Electroanal. Chem.* **240** (1988) 133.
[382] J. B. Goodenough, A. Hamnett, B. J. Kennedy, and S. A. Weeks, *Electrochim. Acta* **32** (1987) 1233.
[383] R. M. Ishikawa and A. T. Hubbard, *J. Electroanal. Chem.* **69** (1976) 317.
[384] R. W. Revie, J. O'M. Bockris, and B. G. Baker, *Surf. Sci.* **52** (1975) 664.
[385] R. Durand, R. Faure, D. Aberdam, and S. Traore, *Electrochim. Acta* **34** (1989) 1653.
[386] N. Batina, D. G. Frank, J. Y. Gui, B. E. Kahn, C. H. Lin, F. Lu, J. W. McCargar, G. N. Salaita, D. A. Stern, D. C. Zapien, and A. T. Hubbard, *Electrochim. Acta* **34** (1989) 1031.
[387] B. G. Baker, D. A. J. Rand, and R. Woods, *J. Electroanal. Chem.* **97** (1979) 189.
[388] C. Lamy, J. M. Léger, J. Clavilier, and R. Parsons, *J. Electroanal. Chem.* **150** (1983) 71.
[389] R. S. Cameron, G. A. Hards, B. Harrison, and R. J. Potter, *Platinum Metals Review* **31** (1987) 173.
[390] B. V. Tilak, R. S. Yeo, and S. Srinivasan, in *Comprehensive Treatise of Electrochemistry*, Vol. 3, Ed. by J. O'M. Bockris, B. E. Conway, E. Yeager, and R. E. White, Plenum Press, New York, 1981, p. 39.
[391] V. S. Bagotsky and Yu. B. Vassiliev, *Electrochim. Acta* **12** (1967) 1323.
[392] T. Biegler and D. F. A. Koch, *J. Electrochem. Soc.* **114** (1967) 904.
[393] P. A. Christensen, A. Hamnett, and S. A. Weeks, *J. Electroanal. Chem.* **250** (1988) 127.
[394] T. Iwasita and W. Vielstich, *J. Electroanal. Chem.* **242** (1988) 451.
[395] B. Beden, M.-C. Morin, F. Hahn, and C. Lamy, *J. Electroanal. Chem.* **229** (1987) 353.
[396] B. Beden, F. Hahn, S. Juanto, C. Lamy, and J.-M. Léger, *J. Electroanal. Chem.* **225** (1987) 215.
[397] A. Papoutsis, C. Lamy, and J.-M. Léger, 38th ISE Meeting, Maastricht, The Netherlands, 1987, Extended Abstracts, p. 282.
[398] S. Juanto, B. Beden, F. Hahn, J.-M. Léger, and C. Lamy, *J. Electroanal. Chem.* **237** (1987) 119.
[399] B. Beden, S. Juanto, J.-M. Léger, and C. Lamy, *J. Electroanal. Chem.* **238** (1987) 323.
[400] P. S. Fedkiw, C. L. Traynelis, and S. R. Wang, *J. Electrochem. Soc.* **135** (1988) 2459.
[401] R. R. Adžić, in *Modern Aspects of Electrochemistry*, No. 21, Ed. by R. E. White, J. O'M. Bockris, and B. E. Conway, Plenum Press, New York, 1990, p. 163.

[402] C. Lamy, J.-M. Léger, and J. Clavilier, *J. Electroanal. Chem.* **135** (1982) 231.
[403] S. G. Sun and J. Clavilier, *J. Electroanal Chem.* **236** (1987) 95.
[404] A. K. Vijh, *J. Catal.* **37** (1975) 410.
[405] L. D. Burke and K. J. Dwyer, *Electrochim. Acta* **34** (1989) 1659.
[406] L. D. Burke and K. J. Dwyer, *Electrochim. Acta* **35** (1990) 1821, 1829.
[407] M. R. Andrew, J. S. Drury, B. D. McNicol, C. Pinnington, and R. T. Short, *J. Appl. Electrochem.* **6** (1976) 93.
[408] V. B. Hughes and R. Miles, *J. Electroanal. Chem.* **145** (1983) 87.
[409] M. Watanabe, M. Uchida, and S. Motoo, *J. Electroanal. Chem.* **229** (1987) 395.
[410] M. A. Quiroz, I. González, Y. Méas, E. Lamy-Pitara, and J. Barbier, *Electrochim. Acta* **32** (1987) 289.
[411] A. Hamnett and B. J. Kennedy, *Electrochim. Acta* **33** (1988) 1613.
[412] M. Beltowska-Brzezinska, J. Heitbaum, and W. Vielstich, *Electrochim. Acta* **30** (1985) 1465.
[413] G. Kokkinidis and D. Jannakoudakis, *J. Electroanal. Chem.* **153** (1983) 185.
[414] M. Shibata and S. Motoo, *J. Electroanal. Chem.* **229** (1987) 385.
[415] Yu. B. Vassiliev, V. S. Bagotski, N. V. Osetrova, and A. A. Mikhailova, *J. Electroanal. Chem.* **97** (1979) 63.
[416] H. Kita, H. Nakajima, and K. Shimazu, *J. Electroanal. Chem.* **248** (1988) 81.
[417] P. Stonehart and J. T. Lundquist, *Electrochim. Acta* **18** (1973) 349, 907.
[418] J. B. Goodenough, A. Hamnett, B. J. Kennedy, R. Manoharan, and S. A. Weeks, *Electrochim. Acta* **35** (1990) 199.
[419] H. Matsui and A. Kunugi, *J. Electroanal. Chem.* **292** (1990) 103.
[420] S. Sibille, J. Moiroux, J.-C. Marot, and S. Deycard, *J. Electroanal. Chem.* **88** (1978) 105.
[421] M. Beltowska-Brzezinska and J. Heitbaum, *J. Electroanal. Chem.* **183** (1985) 167.
[422] M. Beltowska-Brzezinska, *Electrochim. Acta* **30** (1985) 1193.
[423] M. Avramov-Ivić, R. R. Adžić, A. Bewick, and M. Razak, *J. Electroanal. Chem.* **240** (1988) 161.
[424] M. Enyo, *J. Electroanal. Chem.* **186** (1985) 155.
[425] J. Horkans, *J. Electrochem. Soc.* **131** (1984) 1615.
[426] M. Enyo, *J. Electroanal. Chem.* **201** (1986) 47.
[427] K. I. Machida, K. Nishimura, and M. Enyo, *J. Electrochem. Soc.* **133** (1986) 2522.
[428] M. D. Spasojević, R. R. Adžić, and A. R. Despić, *J. Electroanal. Chem.* **109** (1980) 261.
[429] M. L. Avramov-Ivić and R. R. Adžić, *Bull. Soc. Chim. Beograd* **48** (1983) 357.
[430] S. Motoo and M. Shibata, *J. Electroanal. Chem.* **139** (1982) 119.
[431] R. R. Adžić, M. L. Avramov-Ivić, and A. V. Tripković, *Electrochim. Acta* **29** (1984) 1353.
[432] K. Nishimura, R. Ohnishi, K. Kunimatsu, and M. Enyo, *J. Electroanal. Chem.* **258** (1989) 219.
[433] A. Capon and R. Parsons, *J. Electroanal. Chem.* **44** (1973) 239.
[434] A. Capon and R. Parsons, *J. Electroanal. Chem.* **45** (1973) 245.
[435] V. E. Kazarinov, G. Ya. Tysyachnaya, and V. N. Andreev, *Elektrokhimiya* **8** (1972) 396.
[436] B. Beden, A. Bewick, and C. Lamy, *J. Electroanal. Chem.* **148** (1983) 147; **150** (1983) 505.
[437] J. Willsau and J. Heitbaum, 35th ISE Meeting, Berkeley, California, 1984, Abstract A8-20, p. 498.
[438] E. Rasch and J. Heitbaum, *J. Electroanal. Chem.* **205** (1986) 151.
[439] A. Hamelin, C. Lamy, and S. Maximovitch, C. R. Acad. Sci. **282** (1976) 403, 1065.
[440] S. Motoo and N. Furuya, *J. Electroanal. Chem.* **184** (1985) 303.
[441] J. Clavilier and S. G. Sun, *J. Electroanal. Chem.* **199** (1986) 471.

[442] R. R. Adžić, A. V. Tripković, and V. B. Vesović, *J. Electroanal. Chem.* **204** (1986) 329.
[443] S. G. Sun, J. Clavilier, and A. Bewick, *J. Electroanal. Chem.* **240** (1988) 147.
[444] R. R. Adžić, A. V. Tripković, and W. O'Grady, *Nature* **296** (1982) 137.
[445] R. R. Adžić, A. V. Tripković, and N. M. Marković, *J. Electroanal. Chem.* **150** (1983) 79.
[446] K. Venateswara Rao and C. B. Roy, *Indian J. Technol.* **19** (1981) 26.
[447] E. Rasch and J. Heitbaum, *Electrochim. Acta* **32** (1987) 1173.
[448] V. L. Muraghar and Hira Lal, *Trans. SAEST* **12** (1977) 255.
[449] A. Pavese, V. Solis, and M. C. Giordano, *J. Electroanal. Chem.* **245** (1988) 145.
[450] C. Lamy, F. Hahn, and B. Beden, *Port. Electrochim. Acta* **7** (1989) 435.
[451] T. Hartung, J. Willsau, and J. Heitbaum, *J. Electroanal. Chem.* **205** (1986) 135.
[452] M. Shibata, N. Furuya, M. Watanabe, and S. Motoo, *J. Electroanal. Chem.* **263** (1989) 97.
[453] M. Shibata, O. Takahashi, and S. Motoo, *J. Electroanal. Chem.* **249** (1988) 253.
[454] M. Watanabe, Y. Furuuchi, and S. Motoo, *J. Electroanal. Chem.* **250** (1988) 117.
[455] M. Shibata and S. Motoo, *J. Electroanal. Chem.* **202** (1986) 137.
[456] M. Shibata and S. Motoo, *J. Electroanal. Chem.* **201** (1986) 23.
[457] V. E. Kazarinov, V. N. Andreev, and A. V. Shlepakov, *Electrochim. Acta* **34** (1989) 905.
[458] B. Beden, C. Lamy, N. R. de Tacconi, and A. J. Arvia, *Electrochim. Acta* **35** (1990) 691.
[459] M. W. Breiter, *J. Electroanal. Chem.* **109** (1980) 243.
[460] S. A. Bilmes, N. R. de Tacconi, and A. J. Arvia, *J. Electrochem. Soc.* **127** (1980) 2184.
[461] S. A. Bilmes, N. R. de Tacconi, and A. J. Arvia, *J. Electroanal. Chem.* **164** (1984) 129.
[462] J. Sobkowski and A. Czerwinski, *J. Phys. Chem.* **89** (1985) 265.
[463] H. Kita and H. Nakajima, *Electrochim. Acta* **31** (1986) 193.
[464] C. Gutiérrez, J. Caram, and B. Beden, *J. Electroanal. Chem.* **305** (1991) 289.
[465] Y. Ikezawa, H. Saito, H. Fujisawa, S. Tsuji, and G. Toda, *J. Electroanal. Chem.* **240** (1988) 281.
[466] K. Kunimatsu, H. Seki, W. G. Golden, J. Gordon II, and M. R. Philpott, *Langmuir*, **2** (1986) 464.
[467] C. Lamy, B. Beden, and J. M. Léger, *Bull. Soc. Chim. Fr.* **3** (1985) 421.
[468] L. W. H. Leung, A. Wieckowski, and M. J. Weaver, *J. Phys. Chem.* **92** (1988) 6985.
[469] L. W. H. Leung, S. C. Chang, and M. J. Weaver, *J. Chem. Phys.* **90** (1989) 7426.
[470] S. C. Chang and M. J. Weaver, *J. Electroanal. Chem.* **285** (1990) 263.
[471] T. K. Gibbs, C. McCallum, and D. Pletcher, *Electrochim. Acta* **22** (1977) 525.
[472] M. Watanabe and S. Motoo, *J. Electroanal. Chem.* **60** (1975) 275.
[473] S. Motoo and M. Watanabe, *J. Electroanal. Chem.* **111** (1980) 261.
[474] S. Motoo, M. Shibata, and M. Watanabe, *J. Electroanal. Chem.* **110** (1980) 103.
[475] M. Shibata and S. Motoo, *J. Electroanal. Chem.* **194** (1985) 261.
[476] M. Watanabe and S. Motoo, *J. Electroanal. Chem.* **202** (1986) 125.
[477] A. M. Baruzzi, E. P. M. Leiva, and M. C. Giordano, *J. Electroanal. Chem.* **158** (1983) 103.
[478] A. M. Baruzzi, E. P. M. Leiva, and M. C. Giordano, *J. Electroanal. Chem.* **189** (1985) 257.
[479] Yu. B. Vassiliev, V. S. Bagotski, N. V. Osetrova, and A. A. Mikhailova, *J. Electroanal. Chem.* **189** (1985) 311.
[480] M. L. Marcos, J.-M. Vara, J. González-Velasco, and A. J. Arvia, *J. Electroanal. Chem.* **224** (1987) 189.

[481] M. L. Marcos, J. González-Velasco, J.-M. Vara, M. C. Giordano, and A. J. Arvia, *J. Electroanal. Chem.* **270** (1989) 205.
[482] M. L. Marcos, J. González-Valasco, J.-M. Vara, M. C. Giordano, and A. J. Arvia, *J. Electroanal. Chem.* **287** (1990) 99.
[483] M. W. Breiter, *Electrochim. Acta* **12** (1967) 1213.
[484] B. Beden, A. Bewick, M. R. Razak, and J. Weber, *J. Electroanal. Chem.* **139** (1982) 203.
[485] E. M. Belgsir, E. Bouhier-Charbonnier, H. L. Essis-Yei, K. B. Kokoh, B. Beden, H. Huser, J.-M. Léger, and C. Lamy, *Electrochim. Acta* **36** (1991) 1157.
[486] K. D. Snell and A. G. Keenan, *Electrochim. Acta* **26** (1981) 1339.
[487] K. D. Snell and A. G. Keenan, *Electrochim. Acta* **27** (1982) 1683.
[488] V. E. Kazarinov, Yu. B. Vassiliev, V. N. Andreev, and G. Horanyi, *J. Electroanal. Chem.* **147** (1983) 247.
[489] R. Holze, *J. Electroanal. Chem.* **246** (1988) 449.
[490] T. Iwasita and W. Vielstich, *J. Electroanal. Chem.* **257** (1988) 319.
[491] M. Shibata, N. Furuya, and M. Watanabe, *J. Electroanal. Chem.* **267** (1989) 163.
[492] M.-C. Morin, C. Lamy, J.-M. Léger, J. L. Vasquez, and A. Aldaz, *J. Electroanal. Chem.* **283** (1990) 287.
[493] F. Cases, M. Lopez-Atalaya, J.-L. Vasquez, A. Aldaz, and J. Clavilier, *J. Electroanal. Chem.* **278** (1990) 433.
[494] M. Novak, S. Lantos, and F. Marta, *Acta Phys. Chem. Hung.* **18** (1972) 151, 155.
[495] G. Horanyi and M. Novak, *Acta Chim. Acad. Sci. Hung.* **75** (1973) 271.
[496] E. I. Sokolova, S. V. Kalcheva, and S. N. Raicheva, *C. R. Acad. Bulg. Sci.* **26** (1973) 383, 387.
[497] A. G. Dzhambova, E. I. Sokolova, and S. N. Raicheva, *C. R. Acad. Bulg. Sci.* **34** (1981) 815.
[498] S. N. Raicheva, M. V. Christov, and E. I. Sokolova, *Electrochim. Acta* **26** (1981) 1669.
[499] M. Fujihira, S. Tasaki, T. Osa, and T. Kuwana, *J. Electroanal. Chem.* **150** (1983) 665.
[500] M. V. Christov and E. I. Sokolova, *J. Electroanal. Chem.* **175** (1984) 183.
[501] P. Ocon, C. Alonso, R. Cedran, and J. González-Velasco, *J. Electroanal. Chem.* **206** (1986) 179.
[502] R. S. Gonçalves, J.-M. Léger, and C. Lamy, *Electrochim. Acta* **33** (1988) 1581.
[503] P. T. A. Sumodjo, E. J. da Silva, and T. Rabockai, *J. Electroanal. Chem.* **271** (1989) 305.
[504] S. G. Sun, D. F. Yang, and Z. W. Tian, *J. Electroanal. Chem.* **289** (1990) 177.
[505] M. Beltowska-Brzezinska, E. Dutkiewicz, and P. Skoloda, *J. Electroanal. Chem.* **181** (1984) 235.
[506] D. Takky, B. Beden, J.-M. Léger, and C. Lamy, *J. Electroanal. Chem.* **193** (1985) 159.
[507] G. Horanyi, P. König, and I. Télés, *Acta Chim. Acad. Sci. Hung.* **72** (1972) 165.
[508] S. N. Raicheva, E. I. Sokolova, S. V. Kalcheva, and W. Vielstich, 29th ISE Meeting, Budapest, 1978, Extended Abstract, p. 780.
[509] S. N. Raicheva, *Z. Phys. Chem.* **264** (1983) 65.
[510] R. Holze and M. Beltowska-Brzezinska, *Electrochim. Acta* **30** (1985) 937.
[511] R. Holze and M. Beltowska-Brzezinska, *J. Electroanal. Chem.* **201** (1986) 387.
[512] B. Beden, I. Cetin, A. Kahyaoglu, D. Takky, and C. Lamy, *J. Catal.* **104** (1987) 37.
[513] K. Kunimatsu, *J. Electroanal. Chem.* **213** (1986) 149.
[514] R. S. Gonçalves, B. Beden, J. M. Léger, and C. Lamy, 39th ISE Meeting, Glasgow, 1988, Extended Abstract 103.
[515] D. Takky, Ph.D. thesis, University of Poitiers, 1987.

[516] J. C. Card, S. E. Lyke, and S. H. Langer, *J. Appl. Electrochem.* **20** (1990) 269.
[517] H. Hitmi, E. M. Belgsiz, J. M. Léger, and C. Lamy, *Electrochim. Acta* (submitted).
[518] S. N. Raicheva, S. V. Kalcheva, M. V. Christov, and E. I. Sokolova, *J. Electroanal. Chem.* **55** (1974) 213.
[519] S. V. Kalcheva, M. V. Christov, E. I. Sokolova, and S. N. Raicheva, *J. Electroanal. Chem.* **55** (1974) 223.
[520] S. V. Kalcheva, M. V. Christov, E. I. Sokolova, and S. N. Raicheva, *J. Electroanal. Chem.* **55** (1974) 231.
[521] S. N. Raicheva, E. I. Sokolova, and S. V. Kalcheva, *C. R. Acad. Sci. Bulg.* **28** (1975) 1081, 1219.
[522] F. Cases, J.-L. Vasquez, J.-M. Perez, A. Aldaz, and J. Clavilier, *J. Electroanal. Chem.* **281** (1990) 283.
[523] S. N. Raicheva, E. I. Sokolova, and S. V. Kalcheva, *J. Electroanal. Chem.* **61** (1975) 325.
[524] G. Horanyi and M. Novak, *Acta Chim. Sci. Hung.* **75** (1973) 369.
[525] B. Rasch and T. Iwasita, *Electrochim. Acta* **35** (1990) 989.
[526] M. Shibata and S. Motoo, *J. Electroanal. Chem.* **187** (1985) 151.
[527] B. Bänsch, T. Hartung, H. Baltruschat, and J. Heitbaum, *J. Electroanal. Chem.* **259** (1989) 207.
[528] G. Horanyi, *J. Electroanal. Chem.* **51** (1974) 163.
[529] A. Wieckowski, J. Sobkowski, P. Zelenay, and K. Franaszczuk, *Electrochim. Acta* **26** (1981) 1111.
[530] A. Wieckowski, *Electrochim. Acta* **26** (1981) 1121.
[531] A. G. Kornienko, L. A. Mirkind, and M. Y. Fioshin, *Elektrokhimiya* **3** (1967) 1370.
[532] G. Inzelt and G. Horanyi, *Acta Chim. Acad. Sci. Hung.* **101** (1979) 215.
[533] G. Horanyi and E. M. Rizmayer, *Electrochim. Acta* **34** (1989) 197.
[534] A. K. Vijh, *Can. J. Chem.* **49** (1971) 78.
[535] K. Venkateswara Rao and C. B. Roy, *Ind. J. Chem.* **24A** (1985) 820.
[536] E. Santos and M.-C. Giordano, *Electrochim. Acta* **30** (1985) 871.
[537] E. P. Leiva, E. Santos, R. M. Cerviño, M.-C. Giordano, and A. J. Arvia, *Electrochim. Acta* **30** (1985) 1111.
[538] G. Pierre, A. Ziade, and M. El Kordi, *Electrochim. Acta* **32** (1987) 601.
[539] W. Hauffe and J. Heitbaum, *Electrochim. Acta* **23** (1978) 299.
[540] G. Kokkinidis and D. Jannakoudakis, *J. Electroanal. Chem.* **133** (1982) 307.
[541] N. W. Smirnova, O. A. Petrii, and A. Grzejdziak, *J. Electroanal. Chem.* **251** (1988) 185.
[542] F. Kadirgan, E. Bouhier-Charbonnier, C. Lamy, J.-M. Léger, and B. Beden, *J. Electroanal. Chem.* **286** (1990) 41.
[543] R. R. Adžić and M. Avramov-Ivić, *J. Catal.* **101** (1986) 532.
[544] J.-M. Orts, A. Fernandez-Vega, J.-M. Feliu, A. Aldaz, and J. Clavilier, *J. Electroanal. Chem.* **290** (1990) 119.
[545] F. Hahn, B. Beden, F. Kadirgan, and C. Lamy, *J. Electroanal. Chem.* **216** (1987) 169.
[546] H. Ewe, E. Justi, and M. Pesditschek, *Energy Conversion* **15** (1975) 9.
[547] C. H. Hamann and P. Schmöde, *J. Power Source* **1976/77**(1) 2.
[548] E. M. Belgsir, Ph.D. thesis, University of Poitiers, 1990.
[549] G. Horanyi and K. Torkos, *J. Electroanal. Chem.* **125** (1981) 105.
[550] H. Huser, J.-M. Léger, and C. Lamy, *Electrochim. Acta* **30** (1985) 1409.
[551] P. Ocon, B. Beden, H. Huser, and C. Lamy, *Electrochim. Acta* **32** (1987) 387.
[552] P. Ocon, B. Beden, and C. Lamy, *Electrochim. Acta* **32** (1987) 1095.
[553] C. Alonso and J. González-Velasco, *J. Electroanal. Chem.* **248** (1988) 193.
[554] E. I. Sokolova, *Electrochim. Acta* **24** (1979) 147.
[555] R. S. Gonçalves and T. Rabockai, *Z. Phys. Chem.* **131** (1982) 181.

[556] A. Kahyaoglu, B. Beden, and C. Lamy, *Electrochim. Acta* **29** (1984) 1489.
[557] R. S. Gonçalves, W. E. Triaca, and T. Rabockai, *Anal. Lett.* **18** (1985) 957.
[558] G. Horanyi and E. M. Rizmayer, *Acta Chim. Scand.* **37** (1983) 451.
[559] J. P. Ango, B. Beden, O. Enea, H. Essis-Yei, C. Lamy, and J.-M. Léger, Biomass for Energy and Industry, 4th European Conference, Orléans, France, 1987, Extended Abstract 4-2, p. 339.
[560] O. Enea and J.-P. Ango, *Electrochim. Acta* **34** (1989) 391.
[561] J.-M. Léger, I Fonseca, F. Bento, and I. Lopes, *J. Electroanal. Chem.* **285** (1990) 125.
[562] M. M. Baizer, T. Nonaka, K. Park, Y. Saito, and K. Nobe, *J. Appl. Electrochem.* **14** (1984) 197.
[563] G. Pierre, M. El Kordi, and G. Cauquis, *Electrochim. Acta* **30** (1985) 1219, 1227.
[564] G. Pierre, M. El Kordi, G. Cauquis, G. Mattioda, and Y. Christidis, *J. Electroanal. Chem.* **186** (1985) 167.
[565] E. M. Belgsir, H. Huser, C. Lamy, and J. M. Léger, in *Heterogeneous Catalysis and Fine Chemicals II*, Poitiers, M. Guisnet, J. Barrault, C. Bouchoule, D. Duprez, G. Pérot, R. Maurel, and C. Montassièr (eds.), Elsevier, Amsterdam, 1991, p. 463.
[566] G. Horanyi, V. E. Kazarinov, Yu. B. Vassiliev, and V. N. Andreev, *J. Electroanal. Chem.* **147** (1983) 263.
[567] G. Horanyi, *J. Electroanal. Chem.* **117** (1981) 131.
[568] P. Gonzalo, A. Aldaz, and J.-L. Vásquez, *J. Electroanal. Chem.* **130** (1981) 209.
[569] J.-M. Feliu, J. Claret, C. Müller, J.-L Vásquez, and A. Aldaz, *J. Electroanal. Chem.* **178** (1984) 271.
[570] R. Albalat, J. Claret, E. Gómez, C. Müller, and M. Sarret, *Electrochim. Acta* **34** (1989) 611.
[571] A. Fernandez-Vega, J.-M. Feliu, A. Aldaz, R. Albalat, J. Claret, and C. Müller, *J. Electroanal. Chem.* **266** (1989) 137.
[572] M. L. R. Rao and R. F. Drake, *J. Electrochem. Soc.* **116** (1969) 334.
[573] E. Skou, *Electrochim. Acta* **22** (1977) 313.
[574] S. Ernst, J. Heitbaum, and C. H. Hamann, *J. Electroanal. Chem.* **100** (1979) 173.
[575] H. Lerner, J. Giner, J. S. Soeldner, and C. K. Colton, *J. Electrochem. Soc.* **126** (1979) 237.
[576] S. Ernst, J. Heitbaum, and C. H. Hamann, *Ber. Bunsenges, Phys. Chem.* **84** (1980) 50.
[577] M. L. F. de Mele, H. A. Videla, and A. J. Arvia, *J. Electrochem. Soc.* **129** (1982) 2207.
[578] M. L. F. de Mele, H. A. Videla, and A. J. Arvia, *Bioelectrochem. Bioenerg.* **9** (1982) 469.
[579] M. Sakamoto and K. Takamura, *Bioelectrochem. Bioenerg.* **9** (1982) 571.
[580] N. Xonoglou and G. Kokkinidis, *Bioelectrochem. Bioenergy.* **12** (1984) 485.
[581] M. L. F. de Mele, H. A. Videla, and A. J. Arvia, *Bioelectrochem. Bioenerg.* **16** (1986) 213.
[582] N. Xonoglou, I. Noumtzis, and G. Kokkinidis, *J. Electroanal. Chem.* **237** (1987) 93.
[583] L. H. Essis-Yei, B. Beden, and C. Lamy, *J. Electroanal. Chem.* **246** (1988) 349.
[584] L. A. Larew and D. C. Johnson, *J. Electroanal. Chem.* **262** (1989) 167.
[585] I. T. Bae, Xuekun Xing, C. C. Liu, and E. Yeager, *J. Electroanal. Chem.* **284** (1990) 335.
[586] R. R. Adžić, M. W. Hsiao, and E. B. Yeager, *J. Electroanal. Chem.* **260** (1989) 475.
[587] H. L. Chum and M. M. Baizer, *The Electrochemistry of Biomass and Derived Material*, ACS Monograph, American Chemical Society, Washington, D.C., 1985, p. 225.
[588] A. Malki, Ph.D. thesis, University of Poitiers, 1988.

3

Surface States on Semiconductors

R. A. Batchelor and A. Hamnett

Department of Chemistry, The University, Newcastle-upon-Tyne NE1 7RU, U.K.; present address for R.A.B.: *Pilkington Technology Centre, Lathom, Ormskirk, Lancashire L40 5UF, U.K.*

I. INTRODUCTION

The dominant interpretative framework within which semiconductor electrochemical studies have been carried out since the very earliest papers of Brattain and Garrett[1] is that of the ideal metal/semiconductor interface. This framework has been tested most thoroughly for extrinsically doped semiconductors held at reverse bias, for which the majority carriers are attracted away from the interface toward the bulk of the semiconductor. The result is a layer, usually many hundreds of angstroms thick, which possesses an electrostatic charge owing to the uncompensated ionized extrinsic dopant sites and in which the potential distribution is described in terms of a classical Schottky barrier. Any *change* in applied potential is then postulated to be accommodated almost entirely within this space-charge layer, provided the semiconductor remains in reverse bias and the electrolyte has a concentration exceeding approximately 0.01 *M*. The central difference between this situation and that found in the more familiar electrochemistry of the *metal*/electrolyte interface is that the very high charge carrier density in the metal allows the buildup of a high uncompensated charge at the surface, ensuring that the change in applied potential is accommodated not within the metal, but in the electrolyte itself.

Modern Aspects of Electrochemistry, Number 22, edited by John O'M. Bockris *et al.*
Plenum Press, New York, 1992.

If the electrolyte in turn has a high concentration of ions, then the potential change will be accommodated across an extremely narrow region just outside the metal surface termed the Helmholtz layer; this layer will normally be only a few angstroms thick and experience very high electric fields. By contrast, at the ideal semiconductor/electrolyte interface, fields are much lower, and the change in the voltage dropped across the Helmholtz layer will be very small.

However, this semiconductor-electrolyte model has been found to be scarcely more adequate than the corresponding ideal metal-semiconductor model. As in the lattter case, it has proved necessary to postulate the existence of electronic states localized at the interface, whose population is controlled by the position of the Fermi level. Such states, originally introduced by Bardeen[2] to account for the deviation from ideality encountered in the metal/silicon interface, can fill or empty as the potential difference across the interface (and hence the semiconductor Fermi level) is altered. This alteration in surface-state population is accompanied by a change in the surface charge and hence in the potential distribution, with the change in applied potential now being partitioned between the space-charge region within the semiconductor and the very thin Helmholtz layer, provided the electrolyte concentration is much larger than the concentration of majority carriers in the semiconductor.

The introduction of these interfacial states greatly complicates our analysis of experimental data. As we shall see below, both the *origin* of these states and their *number density* are objects of considerable controversy. Bardeen's original analysis showed that even a surface-state density corresponding to less than one percent of the surface atomic population could have a major impact on the potential change at the interface, particularly if the *energy* distribution of such states were narrow (that is, if all the surface electronic states were close in energy). The very small coverage of states needed to give rise to observable macroscopic effects has meant that identification of such states by spectroelectrochemical means has proved extremely difficult, and indirect evidence has usually been presented to establish their chemical origin. It is a major goal of modern semiconductor electrochemical research to establish more direct means of identifying interfacial states and to relate them to the chemical constitution of the interface.

However, the difficulty of this task should not be understated; in addition to the "intrinsic" surface states originally introduced by Bardeen, a number of extrinsic effects have also been identified as leading to surface states. These include the possibility of chemical attack from solution that can lead to the presence of corrosion intermediates such as hydridic or hydroxylic bonds on the surface of silicon, and even complete oxide layers in extreme cases. Adsorption of electrochemically active species onto the surface can also give rise to surface states, as can crystal defects, which may be similar to defects found in the bulk semiconductor or may exist by virtue of surface preparation procedures such as polishing or etching. Another source of surface states is the intermediates that may form during photochemically induced reactions: a major thrust in the development of semiconductor electrochemistry has been the possibility of using devices based on the semiconductor/electrolyte interface for the conversion of solar to electrical energy, and an important limitation on the efficiency of these systems is the presence of surface states that may act not merely to give a less than optimal potential distribution but may also give rise to electron–hole recombination effects, greatly lowering the efficiency close to the flat-band potential.

This additional role of surface states, allowing routes for electron transfer processes that might otherwise take place with very low efficiency, has recently received much attention. Both hole and electron transfer processes at the surface of wide-bandgap semiconductors are now believed to be mediated by surface states, and considerable progress has been made in the understanding of the surface chemistry of materials such as TiO_2 in recent years.

It is clear from the above that any review of the importance of surface states in semiconductor electrochemistry will be almost coterminous with the entire subject. Although the last few years have seen a decrease from the spate of papers that appeared at the beginning of the previous decade, interest in the fundamental aspects of the subject is still strong, and it seems opportune to review the various developments that relate to the concept of surface state in the light of the most recent insights. In order to maintain some coherence, we have restricted our discussion to single-crystal samples of III–V semiconductors, cadmium chalcogenides, silicon, and TiO_2. Detailed accounts of the electrochemistry of all of these materials have appeared recently,[3] allowing us to focus on those aspected that have been identified as likely to originate with surface states and their varying populations. We shall see that for all these materials progress has been made from the original postulates, designed to account for deviations from the ideal model, to a real molecular understanding of the surface chemistry. Much, however, yet remains to be done.

II. MODELS OF THE POTENTIAL DISTRIBUTION AT THE SURFACE

1. The Basic Model

The mathematical description of the potential distribution at the surface has been reviewed elsewhere,[4,5] and only a brief overview will be presented here. The potential distribution shown in Fig. 1 for a wide-bandgap *n*-type semiconductor in contact with an electrolyte consists of three regions, the interior of the semiconductor, the Helmholtz layer at the surface, and the electrolyte itself. The potential drop in the latter can be made small by increasing the ionic concentration to above ca. 10^{-2} *M*, assuming a donor density

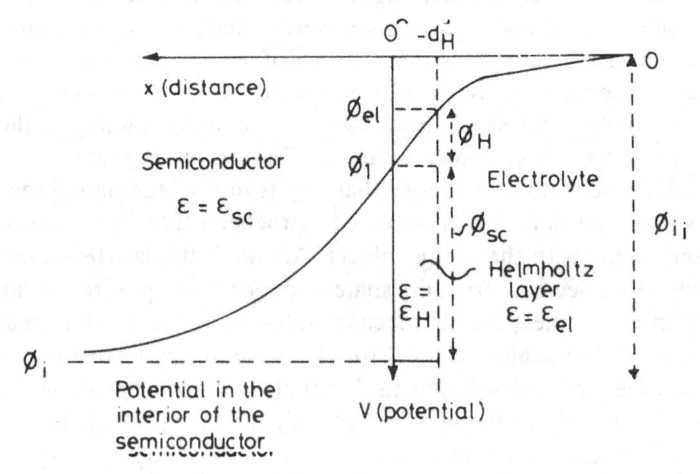

Figure 1. Detailed potential distribution for a semiconductor/electrolyte interface.

of less than 10^{19} cm^{-3}, and we shall not be concerned with it. The central question is the partition of the potential between the Helmholtz layer and the interior of the semiconductor. If the donor levels in the depletion layer of the semiconductor are fully ionized, then the density of *positive* charge in this region will be just N_D, the donor density. The density of *negative* charge will be given by the Boltzmann formula:

$$n(x) = N_D \exp\{e_0[\phi(x) - \phi_b]/k_BT\} \tag{1}$$

where $\phi(x)$ is the potential at distance x from the surface of the semiconductor, and ϕ_b is the potential in the bulk of the semiconductor, and we note that $\phi(x) - \phi_b < 0$. This gives us one relationship between ϕ and x; a second is Poisson's equation:

$$\partial^2\phi/\partial x^2 = -\rho/\varepsilon\varepsilon_0 \tag{2}$$

where ρ is the net charge density and is given by $e_0[N_D - n(x)]$, and ε is the static dielectric constant of the semiconducting medium. Combining these gives the Poisson–Boltzmann equation appropriate to our system:

$$\partial^2\phi/\partial x^2 = -e_0N_D\{1 - \exp[e_0(\phi - \phi_b)/k_BT]\}/\varepsilon\varepsilon_0 \tag{3}$$

Integration gives

$$(\partial\phi/\partial x)^2 = -2e_0N_D((\phi - \phi_b) - (k_BT/e_0) \times \{\exp[e_0(\phi - \phi_b)/k_BT] - 1\})/\varepsilon\varepsilon_0 \tag{4}$$

We can simplify this since k_BT/e_0 is small (ca. 25 mV at room temperature), and because $\phi - \phi_b$ is negative, the exponential term becomes vanishingly small and

$$\partial\phi/\partial x \approx \pm(2e_0N_D/\varepsilon\varepsilon_0)^{1/2}(\phi_b - \phi - k_BT/e_0)^{1/2} \tag{5}$$

If the potential difference across the depletion layer, $\phi_b - \phi$, is written as ϕ_{sc}, then integration of Eq. (5) leads to

$$W_{sc} = (2\varepsilon_{sc}\varepsilon_0/e_0N_D)^{1/2}(\phi_{sc} - k_BT/e_0)^{1/2} \tag{6}$$

where W_{sc} is the width of the depletion layer. The *charge* in this layer is given, to the same approximation, by the Gauss equation:

$$Q_{sc} = \varepsilon_0\varepsilon_{sc}(\partial\phi/\partial x)_{x=0} = (2e_0N_D/\varepsilon_0\varepsilon_{sc})^{1/2}(\phi_{sc} - k_BT/e_0)^{1/2} \tag{7}$$

Provided that the charge density in the semiconductor is much smaller than that in the electrolyte, a more complete analysis shows that the great majority of any *change* in potential applied to the semiconductor will appear across the semiconductor depletion layer, since the double layer external to the semiconductor is far smaller than the width W_{sc}. It follows that the position in energy of the semiconductor band edges at the interface should also be unaltered on change of the applied potential. As a result, we can write ϕ_{sc} for a given applied potential V as $V - V_{FB}$, where the flat-band potential V_{FB} is that at which the depletion layer disappears and there is no electric field within the semiconductor. From Eq. (7), we can therefore calculate the differential capacitance of the semiconductor as a function of applied potential:

$$C_{sc} = \partial q_{sc}/\partial V = \tfrac{1}{2}(2\varepsilon_{sc}\varepsilon_0 e_0 N_D)^{1/2}(V - V_{FB} - k_B T/e_0)^{-1/2} \quad (8)$$

and

$$C_{sc}^{-2} = 2(V - V_{FB} - k_B T/e_0)/(e_0 N_D \varepsilon_0 \varepsilon_{sc}) \quad (9)$$

This is the celebrated Mott–Schottky relationship; assuming that the capacitance of the Helmholtz layer is much larger than C_{sc}, Eq. (8) should yield the dominant impedance of the interface, and the fact that this relationship is frequently observed was originally thought to constitute good evidence for the validity of the model.

It should be noted that the actual measurement of the capacitance associated with the semiconductor/electrolyte interface is fraught with difficulty. The capacitance in Eq. (8) is frequency independent, since our model assumes that the establishment of electronic equilibrium is extremely fast. Even in cases where there are no other contributions to the interfacial capacitance, this may not be true, and the problems are discussed elsewhere.[5] Provided Eq. (8) is obeyed, it should be possible to calculate the semiconductor dopant density from the slope of the plot of C_{sc}^{-2} versus applied potential. Figure 2 shows such a plot for ⟨100⟩ *p*-GaAs in 0.5 *M* H_2SO_4 for various frequencies, and it is evident that a frequency-independent capacitance is a good approximation to the interface impedance. However, the "ideal" behavior of Fig. 2 is far from common, and the analysis of the deviation of the capacitive behavior

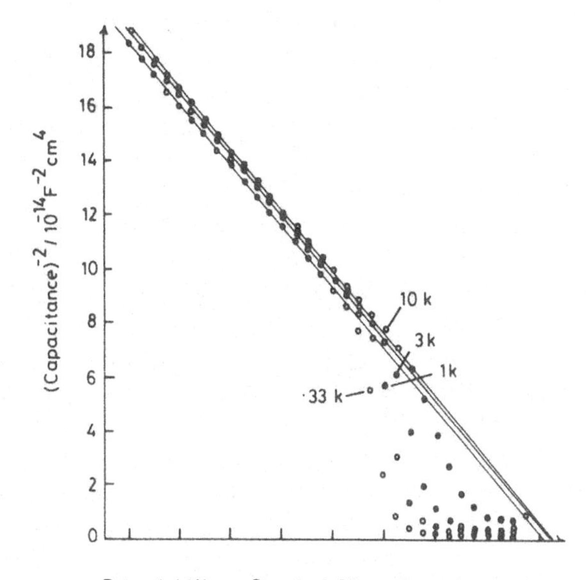

Figure 2. Series capacitance measurements for 100 *p*GaAs in 0.5 *M* H_2SO_4 at the frequencies shown. The electrode was pre-etched in Br_2/MeOH.

from ideality has played an important role in characterizing non-ideal surfaces.

2. The Effect of an Additional Charged Layer at the Surface

If a film of surface charge is inserted between the semiconductor and the electrolyte, it may have a major influence on the potential distribution. The resulting changes in this distribution can be calculated assuming that the surface film is of small width, and typical results are shown in Fig. 3 for the case of 10^{18} cm^{-3} doped *p*-GaP for which the *total* interfacial potential was 1.0 V. It can be seen from Fig. 3 that a surface charge density of 2×10^{-6} C cm^{-2} will have a marked effect on the potential distribution, and this charge density corresponds to approximately 10^{13} electronic charges per square centimeter. This number enters into a useful rule of thumb: uncompensated charge densities of this magnitude can be expected to lead to significant deviation from the ideal model.

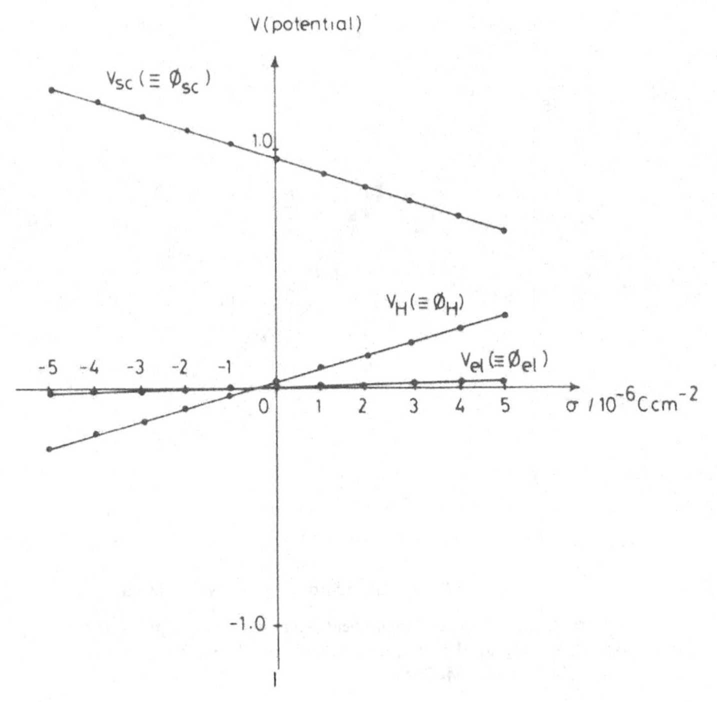

Figure 3. Calculated potential differences ϕ_{sc} and ϕ_H for n-GaP (10^{18} cm^{-3} dopant density) as a function of surface-charge density, σ, for a fixed value of ϕ_i = 1.0 V.

A high density of surface charge will therefore displace the band edges relative to an external reference. In the gas phase, this will manifest itself, for example, in shifts of the photoelectron peaks associated with Ga and As in GaAs. In electrolyte, the main effect will be a shift in the flat-band potential.

A particularly important case arises if the surface charge *varies with applied potential.* If localized electronic levels are considered to exist within the bandgap, and if they are in electronic equilibrium with the bulk of the semiconductor, their population will change with applied potential. This is illustrated in Fig. 4d and e: a narrow band of *localized* states in the midgap region at the surface, which is uncharged when filled, empties and becomes positively charged when the bulk Fermi level passes through the energy of the surface state. The effect on the potential distribution is shown in Fig. 5 for

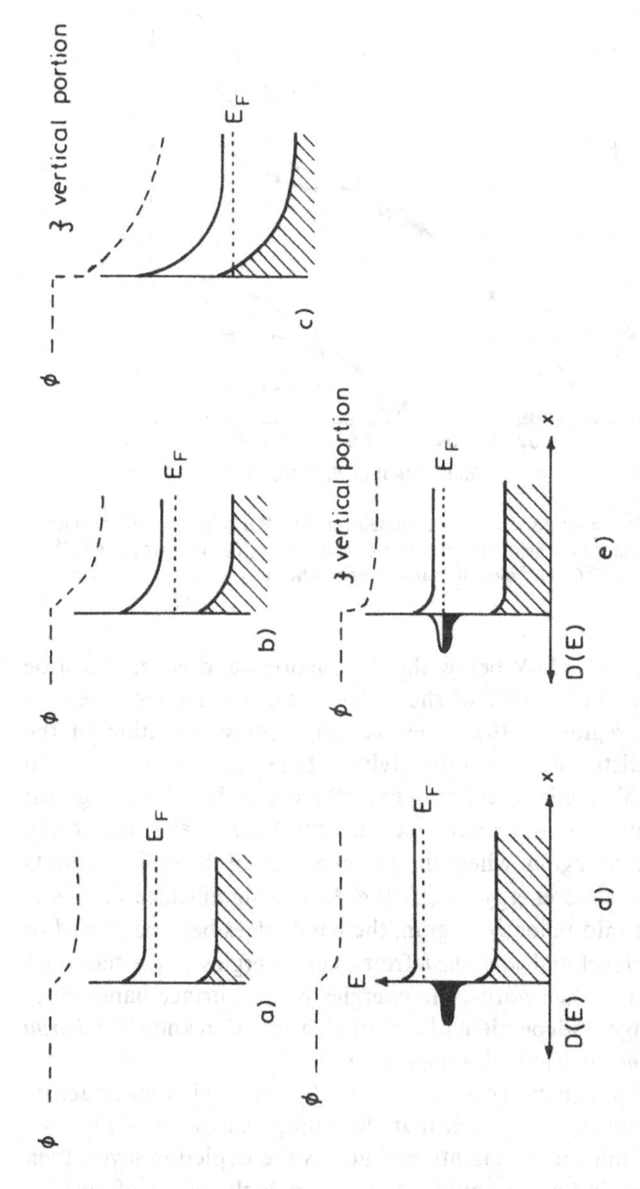

Figure 4. Potential distribution in a wide-bandgap *n*-type semiconductor (a) for a small reverse bias, (b) for an intermediate bias, (c) for a larger bias sufficient to cause inversion, (d) for a small bias with a filled set of surface states, (e) and for a larger bias in which the surface states have been partially ionized.

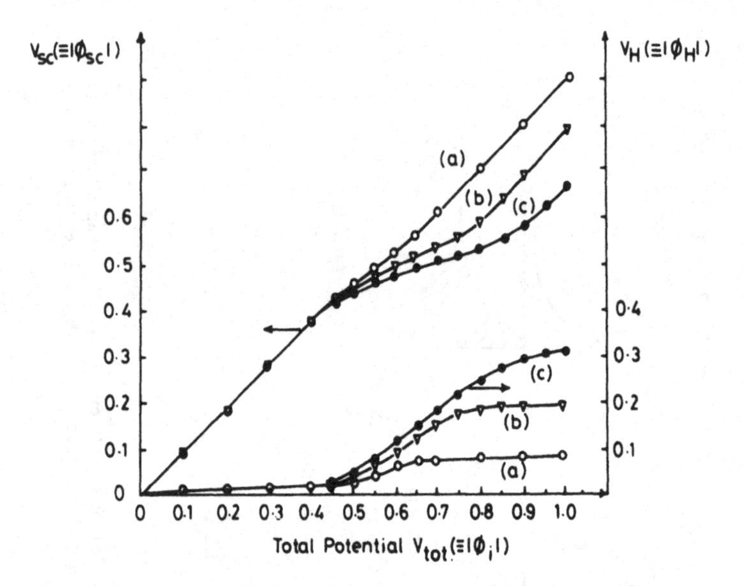

Figure 5. Potential distribution calculated for *n*-GaP in which a surface state is introduced 0.5 eV below the conduction-band edge; (a) 10^{-6}, (b) 3×10^{-6}, and (c) 5×10^{-6} C cm^{-2} overall charge density when empty; $d_H = 3$ Å, $\varepsilon_H = 6$, $\varepsilon_{el} = 80$.

a state located 0.5 eV below the conduction-band edge. It can be seen that as the *density* of the states at the surface increases, an increasing region of the potential range shows partition of the potential distribution into the Helmholtz layer. It is of interest to note that this crude calculation nevertheless leads to three regions: for high and low potentials, we see that $\partial\phi_{sc}/\partial V \approx 1$, but in the mid-potential region, where the Fermi energy of the bulk intersects the surface state energy, $\partial\phi_{sc}/\partial V \ll 1$. The significance of this is that in this mid-potential region, the band edges become *pinned* to the Fermi level and *unpinned* from their energies at the flat-band potential. In other words, the energies of the surface band edges are not longer independent of potential, a situation known as *Fermi level pinning* or *band-edge unpinning.*

A similar situation can occur when the potential dropped across the depletion layer is greater than the bandgap, as shown in Fig. 4c. If equilibrium can be maintained across the depletion layer, then the situation in Fig. 4c would lead to a very high density of positive holes at the surface, and the semiconductor becomes effectively degenerate, with the potential behaving essentially as at a metal/electrolyte interface. This process is termed *inversion* and may be encountered in narrow-bandgap semiconductors; in wide-bandgap materials, however, it is normally not possible to establish thermal equilibrium, and surface carrier density remains small.

Although, as we shall see, the models above have been applied with much success, the question of the molecular identity of the surface states has often remained unanswered for reasons that are now clear; given the very large effect of a surface state density of 10^{13} charges per square centimeter, which corresponds to a coverage of substantially less than 1%, the detection of these states remains beyond the limit of most of our available *in situ* spectroscopic techniques. Even so, progress has been made recently and is reviewed below.

III. SURFACE STATES AT THE SEMICONDUCTOR/VACUUM INTERFACE

The concept of the surface state was introduced in 1932,[6] when Tamm predicted the presence of localized electronic states at the semiconductor surface using an essentially one-dimensional model. The Kronig-Penney potential, on which Tamm based his analysis, is a period linear chain of square potential wells for which the Schrödinger equation can be solved in terms of Bloch functions, $u_{k,n}(z)$. The wave functions will take the form

$$\psi_{k,n} = u_{k,n}(z)\, e^{ikz} \tag{10}$$

where $u_{k,n}(z)$ has the periodicity of the lattice, z is the one-dimensional axis, and k and n are quantum numbers. If the surface of this one-dimensional crystal is represented by an abrupt step in the potential, then, as Tamm showed, electronic states localized at the surface should result, with energies within the bandgap. Shockley extended this work,[7] showing that for a generalized periodic potential which is terminated at a potential maximum in the crystal, surface states would also result. However, whereas Tamm states would be expected to appear for a suitably large surface perturbation, the Shockley states should be present for small perturbations at the surface and then disappear as the perturbation becomes large. The differences between these approaches and the general theory of surface states are both discussed elsewhere.[8,9] However, the sheer complexity of real surface structures renders the distinction between Tamm and Shockley states rather artificial.

Bardeen proposed that such surface states were responsible for the experimentally observed insensitivity of metal-semiconductor barrier heights to the work function of the metal[10] For an ideal Schottky barrier, the situation is shown in Fig. 6a: as the metal and semiconductor come into equilibrium, their Fermi levels must adjust to be the same, and, in the case of an n-type semiconductor with Fermi level above that of the metal, the resulting potential

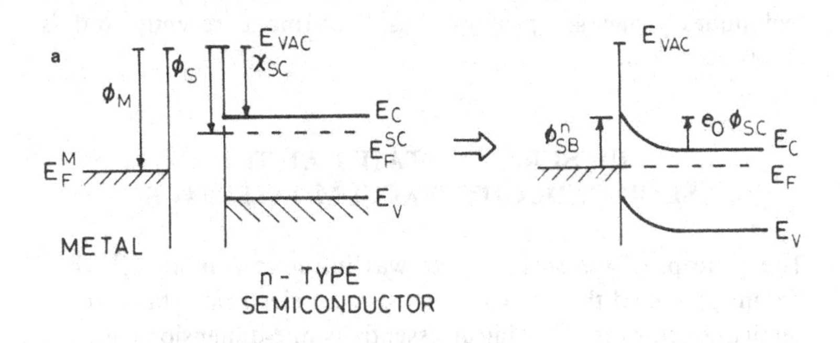

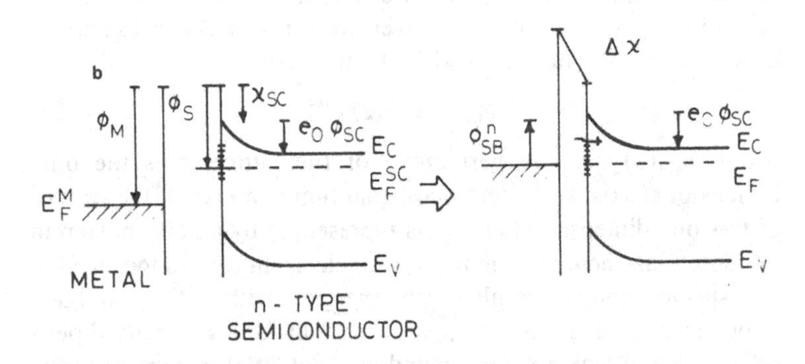

Figure 6. (a) Ideal Schottky barrier formation; ϕ_M is the metal work function, ϕ_s the energy of the semiconductor Fermi level with respect to the vacuum level, ϕ_{sB} the Schottky barrier height, and ϕ_{sc} the potential dropped across the space-charge region. (b) Schottky barrier formation in the presence of surface states. $\Delta\chi$ is the change in potential across the interface due to surface-state charging.

drop should occur across the depletion layer in the semiconductor. This is a similar type of argument to that advanced above for the semiconductor/electrolyte interface, where the majority of the potential drop takes place across the medium with the lower charge carrier density. The analysis leads to the conclusion that there should be a direct relationship between the work function (and hence the Fermi level) of the metal, ϕ_M, and the Schottky barrier height, ϕ_{SB}. In fact,

$$\phi_{SB} = \phi_M - \phi_S \tag{11}$$

where ϕ_S is the work function of the semiconductor, measured from the Fermi level. Commonly, in fact, the variation of the barrier height for a given semiconductor is far less than that implied by the above equation, as can be seen from Fig. 7,[11] which shows that for most metals forming barriers with either *p*- or *n*-type GaAs, the barrier height is pinned close to the midgap. This pinning can be explained by assuming a high density of surface states. On barrier formation, as shown in Fig. 6b, the semiconductor Fermi level will drop, as previously described, until it intersects the surface state energy. The surface states will start to ionize, giving rise to a narrow region of surface charge over which a substantial fraction of further potential change will take place. However, the potential change across the narrow surface charge region will not contribute to the barrier height, as electrons can tunnel across this region readily. If

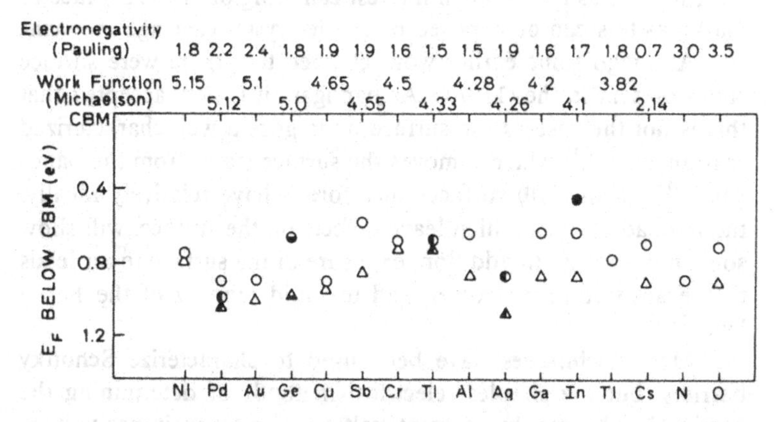

Figure 7. Fermi level pinning positions at *n*- and *p*-GaAs[11]; different symbols represent the work of different groups. (CBM is Conduction-Band Minimum.)

the density of surface states is sufficiently high, they will pin the Fermi level, so that changes in the metal work function will alter the surface-state population but not the barrier height.

The effect of surface charge on the potential distribution is rather more complicated than that discussed earlier, because the charge on the semiconductor will be compensated by the attraction of oppositely charged carriers in the metal or electrolyte.[12,13] In principle, charged surface states located *right at the surface* should have no effect on the potential distribution, as their expected effect should be canceled by equal and opposite charges at the metal interface. Only when the finite width of the interface region and the presence of surface states some angstroms into the semiconductor are taken into account do calculations show that Fermi level pinning will occur.[12,13]

The extensive studies on the metal/semiconductor interface have been reviewed by Brillson[14] and by Lindau and Kendelwicz,[11] The model of Bardeen remains predominant in interpretation, but research has focused recently on determining the origin of the surface states so that they can be eliminated or controlled. A particularly important example is GaAs, because it shows a high density of surface states at interfaces with all materials apart from lattice-matched $GaAl_xAs_{1-x}$; this greatly reduces its considerable device potential. Studies are often carried out under ultra-high-vacuum conditions in order to reduce the possibility of contamination, which has led to much interest centering on the ⟨110⟩ face of GaAs, as this can be exposed by single-crystal cleavage *in vacuo.*

Although some earlier work claimed that there were surface states present in the ⟨110⟩ GaAs bandgap, it is now accepted that this is not the case as the surface undergoes a well-characterized rearrangement,[15] which removes the surface states from the bandgap.[16-18] Clean ⟨110⟩ surfaces therefore behave relatively ideally, though bad cleaves, which leave defects on the surface, will show some nonideality. In addition, exposure of the surface to air leads to the adsorption of oxygen and to rapid pinning of the Fermi level.[19,20]

Many techniques have been used to characterize Schottky barriers, but the simplest electrical methods of determining the barrier height are by current-voltage and capacitance-voltage measurements, the latter being interpreted within the Mott-

Schottky model discussed above.[21] Another technique, which is very reliable, is internal photoemission, which measures the photocurrent arising from excitation from the metal Fermi level to the conduction-band edge. By varying the wavelength, the energy difference can be obtained directly.[14]

A variety of models have been developed to describe the formation of nonideal Schottky barriers, of which four will be discussed here: metal-induced gap state model, unified defect model, disorder-induced gap state model, and the effective work function model.

In 1965, Heine pointed out that in an intimate metal-semiconductor contact, wave functions should be matched from metal to semiconductor, so that *no* localized states would result.[22] Instead, the metal wave functions, which tail into the semiconductor, will affect the charge balance at the interface in a manner similar to the distribution of surface states. Tersoff treated the model more quantitatively and has successfully derived predictions for the Schottky barrier heights that are independent of the properties of the metal.[23] The metal-induced gap state model, however, fails by itself to explain the variation of barrier height with metal or with pretreatment, so that other aspects must be considered, such as the presence of defect and impurity states.[24,25] In addition, this model is clearly not applicable to semiconductor/insulator or semiconductor/electrolyte interfaces, where no delocalized metallic states are present.

Spicer *et al.* proposed that pinning might result from the formation of adsorption-induced defects,[26] but that the *nature* of the defect would not be especially sensitive to the identity of the adsorbing species. Photoelectron and photoemission spectra have allowed band edges to be studied for semiconductors coated in gases and insulators, and this has shown that the properties of the overlayer are not the determining factor in the band bending. The *unified defect model* (UDM) can, therefore, explain the similarity in pinning between metal-oxide-semiconductor, metal–insulator–semiconductor, and Schottky junctions, and this generality leads one to ask whether the defects involved might not also be important at the semiconductor–electrolyte junction. As a result, it is of considerable interest to consider the chemical identity of the defects invoked in the UDM. Weber *et al.* have suggested that for GaAs the defect is the As_{Ga} antisite, that is, an arsenic atom on a gallium site, since calculations indicate that the energy of this deffect is close to the pinning position on this semiconductor.[27] As_{Ga} will behave as a double donor, so that it can pin *p*-GaAs faces, and Spicer *et al.* have suggested that it will be associated with a Ga_{As} antisite defect which will provide two acceptor levels between the donor levels and the valence band.[28] The projected situation in the advanced unified defect model (AUDM) is shown in Fig. 8.

Calculations on the Fermi levels that have been observed imply that there are twice as many As_{Ga} antisites as Ga_{As} antisites, which is consistent with the fact that Bridgeman grown GaAs is As rich. Surface treatments will affect the barrier height by altering the amount of excess As or Ga present, and therefore the relative concentrations of the two types of defect. Excess As should shift the Fermi level toward the conduction band whereas excess Ga will shift it toward the valence band.

Woodall and Freeouf[29,30] have criticized the defect models because they require a much higher surface defect density than that observed in the bulk of molecular-beam epitaxy (MBE)-grown GaAs and instead suggest that the Fermi level is determined by the work functions of interface phases. However, the difficulty in estimating the work functions of microcrystalline interface phases makes this model difficult to apply, and the very large variation in

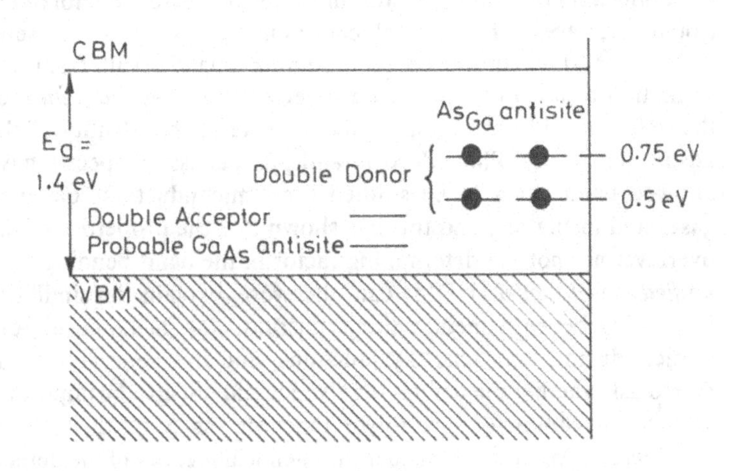

Figure 8. Energy-level diagram from GaAs in the AUDM model.[28] (VBM is Valence Band Maximum.)

work function which is expected on varying the interface is not reflected in the experimental barrier heights. For example, some GaAs and InP interfaces undergo reaction with the metal, so that the work functions of the known metal arsenides and phosphides might be expected to be important, but there is no experimental indication of such large work functions.

The defect-induced gap state (DIGS) concept was introduced from the theory of semiconductor-insulator structures because of the similarity in Fermi level pinning at these interfaces and Schottky barriers.[31,32] The gap states in this model result from a thin disordered layer of semiconductor in which the perturbation from the perfect underlying solid causes mixing of the bonding and antibonding states, and a continuum of states is expected across the bandgap, as in Fig. 9. Surface disorder of the type envisaged here could well be present at the semiconductor/electrolyte interface as well; however, although surface-state distributions of the type predicted have been measured by impedance at metal-insulator-semiconductor structures, it is not clear that such a distribution has been measured at the semiconductor/electrolyte interface.

There is still little unchallenged understanding of the underlying causes of the Fermi level pinning of Schottky barriers and metal-insulator-semiconductor structures even in the gas phase. However, recent work by Brillson and co-workers[33,34] on MBE-grown ⟨100⟩ GaAs has revealed a variation of barrier height with

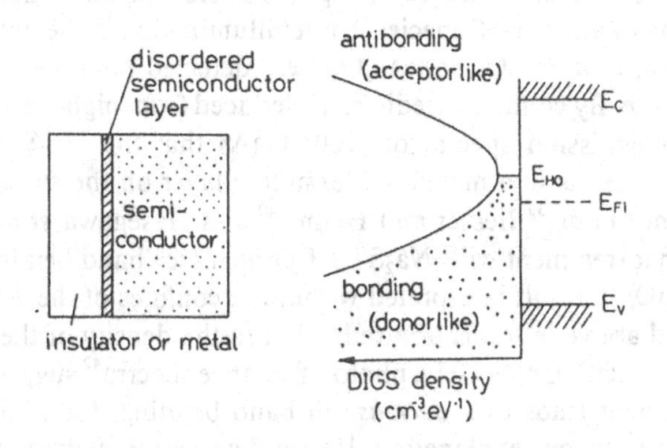

Figure 9. Energetics of the defect-induced gap state model.[32] E_{HO} is the charge neutrality point of the DIGS density spectrum.

metal work function which approaches ideal Schottky behavior. The pinning observed in earlier work is assigned to the very much higher density of defects that can act as deep traps in cleaved melt-grown GaAs as compared to MBE-grown GaAs.

For practical purposes, it has been observed that some chemical pre-treatments appear to much reduce the high surface recombination velocity and possibly alter the surface-state density. One of the most successful of these treatments appears to be the spin coating of films of $Na_2S \cdot 9H_2O$ onto GaAs as used by Sandroff *et al.*[35] Recombination velocity reduction by a factor of 60 resulted at a GaAs/air interface, when the surface was coated in this way, and the surface remained passivated for several days. Photoluminescence showed a 250-fold increase in intensity on treatment, a result again indicative of a large decrease in surface recombination. In order to characterize such processes, it is necessary to observe chemical changes in the first few monolayers of the sulfide/semiconductor interface. Sandroff *et al.*[36,37] used X-ray photoelectron spectra (XPS) to show that Na_2S and $(NH_4)_2S$ solutions will remove the Ga_2O_3 and As_2O_3 layers and result in As XPS peaks at different positions. For the Na_2S-treated surface, the peaks are at the same position as in As_2S_3, but for GaAs treated with $(NH_4)_2S$, the As peak position implied a sulfur oxidation state of -1. It was suggested that this was the result of $Na_2S \cdot 9H_2O$ terminating the ⟨100⟩ surface by S-bridging between arsenide ions but that $(NH_4)_2S$ gives bridging by S—S species as indicated in Fig. 10. There was no evidence of formation of any Ga—S species. Under illumination in the presence of oxygen, both Ga_2O_3 and As_2O_3 were detected after 40 minutes of exposure. By contrast, Tiedje *et al.* deduced from high-resolution UV photoemission spectra of ⟨100⟩ GaAs that the $Na_2S \cdot 9H_2O$ treatment results in a mixed-oxide/sulfide layer on the surface.[38]

Spindt *et al.*,[39] Besser and Helms,[40] and Hasegawa *et al.*[41] all report that treatment with $Na_2S \cdot 9H_2O$ *increases* band bending on GaAs ⟨100⟩, a result interpreted within the confines of the AUDM discussed above in terms of a reduction in the density of the As_{Ga} antisite defect.[39] Conversely, photoreflectance spectra[42] suggest that the treatment leads to a *decrease* in band bending, for which we can see no obvious explanation. It would appear that although the recombination velocity *is* decreased quite drastically on treatment with the sulfide, and the band bending *is* shifted (though the

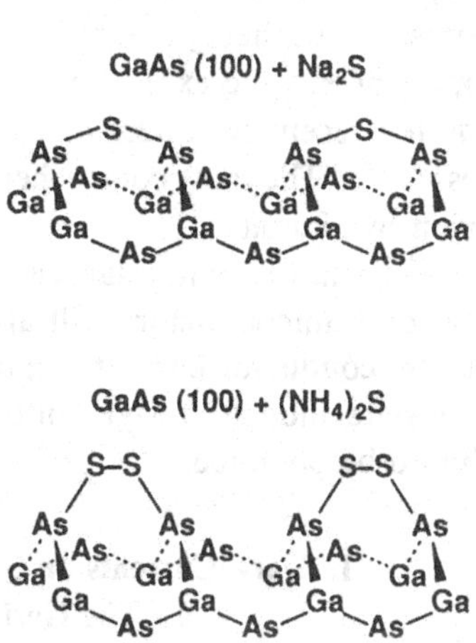

Figure 10. (a) Model for the GaAs surface treated with Na_2S; (b) model for GaAs surface treated with $(NH_4)_2S$.[37]

direction remains controversial), the Fermi level remains pinned. The other conclusion that might, perhaps less charitably, be drawn is that our understanding of even so well studied a surface as ⟨100⟩ GaAs is still quite primitive.

IV. SURFACE STATES AND ELECTROCHEMICAL BEHAVIOR IN THE DARK

The presence of surface states has usually been diagnosed when it has not proved possible to explain experimental electrochemical results in terms of classical models of the semiconductor/electrolyte interface. Where results can be interpreted in terms of surface-state models, this is usually taken to demonstrate the existence of such states, provided that no other explanation, such as oxide growth, appears reasonable. Although the logical weakness of this approach is clear, the paucity of techniques available to the electrochemist has left little alternative.

In order to understand real interfaces, it has been normal to compare their properties to the predictions of models that both include and exclude surface-state effects. Many of the fundamental models have been established for some time, and there are a number of reviews covering both theoretical and experimental aspects,[5,43–56] These reviews often have slightly different emphases; in spite of its age, Myamlin and Pleskov's book[46] remains extremely useful as a repository of theory and experiment on the elemental semiconductors, whereas the explosion of interest in solar energy conversion has led recent reviewers to concentrate on photoconversion processes.[54,56] The most comprehensive recent review of theory is that given by Hamnett.[5]

An analysis of faradaic currents in the dark and under illumination on semiconductors will allow us to test the classical models of semiconductor current response. By comparing successful surface-state models, insight into the underlying surface processes should be obtained.

1. Dark Currents on Semiconductor Electrodes—The Gerischer Model

The major difference between our models of semiconductor electrodes and of metal electrodes in reasonably concentrated electrolyte is that any change in potential applied to a metal is dropped across the very thin Helmholtz layer, and the energy of the electrons at the metal surface is therefore under potential control. Following our discussion above of the potential distribution at the semiconductor/electrolyte interface, it is apparent that for a reverse-biased semiconductor, change in the applied potential will not affect the energy of the electrons at the surface, but will affect the surface *concentration* of charge carriers.

The potential distribution at the semiconductor/electrolyte interface is very similar to that at a metal–semiconductor Schottky barrier. However, it must be emphasized that the similarity between the two types of junction does not extend to their dark-current properties owing to three principal differences between metal and electrolyte[52]:

(i) The metal has a very high density of electronic states distributed across a broad continuum compared to which dissolved or adsorbed redox species offer in general a far lower density of states across a restricted range of energies.

(ii) The redox centers adsorbed on a semiconductor in electrolyte are able to migrate from the surface, but the electrons are in localized energy levels on the species and their energy is changed by the movements of solvent molecules and ions in solution.

(iii) Both reversible and irreversible chemical changes are likely to take place at the semiconductor/electrolyte interface, such as corrosion or oxide layer formation, and the result of such changes will be a drastic change in dark-current properties.

Dark current at a Schottky barrier can be described by the Richardson equation[21] for thermionic emission, which estimates a rate of transport of charge carriers across the depeletion layer. Although Memming has been able to fit some results for semiconductor/electrolyte behavior to the currents predicted for thermionic emission,[57] in general, because of the differences listed above, it is usually the kinetics of electron transfer across the interface that determines the current flow in the dark.

The theory of electron transfer at the semiconductor/electrolyte interface was developed by Marcus[58] and Gerischer[47] among other workers, and we follow the treatment of Gerischer here. The electronic energy level in a redox ion is considered to vary with the thermally fluctuating polarization of its electrolyte medium, which induces a large broadening of the electronic energy levels. The probability of a given electronic energy being adopted at any one time is given by[59]:

$$D_{ox}(E) = (4\pi\lambda k_B T)^{-1/2}\exp[-(E - E^0_{ox})^2/4\lambda k_B T] \quad (12)$$

$$D_{red}(E) = (4\pi\lambda k_B T)^{-1/2}\exp[-(E - E^0_{red})^2/4\lambda k_B T] \quad (13)$$

for the oxidized and reduced forms of the redox couple, respectively, within a harmonic approximation, where λ is the *reorganization* energy and corresponds to the change in energy of the outer electronic level to or from which the electron is being transferred as the redox species is deformed to the transition state. The assumption that we have made is that λ is the same for both oxidized and reduced states; from this, the separation in energy between the maxima of D_{ox} and D_{red} is easily shown to be 2λ, and a plot of $D(E)$ is shown in Fig. 11. It should be noted that $D(E)$ is not directly analogous to an electronic density of states as used in solid-state physics because the broadening is caused by a potential slowly varying in *time* rather than *space*, and the different states in it do not overlap in space and time.

The fundamental assumption of Gerischer's theory is that electron exchange should take place between electronic states of energies within $k_B T$ of each other. The electron is much lighter

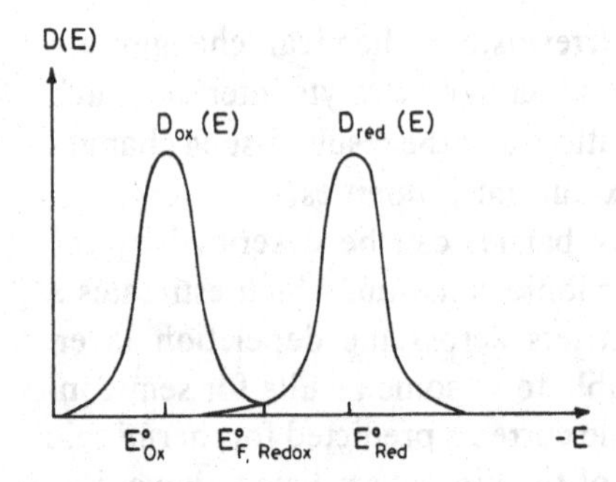

Figure 11. Distribution function for electronic redox energy levels in electrolyte.

than the nucleus, and this mass difference allows us to separate electron and nuclear wave functions. Electron transfer can then be considered to take place essentially vertically on the potential energy diagram, a result familiar as the Franck–Condon principle in spectroscopy. Within the theory of electron transfer between molecules, this process should take place without any significant change in the positions or momenta of the atoms constituting the molecular framework. If energy cannot be dissipated or absorbed by the motion of the atoms, then, within the one-electron model, conservation of energy will require the electron energy to remain unchanged on transfer. Thus, for electron transfer to occur, the electronic energies of donor and acceptor must fluctuate to equality, or, at an electrode, the electronic energy of the redox level must fluctuate to that of the conduction or valence band levels of the semiconductor.

The rate of electron transfer at any one energy can be written as proportional to the density of electronic states in the solid at that energy and to the probability of the redox ion having that electronic energy and also to a transmission factor $\kappa(E)$, which includes the quantum mechanical probability for electron transfer summed over all distances while the redox couple is near the surface. The total current in the case of a cathodic process can therefore be written as such an expression integrated over all possible energies at the surface:

$$j^- \propto C_{\text{ox}} \int_{-\infty}^{0} \kappa(E) \cdot D_-(E) \cdot D_{\text{ox}}(E)\, dE \qquad (14)$$

A similar expression can be written for anodic processes, where j^+ is the electron transfer flux from the reduced species in solution to the electrode surface:

$$j^{+} \propto C_{\mathrm{red}} \int_{-\infty}^{0} \kappa(E) \cdot D_{+}(E) \cdot D_{\mathrm{red}}(E)\, dE \tag{15}$$

and the integration is over all bound states. $C_{o}x$ and C_{red} are the concentrations of oxidizing and reducing agent in solution, and $D_{-}(E)$ and $D_{+}(E)$ are the densities of unoccupied and occupied electronic states on the electrode.

The formulas above are correct for the metal/electrolyte interface, and the change in current with potential can be traced clearly within this model to the changing relative positions of metal and redox energy levels. For the semiconductor/electrolyte interface, the situation is more subtle; the integral above can, in principle, be partitioned into contributions from the valence band (VB) and the conduction band (CB). In practice, only one of these contributions will normally be significant. The second difference is that the actual concentration of carriers in the semiconductor at the surface must be included. The final difference is that we must, at least in principle, include the possibility of tunneling across the depletion layer, particularly for highly doped semiconductors. The general situation is illustrated in Fig. 12, where the densities of filled and

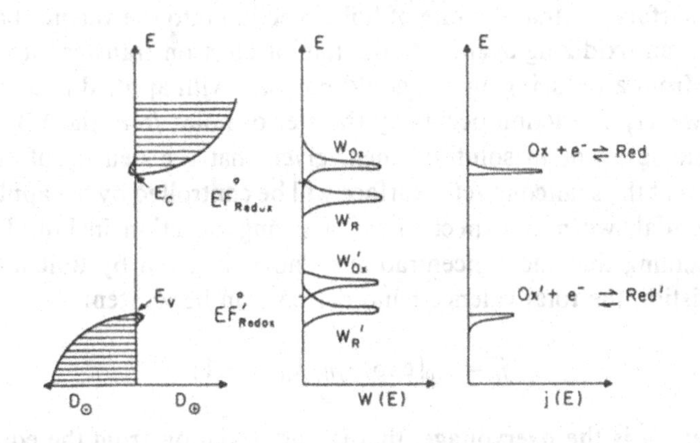

Figure 12. Distribution functions for electrons, $D_{-}(E)$, and holes, $D_{+}(E)$, and for the oxidized and reduced forms of the redox couples. The rates of electron transfer are also plotted and show that the couple W interacts with the CB, and W' with the VB.[47]

empty states in the semiconductor are plotted, as well as the distribution functions for two separate redox couples and the resulting rate of electron transfer.

Figure 12 hides a number of conceptual problems; it should be remembered that the energies of the CB and the VB in the semiconductor are measured relative to the vacuum electron level, whereas the energies of the redox couple are measured relative to a standard redox couple in solution. Relating the one to the other must be done through theoretical estimates of solvation energies, and the position of the standard hydrogen electrode has now been estimated by a variety of such approaches to lie between 4.3 and 4.8 eV below the vacuum level.[60] Using this type of estimate, it is possible to position the redox levels with respect to the energy bands of the semiconductor and hence to estimate the likely rate of electron transfer. Qualitatively, we can say that if the distribution function $D(E)$ of the redox couple lies mainly between the VB and the CB, then electron transfer between semiconductor and redox couple should be severely inhibited. Of course, the likelihood of this occurring will depend on both the bandgap and the size of the reorganization parameter λ; given that the latter can be estimated to be of the order of at least 1 eV,[49] then it is clear that this condition will not be met for narrow-bandgap semiconductors.

For the classical Schottky barrier, the band edges are fixed at the surface, so that the rate of hole injection into the valence band from an oxidizing agent and the rate of electron transfer into the CB from a reducing agent should not vary with applied potential. However, if reaction occurs by transfer of holes *from* the VB to a reducing agent in solution, then, given that the density of such holes at the semiconductor surface will be controlled by the applied potential, we may expect to see a strong variation in hole flux. Assuming that the concentration of holes is given by Boltzmann statistics, the total valence band current can be written:

$$j_v = j_{v,0}[\exp(e_0\eta/k_BT) - 1] \tag{16}$$

where η is the overvoltage, that is, the deviation from the equilibrium potential, E^0, where no net current flows, and $j_{v,0}$ is the exchange current density at E^0 and is equal to the magnitude of the hole injection current into the VB.

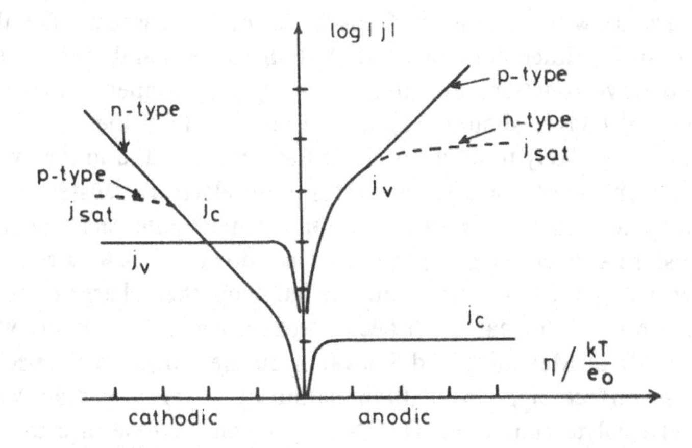

Figure 13. Current-voltage curves for the partial currents in the VB and the CB in the Gerischer model.

Similarly, for an *n*-type semiconductor, the density of electrons in the CB is controlled by the applied potential and

$$j_c = j_{c,0}[1 - \exp(-e_0\eta/k_B T)] \tag{17}$$

With these expressions, the current/overvoltage behavior of both *n*- and *p*-type semiconductors can be predicted provided the effect of ionic transport in the solution can be neglected, as shown in Fig. 13.

When an *n*-type semiconductor is held at very anodic potentials, so that the Fermi level drops below the VB edge, a very large density of holes is expected to develop at the surface, as in Fig. 4c. In fact, the presence of the depletion layer will act as a barrier to the migration of holes to the surface, so that the rate of reaction will no longer be controlled by the rate at which charge transfer occurs at the interface, but by the rate of thermal generation of holes in the depletion layer. We anticipate, therefore, that anodic dark currents in the inversion region should reach a saturation value at sufficiently high anodic potentials.

2. Difficulties with the Gerischer Approach

Although the model described above gives an important insight into dark currents on semiconductors, its predictions are frequently at variance with experiment. Gerischer's approach assumes that the main surface interactions are with the bulk valence- and conduction-band wave functions and that the potential dropped across the Helmholtz layer should remain constant. In fact, the electronic structure is likely to be distorted at the interface, and in the event of a high density of intergap states, then electron transfer could take place via these states. Certainly, when quite facile charge transfer has been observed to and from redox couples whose redox energy is predicted to lie within the bandgap, then charge transfer via surface states has often been invoked. An early example was the work of Memming and Schwandt on the reduction of species at the surface of *p*-GaP.[61] Illumination of *p*-GaP in contact with an electrolyte containing $Fe(CN)_6^{3-}$ gave an increase in cathodic current as electrons excited into the CB were transferred to the redox species. However, the dark cathodic current was too large to be explained by the transfer of minority carriers through the CB, a process that will be severely limited by thermal excitation in a semiconductor with a bandgap of 2.3 eV. Given the fact that the value of λ for the ferro/ferricyanide couple is believed to be only ca. 0.5 eV, overlap of $D_{ox}(E)$ and $D_{red}(E)$ with both the VB and the CB is not possible, and charge transfer was postulated to take place through surface states in the bandgap after inelastic transfer from both the VB (in the dark) and the CB (on illumination) as shown in Fig. 14.

As the concentration of Fe(III) in solution was increased, the reduction current in the dark eventually saturated, an effect ascribed to the transfer of electrons from the VB to the surface state becoming rate-limiting, instead of transfer of charge from the surface state to Fe(III). In this system, illumination also results in hydrogen evolution at the CB edge, which will compete with the surface state for the photogenerated electrons.

In another early paper, Tyagai and Kolbasov[62] investigated the current-voltage characteristics of CdS and CdSe in the presence of Fe^{III}/Fe^{II} couples and obtained *fractional* power relationships for the dependence of the current on the redox species concentrations. To interpret this result, Tyagai and Kolbasov assumed an exponential distribution of interface states and, by using Schottky-Read kinetics for the surface-state population,[4,5] were able to model the concentration dependence found. However, the system studied

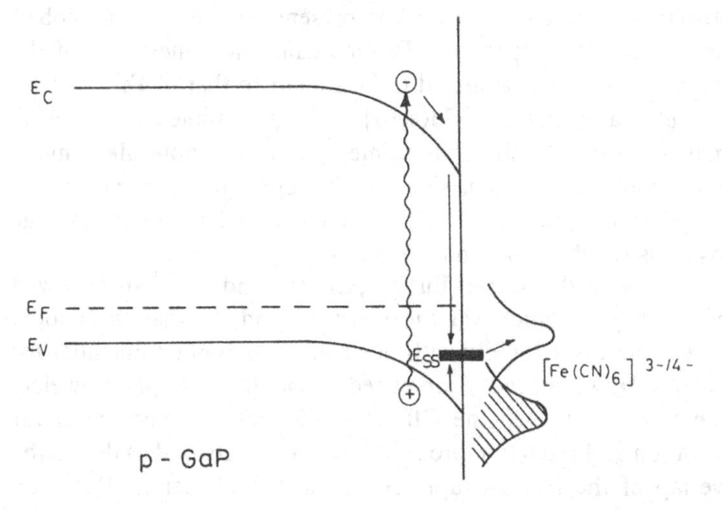

Figure 14. Energy diagram for *p*-GaP in the presence of $[Fe(CN)_6]^{3-/4-}$. E_{ss} is the postulated surface-state energy.[61]

is clearly complex: the presence of the redox couples led to surface pitting on the electrodes, and layers of sulfur or selenium were observed to form. The results also depended in a complex way on the amount of current passed by the electrode, and later workers have sought more compelling evidence for the mediation of surface states in this system.

Memming *et al.*[63,64] in more recent investigations of CdS, support the assertion that charge transfer must be capable of taking place through surface states. They observed efficient hole transfer from CdS to $[Fe(CN)_6]^{4-}$, hydroquinone, and polysulfide redox couples, despite the fact that the energy overlap of these couples with the VB is expected to be very small. Changes in the degree of adsorption were considered as possible sources of the deviation from the expected pattern, but the most convincing explanation was the presence of surface states lying close in energy to the VB edge, which facilitate electron transfer from the redox couple to the VB.

It becomes more difficult to predict the route of charge transfer in narrower bandgap semiconductors such as GaAs and InP, since the thermal broadening of the majority of redox levels is similar in size to the bandgap itself. Meissner *et al.*[65] considered the dark currents on InP and GaAs in the presence of $Eu^{3+/2+}$ and cobalt sepulchrate, $[Co(sep)]^{3+/2+}$. The reorganization energy, λ, of the $Eu^{3+/2+}$ couple was assumed to be similar to that of $Fe^{3+/2+}$ (i.e., ca. 1 eV), and that for $[Co(sep)]^{3+/2+}$ was assumed to be rather smaller (ca 0.4 eV) since the framework of this molecule is much more rigid. The two Gaussian distributions, $D(E)$, are plotted on a logarithmic scale in Fig. 15 and compared to the band-edge positions for the two semiconductors.

On GaAs, the couples $Eu^{3+/2+}$, $Cr^{3+/2+}$, and $V^{3+/2+}$ are observed to undergo oxidation via the valence band, so that the anodic current increases rapidly with more positive applied potential for *p*-GaAs. However, reduction of redox couples takes place by electron transfer through the CB. This difference in mechanism for oxidation and reduction processes can be understood as due to the overlap of the redox couple energy with both bands. However, although the distribution of electron states is much narrower for the $[Co(sep)]^{3+/2+}$ couple, and energy overlap with the VB is expected to the very small, it was found that oxidation still occurred, even for this couple, by hole transfer from the VB. This is clearly incompatible with Gerischer's model, and Meissner *et al.* concluded

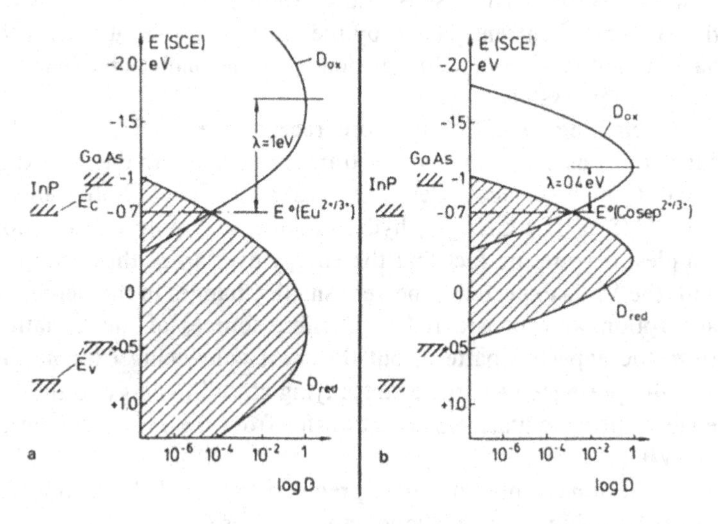

Figure 15. Expected distribution of electron states for $Eu^{3+/2+}$ (a) and $[Co(sep)]^{3+/2+}$ (b) compared to the position of the band edges in GaAs and InP.[65]

that either the electron state distribution must be different from Gaussian or electron transfer takes place via surface states.

The behavior of InP is rather different from that of GaAs despite its similar bandgap and band-edge positions.[66] Both oxidation and reduction of the $Eu^{3+/2+}$ couple takes place via the CB in InP, so that, in *n*-type InP, electrons are injected into the CB as Eu^{2+} is oxidized. This is surprising, as far better energy overlap is expected for the filled Eu^{2+}(aq) states with the VB than with the CB. For *p*-InP, electron injection into the CB is not detected because electron transport into the bulk is opposed by the depletion-layer field. No additional dark current is therefore observed on addition of reducing agents that overlap energetically with the VB, and the implication is that VB electron transfer is inhibited for InP. Meissner and co-workers[66] suggested that this inhibition results from a thin tenacious layer of oxide on InP; the oxide In_2O_3 has a bandgap of 3.5 eV, and transport of charge carriers in In_2O_3 with energies in the forbidden gap is likely to be highly unfavorable. If the band edges at the InP/In_2O_3 junction are considered to match as in Fig. 16, then electron transfer to the CB should occur readily but that to the VB will require considerable activation.

The parallels between this argument and the surface-state model are strong, and they become stronger as the oxide layer approaches monolayer thickness, and it is no longer meaningful to talk of an oxide band structure. Then we would say that electron transfer is most probable where the surface electronic density of states is highest.

In addition to the problems posed for the Gerischer model by the presence of surface states, we have also neglected the effects of charge carrier tunneling through the depletion layer either directly to the electrolyte or to surface states. This problem has been considered by several authors.[67-72] Calculations indicate that tunneling is only likely to be significant for very highly doped semiconductors, with $N_D \geq 10^{19}\ cm^{-3}$, where the depletion layer is narrow and the fields are high. Anodic potential scans on ⟨100⟩ *n*-GaAs in 1 *M* NaOH(aq) show that breakdown of the barrier and subsequent anodic dissolution takes place at higher anodic potentials, the lower the donor density, as shown in Fig. 17. However, even for $N_D \approx 10^{15}\ cm^{-3}$, *n*-GaAs avalanche breakdown occurs at a potential of about 25 V, a factor of ten lower than that predicted

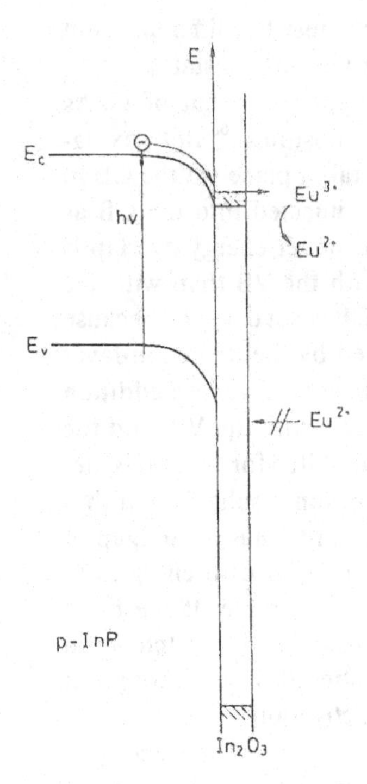

Figure 16. Model of energy levels for a thin layer of In_2O_3 on InP.[66]

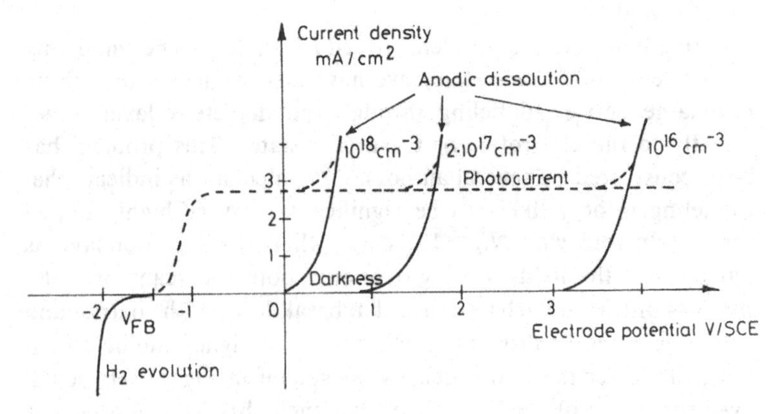

Figure 17. Current-potential curves for ⟨100⟩ *n*-GaAs in 1 *M* NaOH.[73]

or indeed observed on Schottky diodes. Tranchart *et al.*[73] ascribed these results to the presence of a surface defect at which microbreakdown can occur, leading to the propagation of an etch pit into the depletion layer. In a theoretical study, Schmickler[71] has shown that the presence of intragap states, such as donor levels, can considerably increase the rate of tunneling through the depletion layer, and this observation is likely to be correlated with the early breakdown of the semiconductor-electrolyte barrier.

A final problem in the application of Gerischer's model is that the presence of surface states, so freely invoked to account for discrepancies between theory and practice, leads to the possibility of complete or partial Fermi level pinning. If the energies of the CB and the VB are *not* fixed at the interface, then the electron transfer rates will all alter with applied potential since at least some of this potential will appear across the Helmholtz layer. The fact that this possibility will always exist makes *any* deduction purely from dark-current measurements very hazardous. We shall see below that partial Fermi level pinning can explain many of the apparent deviations from Gerischer's model.

3. Corrosion Reactions in the Dark—Strong Interactions at the Interface

(*i*) *Mechanism and Thermodynamics of Anodic Corrosion*

All semiconductor electrodes in electrolyte will undergo either reductive or oxidative decomposition if a suitable potential is applied or a suitable redox couple is present. Under these circumstances, the ideal model of the semiconductor surface, essentially unaffected by the electrolyte, no longer applies. The electronic structure of the surface is drastically altered, and the language of chemical change must be applied to the semiconductor; in turn, we would hope to reconcile this with our band energy level approach as outlined above. Gerischer has considered the mechanisms and thermodynamics of a variety of corrosion processes,[74–76] and we consider in particular the process of anodic corrosion in this section.

In general, decomposition processes take place in two steps:

(i) A surface bond is broken by chemical reaction with a species from the electrolyte under the influence of the applied potential. This will result in one surface atom being bonded to the electrolyte species and the other atom having a radical character.

(ii) It is likely that this second atom will also rapidly react with the electrolyte to form a second surface covalent bond.

Each step involving the formation of a covalent bond between a radical and an electrolyte species requires the loss or gain of an electron from the solid. For a semiconductor, the electron can be transferred to or from either the valence or the conduction band, and it is clearly important which band is involved, because the energy difference between them is large. The oxidative decomposition of an elemental semiconductor can proceed by a mechanism involving charge carriers or by a chemical decomposition. The two steps in a mechanism involving charge carriers can take place in one of the following ways:

- Step 1a

$$(\mathrm{A{-}A})_s + \mathrm{X} \rightarrow (\mathrm{AX})^+ + \mathrm{A}_s^{\bullet} + e_{\mathrm{cb}}^-$$

- Step 1b

$$(\mathrm{A{-}A})_s + p_{\mathrm{vb}}^+ \rightarrow (\mathrm{AX})^+ + \mathrm{A}_s^{\bullet}$$

- Step 2a

$$\mathrm{A}_s^{\bullet} + \mathrm{X} \rightarrow (\mathrm{AX})^+ + e_{\mathrm{cb}}^-$$

- Step 2b

$$\mathrm{A}_s^{\bullet} + p_{\mathrm{vb}}^+ + \mathrm{X} \rightarrow (\mathrm{AX})^+$$

Steps 1a and 1b are alternative initial steps in which a surface bond is broken by loss of an electron to the CB or gain of a hole from the VB. The process represented by step 1b should become more favored as hole density increases near the surface and should therefore be distinguishable from step 1a. If the initial step is rate-limiting because it requires the rupture of a covalent bond rather than a radical reaction, then it is more difficult to determine whether the second step is of the form in step 2a or 2b. The VB in most semiconductors apart from the transition-metal chalcogenides is made up of bonding-type electronic states, so that the presence of a hole will lead to a general weakening in binding. Step 1b is, therefore, the transition from the delocalized absence of a binding electron to the localization of electron loss on an atom–atom bond.

Alternatively, oxidative decomposition may take place by the following mechanism:

$$(A{-}A)_s + X{-}X \rightarrow AX + AX$$

This mechanism is a chemical decomposition in which the semiconductor's charge carriers play no part, and the applied potential should not affect the rate.

The energy profiles for oxidative decomposition for the mechanism involving charge carriers are shown in Fig. 18, which is taken from the work of Gerischer and Mindt.[75] The process of electron loss to the conduction band (step 1a) is expected to be energetically less favorable than electron loss to the valence band, since the latter lies much lower in energy. As a result, the activation energy for step 1a will be larger than that for step 1b by an amount roughly equal to the bandgap. The surface radical energy is likely to lie within the bandgap and, at first sight, might be expected to be closer to the VB than the CB because the radical orbital will retain some bonding character. However, the energy of the radical orbital will be profoundly influenced by interaction with the electrolyte, and any predictions are likely, at best, to be highly qualitative. These surface radicals will, of course, constitute electronic

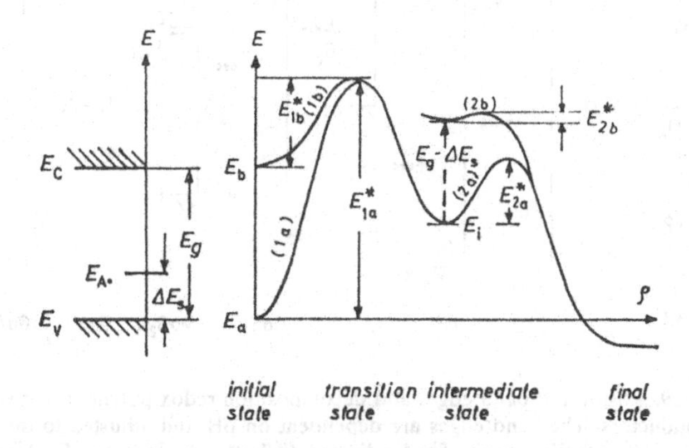

Figure 18. Energy profiles for oxidative decomposition with respect to the energy-band model.[75]

surface states, and much work has been carried out to determine the energies of such states.

Thermodynamically, it is possible to associate a redox potential with each decomposition reaction; for example,

$$GaAs + 6H_2O \rightarrow Ga(OH)_3 + H_3AsP_3 + 3H_2$$

$$GaAs + \tfrac{3}{2}H_2 \rightarrow Ga + AsH_3$$

Oxidation will be thermodynamically feasible when the Fermi level at the surface drops below the redox potential of the oxidation reaction. The decomposition potentials for oxidation and reduction have been calculated for various semiconductors by Gerischer[76] and are shown in Fig. 19 together with the corresponding band edges, all at pH 7. As can be seen, all the semiconductors shown might be expected to undergo some oxidative decomposition while the Fermi level remains within the depletion layer. The decompositions are, however, frequently kinetically unfavorable, especially if the surface is stabilized by the presence of a redox couple to which

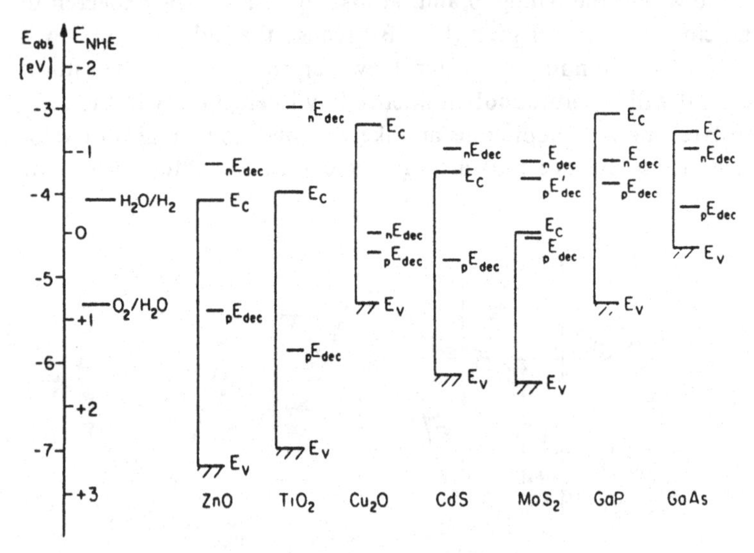

Figure 19. Position of band edges and decomposition redox potentials for various semiconductors. The band edges are dependent on pH and adjusted to the point of zero zeta potential (pzzp) [and adjusted to both an absolute (E_{abs}) and an electrochemical (E_{NHE}) scale]. The decomposition potentials are calculated for the assumed decomposition reactions; they should be correct to ±0.2 eV.[76]

rapid transfer of holes can take place at the surface. Stabilization has been particularly closely sutdied in phtoeffects and is duscussed in more detail below.

(ii) Anodic Decomposition and Etching of GaAs

The electrochemical oxidation and etching of gallium arsenide is of particular interest because of its use in device preparation and has therefore been carefully studied.[77-79] Current-voltage curves by themselves are not sufficient to characterize the oxidation process since, although oxidizing agents may inject a substantial number of holes into the VB, the holes that lead to surface oxidation and dissolution will not contribute to the nett current measured. Instead, the rate of oxidation of the redox couple at the surface must be measured independently to obtain the total rate of hole injection. The most commonly used technique is that of the rotating ring-disk electrode (RRDE),[80] in which the working (disk) electrode is a suitably turned piece of semiconductor crystal, and the change in concentration of the redox species is monitored at the ring. Fortunately, the products of GaAs oxidative dissolution are not electroactive, so they do not interfere with the detection of the redox species at the ring. Reactions where the rate is limited by diffusion of redox species to the surface are readily distinguished through the linear variation of the disk current with the square root of the rotation speed described by the Levich equation.

Kelly and Notten[81] have studied the corrosion of *p*-GaAs in the presence of acidic Ce(IV): as can be seen from Fig. 20a, the rate of reduction of Ce(IV) is independent of the applied potential. The limiting cathodic current in Fig. 20b corresponds to the situation in which all injected holes are transported to the bulk of the electrode, and this current obeys the Levich equation. It follows that the rate of hole injection is controlled by the rate at which Ce(IV) ions can diffuse to the electrode surface and is independent of the applied potential. Now, at more anodic potentials, the rate of transport of holes from the surface falls, and some of the injected holes start to take part in corrosion reactions, so the observed disk current falls. When the applied potential becomes sufficiently positive, the majority holes will migrate more rapidly from the bulk to the surface, and the current flow becomes anodic.

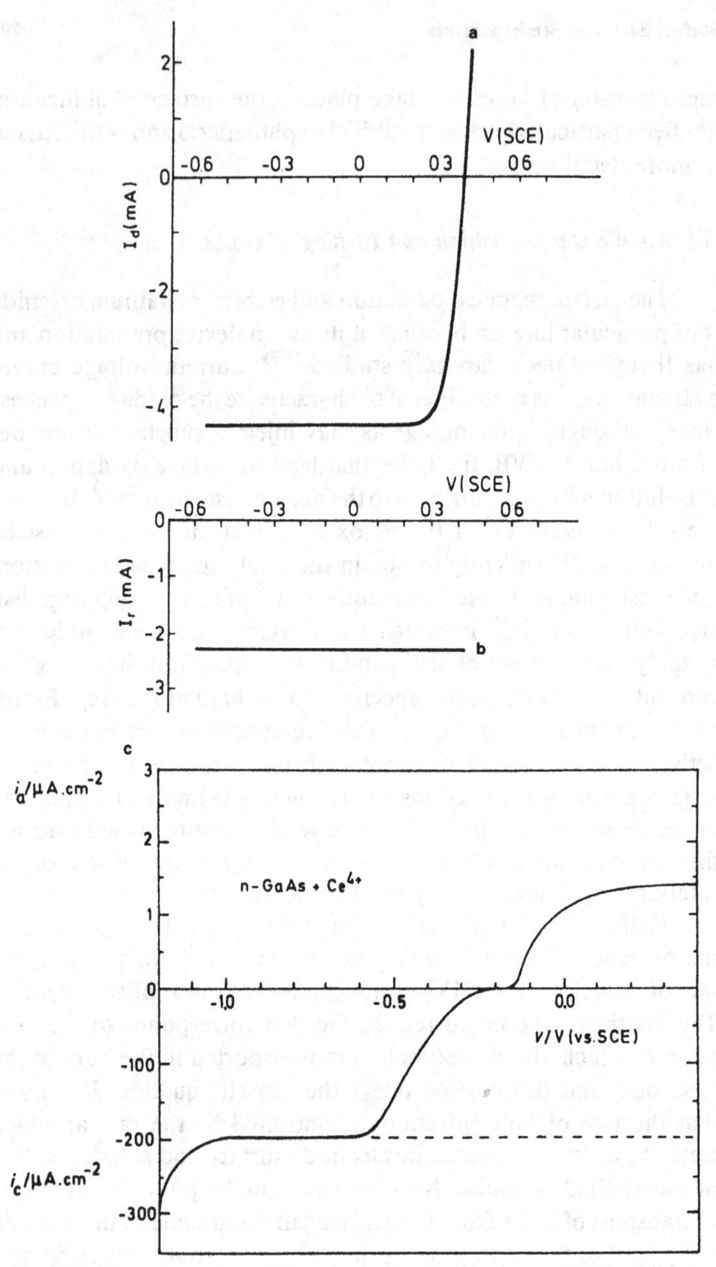

Figure 20. Ring current (a) and disk current (b) for *p*-GaAs in 0.05 *M* $Ce^{4+/0.5}$ *M* H_2SO_4 swept at 50 mV s^{-1} and rotated at 83 Hz[81] and disk current (c) for *n*-GaAs in 10^{-2} *M* Ce^{4+}/2 *M* H_2SO_4 at 17 Hz.[84]

Decker *et al.*[82] have measured the ring and disk currents of *n*-GaAs electrodes in acidic Ce(IV) and alkaline $[Fe(CN)_6]^{3-}$ electrolytes. In this case also, the ring currents are independent of applied potential, and the disk current obeys the Levich equation, indicating that the rate of hole injection is limited by the rate of transport of oxidizing agent to the electrode surface. The cathodic disk current tends to a limiting value in *n*-GaAs, but at sufficiently negative potentials reductive decomposition of the electrode will start to take place. As the applied potential is swept positive, the cathodic disk current decreases as injected holes become involved in corrosion reactions. Unlike that for *p*-GaAs, the disk current for *n*-GaAs should become negligible at very positive potentials since the electrode is in deep depletion and all the injected holes will be trapped at the surface and take part in corrosion. In fact, Gerischer and Wallem-Mattes[83] observed a limiting anodic current a factor of 400 times less than the limiting cathodic current. The anodic current must result from electrons excited into the conduction band at the surface and then accelerated into the bulk by the electric field. Vanmaekelbergh *et al.*[84] have studied the effect of rotation rate on the relationship between limiting anodic and cathodic currents on *m*-GaAs for acidic Ce(IV) and alkaline $[Fe(CN)_6]^{3-}$ electrolytes; the results, shown in fig. 20c, are interpreted according to a corrosion mechanism already suggested for photoanodic attack,[85] with an additional step due to thermal excitation of an electron from a decomposition intermediate:

$$(GaAs)_{surf} + p^{+}_{vb} \rightarrow X_1; \qquad \text{rate constant } k_1$$

$$X_1 + X_1 \rightarrow X_2 + (GaAs)_{surf}; \qquad \text{rate constant } k_2$$

$$X_i + X_1 \rightarrow X_{i+1} + (GaAs)_{surf}; \qquad \text{rate constant } k_{i+1}$$

$$X_5 + X_1 \rightarrow \text{products} + 2(GaAs)_{surf}; \qquad \text{rate constant } k_6$$

$$X_1 \rightarrow X_2 + e^{-}_{cb}$$

Initial hole injection into a GaAs surface unit leads to an intermediate X_1 which is taken to be mobile on the surface so that subsequent oxidation of other intermediates can take place by reaction with X_1. Further oxidation of X_1 can also take place by thermal excitation of an electron to the conduction band to give an anodic current.

The nature of the electrically active decomposition intermediate or surface state on GaAs is not known, but Vanmaekelbergh *et al.*[85] suggested that it reacts with the solvent to give a neutral species:

$$X_1 + H_2O \rightleftarrows X_1OH + H^+$$

It is more likely for such a neutral species to be able to combine with a positively charged intermediate than for two positively charged intermediates to combine:

$$X_1 + X_1OH + H^+ \rightarrow X_2 + H_2O$$

Heller[86] has considered the way in which chemical species are likely to alter the surface-state distribution on semiconductor electrodes. He associated the high surface recombination velocity of GaAs with the presence of elemental arsenic, which induces surface states in the bandgap. The origin of the arsenic may lie in the thermodynamically very favorable reaction between As(III) and the GaAs surface:

$$As_2O_3 + 2GaAs \rightarrow Ga_2O_3 + 4As; \qquad \Delta H = -260\ \mathrm{kJ\ mol^{-1}}$$

Solomun *et al.*[87] observed elemental As by photoelectron spectroscopy (PES) on the surface of $\langle 110 \rangle$ *n*-GaAs electrodes that had been anodically oxidized in acidic or neutral solution, although accumulation did not seem to take place in alkali. The gallium oxides dissolve into the electrolyte after formation.

It is reasonable to expect the decomposition intermediate to have an unpaired electron on As rather than Ga, as As is the more electronegative atom. The *n*-GaAs electrodes all give significant anodic currents, which are particularly large for surfaces that have been heavily disordered by ion implantation. Disordered surfaces are expected to contain a high density of electronic states in the bandgap.

The anodic current transients are larger for *n*-GaAs in alkaline than in acid solutions, which is interpreted in terms of surface states lying closer to the CB edge in the case of alkaline electrolytes. Thermal excitation of electrons into the CB may then take place. Solomun *et al.*[87] used an argument similar to that of Heller[86] to explain why the defect state should lie closer to the CB in alkali. The As · radical will interact with nucleophile from solution to give bonding and antibonding states as shown in Fig. 21. In alkaline

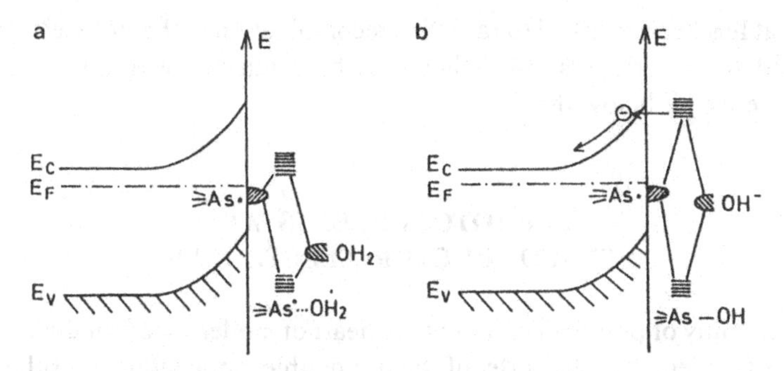

Figure 21. Molecular-orbital-type energy scheme for the interaction of an As · intermediate with H_2O (a) and OH^- (b).[87]

electrolyte, the nucleophile reagent will be OH^- rather than H_2O, so that the energy matching and overlap are likely to be larger, and the surface-state antibonding character level is likely to be high in energy.

Such arguments should be regarded with some caution since, as we have already indicated, the molecular origin of surface states on ⟨110⟩ GaAs is not understood even for adsorption under ultra-high-vacuum (UHV) conditions. Changes at the submonolayer level, particularly at active sites on the electrode, are likely to play a dominant role, possibly in ways that cannot easily be accommodated within one-electron energy level schemes of the sort typified in Fig. 21.

A study[88] by Vanmaekelbergh and Kelly of the anodic decomposition of *n*-GaP in aqueous acidic Ce(IV) showed that hole injection was diffusion-limited only for lower rotation rates and higher temperatures. Results for the limiting anodic current were analyzed in this case on the basis of the model used in an earlier paper.[84] It was assumed that the rate depended on the temperature only by virtue of the thermal excitation of electrons from the first decomposition intermediate to the CB according to:

$$k_2^{\mathrm{ex}} = k_{2,0}^{\mathrm{ex}} \exp(-E_a/k_B T) \tag{18}$$

where E_a is the activation energy of the process, which is just taken to be the energy difference between the intermediate and the CB; this is found from the analysis to be 0.6 eV. The authors pointed out that this energy corresponds to a luminescence peak at 1.6 eV that has been ascribed to radiative recombination at the surface[89-91] and whose origin is also believed to be connected with a surface state 0.6 eV below the CB.

V. PHOTOCURRENTS AT SEMICONDUCTOR ELECTRODES

The study of photoeffects lies at the heart of modern semiconductor electrochemistry, because of their possible application to solar energy conversion. Figure 22 shows the situation schematically for illumination of an *n*-type semiconductor, where the holes photogenerated in the VB are accelerated toward the surface, at which they may react with one or more redox couples, while the photogenerated electrons are swept into the interior of the semiconductor. Theoretical treatment of photocurrent behavior is difficult since:

(i) diffusion of charge carriers cannot be separated from migration under the influence of the space-charge field;

(ii) both minority and majority carriers must be considered under nonequilibrium conditions;

(iii) recombination of electrons and holes may take place in the space-charge layer as well as the bulk;

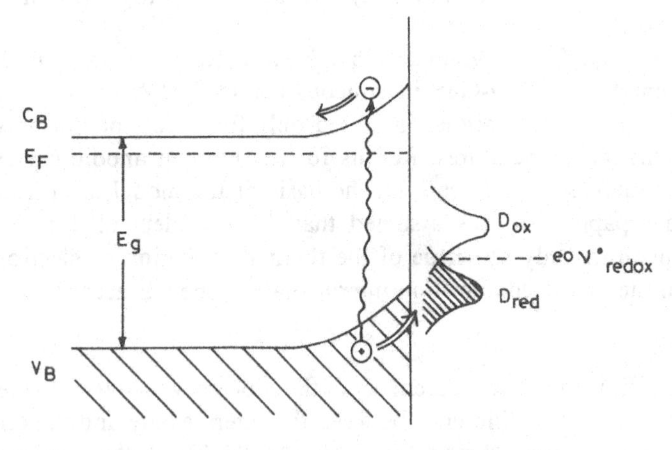

Figure 22. Energetic parameters for photoexcitation at an *n*-type semiconductor electrode.

(iv) surface boundary conditions are difficult to define if surface recombination does take place, or if the rate of charge transfer to the electrolyte is slow compared to the rate of generation;

(v) large concentrations of photogenerated minority charge carriers within the depletion layer or trapped at the surface will change the potential distribution across the interface.

Further complications will be found in real systems as a result of photoinduced chemical change at the interface leading to hysteresis, surface inhomogeneity, and possibly complex adsorption/reaction kinetics on the surface.

From the list of theoretical difficulties, it should be clear that localized electronic states could have a major effect on points (ii)–(v). Within the depletion layer, deep trap states in the bandgap act as a route for electron–hole recombination and can significantly change the concentration and flux of both types of carrier. Surface electronic states are also likely to act as recombination centers and may also act as intermediates for charge transfer to solution. Furthermore, high densities of photogenerated states trapped in surface states would significantly change the potential drop across the Helmholtz layer and the depletion layer, an effect which can be explored by impedance, photocapacitance, and electroreflectance techniques.

Modern photocurrent models have been reviewed by Peter[54] and, with particular emphasis on the theoretical aspects, by Hamnett.[5] The transport of minority carriers (holes of concentration p) within the (n-type) semiconductor is described by the transport equation

$$D_p \cdot d/dx(dp/dx + e_0 Ep/k_B T) + g_p = R_p \qquad (19)$$

where D_p is the diffusion coefficient, g_p is the rate of hole generation, and R_p is the rate of loss through recombination at point x, and it is assumed that the mobility is given by the Einstein relationship $\mu = e_0 D/k_B T$. The first term in the parentheses results from hole diffusion and the second from migration under the influence of the depletion-layer electric field E. The generation rate, g_p, will decrease with increasing distance x into the semiconductor, since the light flux decreases with absorption following Beer's law. Hence,

$$g_p = \alpha\phi_0\theta\, e^{-\alpha x} \qquad (20)$$

where α is the absorption coefficient, and ϕ_0 is the light intensity at the surface. The quantity θ is the quantum efficiency for the generation of mobile electron-hole pairs in absorption and is usually assumed to be unity. The recombination rate depends in a complex manner on the density and population of interband states and will be discussed in detail later.

Gärtner Equation. An early approximate solution to Eq. (19) was given by Gärtner,[92] who derived an expression for the flux of minority carriers toward the surface of an illuminated reverse-biased p-n junction. When transport of photogenerated carriers to the electrode surface and not charge transfer at the surface is rate-limiting, then the Gärtner expression should be applicable to semiconductor photoelectrochemistry as well.

Gärtner assumed that any holes formed in the depletion layer would be swept to the surface without any recombination taking place, so that the depletion-layer contribution to the surface flux can be calculated simply by integrating the generating function over the depletion layer. In the bulk of the semiconductor, beyond the depletion layer, photoexcitation can also take place, but here the transport equation takes the simplified form

$$D_p \cdot d^2p/dx^2 + \alpha\phi_0\theta\, e^{-\alpha x} = kp \tag{21}$$

where the recombination law is assumed to be first order, with a rate constant k. This is related to the physically important variable L_p, the minority carrier diffusion length, and to D_p, by the expression $k = D_p/L_p^2$. By further assuming that the concentration of holes at the boundary between bulk and depletion layer is zero (i.e., that all the holes in the depletion layer are immediately swept to the surface) and that the hole concentration tends to zero in the semiconductor bulk, the current due to minority holes crossing the interface is given by

$$J = e_o\phi_0\theta[1 - e^{-\alpha W}/(1 + \alpha L_p)] \tag{22}$$

where W is the depletion-layer width defined above.

This expression has been widely applied in the interpretation of photocurrents because of its relative simplicity. By fitting the photocurrent at high depletion potentials to the formula, and measuring W, α, and ϕ_0 independently, the value of L_p can be estimated. Unfortunately, it is far from easy to measure the incident light intensity, since a correction must be made for that light reflected from the surface, and ϕ_0 refers to the intensity of light that is transmitted across the interface. As a result, quantitative estimates of the quantum efficiency for photocurrent conversion are far from easy.

Lemasson and co-workers[93,94] have measured the photoelectrochemical response of a variety of semiconductors and interpreted them in terms of the Gärtner equation. They observed that at any potential the photocurrent increases linearly with the incident photon flux, which implies that the photogeneration and transport of carriers must be the rate-limiting step, a result consistent with the Gärtner approach. For higher wavelengths, near the absorption threshold, the incident light should not be greatly attenuated within the depletion layer ($\alpha W \ll 1$), and the Gärtner equation will reduce to

$$J \approx \frac{(L_p + W)e_0\phi_0\alpha\theta}{(1 + \alpha L_p)} \approx \alpha\phi_0\theta e_0(L_p + W) \tag{23}$$

the latter approximation assuming that $\alpha L_p \ll 1$ as well.

From this, plots of photocurrent versus space-charge width W, itself proportional to $(V - V_{fb})^{1/2}$, where V_{fb} may be obtained from impedance measurements, should be linear. This has been found to be the case at high depletion for the materials investigated by Lemasson and co-workers, and extrapolation yields the minority carrier diffusion length. For ZnSe, CdSe, CdTe, GaP, and InP, a reverse bias of at least 1 V proved necessary to obtain linear plots, even at wavelengths close to threshold, and deviations for potentials closer to the flat-band potential were attributed to recombination in the depletion layer and at the interface; this recombination was found to be strongly dependent on the pretreatment of the electrodes.

A further problem with the Gärtner analysis is the boundary condition that the hole concentration at the boundary between depletion layer and bulk should be zero. This assumes a very high rate of transfer at the electrode surface, so that minority carriers have no opportuntiy to accumulate in the depletion layer. Tyagai[95] was the first to address this problem by using a flux condition at this boundary rather than the assumption of $p = 0$. Wilson[96] modified Tyagai's approach by determining the flux condition from the surface boundary conditions and by allowing minority carrier recombination at the surface. Wilson's approach was the first to concentrate attention on the photocurrent onset potential region, an important step forward since it is this region that is technologically the most important for efficient solar conversion.

1. The Photocurrent Onset Region

In the absence of surface or depletion-layer recombination, the theories of Gärtner, Tyagai, and Wilson all predict a rapid rise in the photocurrent as the depletion-layer width is increased. However, experimentally it is well established that unless special surface pretreatments are carried out,[97] the photocurrent usually rises rather slowly as the potential moves into reverse bias and indeed may show a shift of 0.5–1 V before any appreciable photocurrent flows. This effect has been observed by Wilson on TiO_2,[96] by Memming and co-workers[65,98,99] on *n*- and *p*-type III–V semiconductors, by Butler and Ginley[100] on *p*-GaP, and by Wheeler and Hackerman[101] on *n*-CdSe.

Typical calculated hole fluxes are shown in Fig. 23[96] as a function of space-charge voltage for a range of surface reaction parameters. Although it is possible to simulate the delayed photocurrent onset by using very small rate constants for hole consump-

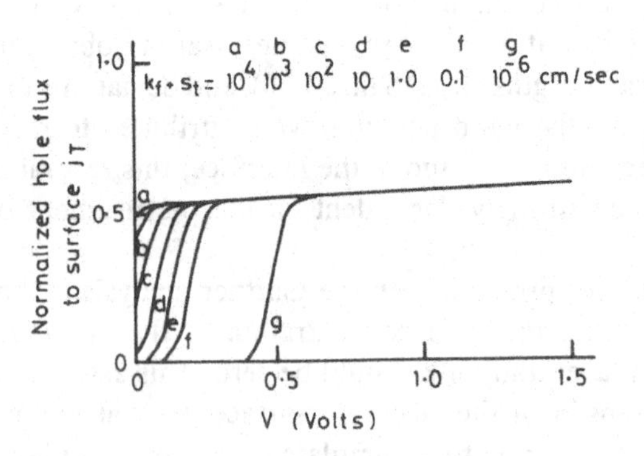

Figure 23. Hole flux calculated by Wilson for $L_p = 10^{-4}$ cm,, $\alpha = 10^4\ cm^{-1}$, $N_D = 10^{18}\ cm^{-3}$ and surface reaction parameters as shown.[96]

tion at the surface, the use of any reasonable rate constants predicts the onset of photocurrent to take place with 0.1 V of V_{fb}. It follows that even the incorporation of realistic boundary conditions between the depletion layer and the bulk does not result in a significant delay in photocurrent onset as compared to the model of Gärtner, and explanations of observed behavior must be sought elsewhere.

Wilson[96] incorporated into his model a simple surface recombination flux, and the results of this are shown in Fig. 24. The dashed line in this figure shows the experimental photocurrent curve plotted versus voltage from the (impedance-measured) flat-band potential for polished TiO_2. Plotted as well are two calculated curves, the first showing the effects of a hypothetical recombination center 0.7 eV below the CB edge, and the second showing the effects of a hypothetical exponential distribution of surface states extending down from the CB edge into the bandgap region. It is clear that this second hypothesis does account for the observed photocurrent; physically, the hole flux is given by Fig. 23, but recombination does

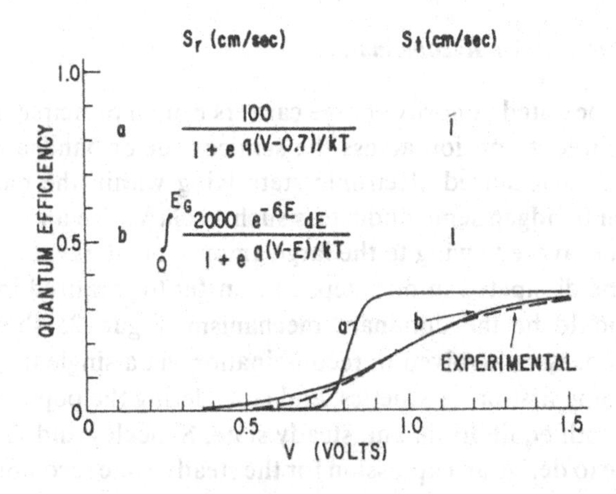

Figure 24. Quantum efficiency versus voltage across the TiO_2 depletion layer. The dashed line is the experimental result, and the solid lines are those calculated from Wilson's model for $\alpha = 10^4$ cm^{-1}, $N_D = 10^{20}$ cn^{-3}, and $L_p = 5 \times 10^{-5}$ cm. Curve a is for a single surface-state level 0.7 eV below the CB edge, and curve b is for an exponentially decreasing density of surface states from the CB into the bandgap.

not lead to any observed photocurrent, and the hole flux is partitioned between recombination and faradaic transfer, with the former being dominant near the flat-band potential and the latter becoming significant as the recombination sites are exhausted.

Wilson's paper was of great importance in showing that surface states could account quantitatively for the delayed photocurrent onset. Since 1978 it has been recognized that, in addition to surface recombination, two other effects might play a major role in delayed photocurrent onset:

(i) recombination within the depletion layer;

(ii) shifts of the semiconductor band edges due to illumination, which will always tend to reduce the potential actually dropped across the depletion layer.

In principle, all three effects might be operative simultaneously, though it has usually been the case that individual authors, in considering specific cases, have concentrated on one or two aspects of the problem. In the following sections we shall consider these different approaches in more detail. Before that, we shall consider recombination more generally.

(i) Charge Carrier Recombination

Photoexcited minority charge carriers can, in principle, recombine by direct transition across the semiconductor bandgap or via transfer to a localized electronic state lying within the bandgap. For wider bandgap semiconductors such as GaAs, the direct transition is not favored owing to the large amount of energy that would have to be dissipated in one step, so transfer to localized intergap states should be the dominant mechanism. Figure 25 shows the kinetic processes involved in recombination via a single trap state. By assuming first-order kinetics, and considering the population of traps at both equilibrium and steady state, Shockley and Read[7,102] were able to derive an expression for the steady-state recombination rate:

$$-R_p = \frac{dp}{dt} = -\frac{C_n C_p\{np - N_c N_v \exp[(E_v - E_c)/k_B T]\}}{C_n(n + n_t) + C_p(p + p_t)} \tag{24}$$

where

$$n_t = N_c \exp[(E_t - E_c)/k_B T]$$

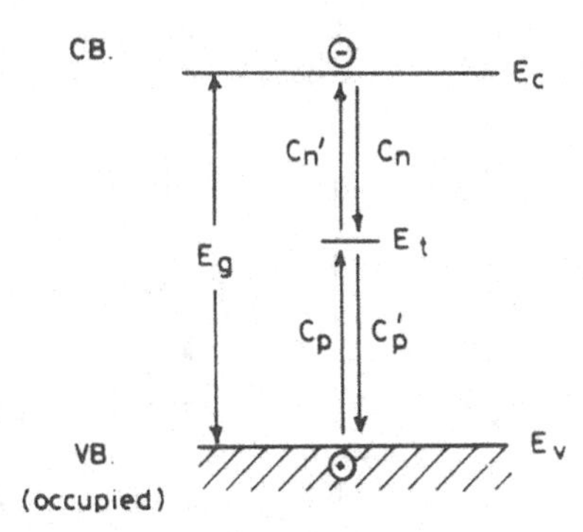

Figure 25. Kinetic parameters in the Shockley–Read model.

and $p_t = N_v \exp[(E_v - E_t)/k_B T]$, E_t is the trap energy, and N_c and N_v are the conduction- and valence-band densities of state, respectively.

For the semiconductors that are considered here, the bandgap is sufficiently large for the term $N_c N_v \exp[(E_v - E_c)/k_B T]$ to be neglected. However, the expression remains involved, and further approximations are usually sought for particular cases. Inverting Eq. (24) gives

$$\frac{1}{R_p} = -\frac{1}{(\partial p/\partial t)} = \underset{\textbf{I}}{\frac{1}{C_p p}} + \underset{\textbf{IIA}}{\frac{n_t}{C_p n p}} + \underset{\textbf{IIB}}{\frac{p_t}{C_n n p}} + \underset{\textbf{III}}{\frac{1}{C_n n}} \tag{25}$$

Outside the depletion region, in the bulk of the semiconductor, the minority carrier density, p, should be small ($p \ll n$), and the first term will dominate, so in this region

$$R_p \approx C_p p \equiv p/\tau_p \tag{26}$$

where τ_p is the hole recombination time. It is this relationship that we have used in the derivation of the expressions for photocurrent given above. In the depletion layer, and for electronic states at the surface, the behavior is likely to be significantly more complex, but based on the above analysis, some of the possible rate-limiting cases are illustrated in Fig. 26. The theories of photocurrent can then be considered in the light of these relationships.

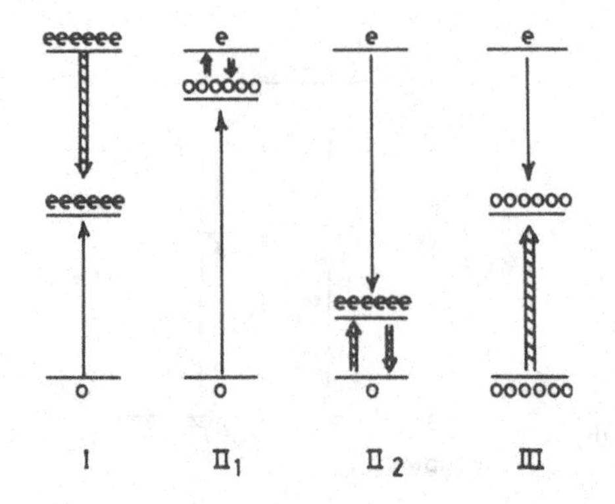

Figure 26. Rate-limiting regimes for Shockley–Read statistics. The roman numerals show which term in Eq. (25) is dominant: **I**, $n \gg n_t, p, p_t$; **IIA**, $n_t \gg n, p, p_t$; **IIB**, $p_t \gg p, n, n_t$; **III**, $p \gg n, n_t, p_t$. The single arrows indicate the rate-limiting step for each case.

(*a*) *Depletion-layer recombination*

The theories of photocurrent that we have considered all deviate most from experiment under low depletion, when the carrier densities in the depletion layer will be higher, and the assumption of no space-charge recombination correspondingly less satisfactory. Sah *et al.*[103] treated the problem of depletion-layer recombination in a p–n junction by assuming that thermal equilibrium was established between *minority* carriers in the depletion layer, but that these were not necessarily in equilibrium with *majority* carriers. The consequence is that *quasi* Fermi levels can be defined for both electrons and holes separately; these quasi Fermi levels are at constant energy in the depletion layer and are separated by an energy denoted e_0V. The advantage of the quasi-equilibrium approach is that it allows minority carrier density across the depletion layer to be related to the density at the surface, so that the rate of recombination can be calculated readily.

Sah *et al.*[103] considered the case of a single set of traps at the intrinsic Fermi level of the semiconductor with equal rates of electron and hole capture, where the recombination rate should reduce to

$$R_p \approx \sigma \nu_{\text{th}} N_t \cdot np/(n+p) \tag{27}$$

where σ is the capture cross section of the trap for electrons and holes, ν_{th} is the carrier thermal velocity, and N_t is the trap density, and $\sigma\nu_{\text{th}}N_t$ is an explicit expression for C_n or C_p. Integration of this expression across the depletion layer will then yield the current *loss* due to recombination.

Reichman[104] has used the approach of Sah *et al.* to show that depletion-layer recombination can be used to explain the delayed photocurrent onset. The variation in collection or quantum efficiency with surface charge transfer rate constant is illustrated in Fig. 27a. Reichman and Russak[105] applied their theory to the photocurrent behavior of CdSe in polysulfide electrode, but conclusions drawn from this system must be treated as qualitative owing to the susceptibility of the surface to corrosion and chemical change.[106-108] Sulfur is likely to build up on the electrode, and the system is unstable with respect to the formation of a surface layer

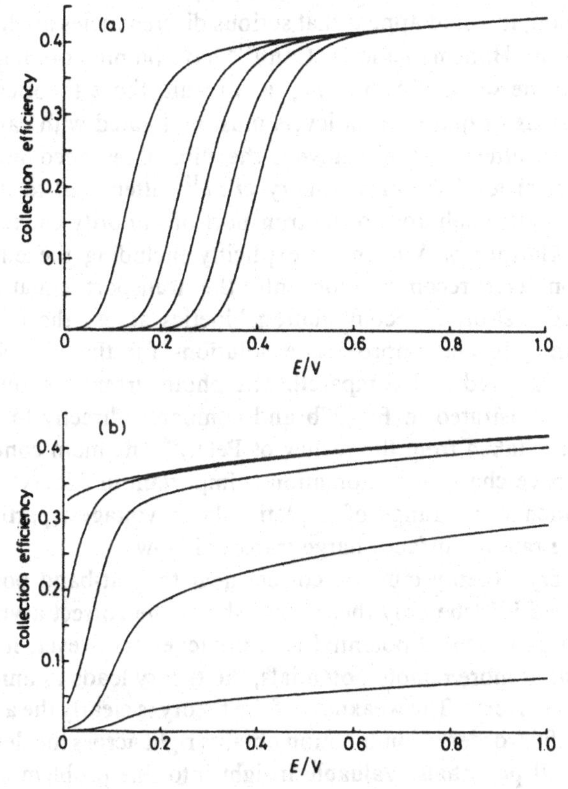

Figure 27. Quantum efficiencies predicted by Reichman (a) and Albery (b) theories for identical conditions: photon flux = 10^{15} cm^{-2} s^{-1}1, charge transfer rate constants 10^4, 10^3, 10^2, 10 and 10 cm s^{-1} from left to right, $\alpha = 5 \times 10^4$ cm^{-1}, $L_p = 10^{-5}$ cm, $N_D = 10^{18}$ cm^{-3}, $\tau = 10^{-9}$ s, $\Delta\phi_{sc} = 0.6$ V.[54]

of CdS. Horowitz[109] has fitted Reichman's theory to the quantum yield of *p*-GaP in 0.5 *M* H_2SO_4 by using a very low value of the hole exchange current at the surface to obtain the observed delay in photocurrent onset. However, this interface has been intensively investigated by others, and more convincing explanations involving surface-state recombination have been proposed.

El Guibaly *et al.*[110] have also followed Sah *et al.* in calculating depletion-layer recombination, but they have included a term for hole recombination at the surface. However, the assumption of flat quasi Fermi levels in the depletion layer has been criticized by several workers. For light intensities appropriate to solar energy conversion, it would appear that serious discrepancies are likely.[111] In addition, Haneman and McCann[112] have pointed out important errors in the work of Sah *et al.*; as a result, those theories based on the ideas of quasi Fermi levels must be treated with caution.

In an attempt to circumvent the difficulties encountered in quasi Fermi level theories, Albery *et al.*[113] attempted to derive a more complete solution to the transport of minority carriers than that of Gärtner or Wilson by explicitly including the effects of depletion-layer recombination into the transport equation. By assuming first-order recombination kinetics across the depletion layer, analytic and approximate solutions for the photocurrent could be derived and compared. The photocurrent response predicted is illustrated in Fig. 27b and compared directly to that of Reichman, taken from the review of Peter.[54] The main conclusion is that space-charge recombination is important in Albery's theory for a much wider range of depletion-layer voltages, particularly when the rate of surface charge transfer is low.

Albery's theory must be correct near the flat-band potential, and indeed it is the only theory that shows the correct asymptotic behavior as flat-band potential is approached. However, it is also clear that at more remote potentials, the theory leads to unrealistically low currents. The weakness of the theory is clearly the assumption that first-order recombination obtains right across the depletion layer at all potentials. Valuable insight into this problem and the nature of photoprocesses at semiconductor electrodes can be obtained from the form of the minority carrier distribution that the model predicts. Figure 28 shows the predicted variation of the logarithm of minority carrier density as a function of distance from

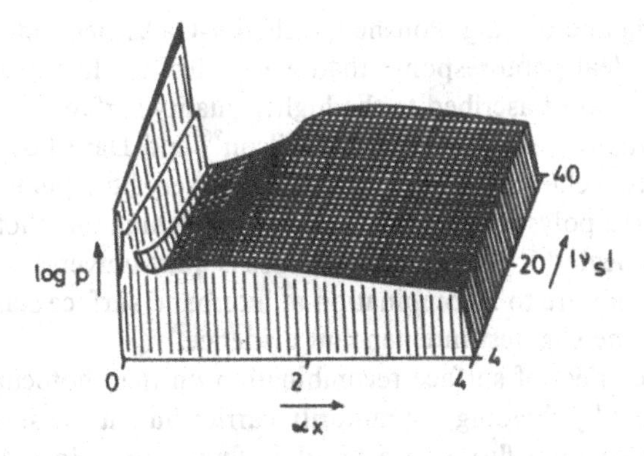

Figure 28. Form of minority carrier concentration, log p, as a function of the normalized distance into the semiconductor from the surface, αx, and the depletion-layer potential, $|v_s|$. The parameters used in this calculation were $(\alpha L_D)^{-1} = 6.90$, $L_p/L_D = 6.97$, and $k'_\Sigma L_D/D_p = 1.45 \times 19^{-2}$, where L_D is the semiconductor screening length $(=(\varepsilon_0 \varepsilon_{\text{sck}} T/e_0^2 N_D)^{1/2})$, k'_z is the total rate constant for all hole (minority carriers) reactions at the surface and the other symbols are defined in the text.[5]

the electrode surface, and the depletion-layer potential $|v_s| \equiv |e_0 \Delta\phi_{sc}/k_B T|$. As expected, the region depleted of minority carriers extends further into the semiconductor as the surface electric field increases. However, close to the surface, the concentration of minority carriers rises steeply as carriers await transfer to the electrolyte.

In a later paper, Albert and Bartlett[114] analyzed in detail the effect that the variation of carrier concentrations across the depletion layer will have on the rate-limiting step for recombination rate law is likely to shift from case **I** in Fig. 26 to cases **II** and **III**. The kinetic laws that operate will also depend on the incident light intensity and depletion-layer potential as well as the energy of the trap state in the bandgap.

(*b*) *Surface recombination*

Recombination is also likely to take place at the semiconductor/electrolyte interface through surface states that lie in the bandgap. The photocurrents for potentials close to the flat-band potential are observed to be very sensitive to such electrode pretreatment as polishing and etching. Polished, etched n-GaAs electrodes showed a more ideal photoresponse than electrodes that had merely been etched, a result ascribed to the higher quality surface.[115] A similar result was reported for TiO_2 by Wilson[96] and Dare-Edwards and Hamnett.[97] Unetched CdS and CdSe gave very little photoresponse in alkaline polysulfide electrolyte, but on etching the photocurrent onset appeared.[116] Poor photoresponse is quite universally ascribed in the literature to recombination at interband surface states, often without the slightest independent evidence.

The effect of surface recombination on the photocurrent can be treated by dividing the minority carrier flux at the surface into a recombination flux and a faradaic flux. Using Eq. (24), Gerischer[117] has derived limiting cases for rates of surface recombination in a similar manner to that used in the discussion illustrated in Fig. 26. Wilson[96,118] investigated the case in which the surface concentration of photogenerated minority carriers, p_s, is much smaller than the majority carrier concentration at the surface, n_s, and the value of n_t^s, defined analogously to n_t, to obtain the recombination rate:

$$R_p \approx v\sigma_R N_{ss}[n_s p_s/(n_s + n_t^s)] \equiv S_r p_s$$

where N_{ss} is the density of surface states at a particular energy E_{ss}, $n_t^s = N_c \exp[(E_{ss} - E_c)/k_B T]$, σ_R is the hole capture cross section of the surface states, and v is the thermal velocity of the holes. By expressing the hole recombination flux in the form $S_r p_s$ and the faradaic flux as $S_t p_s$, Wilson was able to derive an expression for the total hole flux to the surface which avoided the restrictive assumptions of the more elementary transport models referred to above. S_r can then be evaluated as a function of applied potential by assuming an equilibrium population of minority carriers at the surface. For surface states at an energy E_{ss}:

$$\begin{aligned} S_r &\approx \frac{v\sigma_R N_{ss} n_s}{n_s + n_t} \\ &= \frac{v\sigma_R N_{ss} N_c \exp[(E_F - E_c - e_0\phi_{sc})/k_B T]}{N_c \exp[(E_F - E_c - e_0\phi_{sc})/k_B T] + N_c \exp[(E_{ss} - E_c))/k_B T]} \\ &= \frac{v\sigma_R N_{ss}}{1 + \exp(e_0\phi_{sc}/k_B T)\exp[(E_{ss} - E_F)/k_B T]} \end{aligned} \tag{28}$$

where E_F is the bulk Fermi level, ϕ_{sc} is the space-charge voltage, and E_{ss} is the surface-state energy as above. Curve a in Fig. 24 shows that a single set of surface states at 0.7 eV below the conduction-band edge in TiO_2 can effectively prevent photocurrent for space-charge voltages below 0.7 V, since in this region the surface states are populated with electrons and recombination is facile. Then, as the potential increases, the majority carrier Fermi level drops below the surface states, empty the surface states of electrons, and the recombination rate falls, and faradaic hole transfer begins to compete effectively. The experimental result for a polished TiO_2 electrode is shown by the dashed line in Fig. 24; the photocurrent onset region is far broader than that predicted from the single-energy surface-state model, and a distribution of surface-state energies is needed to account for the overall behavior, in agreement with impedance measurements and other techniques that are discussed in more detail below. Incorporation of a surface-state distribution is straightforward within Wilson's formalism: integration of S_r over all surface-state energies for a simple exponential distribution of surface states leads to curve b in Fig. 24, which is in excellent agreement with the experimental data. Of course, it must be emphasized again that *any* photocurrent behavior can be modeled by appropriate choice of a suitable surface-state distribution, and independent confirmation of such a distribution is needed before this approach can be validated.

Peter *et al.*[119] have developed a model for steady-state and transient photocurrent response using the kinetic parameters shown in Fig. 29. Minority carrier flux is calculated by the Gärtner equation, and the deviation of surface-state population from the Fermi level-controlled equilibrium value can be calculated in the steady state. The concentration of *majority* carriers at the surface is assumed always to be close to the equilibrium value, in common with the approach of Wilson.

Normalized photocurrents calculated for surfaces states having a density of 10^{13} cm^{-2} and situated 0.3 eV above the valence band are shown in Fig. 30 for a variety of photon fluxes. The curves are quite similar to those predicted by the Reichman treatment of depletion-layer recombination given above: both give an earlier photocurrent onset with potential at higher light intensities, since the increasing flux of minority carriers reduces the concentration of majority carriers in the traps and the rate of minority carrier

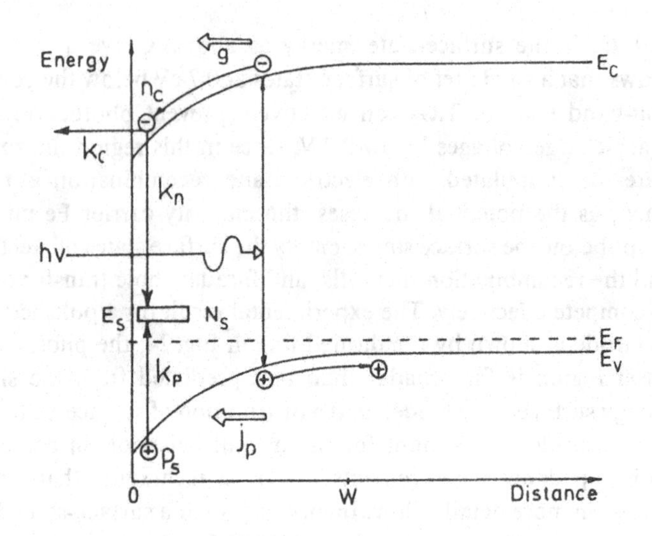

Figure 29. The model of Peter *et al.* for surface recombination for a *p*-type semiconductor; k_n and k_p are the rate constants for electron and hole capture by the surface state, and k_c is the rate constant for electron transfer from the CB to the electrolyte.[119]

recombination drops. Similarly to Wilson's model, the Fermi level is observed to exert a controlling influence on the surface-state population for the parameters chosen, so that as the space-charge voltage is increased above 0.3 V, the electron population in the surface states increases, the low of electron flux to recombination decreases, and the cathodic photocurrent rises. In fact, with some notable exceptions, the experimental evidence for the type of photocurrent onset shift with light intensity illustrated in Fig. 30 is not especially convincing, and little variation is usually observed. A general problem with the different photocurrent models proposed above is that parameters can usually be chosen so that they give similar predictions, and it becomes difficult to distinguish which model truly reflects the underlying phenomena, unless the parameters can be obtained independently.

A closely related model to Peter's has been developed by Memming, who used a steady-state analysis that also included explictly both conduction electron transfer and surface-state recombination. Unlike Peter, who dealt in changes of conduction-band and surface-state population from equilibrium, Kelly and

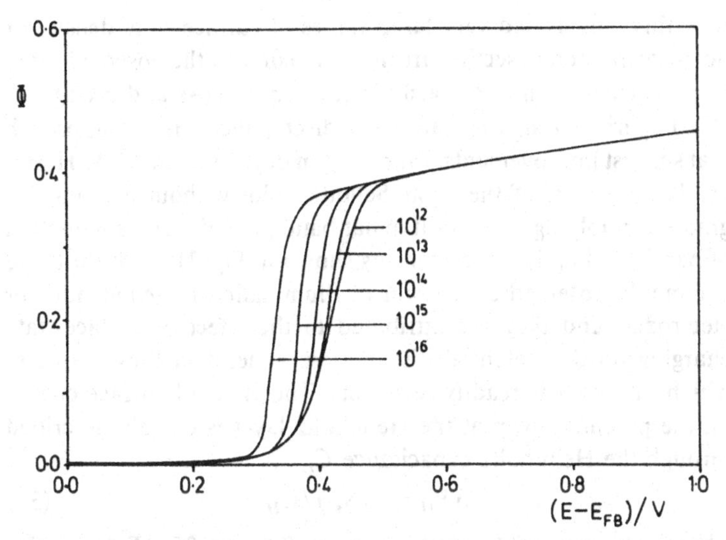

Figure 30. Quantum efficiencies for various photon fluxes calculated from the model of Peter *et al.* for the photoelectrochemical response of a *p*-type semiconductor. Parameters used were $N_{ss} = 10^{13}$ cm^{-2}, $E_{ss} = 0.3$ eV, $\alpha = 3.3.10^{4}$ cm^{-1}, $k_n = 10^{-7}$ cm s^{-1}, $k_p = 10^{-8}$ cm s^{-1}, $k_c = 10^{4}$ cm s^{-1}, $N_A = 8 \times 10^{16}$ cm^{-3}, $\varepsilon = 11$, and $L_n = 7 \times 10^{-6}$ cm.[119]

Memming[98] derived expressions for surface-state population and photocurrent by using the steady-state condition in the rate equations:

$$dn_s/dt = g - k_n n_s N_{ss}(1-f) - k_c c n_s \tag{29}$$

$$N_{ss}\, df/dt = k_n n_s N_{ss}(1-f) - k_p p_s N_{ss} f \tag{30}$$

where f is the fractional surface-state population by electrons, g is the minority carrier flux to the surface, c is the concentration of reducible electroactive species in solution at the electrode surface, and other parameters are defined in Fig. 29. A consequence of this approach is that surface states will not approach their equilibrium, Fermi level-controlled population as the illumination intensity tends to zero. A full kinetic treatment would require a Shockley-Read analysis in order to include the loss of carriers from the surface states by thermal excitation. However, a significant advantage of Memming's approach is that no energy need be specified for the surface states as their population is determined solely by the rates of transfer from conduction and valence bands. The delay in photocurrent onset is, however, still estimated at about 0.3 V with this model, and very large values of surface-state density or hole-acceptor cross section are needed to predict the observed delay in photocurrent onset of ca. 0.5 V seen on *p*-GaAs and *p*-GaP.

Experiment also reveals other discrepancies in the approach that suggest improvements. Illumination of *p*-GaAs in 0.5 *M* H_2SO_4 results in a shift of the Mott–Schottky plot without a change in gradient, implying that a shift in flat-band potential and a reduction of band bending has occurred. As shown in Fig. 31, such shifts are commonly observed as a result of illumination of semiconductor electrodes, and they are attributed to the effect of surface-state charging on the potential distribution, at least in those cases in which the effect is readily reversible. The effect of surface charge on the potential drop at the Helmholtz layer is usually described through the Helmholtz capacitance C_H:

$$\Delta V_H = e_0 N_{ss} f / C_H \tag{31}$$

where estimates of C_H vary between 2.5 and 25 $\mu F\,cm^{-2}$. The flat-band shift of 0.25 V led Kelly and Memming[98] to estimate a surface-state density of ca. $10^{13}\ cm^{-2}$.

The original model predicted the surface-state population f to tend to unity in the region of limiting photocurrent, so that no illumination-dependent flat-band shift would be expected in this

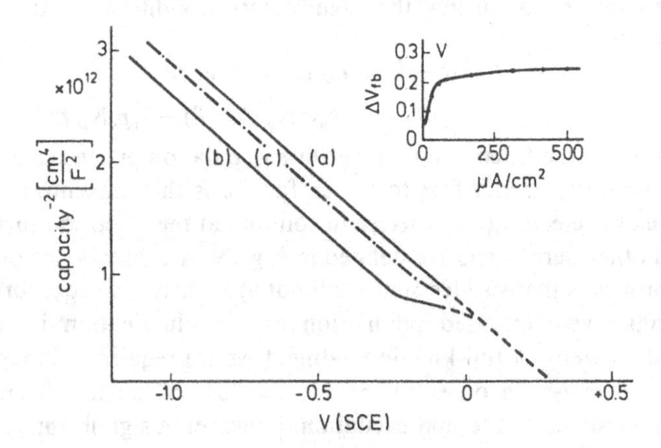

Figure 31. Mott–Schottky plots for *p*-GaAs in 0.5 *M* H_2SO_4 (aq): (a) in the dark; (b) illuminated to give a limiting photocurrent of 50 $\mu A\,cm^{-2}$; (c) as for (b) but with 0.05 *M* $K_3Fe(CN)_6$. The inset shows the flat-band potential as a function of the limiting photocurrent density, which, in turn, is determined by the incident light intensity.[98]

region, contrary to the results shown in Fig. 31. The flat-band shift is also much reduced at low light intensities in the presence of redox couples whose potential lies close to the lower part of the p-GaAs bandgap; the effects of the $[Fe(CN)_6]^{3-/4-}$ couple are illustrated in Fig. 31c. These redox couples also give a much earlier photocurrent onset.

The kinetic model can be extended to include electron transfer to solution from surface states, a phenomenon already encountered above. The variation of space-charge potential with surface-state population will affect the photocurrent in two important ways. Firstly, in a p-type semiconductor, the surface hole concentration, $p_s = p_0 \exp(-e_0\phi_{sc}/k_BT)$, is very sensitive to the extent of band bending and has a major effect on the recombination rate. The illumination-induced reduction in depletion-layer potential increases the surface hole concentration and therefore the hole concentration in surface states, so that the onset in photocurrent will be further delayed. The change in Helmholtz layer potential will also shift the band-edge and surface-state energies with respect to the redox couple in the electrolyte. Energy overlap between the surface and redox couples will therefore be changed, and this will, in turn, cause changes in the rate constants for transfer which can be estimated from the theory of Gerischer[47] that we presented earlier. For electron transfer from the CB,

$$k_c = k_c^0 \exp\left(\frac{-(E_{\mathrm{CB}} - E^0_{F,\mathrm{redox}} - e_0\Delta V_H + \lambda)^2}{4k_BT\lambda}\right) \tag{32}$$

where $E^0_{F,\mathrm{redox}}$ is the absolute chemical potential of the redox couple in solution, and E_{CB} is the energy of the conduction band in the absence of surface-state charging. A similar expression can be derived for the rate constant for electron transfer from the surface state:

$$k_s = k_s^0 \exp\left(\frac{-(E_{\mathrm{SS}} - E^0_{F,\mathrm{redox}} - e_0\Delta V_H + \lambda)^2}{4k_BT\lambda}\right) \tag{33}$$

where E_{SS} is the surface-state energy level.

Kelly and Memming[98] calculated numerically the photocurrent as a function of potential and photon flux, and the results are shown in Fig. 32; in these calculations, the flux of electrons to the surface, g, has been assumed equal to the incident photon flux for simplicity. The inclusion of band-edge shift and the efficient surface-state-mediated recombination now lead to a calculated delay in photocur-

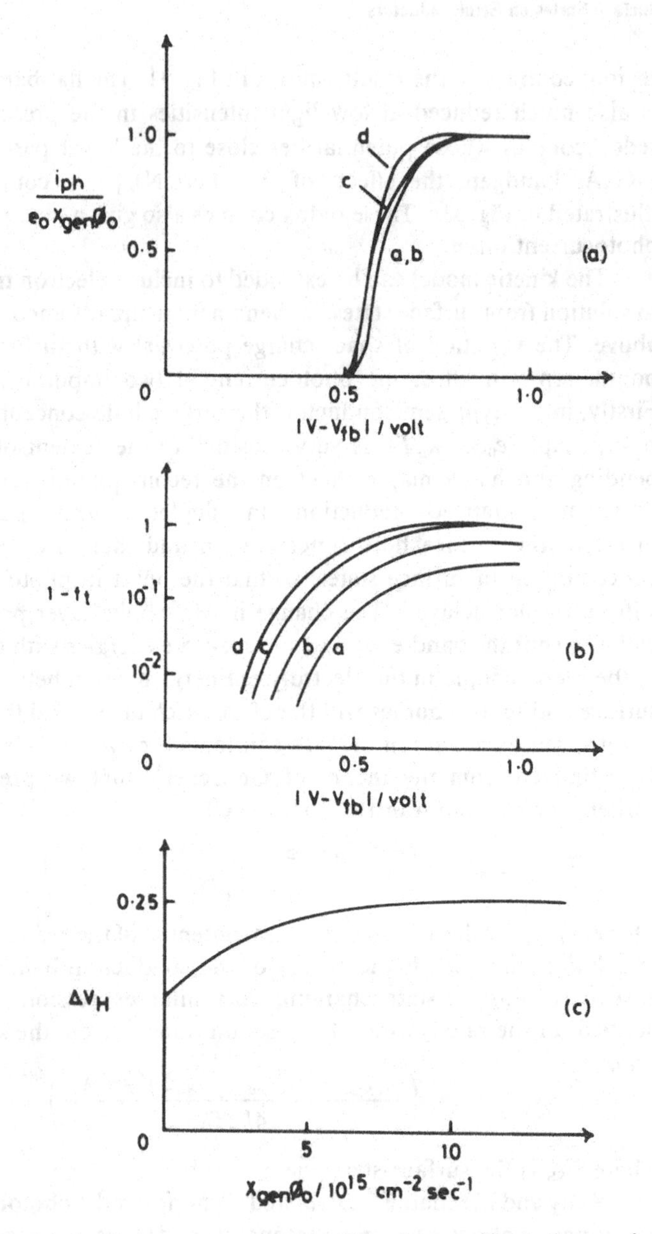

Figure 32. Calculated photocurrent quantum efficiencies according to the extended Kelly–Memming model: (a) normalized photocurrent-voltage curves for 10^{14} (curve a), 10^{15} (curve b), 10^{16} (curve c), and 10^{17} cm^{-2} s^{-1} (curve d) photon flux, with the parameters $k_c^0 c = 10^3$ cm s^{-1}, $k_s^0 c = 10^8$ s^{-1}, $N_{ss} = 10^{13}$ cm^{-2}, $k_n = 10^{-7}$ cm^3 s^{-1} and $k_p = 10^{-9}$ cm^3 s^{-1}. $C_H = 6.4\ \mu$F cm^{-2}, and $\lambda = 1.0$ eV; (b) f as a function of applied potential in the calculation in (a); (c) dependence of ΔV_H at limiting photocurrent illumination flux, using parameters from (a).[98]

rent onset of ca. 0.5 V, close to that observed experimentally for *p*-GaAs in 0.5 *M* H_2SO_4. The upward band-edge shift that occurs on *p*-type semiconductors as the surface states fill with electrons should enhance the rate constants for transfer to such redox couples as H^+/H_2, $Eu^{3+/2+}$, and $Cr^{3+/2+}$, though within the model presented here, it is clearly the effect of recombination that predominates.

A stroking feature of Fig. 32a is the smaller dependence of the normalized photocurrent on the incident light intensity than is found in other models, a result in greater accord with the bulk of available experimental data. This is a direct result of the inclusion of surface-state electron transfer and band-edge shift in one model. At low levels of light intensity, there will be few electrons in the surface states, so that transfer to the surface states will be favored. Electrons in the surface states will then either undergo recombination with surface holes or transfer into solution, the balance between these two processes being determined by the depletion-layer potential, which controls the surface *hole* concentration. The transfer of photogenerated electrons to empty surface states that can occur at low light intensity does not now result in such a drop of photocurrent efficiency in the onset regime, since electron transfer can take place from the surface states to the electrolyte. As the illumination intensity in increased, the photogenerated electron flux to the surface is increased, the surface states start to fill with electrons, and the net fractional flux of electrons to the surface states decreases. Hence, a greater proportion of the electrons should be transferred directly from CB to electrolyte, and the recombination rate should *decrease.* However, the filling of the surface states also causes a decrease in band bending on a *p*-type semiconductor, and this increases the surface hole concentration and hence the surface recombination rate. In Fig. 32, the rate constants have been chosen so that these two effects cancel, and the normalized photocurrent varies little with illumination intensity. Also shown are the calculated surface-state occupancies as a function of potential and illumination intensity in Fig. 32b and the predicted band-edge shift at the limiting photocurrent as a function of light intensity. The latter is in very good agreement with the experimental results in Fig. 31.

The improved photocurrent onset with redox couples such as $[Fe(CN)_6]^{3-/4-}$, $Fe^{3+/2+}$, and I_2/I^- can be explained as the result of good energy overlap with surface states lying in the lower half of the bandgap. This can be modeled by using a large value for the surface-state electron transfer rate constant, which is found to improve the predicted onset, as observed experimentally. The more rapid rise of electron capture from surface states will reduce the band-edge shift with illumination, by removing electrons from the surface states.

The theory of Kelly and Memming can, therefore, give an attractive, semiquantitative description of photocurrent onset delay. However, problems in applying the theory quantitatively lead to some difficulties since the treatment of electron transfer from both semiconductor bulk and electrolyte to the surface states has been considerably oversimplified. Essentially, the theory presupposes that electron transfer is irreversible, an assumption embodied in the flux equations discussed earlier. Clearly, electron injection into the surface states from solution is also possible, and if the rate constant for this process is sufficiently large, then the surface states will, in effect, equilibrate with the electrolyte redox couple, and this process will determine the surface-state population. For systems where the surface-state density is high, control of surface-state population by redox chemical potential will have a major effect on the Helmholtz potential drop. This phenomenon is discussed in the section on Fermi level pinning (Section VI), along with the possibility that not all change in applied potential is dropped across the depletion layer, even under dark conditions. Illumination can also result in considerable photocorrosion and chemical change, leading to hysteresis in photocurrent curves and to interpretation by the simple models above becoming impossible.

Gobrecht and Blaser[220] proposed, some years ago, a mechanism for surface recombination that does not require the presence of surface states. Instead, they pointed out that if rapid hole and electron transfer could occur, respectively, from the VB to the oxidized form and from the CB to the reduced form of a redox couple in solution, then recombination mediated by the solution species would have taken place. Both Memming and Gerischer questioned the applicability of such a model to the photoelectrochemistry of Ge, which Gobrecht and Blaser used as an example. However, more recently, Meissner *et al.*[65] have considered this process as a possible source of the poor photocurrent onset of *p*-GaAs. In practice, it is difficult to determine whether redox reactions or a surface-bound species is responsible for recombination, as both mechanisms depend on surface hole concentration in a similar manner. Rajeshwar[121] has proposed a photocurrent model in which charge transfer is *exclusively* via surface states that are identified as adsorbed redox species capable of existing in both oxidized and reduced forms on the electrode surface. This is obviously a broad view of what constitutes a surface state, but it must be emphasized that the kinetic models that we have been considering are equally applicable here.

(ii) *Photocurrent Onset and Surface States on p-GaP*

The photoelectrochemistry of *p*-GaP in acidic electrolyte has received considerable attention owing to the comparative stability of the surface and the possibility of fabricating useful solar energy conversion devices. Nakato *et al.*[122] observed that the photocurrent onset of *p*-GaP occurred ca. 0.7 V negative of the flat-band potential, and Butler and Ginley[123] concluded that this resulted from recombination at centers near the electrode surface.

Uosaki and Kita[124] ascribed the late onset in photocurrent at *p*-GaP in p.5 *M* H_2SO_4, 1 *M* NaOH, and 1 *M* NaOH with methyl viologen to the oxidation by valence-band holes of reduced species that are then reduced by photogenerated CB electrons. This is essentially the recombination mechanism of Gobrecht and Blaser[120] and is supported by the fact that the anodic *dark* current has an onset very close in potential to the photocurrent onset and by the fact that in the presence of methyl viologen, the delay in onset is considerably smaller since rapid reduction kinetics compete more effectively with the rate of recombination.

Dare-Edwards *et al.*[125] measured the photocurrent of *p*-GaP in 0.5 *M* H_2SO_4 as a function of light intensity and observed a photocurrent onset much closer to flat-band potential at low light intensity. Both Reichman's theory of depletion-layer recombination and the surface-recombination theories of Memming and Peter predict the reverse, that is, that the onset should be *more delayed* for low levels of illumination. The implication is that the density of surface-recombination levels is increased by illumination, and it was proposed that the photogenerated electrons reduce hydrogen ions at the surface to give hydrogen atoms, $H^{\cdot}_{ads}$, adsorbed on the surface. These will constitute filled electronic states that lie within the bandgap, and Dare-Edwards *et al.* proposed that recombination can then take place by tunneling of holes across the depletion layer to reoxidize the adsorbed radicals, a process that will be favored for smaller depletion-layer potentials. Close to the flat-band potential, the hydrogen atoms will be oxidized leading to the earlier photocurrent onset on the sweeps *away* from the flat-band potential as observed in Fig. 33. Support for this model came from impedance measurements that confirmed the presence of a surface state in the depletion-layer potential range of 0.2 to 0.8 V.

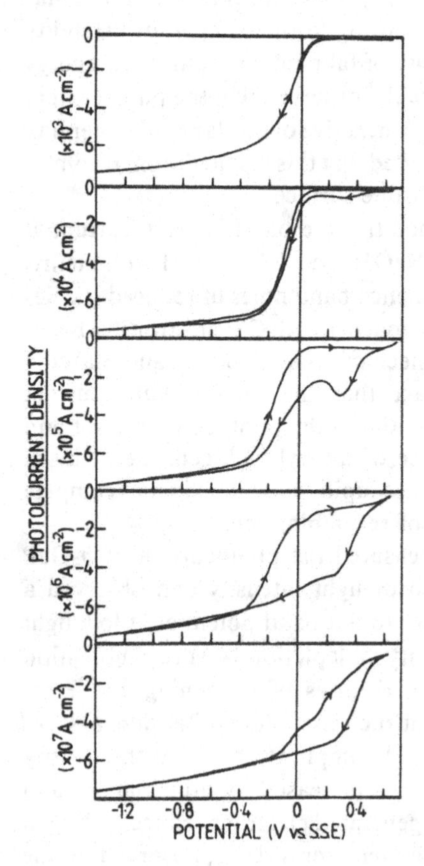

Figure 33. Variation of photocurrent with potential for various incident light intensities for *p*-GaP in 0.5 *M* H_2SO_4.[125]

Albery and Bartlett[126] studied photoreduction at *p*-GaP in acidic electrolyte in the presence and absence of Fe(III) and methyl viologen using rotating ring-disk electrodes. They also concluded that adsorbed hydrogen atoms are important in surface recombination, but did not accept that they are responsible for the surface state just above the VB edge, since they expect H to be of higher energy. A more complex kinetic model was therefore developed.

Peter and co-workers[127] measured photocurrent conversion efficiencies of *p*-GaP in 0.5 *M* H_2SO_4 and were able to show that the results, for a variety of wavelengths, were consistent with the full Gärtner equation for potentials more than 0.7 V from the flat-band potential. Illumination of the surface for 100 s at a reducing potential leads to an increase in the anodic dark current passed on a potential sweep to +0.4 V. If this is wholly ascribed to reoxidation of $H^{\cdot}_{ads}$, then a surface coverage approaching 10^{14} cm^{-2} is present. Peter and co-workers also reported that the surface concentration of $H^{\cdot}_{ads}$ appears to increase with the square root of the time at which the electrode is prepolarized at cathodic potentials, which is consistent with a diffusion-limited process. As Butler and Ginley[128] had already observed injection of protons into *p*-GaP under high electric fields, Peter and co-workers therefore suggested that the adsorbed hydrogen atoms could diffuse up to 1.5 nm into the electrode surface. They no longer, therefore, constitute true surface states and are classified as "near surface states," though they will certainly act as recombination sites in a similar manner to true surface states.

2. Photocorrosion at Semiconductor Electrodes

Photoprocesses at semiconductor electrode surfaces will often result in surface corrosion, as the semiconductor undergoes reaction with the photogenerated minority carriers. The intermediates in the corrosion reaction will generally be stabilized by interaction with the electrolyte, so that photocorrosion can often compete for the photogenerated carriers with redox couples in solution. Control of photocorrosion is particularly important in systems that could be used for solar energy conversion, where if even a very small proportion of the carrier flux to the surface results in corrosion, then long-term degradation of the electrode will occur. Furthermore, as has already been discussed for dark corrosion, the reaction intermediates of photocorrosion resulting from partial oxidation or reduction are likely to be electronically highly active, and to act as photogenerated surface states.

Gomes and co-workers[85,129-134] have considered, in a series of papers, possible general schemes for the photocorrosion of an n-type semiconductor. For a material that required n equivalents for complete electrochemical oxidation, the process is assumed to consist of the following n elementary steps:

$$(\mathrm{SC})_{\mathrm{surface}} + p^+ \underset{k_{-1}}{\overset{k_1}{\rightleftharpoons}} \mathrm{X}_1$$

$$\mathrm{X}_1 + \mathrm{M} \underset{k_{-2}}{\overset{k_2}{\rightleftharpoons}} \mathrm{X}_2$$

$$\mathrm{X}_i + \mathrm{M} \underset{k_{-(i+1)}}{\overset{k_{i+1}}{\rightleftharpoons}} \mathrm{X}_{i+1}$$

$$\mathrm{X}_{n-1} + \mathrm{M} \xrightarrow{k_n} \text{decomposition products}$$

Further oxidation of decomposition intermediates is assumed to be via M, which is either another photogenerated hole or the first oxidation intermediate, X_1, if this is sufficiently mobile. Oxidation of a reducing agent Y in solution could take place directly by hole capture at the surface:

$$p^+ + \mathrm{Y} \rightarrow \mathrm{Z}$$

or by any of the oxidation intermediates:

$$\mathrm{X}_i + \mathrm{Y} \xrightarrow{k'_{-i}} \mathrm{X}_{i-1} + Z \qquad (i = 1, 2 \ldots, n-1)$$

As well as the surface states that are present in the dark, the intermediates will also be able to act as recombination centers for the minority carriers by capturing electrons from the CB:

$$\mathrm{X}_i + e^- \xrightarrow{k^r_{-i}} \mathrm{X}_{i-1} \qquad (i = 1, 2 \ldots, n-1)$$

It will be apparent that for the six-equivalent oxidation of III–V semiconductors, this scheme is too complex for the analysis of results. Instead, various simplified forms have been analyzed, and the results compared with experiment to determine the most suitable model. Figure 34 shows the variation of stabilization ratio, *s*, which is defined as the fraction of the photocurrent that leads to faradaic conversion of the solution redox couple, against the total photocurrent. Three sets of data are presented in which the photocurrent has been varied either by changing the applied potential or the light intensity.[85] As can be seen, the stabilization ratios are similar for a given photocurrent, provided that the photocurrent is sufficiently large. However, for lower photocurrents, *s* increases, though less rapidly for the case where the light intensity is decreased. The fact that *s* decreases as the light intensity increases is clear evidence for a multistep decomposition process, since if decomposition and transfer to the solution redox couple Y/Z were both first order, their ratio should be independent of light intensity. In fact, Vanmaekelbergh *et al.*[85] were able to fit their data for *s* to two types of mechanism. In one type, the holes oxidize the redox couple directly, and the first step of the anodic decomposition process is reversible. In the second type, redox couple oxidation takes place by direct electron transfer from solution to either the first or the thilrd decomposition intermediate, and M in the scheme above is identical to X_1.

Analysis shows that these two processes can be distinguished as the photocurrent increases. At high currents, the decomposition steps will become effectively irreversible, and direct competition betweeen stabilization and decomposition in the first reaction is then expected to lead to a constant value of *s* for the first type of mechanism. However, where decomposition takes place by successive mediation of X_1, it will be of higher order in the surface hole concentration than the rate of electron transfer to X_1 or X_3, and, as a consequence, decomposition will dominate at higher photocurrents, and $s \to 0$. This is, in fact, the situation observed in Fig. 34. The similarity of the derived values of *S* at higher values of photocurrent, whether the change in the latter was induced by potential or light intensity, was taken as evidence that recombination by "normal" surface states dominated at high photocurrent. The deviation of the curves in Fig. 34 at lower photocurrents was ascribed

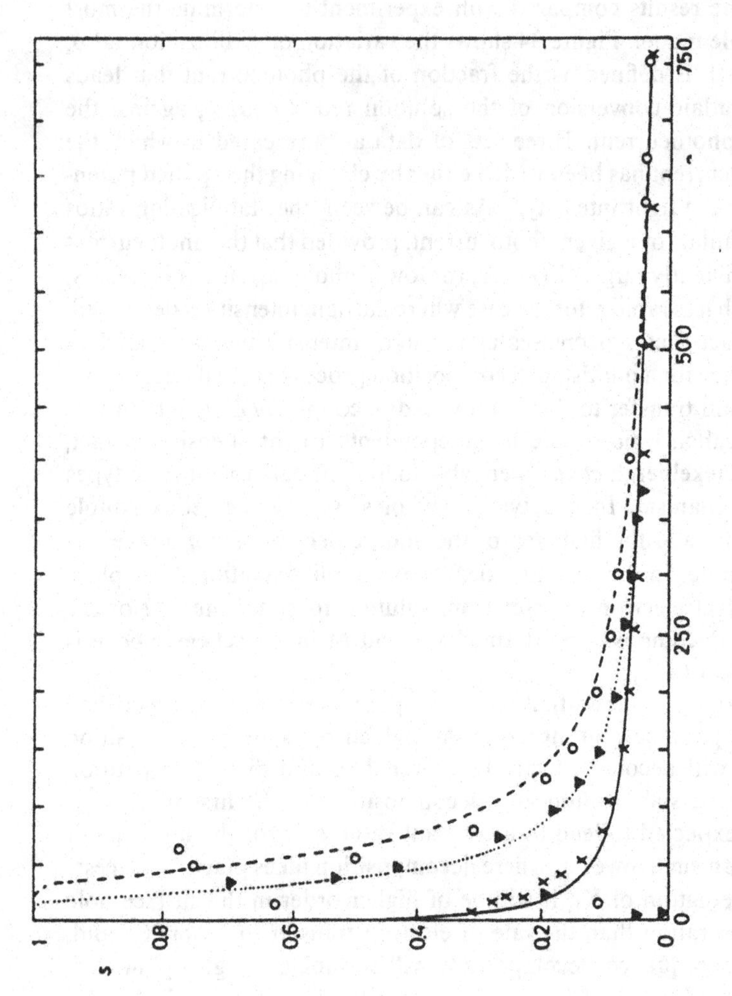

Figure 34. Stabilization ratio, s, as a function of photocurrent density, i, for $\langle\bar{1}\bar{1}\bar{1}\rangle$ n-GaAs in 1.5 M Fe^{2+} at pH1: ——, varying light intensity at +0.5 V vs. SSE; · · ·, varying potential at constant light intensity corresponding to 375 $\mu A\ cm^{-2}$ at +0.5 V vs. SSE; – – –, varying potential at constant light intensity corresponding to 700 $\mu A\ cm^{-2}$ at +0.5 V vs. SSE.[85]

to the increased importance of recombination through the first two decomposition intermediates. It is important that independent evidence of decomposition-induced surface states can be found, other than from an analysis of kinetic data, and we will discuss capacitance measurements on illuminated electrodes below.

Gerischer and Lübke[135] have measured photocorrosion for $MoSe_2$, WSe_2, and GaP in contact with solutions containing $[Fe(CN)_6]^{4-}$. The stabilization ratio is observed to be considerably lower for damaged or corroded $MoSe_2$ and WSe_2 although s did not vary significantly with a change in light intensity. As a result, a simpler corrosion mechanism could be proposed, in which the competing corrosion and redox reactions were both first order in hole concentration.

Frese *et al.*[136,137] developed a kinetic model for semiconductor photocorrosion similar in outline to that of Gomes and co-workers. Like Gerischer and Lübke, they observed that stability was considerably lower on damaged crystalline surfaces. On the basis of binding energy arguments, Frese *et al.* pointed out that atoms on step sites of the semiconductor surface are likely to undergo bond breaking more easily and to act as centers for further corrosion. Atoms lying at kink sites are likely to be correspondingly more reactive still, and their removal by corrosion will lead to the generation of further active sites. The concentration of corrosion at such defect sites on the surface raises some questions about the applicability of the kinetic schemes above, which assume well-defined intermediates for each stage, whereas it may be that the properties of each intermediate will critically depend on the nature of the defect site it occupies as well as its oxidation state. For this reason, kinetic schemes must be treated with some caution, and the predicted states characterized independently before an unequivocal model can be established.

VI. FERMI LEVEL PINNING IN ELECTROCHEMISTRY

Unless there is unequivocal evidence to the contrary, most authors have implicitly or explicitly assumed that very little of the potential at the semiconductor/electrolyte interface is dropped across the Helmholtz layer. In particular, it is usually assumed that *changes* in the applied potential take place within the depletion layer. However, as we have discussed earlier, relatively small amounts of charge will lead to a significant drop in potential in the Helmholtz layer, and such surface charges may result from either surface oxidation or from the presence of surface states. By analogy with the semiconductor-metal junction, the phenomenon of surface charge controlling the potential distribution is termed Fermi level pinning.

For our purposes, it is useful to distinguish two types of Fermi level pinning[5]:

(i) Type I: Equilibrium is established between the semiconductor surface and the electrolyte, but not with the semiconductor bulk. As a result, the surface-state population is determined by the electrochemical potential of the redox couple present so that the flat-band potential will change with redox couple. The electrode will behave ideally under changes in applied potential, and linear Mott-Schottky plots of the correct gradient will be obtainable.

(ii) Type II: The surface states are in equilibrium with the semiconductor bulk but not with the electrolyte. Changes in potential applied to the electrode will alter the surface-state population, which will reduce the resulting change in depletion-layer voltage. Mott–Schottky plots will therefore not give the expected gradients and may not even be straight for nonuniform surface-state distributions.

The effect of changes in surface charge on the potential distribution is often described by invoking the Helmholtz layer capacitance, C_H, which relates the voltage across the Helmholtz layer to the charge stored by the equation $V_H = Q_H/C_H$. By treating the Helmholtz layer as a parallel-plate capacitor, and estimating layer thickness and dielectric constant, C_H can be estimated to lie in the range 2.5 to 30 μF cm^{-2}. In fact, rigorous calculation of the effect of surface charge on the potential distribution is quite involved even in the case of a junction with a metal[12,13] and is more so for a semiconductor–electrolyte junction where adsorption, chemical interaction, and finite charge-carrier size are important. However, taking C_H as 20 μF cm^{-2} and assuming that there is a surface-state density of 10^{13} cm^{-2}, which corresponds to only ca. 1% coverage of the surface in atomic terms, complete filling or emptying of such states would lead to a change in potential of ca. 0.1 V across the

Helmholtz layer. In the literature, it is therefore generally accepted that surface-state densities of the order of 10^{13} cm^{-2} are required for a significant effect on the potential distribution.

Although both types of Fermi level pinning were considered in the early work by Green,[44,138] its importance in photoelectrochemistry has been reemphasized more recently by Bard and co-workers.[139-142] These authors studied the variation of photovoltage and flat-band potential with redox couple for several semiconductors and concluded that changes in redox couple led to changes in the voltage across the Helmholtz layer. Figure 35 relates the semiconductor energies and barrier height ϕ_B to the redox couple and its position vis-à-vis the vacuum level. On most electrodes, ions will adsorb onto the surface and change the potential distribution; as a result, flat-band potentials are often strongly dependent on pH as well as on the redox couple itself. In Fig. 35, we consider only the case where the surface charge due to adsorbed protons (and other ions) is zero. This point is termed the point of *zero zeta potential* (pzzp), a concept derived from colloid chemistry.[143] The

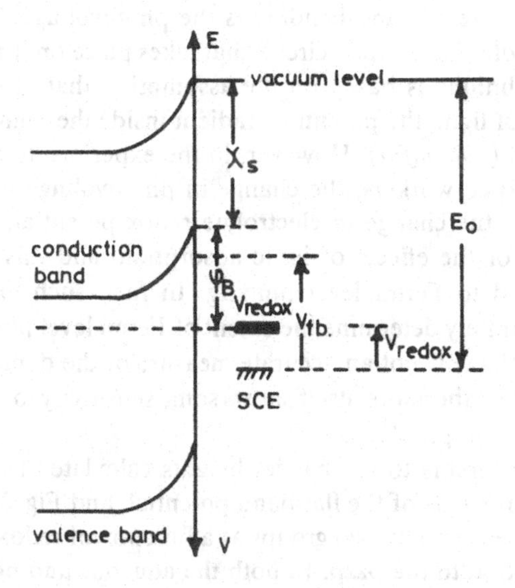

Figure 35. Energy diagram for an n-type semiconductor electrode with zero surface charge. χ_s is the semiconductor electron affinity, and E_0 relates the SCE reference electrode to the vacuum level.[145]

flat-band potential at the pzzp can be determined both experimentally by determining the pzzp separately from electrophoretic measurements on the material and theoretically from the mean electron affinities of the constituent atoms of the semiconductor referenced to the vacuum level. Since the variation in V_{fb} with pH is usually well known, it is always possible, therefore, to correct to the pzzp value for an aqueous electrolyte.

The barrier height should therefore be given by the depletion layer voltage $[V_{redox} - V_{fb}(\text{pzzp})]$ together with a term ΔV_{fc} for the separation of the Fermi level in the bulk of the semiconductor from the bottom of the CB (in an n-type semiconductor). Reference to Fig. 35 shows that

$$\phi_B = e_0[V_{redox} - V_{fb}(\text{pzzp}) + \Delta V_{fc}] \tag{34}$$

It should, therefore, be possible to control the open-circuit band bending by varying the redox couple. Figure 35 shows the principle of this: in the absence of any surface states, equilibrium should be established between the Fermi level in the semiconductor and that corresponding to the redox couple in solution. A convenient measure of band bending is the photovoltage. This is the change in potential at open circuit that takes place on illumination, and the technique is based on the assumption that at sufficiently high levels of light, the potential gradient inside the semiconductor must vanish (*vide infra*). However, in the experiments carried out by Bard and co-workers, the change in photovoltage was usually smaller than the change in electrolyte redox potential, even after correction for the effects of ionic adsorption, and this difference was assigned to Fermi level pinning. In fact, such experiments cannot accurately determine the extent of Fermi level pinning since the photovoltage is not an accurate measure of the depletion-layer voltage and, furthermore, itself shows some sensitivity to the kinetics of the redox couple.[144]

It is preferable to use barrier heights calculated from impedance measurements of the flat-band potential, and Fig. 36[145] shows ϕ_B values measured by two groups as a function of redox potential, after correction to the pzzp. In both the aqueous and nonaqueous electrolytes, it appears that only a proportion of any change in redox potential is reflected in changes in ϕ_B, and the remainder of the change appears in a shift of the flat-band potential. However,

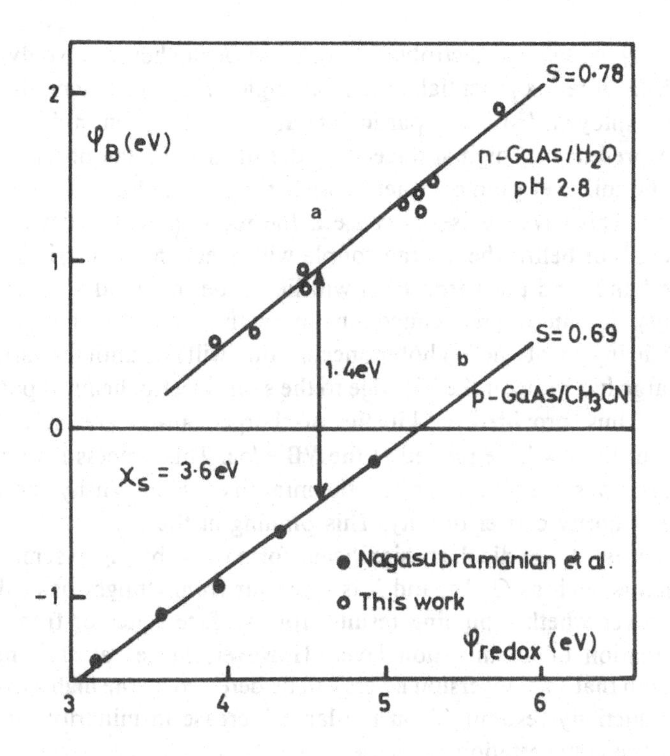

Figure 36. Plot of ϕ_B versus redox potential of the electrolyte. The data are fitted to the expression $\phi_B = S\phi_{redox} + K$.[145]

it is noteworthy that Fig. 36 does show much more ideal behavior than is usually observed for the metal–semiconductor junction, where there appears to be little correlation between the metal work function and the barrier height, as was illustrated in Fig. 7. The reasons for this difference are not clearly understood, but evidently the metal–semiconductor junction is much more prone to significant potential drops within a very short distance of the surface, whereas such potential drops do not occur or are somehow compensated at the semiconductor-electrolyte junction. The observation by Brillson and co-workers[33,34] of near ideal behavior for Schottky barriers formed from very high quality (100) MBE GaAs is particularly interesting as such results are readily obtainable in electrolyte with standard grades of melt-grown semiconductors. A common feature is, however, that the presence of elemental As is associated with high densities of surface states for both types of junction.

Nonaqueous electrolytes have often been chosen to study the effects of redox potential, as a much higher range of potentials can be employed. However, particular care must be taken in deducing the presence of high surface-state densities from the observation of Fermi level pinning when a *wide* range of redox potentials is used. This caveat arises because if the redox potential lies above the CB or below the VB the couple will inject charge carriers into the band, and the Fermi level will lie inside the band as a result. Thus, for an *n*-type semiconductor in the presence of a strong oxidizing agent, the VB hole concentration will rise until the surface charge has brought the VB edge to the same electrochemical potential. Thus, provided the kinetics of charge transfer are rapid, the Fermi level will be pinned at the VB edge. This process is termed *inversion*, since, at the surface, the minority carrier density exceeds the majority carrier density. This pinning at the band edges will, of course, be particularly significant for narrow-bandgap semiconductors such as GaAs, and it is often far from straightforward to discover whether pinning results from surface states or from the formation of an investion layer. However, Jaeger *et al.*[146] have shown that such inversion layers can be detected by the high surface conductivity resulting from the large increase in minority carrier surface concentration.

Type II Fermi level pinning is by its nature more difficult to characterize than type I since it results in nonclassical variation of properties with applied potential. The theories of photocurrent and impedance used to interpret a large corpus of results rely upon the assumption of a classical potential distribution, and it is not easy to separate unambiguously effects due to type II Fermi level pinning from other weaknesses in the models. Alternative methods of characterizing surface charges and the effect on depletion-layer voltage are therefore discussed in a following section.

Changes in the Helmholtz voltage over a range of applied potentials may occur as a result of:

(i) potential-dependent adsorption of charged ionic species;

(ii) surface oxidation or chemical reaction leading to a different surface dipole layer or to the formation of an insulating layer; or

(iii) charge or discharge of electronic states located at the surface.

There is no clear dividing line between (ii) and (iii) as a localized charge carrier can lead to a chemical change which could be reversed at a different potential and the "surface state" emptied. A major difference between the two concepts is that surface states are expected to be electronically active, filling and emptying under potential modulation, whereas chemical reaction can lead to electronic passivation, for example, the formation of an oxide layer on silicon. Often, the terminology used depends on the phenomena to be explained or the preference of the researcher.

The attractive feature of surface-state models is that they are much more amenable to quantitative treatment, as a surface state can be described by its energy and rate constants for loss and gain of carriers in a manner very similar to that used for solid-state devices. For instance, if the surface-state population is in equilibrium with the bulk of the semiconductor, then the density of charged states can be calculated. If the surface states are at an energy E_{ss}^0 and of density N_{ss} in an n-type semiconductor of Fermi level $E_F = E_F^0 + e_0\phi_{sc}$, where $\phi_{sc} < 0$, then the surface-state population can be calculated from

$$n_{ss}/(N_{ss} - n_{ss}) = \exp[-(E_{ss}^0 - e_0\phi_{sc} - E_F^0)/k_BT] \qquad (35)$$

The effect of various surface-state densities at 0.5 eV below the flat-band Fermi level on the potential distribution has already been illustrated above in Fig. 5. As the Fermi level passes through the surface-state energy, the changing potential causes much of the change in potential to be dropped across the Helmholtz layer. Where the Fermi level is remote from the surface states ($|E_{ss} - E_F| \gg 3k_BT$), the surface-state population is not sensitive to the applied potential, and the electrode will appear to behave normally.

Results comparable to those in Fig. 5 were observed by Allongue and Cachet[147] when the capacitance of n-GaAs in contact with 1 M KOH, 0.5 M H_2SO_4 and intermediate-pH electrolytes was measured at steady state. By using a frequency of 24.7 kHz, it was hoped that the space-charge capacitance could be extracted from the impedance and then be related directly to the potential distribution in a Mott-Schottky plot as shown in Fig. 37. The first result, curve a, shows a horizontal region similar to that shown in Fig. 5, indicative of a large proportion of the change in applied potential in this region taking place across the Helmholtz layer

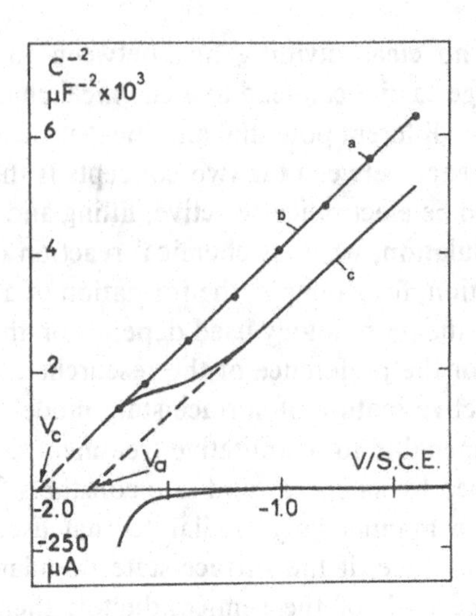

Figure 37. Mott-Schottky plot of *n*-GaAs in 1 *M* KOH at steady state (a) in 1 *M* KOH/1 *M* K_2Se at steady state (b), and in 1 *M* KOH with the potential stepped every 10 s (c).[147]

rather than the depletion layer. If 1 *M* K_2Se is added to the alkaline electrolyte, a linear Mott-Schottky plot is obtained, and this can be mimicked even without the selenide in solution provided ac impedance measurements are taken rapidly, as shown by curve c. All this implies that the effect responsible for the horizontal part of curve a must be a slow change at the electrode surface, which can, perhaps, be better described as a chemical process than as the emptying and filling of a simple surface state. This conclusion is consistent with that of Schröder and Memming[99] and from our own work[148] that reliable flat-band potentials of GaAs can only be obtained by using high scan rates.

Allongue and Cachet assigned the deviation from ideal behavior to formation of an oxide layer, although we could see no evidence for this in a recent ellipsometric study of *n*-GaAs in 1 *M* KOH.[149] Oxide and other insulating layers can, however, have a major effect on potential distribution, particularly for Si and InP, which both form particularly tenacious surface oxides. The insulating layer will separate surface charge on the semiconductor from the electrolyte, thereby greatly decreasing the Helmholtz capacitance and allowing a significant proportion of the interface potential to be dropped across this layer.

The observation that the potential distribution for GaAs appears only to be ideal for measurements taken at rapid scan rates is of considerable practical moment, since any photoelectrochemical cell will be required to function for long periods, and data derived from short-term experiments will give quite misleading predictions of long-term behavior.

Surface states are often thought to be distributed across the bandgap rather than located at a single energy. Theories can, therefore, often be quantitatively fitted to experimental data by choosing a suitable surface-state distribution, although, as we have already emphasized, almost any results can, in principle, be fitted in this way. A good example of this type of fitting procedure comes from the work of Wilson[96] described in Fig. 24 above. Wilson's distribution is very plausible but, as we will see, it has proved far from easy to verify this distribution in any independent way.

The shift in measured flat-band potential on illumination has already been illustrated in Fig. 31 for the case of *p*-GaAs in acid solution. This shift was ascribed to the trapping in surface states of photogenerated charge carriers. In the presence of 0.05 *M* $K_3Fe(CN)_6$ the observed shift on illumination was much reduced since rapid electron transfer kinetics reduced the surface electron concentration. Notten[150] conclusively demonstrated the importance of surface kinetics in the occurrence of flat-band shifts by observing a shift of 0.5 V when a *p*-GaAs disk immersed in 0.05 *M* NaOCl/0.1 *M* KOH was rotated at the very slow rate of 100 rpm; this shift disappeared on rotation at 5000 rpm since the more efficient supply of OCl^- to the surface prevented the buildup of surface-trapped charge.

Flat-band shifts under illumination have now been observed at a range of semiconductor electrodes, for example, GaAs,[99,147,151] InP,[152] GaP,[153] and CdS.[154,155] Allongue *et al.*[151] observed a reversible shift of 350 mV to more positive potential for *n*-GaAs in 1 *M* (Se^{2-}/Se_2^{2-}) electrolyte, but no shift occurred if the electrode was treated with Ru(III). The shift of *n*-CdS flat-band potential under illumination was studied by Meissner *et al.*[154,155]: Mott-Schottky plots were found to be dependent on electrode pretreatment, scan rate, and crystal face, accounting for the large range of flat-band potentials reported in the literature. Illumination of *n*-CdS in 0.1 *M* K_2SO_4 (aq) resulted in a reversible flat-band shift of +0.4 to +0.6 V plus a shift of several hundred millivolts which remained after the light was switched off, presumably as a result of a chemical change to the surface. In the presence of S^{2-} (aq) and other redox couples, no flat-band shift is observed under illumination.

VII. THE CHARACTERIZATION OF SURFACE STATES

Surface states remain difficult to characterize in spite of the ubiquity with which they are invoked to account for deviations from expected behavior. The main reason for this is that relatively low coverages of surface states, lower than those detectable by most spectroscopic techniques, are expected to have a large effect on the potential distribution. As a result, classical electrochemical techniques have dominated the investigation of the semiconductor/electrolyte interface, with impedance measurements playing a central role.

Spectroscopic techniques must be highly surface specific to pick out the surface interband states. Unfortunately, if absorption due to surface states is to be measured, then light of lower energy than the bandgap must be employed or else the transition will be obscured by interband excitation. The absorption of this subbandgap radiation may be followed directly, but it is usually more effective to measure the effect of excitation either on the photocurrent or on the capacitance. However, subbandgap radiation will penetrate through the semiconductor and lead to deep trap excitation, which can be confused with surface-state effects.

Nevertheless, modern techniques have had a considerable impact on our understanding of the semiconductor/electrolyte interface. We have chosen below a selection of both electrochemical and spectroscopic techniques which have proved particularly effective in the study of surface states and their behavior.

1. Impedance of Electrodes in the Dark

The principal use of ac impedance measurements has been in the determination of flat-band potentials and dopant densities from the Mott–Schottky equation given above. However, electrodes rarely show ideal behavior, and space-charge capacitances fitted to the simplest model

R_{sc} C_{sc}

are often dependent on both the direction and rate of scan of the potential sweep, as well as any prepolarization.[99] In addition, considerable frequency dispersion of the capacitance can usually be observed, and Fig. 38 shows the case of ⟨100⟩ *p*-GaP in 0.5 *M* H_2SO_4 (Ref. 156). Essentially linear Mott–Schottky plots are observed with a common intercept for each frequency, but with the gradients increasing with increasing frequency until a maximum value is

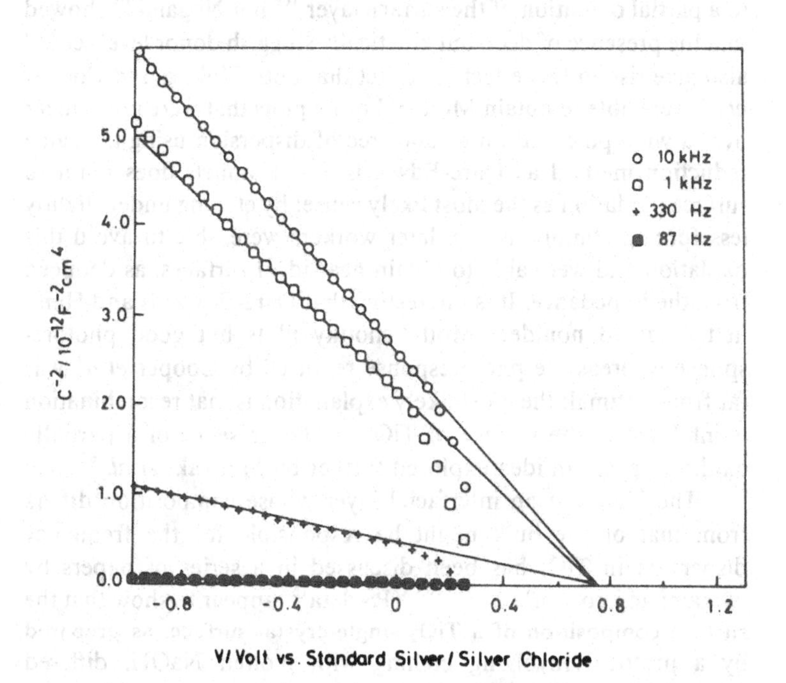

Figure 38. Inverse square capacitance of ⟨100⟩ *p*-GaP in 0.5 *M* H_2SO_4 at different frequencies.[156]

reached. More complex behavior is also observed, with nonlinear plots and frequency-dependent intercepts. Hamnett[5] has recently reviewed the present understanding of impedance response, and in many cases it is apparent that several models can equally well account for the data. This is particularly true for TiO_2, whose ac behavior has been investigated exhaustively.

Preparation of the surface of TiO_2 in a state essentially free from electronic surface states is difficult. Using near-ideal photoresponse as an indicator led Dare-Edwards and Hamnett[157] to a protocol in which polishing, etching with boiling concentrated sulfuric acid, and reduction in hydrogen were carefully ordered. Rather similar protocols were reported by Finklea[158] and Cooper *et al.*,[159] though in both cases slightly milder temperatures for the etchant were used. Dare-Edwards and Hamnett had found that surfaces prepared by their protocol gave Mott–Schottky plots consisting of two linear portions; this behavior was interpreted as due to a partial oxidation of the surface layer,[160] but Nogami[161] showed that the presence of deep but kinetically sluggish donor levels could also give rise to this effect. The fact that both Finklea and Cooper *et al.* were able to obtain Mott–Schottky plots that were both linear over a wide potential range and free of dispersion using the same reduction method as Dare-Edwards and Hamnett does point to surface oxidation as the most likely cause: by etching under slightly less forcing conditions, the later workers were able to avoid this oxidation and were able to obtain near-ideal surfaces, as deduced from the impedance. It is interesting that Dare-Edwards and Hamnett obtained nonideal Mott–Schottky plots but good photoresponse whereas the photoresponse reported by Cooper *et al.* was far from optimal: the most likely explanation is that recombination is inhibited at the surface of TiO_2 by the presence of a partially oxidized layer, an idea explored further by Morisaki *et al.*[162]

The idea that an interfacial layer whose composition differs from that of the bulk might be responsible for the frequency dispersion in TiO_2 has been discussed in a series of papers by Nogami and co-workers.[163-165] XPS data[163] appear to show that the surface composition of a TiO_2 single-crystal surface, as prepared by a protocol involving etching with molten NaOH, differed appreciably from the bulk composition, when this was revealed by argon-ion etching of the sample. Nogami also showed that the surface O 1*s* peak is split; although no comment was made regarding this splitting, it is probably due to carbonate, likely to be present in large concentration in molten NaOH. The only oxidation state of Ti that appears on the surface is Ti(IV); some Ti(III) appears on ion etching, but the process of ion etching is itself reducing in nature since lighter atoms are preferentially removed. The XPS evidence is, therefore, less compelling than at first might appear to be the case, though it is certainly consistent with the postulate of an oxidized surface region. Our own experience of molten NaOH is that it is a highly oxidizing etchant.

More cogent evidence for the existence of an interfacial layer comes from impedance data on these electrodes. Nogami showed that over a substantial frequency range, $|Z|$ decreases as $\omega^{-1/2}$, consistent with diffusion in the electrolyte (perhaps in a porous film) as the rate-limiting step. Using a model that combines diffusion in the electrolyte with deep donor levels in the semiconductor that are in slow equilibrium with the bulk Fermi level, in fast equilibrium with the electrolyte, and show a spatial distribution that decays exponentially or in Gaussian form into the interior of the semiconductor, Nogami was able to account for all aspects of his frequency dispersion.[165]

It has to be said that the complexity of the interface ivestigated by Nogami may owe much to his chosen method of surface preparation. Tomkiewicz, using a not dissimilar protocol, interpreted his data in terms of inhomogeneities in the (001) *surface.*[166] The model used was an equivalent circuit of the form

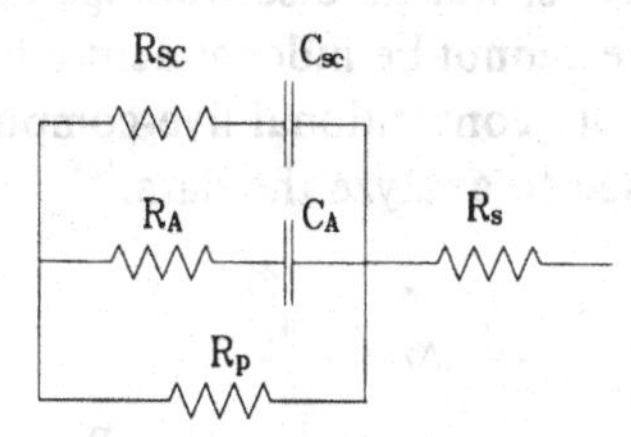

In this equivalent circuit, the parallel components C_{sc} and C_A represent about 80% and 20%, respectively, of the surface area and correspond to different dopant densities. If C_{sc} is extracted from this equivalent circuit, which is normally done by assuming that R_{sc} is very small and extracting C_{sc} at high frequency,[167] then behavior very similar to that discussed above is obtained, with linear Mott–Schottky relationships at high and low potentials, but with a nonlinear central portion lying about 1 V positive of the flat-band potential in 0.1 *M* acetic acid. There is good evidence that this behavior is due to surface effects, since it disappears if the crystal is rinsed in $H_2SO_4:H_2O$ (1:1). The nonlinearity can, then, be attributed to surface states giving rise to a finite drop of potential in the Helmholtz layer. Analysis of the data assuming that the potential can be partitioned between space-charge and Helmholtz layers only gives the straightforward expression

$$\frac{dU}{d(1/C_{sc}^2)} = \frac{\varepsilon\varepsilon_0 e_0 N_D}{2} + \frac{\varepsilon\varepsilon_0 e_0^2 N_D}{2C_H}\left(\frac{dN_{ss}^+}{d\phi_{sc}}\right) \tag{36}$$

where ε is the relative permittivity of the TiO_2 and ε_0 is the permittivity of free space ($8.854 \times 10^{-12}\ Fm^{-1}$).

From this, the variation of N_{ss}^+, the density of *un*occupied surface states, with the potential drop $d\phi_{sc}$ in the space-charge region can be obtained, if C_H is known. The latter was evaluated, approximately, from the variation of flat-band potential with N_D, the donor density, giving a total surface state density of ca. $6 \times 10^{14}\ cm^{-2}$. The position of these states is of considerable interest: their maximum is about 1.8 eV below the conduction-band edge, close to the position postulated to account for the facile photo-oxidation of water as discussed below. Tomkiewicz's analysis is of considerable interest, as it attempts to measure directly the manner in which the applied potential is partitioned between space-charge and Helmholtz layers, but its disadvantage is that the equivalent circuit given above cannot be independently justified. Indeed, in a later paper, the more conventional five-component equivalent circuit below was used to analyze the data.[168]

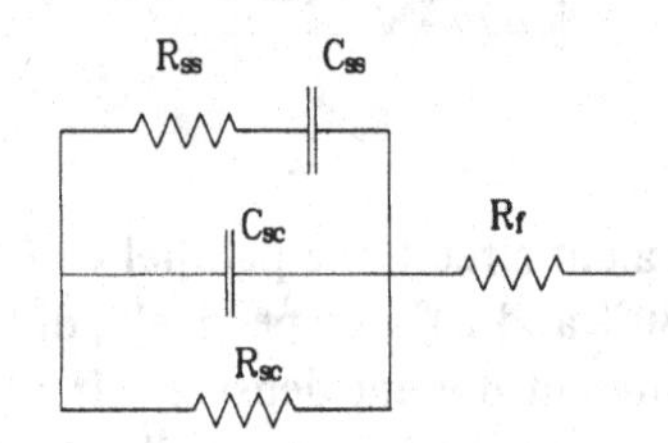

The justification for this type of circuit has been discussed by Dare-Edwards *et al.*[156] It is the simplest possible circuit formed from passive elements that can be used to describe the response of surface states whose time constant falls within the frequency regime of the experiment, and the magnitude of C_{ss} is a direct measure of the density of surface states, as shown elsewhere.[156] For surface states at a single energy, or distributed uniformly across the bandgap, the $R_{ss}C_{ss}$ pair gives a reasonable description for the filling and emptying of the surface states. Clearly, at low frequencies, the pair will behave as a capacitive impedance, but at high frequencies, the *RC* time constant of the pair will prevent complete charging, and the interface will behave more classically. The time constant for surface-state charging is given by $\tau = R_{ss}C_{ss}$, and this will remain fairly constant across the bandgap, provided surface states of the same type are involved. Faradaic current flow across the interface is modeled by the resistor R_f, provided the current remains small. Analysis of the frequency response for TiO_2 in phosphate buffer[168] showed a large band of surface states centered 0.8 eV below the conduction-band edge. From the value of C_{ss}, an estimate of the density of these states can be made, and a value of 1.8×10^{13} cm^{-2} has been reported.[168] A very similar result was reported by Kobayashi *et al.*,[169] whose analysis of similar ac measurements showed surface states at ca. 0.65 eV below the conduction-band edge with a density of 8×10^{12} cm^{-2}; this density is apparently independent of pH in the range 0.4–7.

The five-component model given above has also been used to interpret the ac impedance data for a number of other semiconductor electrodes, usually by fitting the data at 8–12 frequencies at least, using a minimization routine to obtain the best fit.[156,170] Thus, Fig. 39 shows the result of fitting data for ⟨100⟩ *p*-GaP in 0.5 *M* H_2SO_4 in the frequency range 40–10,000 Hz. A linear Mott-Schottky plot is obtained, and a dopant density consistent with Hall measurements can be calculated from the gradient.[156] It can also be seen that the faradaic resistance decreases near the flat-band potential as expected for a smaller depletion layer. The surface-state density, as measured by C_{ss}, appears to peak close to the flat-band potential, whereas R_{ss} displays a minimum consistent with the maintenance of a steady time constant $\tau \approx RC$.

Application of these relatively simple models to silicon has proved far less straightforward, since the surface chemistry of silicon is greatly complicated in aqueous solution by a thin layer of SiO_2

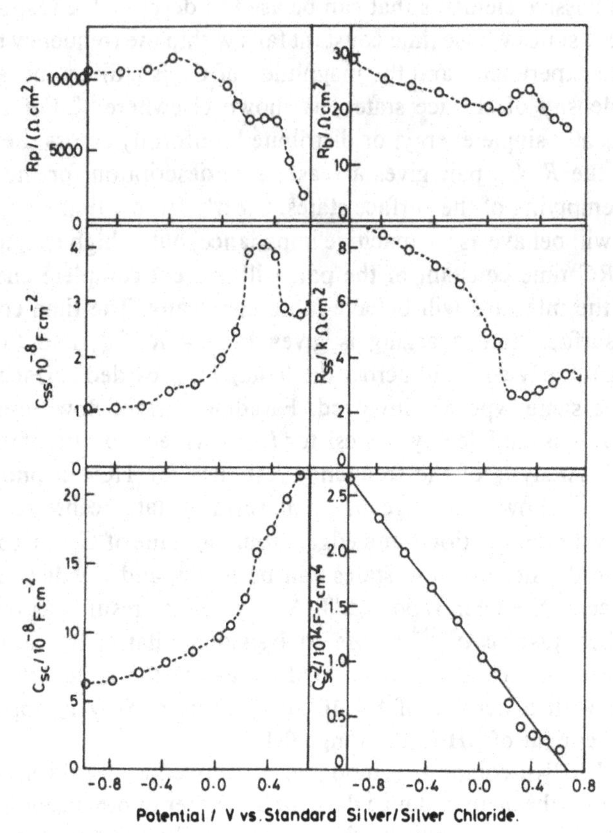

Figure 39. Impedance of ⟨100⟩ *p*-GaP in 0.5 *M* H_2SO_4 from Fig. 38, analyzed according to the five-component model.[156]

that is always present on the surface unless there is a relatively high concentration of fluoride ions in solution. The effect of this layer in non-fluoride-containing solutions was first explored in detail by Memming and Schwandt[171] and subsequently, using transient techniques, by Wolovelsky *et al.*[172] The former workers showed that in sulfuric acid, even at high frequencies, it was not possible to interpret the ac response in terms of a simple series *RC* circuit. Using a surface conductivity method to determine the depletion-layer potential drop and using a surface recombination velocity measurement technique, Memming and Schwandt were able to pinpoint a maximum in surface recombination at a depletion potential of 0.50 V for *p*-type Si (1 Ω cm).

Without interpretable ac data, the exploration of Si electrochemistry is severely hampered, and Wolovelsky *et al.* circumvented this problem with the use of a transient technique that in effect Fourier transforms the data into the time domain. In this way, two distinct rate constants can be seen, corresponding to bulk and surface processes. With this approach, they were able to interpret successfully data for *n*-Si (at a dopant level of ca. 5×10^{15} cm^{-3}) in inert electrolyte solutions of $Ca(NO_3)_2$. For this material, they found good agreement between theory and experiment for potentials well into the depletion region, if it is assumed that all the applied potential is accommodated in the semiconductor. Close to the flat-band potential, however, some charge accumulation is found in surface states, which extend in energy from ca. 0.2 V below to 0.2 V above the flat-band potential, with a total density of ca. 10^{12} cm^{-2}. Analysis of the bulk relaxation in the accumulation region gave a capacitance for the surface oxide of 6 μF cm^{-2}, from which a thickness of ca. 6 Å can be estimated. This last value is quite approximate, but highly reasonable.

In the presence of relatively high fluoride concentrations (>1 *M*), the behavior of *n*-Si can be interpreted more straightforwardly. Working at frequencies near 140 kHz, Memming and Schwandt[171] found quantitative agreement between theoretical and experimental ac response, assuming a roughness factor of about 2 and complete accommodation of the potential change in the depletion layer. Even at this frequency, however, the agreement for *p*-Si was much poorer, and recombination velocity studies revealed a very fast surface state centered at ca. 0.45 eV above the VB edge. This compares with the value of ca. 0.7 eV above the VB edge deduced above for Si in the presence of sulfuric acid. In fact, if fluoride ions are added to the latter solution, both surface states can be seen. Interestingly, recombination velocity studies also showed a surface state on *n*-type Si lying 0.35 eV below the CB edge, but the capture cross section of the latter is considerably smaller, leading to the observation that relatively little effect of these states can be discerned on the capacitance behavior. As a result of their studies, Memming and Schwandt considered it likely that a band of surface states probably extends throughout the midgap region on silicon in the presence of fluoride ions, from ca. 0.35 eV below the CB edge to 0.45 eV above the VB edge, a range of about 0.3 eV.

The effect of these surface states on the anodic polarization of *n*- and *p*-type Si in fluoride solutions is quite spectacular. At low anodic current densities it appears that the surface of silicon dissolves relatively uniformly, but as the current density rises, a remarkable morphological change takes place, with the formation of a porous overlayer. Only at very high anodic current densities is silicon electropolished.[173]

The anodic dissolution of silicon has been extensively investigated.[173-179] A quite remarkable feature of this process is the observation that appreciable quantities of *hydrogen* are liberated, and Turner[173,174] was the first to suggest that at lower potentials, some silicon was formed in the divalent state, and this subsequently disproportionated to give Si(IV) and amorphous Si(0), the latter reacting with water to yield hydrogen. The generation of amorphous silicon was originally postulated as the layer of porous silicon formed was believed to be amorphous. However, more recent data[176,177,180] strongly support the view that the porous silicon formed is in fact crystalline, and it is probable that the hydrogen arises from the direct reaction of the Si(II) first formed with water or HF.[175]

The dissolution model first put forward by Memming and Schwandt[175] invokes an Si atom bound to two surface F and two underlying Si atoms as the reactive site. Capture of a hole by one of the Si—Si bonds leads to an unstable radical cation intermediate, which, in the original mechanism, had the remaining electron from the bond localized on the outer silicon atom. Further hole capture could then take place, leading to scission of the second Si—Si bond and dissolution of $:SiF_2$. Tetravalent dissolution of Si was conceived, in this model, as taking place from hydroxylated parts of the surface, the main difference being that the radical electron left after the first bond scission remains on the inner of the two silicon atoms.

Considerable modification of this model has taken place in recent years. Matsumura and Morrison[181] showed that both current doubling and, at low light intensities, quadrupling could take place if *n*-Si was illuminated in 1 *M* HF at pH 4.9. This suggests that most of the postulated intermediates are oxidized by electron injection rather than hole capture. In particular, intermediates of the form Si—SiF_3 must be sufficiently easily oxidized that their surface energy levels lie in the mid or upper bandgap region. This is of considerable significance, since we know from the results of Memming and Schwandt[171] that the fluorinated silicon surface possesses surface states throughout the midgap region. It would seem very likely that surface crystalline imperfections could easily lead to species of the form Si—SiF_3 and that these could be the surface states found.[182]

Another suggestion[183] is that instead of electron injection, a second hole capture process following the first might lead to dissolution of SiF_4 and formation of Si—H bonds on the surface, the latter by reaction of a surface diradical with HF; this last reaction does not consume either electrons or holes. The postulate of Si—H bonds forming on the surface is an attractive one: it would account both for the known presence of quite large amounts of hydrogen in the porous silicon film that forms[183,184] and also for the evolution of hydrogen. The major difficulty is that analogous chemistry for the Ge and GaAs surfaces would suggest that the energy of the Si—H bond would lie very high, close to the CB, and such a species would rapidly oxidize by electron injection.

A further important result from the work of Matsumura and Morrison is that at higher light intensities, only current doubling is observed. This is of considerable interest: it suggests that the two-electron intermediate Si—SiF_3 does not easily oxidize by electron injection and that if the hole concentration on the surface is sufficiently high, the Si—Si bond will undergo scission by a second hole-capture electron-injection process. The relevance of this to the formation of porous silicon remains controversial: at high local hole concentrations, where the material is strongly biased to positive potentials, the oxidation of the intermediate two-electron product by further hole capture is favored and electropolishing can take place. At lower potentials, initiation of the four-electron dissolution can only take place at preexisting surface states; once initiated, high hole concentrations are favored only at certain morphological sites such as the tips of pores, and hence pore formation takes place. Where the local field is less, at the sides of the pores for example, two-electron oxidation can take place, with competitive chemical oxidation of Si(II) yielding hydrogen and electrochemical oxidation yielding Si(IV) without any hydride intermediate. This model points up the very great importance of the energy of certain intermediates and the fact that surface states can have an importance that transcends simple recombination.

Insulating layers have been invoked for other semiconductors aside from TiO_2 and Si. Braun *et al.*[185,186] have measured the current response of ⟨0001⟩ CdS in KCl (aq) to a potential step and found that the results cannot be fitted to a single exponential decay consistent with one *RC* pair. The authors termed the experiment "time domain spectroscopy" and fitted results to a set of *RC* series pairs in parallel with each other to give a suitable set of exponential decays that can be summed. Each *RC* component is assigned to a different type of surface state, and Braun *et al.* backed this claim with scanning electron microscopy (SEM) photographs of CdS electrodes after use, which show considerable heterogeneity on the 1-μm scale. This approach is very similar to that devised for TiO_2 above, and it has also been assessed by McCann and Badwal for the case of Fe_2O_3 (Ref. 187).

2. Impedance of Illuminated Semiconductor Electrodes

The shifts in flat-band potential that can be measured under illumination for semiconductors have already been mentioned, and an example is shown in Fig. 31. However, a number of groups have also observed peaks in the electrode admittance and capacitance on approach to flat-band potential under illumination,[188-193] an example of which is shown in Fig. 40. In many of these cases, the photoinduced peaks are of considerable interest as they occur very close to the onset of photocurrent and must, therefore, be closely connected to the mechanism that delays photocurrent onset in such semiconductors as GaAs. However, although the dependence of such capacitive peaks on electrolyte, frequency, and illumination intensity might be expected to reveal much about the nature of surface reactions, in practice interpretation is difficult. Both the majority and minority carrier density will be modulated so it becomes difficult to separate the contributions to the total response. Furthermore, illumination-induced flat-band shifts are likely to confound attempts to calculate accurately equilibrium carrier

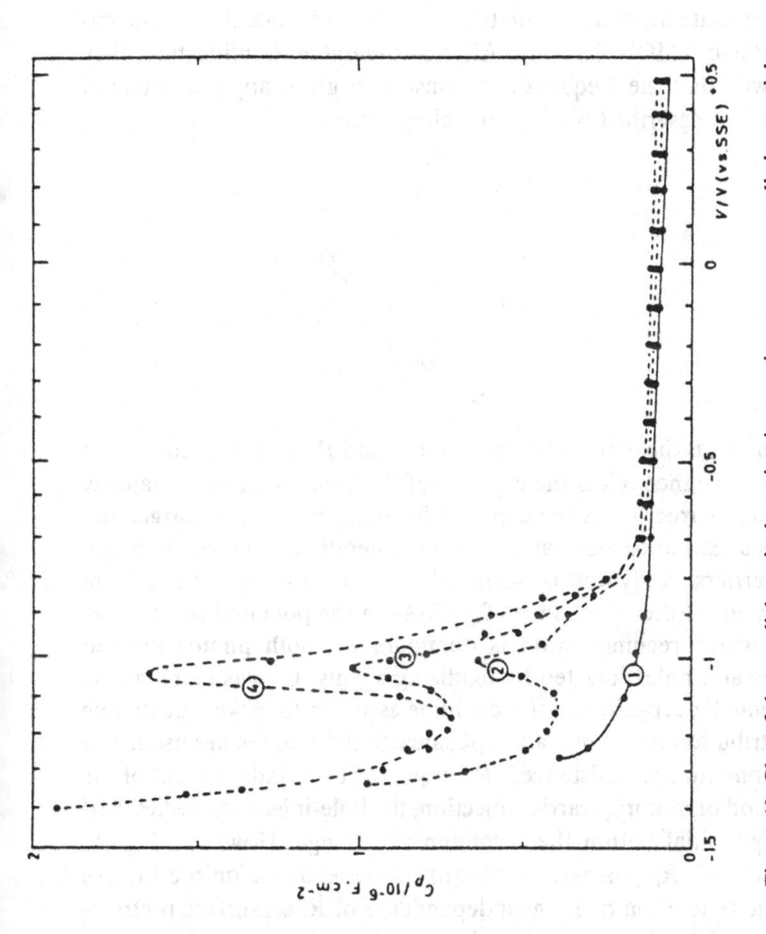

Figure 40. Plot of the parallel equivalent circuit capacitance versus applied potential for $\langle\bar{1}\bar{1}\bar{1}\rangle$ *n*-GaAs in 1 *M* $FeSO_4$, 0.25 *M* K_2SO_4, H_2SO_4 (pH 1) at 1 kHz: (1) in the dark; (2)–(4) at light intensities giving limiting photocurrents of 0.375, 0.75, and 1.5 mA cm^{-2}, respectively.[192]

densities at the electrode surface. Several models for illuminated impedance have, however, now been established, though the understanding they bring is still only semiquantitative.

Pierret and Sah[194,195] developed a theory for the effect of steady-state illumination on the impedance of a metal-oxide-semiconductor (MOS) junction. After considerable simplification, they showed that the frequency response at a given applied potential could be described by the equivalent circuit:

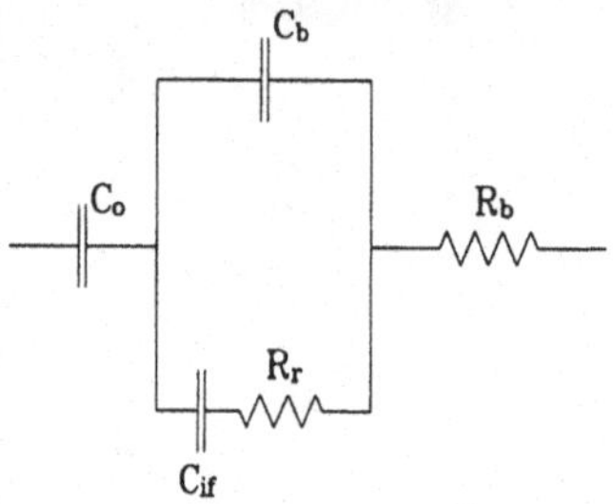

where C_o is the oxide layer capacitance and R_b is the semiconductor bulk resistance, C_b is the capacity of the junction to store majority carrier charge, C_{if} is the capacity for minority carrier charge, and R_r is a resistance associated with the generation and recombination of carriers. Kelly and co-workers[196,197] used a circuit of this form to interpret the impedance of *n*-GaAs in the potential region near V_{fb} where recombination is dominant for both photogenerated holes and holes injected by oxidizing agents. C_o was taken as the Helmholtz capacitance and could be assumed to make a negligible contribution to the overall impedance at the frequencies used. The "recombination resistance," R_r, appeared to be independent of the method of minority carrier injection, the hole-injecting species, and the potential within the recombination range. However, Fig. 41 shows that R_r possesses a near-linear dependance on the inverse of the generation rate. The independence of R_r on surface pretreatment led to the conclusion that recombination must take place within the depletion layer and not at the surface, where surface treatment would modify the recombination centers.

A rather different model for the impedance associated with surface recombination at the illuminated semiconductor/electrolyte interface has been developed by Vanmaekelbergh and Cardon.[198] Illumination is assumed to result in an additional impedance that

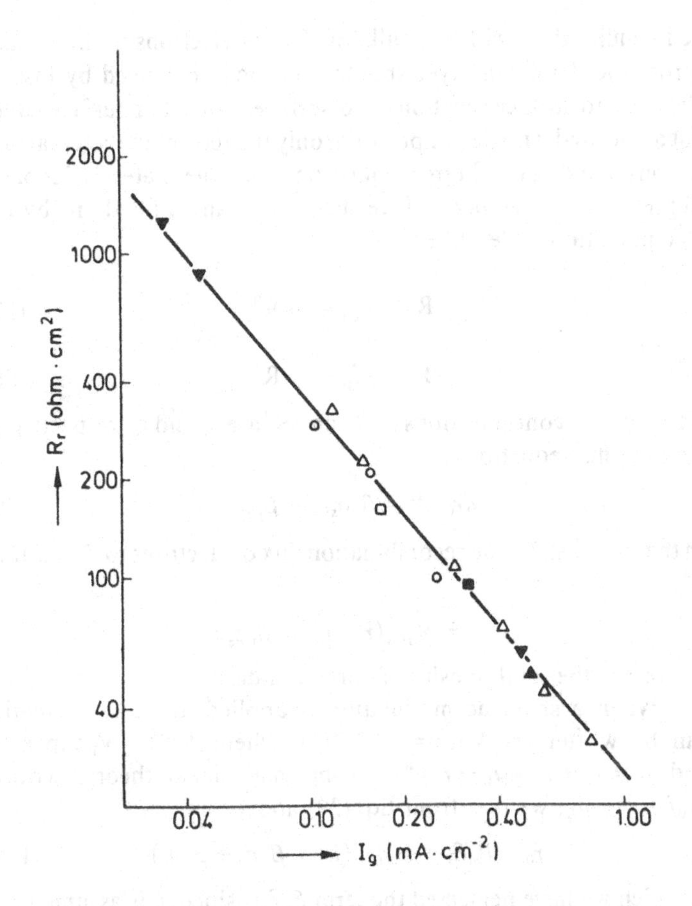

Figure 41. Variation of recombination resistance with generation function for *n*-GaAs in $Ce(SO_4)_2$/0.4 *M* H_2SO_4 at various concentrations and rotation speeds at −0.60 (▼), −0.65 (▲), or −0.70 V (△) vs. SCE, in $K_3Fe(CN)_6$/1 *M* NaOH at different Fe(III) concentrations and rotation rates at −1.5 V vs. SCE (○), in 0.5 *M* K_2SO_4 at pH 5.0 under illumination at −1.0 V vs. SCE (□), and in 0.5 *M* H_2SO_4 under illumination at −0.7 V vs. SCE (■).[197]

is parallel to the equivalent circuit appropriate to conditions in the dark. If it is assumed that the photogenerated hole flux to the surface of an *n*-type semiconductor is unaffected by the applied potential, then the ac current response will be determined entirely by the conduction-band electron flux through the depletion layer. As the majority carrier concentration at the surface is believed to be in equilibrium with the bulk, the flux of electrons to the surface across the depletion layer should be exactly balanced by loss of electrons to surface reactions and surface states. For surface states not associated with decomposition, only the recombination reaction is considered, and thermal emission from the states is ignored. Representing an empty surface state by R^+ and a filled one by R^0, recombination is described by

$$R^+ + e^-_{cb} \xrightarrow{\beta_n} R^0 \tag{37}$$

$$R^0 + p^+_{vb} \xrightarrow{\beta_p} R^+ \tag{38}$$

If the surface concentrations of R^+ and R^0 are r_+ and r_0, respectively, then the flux equation is

$$dr_0/dt = \beta_n n_s r_+ - \beta_p p_s r_0 \tag{39}$$

In the steady state, the recombination flux of electrons to the surface is

$$j_0^r = \beta_n n_s (r - r_0) = \beta_p p_s r_0$$

where r is the total density of surface states.

When a small ac modulation is applied, the total potential can be written as $V'(t) = V + \tilde{V}(t)$, where $\tilde{V}(t) = V_1 \exp(i\omega t)$, and $r_0' = r_0 + \tilde{r}_0 \equiv r_0 + r_1 e^{i\omega t}$. From small-signal theory, writing $dr_0/dt \equiv i\omega\tilde{r}_0$, we find from Eq. (39) above

$$\tilde{r}_0 = \beta_n \tilde{n}_s (r - r_0)/(i\omega + \beta_n n_s + \beta_p p_s) \tag{40}$$

in which we have neglected the term $\beta_p \tilde{p}_s r_0$ since p_s is assumed not to depend on the applied potential.

If n_s is in equilibrium with the bulk, $n_s' = N_D \exp[-e_0(V_{sc} + \tilde{V}_{sc})/k_B T]$, and if $|\tilde{V}_{sc}| \ll k_B T/e_0$, $\tilde{n}_s \approx n_s(-e_0 \tilde{V}_{sc}/k_B T)$, and finally

$$\tilde{r}_0 = -j_0^r (e_0 \tilde{V}_{sc}/k_B T)/(i\omega + \beta_n n_s + \beta_p p_s) \tag{41}$$

The ac component of the electrical current flux due to the capture of conduction-band electrons in surface states, $\tilde{j}_R$, is then given by

$$\tilde{j}_R = -e_0 \beta_n [(r - r_0)\tilde{n}_s - n_s \tilde{r}_0] \tag{42}$$

and the impedance due to recombination, Z_r, is given by

$$Z_r = \tilde{V}_{sc}/\tilde{j}_R = (i\omega + \beta_n n_s + \beta_p p_s)/[(e_0^2/k_B T) j_0^r (i\omega + \beta_p p_s)] \tag{43}$$

which can be represented by an equivalent circuit of the form

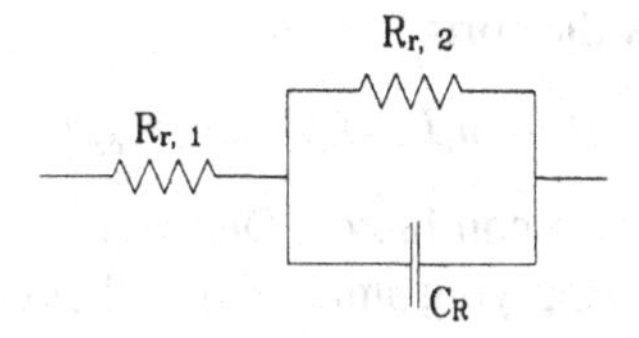

where

$$R_{R,1} = [(e_0^2/k_BT)\beta_n n_s(r - r_0)]^{-1} \tag{44}$$

$$R_{R,2} = [(e_0^2/k_BT)\beta_p p_s(r - r_0)]^{-1} \tag{45}$$

$$C_R = (e_0^2/k_BT)(r - r_0) \tag{46}$$

Vanmaekelbergh and Cardon then considered the consequences of attempting to include the effect of the term $\beta_p \tilde{p}_s r_0$ in Eq. (40), which was neglected in the above analysis. This leads to

$$\tilde{r}_0 = j_0^r[(\tilde{n}_s/n_s) - (\tilde{p}_s/p_s)](i\omega + \beta_n n_s + \beta_p p_s)^{-1} \tag{47}$$

Further progress requires assumptions to be made about the processes consuming surface holes in order to relate $(\tilde{p}_s/p_s)$ to the *majority* carrier modulation. The photogenerated holes which have been swept toward the surface are assumed to occupy a layer where they are in electronic equilibrium with the surface hole density p_s. The total number of holes across the depletion layer in this distribution is taken as $P^\ddagger$ per unit area, and the following rate equation is used:

$$dP^\ddagger/dt = j_h - \beta_p p_s r_0 - \alpha p_s \tag{48}$$

where αp_s is the rate of hole loss arising from redox couple oxidation and surface corrosion reactions, and j_h is the hole flux into the depletion layer from the bulk of the semiconductor. At steady state, $j_0^h = p_s(\alpha + \beta_p r_0)$, and applying small-signal theory to Eq. (48), we have

$$i\omega\tilde{P}^\ddagger = -j_0^h(\tilde{p}_s/p_s) - \beta_p p_s j_0^r(i\omega + \beta_n n_s + \beta_p p_s)^{-1} \times [(\tilde{n}_s/n_s) - (\tilde{p}_s/p_s)] \tag{49}$$

when the expression above for $\tilde{r}_0$ is used.

The total number of holes in the depletion layer can be related to the surface hole concentration by an expression which, for $e_0 V_{sc} \geq 9 k_B T$, takes the form

$$P^{\ddagger} = p_s L_D (k_B T / 2 e_0 V_{sc})^{1/2} \tag{50}$$

where L_D is the semiconductor Debye length, given by $L_D = (\varepsilon\varepsilon_0 k_B T / e_0^2 N_D)^{1/2}$. Applying small-signal theory to this expression gives

$$\tilde{P}^{\ddagger} = (\tilde{p}_s / p_s) \cdot p_s L_D (k_B T / 2 e_0 V_{sc})^{1/2} + (\tilde{n}_s / n_s) \cdot p_s L_D (k_B T / 2 e_0 V_{sc})^{3/2} \tag{51}$$

and $\tilde{P}^{\ddagger}$ can then be eliminated from Eqs. (49) and (51), resulting in an expression where $(\tilde{p}_s / p_s)$ can be written as a function of $(\tilde{n}_s / n_s)$. Equation (47) can then be evaluated and substituted into Eq. (42), from which the impedance can be obtained after some tedious algebra and assuming that $(k_B T / 2 e_0 V_{sc}) \ll 1$:

$$Z_R = \left(\frac{1}{j_0^r}\right)\left(\frac{k_B T}{e_0^2}\right) \times \left[1 + \beta_n n_s \Big/ \left[i\omega \left(1 + \frac{j_0^r \beta_p p_s^2 L_D (k_B T / 2 e_0 V_{sc})^{1/2}}{j_h^2 + \omega^2 p_s^2 L_D^2 (k_B T / 2 e_0 V_{sc})}\right) + \beta_p p_s \left(1 - \frac{j_0^r j_h}{j_h^2 + \omega^2 p_s^2 L_D^2 (k_B T / 2 e_0 V_{sc})}\right)\right]\right] \tag{52}$$

Earlier results on n-GaAs suggest that $p_s < 10^{13}\ \mathrm{cm}^{-3}$ and typically $L_D \approx 10^{-6}$ cm and $j_h \approx 6 \times 10^{14}\ \mathrm{cm}^{-2}\ \mathrm{s}^{-1}$, so that $j_h^2 \gg \omega^2 p_s^2 L_D^2 (k_B T / 2 e_0 V_{sc})$ for the range of frequencies commonly used in ac experiments. Equation (52) can then be simplified to

$$Z_R = \left(\frac{1}{j_0^r}\right)\left(\frac{k_B T}{e_0^2}\right)\left[1 + \beta_n n_s \Big/ \left[i\omega \left(1 + \frac{j_0^r \beta_p p_s^2 L_D (k_B T / 2 e_0 V_{sc})^{1/2}}{j_h^2}\right) + \beta_p p_s \left(1 - \frac{j_0^r}{j_h}\right)\right]\right] \tag{53}$$

Equation (53) gives a frequency response that can be modeled by the same circuit as Eq. (43). C_R and $R_{R,1}$ are then very similar to those defined by Eqs. (46) and (44), and the main difference is the value of $R_{R,2}$. Where $j_0^r / j_h < 1$, the extra factor contributes to the voltage dependence of the impedance, and Z_R can still be written as $R_{R,1} + (R_{R,2}^{-1} + i\omega C_R)^{-1}$.

The model has also been extended to treat recombination via corrosion-induced intermediates according to the kinetic model of Vanmaekelbergh *et al.*[85] Expressions for the effect of the voltage modulation on each stage of the decomposition reaction are obtained by a matrix method, so the resulting equations are of considerable complexity.

Vanmaekelbergh and co-workers[192,199] have reported impedance measurements on illuminated *n*-GaAs by subtracting the bulk ohmic impedance and fitting the resulting impedance to a resistance R_p and capacitance C_p, as shown in Fig. 40. C_p and R_p will therefore include both the normal dark impedance terms and those due to illumination. Figure 40 shows C_p for various illumination intensities for *n*-GaAs in 1 *M* Fe^{2+} (aq). The capacitance maximum shown in this figure is found to be proportional in magnitude to the hole flux to the surface and inversely proportional to the modulation frequency, as shown in Fig. 42.[199] The relationship can be written as

$$C_{\max} = \gamma(i_h/\omega) \tag{54}$$

where i_h is the hole current, and γ is found to be ca. 6 V^{-1}.

To model peaks in C_p, the equation for recombination impedance must be expressed in the form of that for a resistance in parallel with a capacitance. Using a simplified form of Eq. (53):

$$Z_R = \frac{1}{j_0^r} \cdot \frac{k_B T}{e_0^2}\left(1 + \frac{\beta_n n_s}{i\omega + \beta_p p_s[1 - (j_0^r/j_h)]}\right) \tag{55}$$

the equation for the additional parallel capacitance due to illumination can be derived as

$$C_{R,p} = \frac{e_0^2}{k_B T}\frac{j_0^r}{\beta_n n_s}\left[\left(\frac{\omega}{\beta_n n_s}\right)^2 + \left(1 + \frac{\beta_p p_s(1 - j_0^r/j_h)}{\beta_n n_s}\right)^2\right]^{-1} \tag{56}$$

By rewriting Eq. (56) in terms of the surface-state density and population, r and r_0, an expression can be derived for the maximum value of C_p:

$$C_{\max} = \frac{e_0}{2k_B T}\frac{1}{1 + \alpha/\beta_p r}\left(\frac{i_h}{\omega}\right) \tag{57}$$

Equation (57) shows the desired relationship between $C_{\max}$, i_h, the

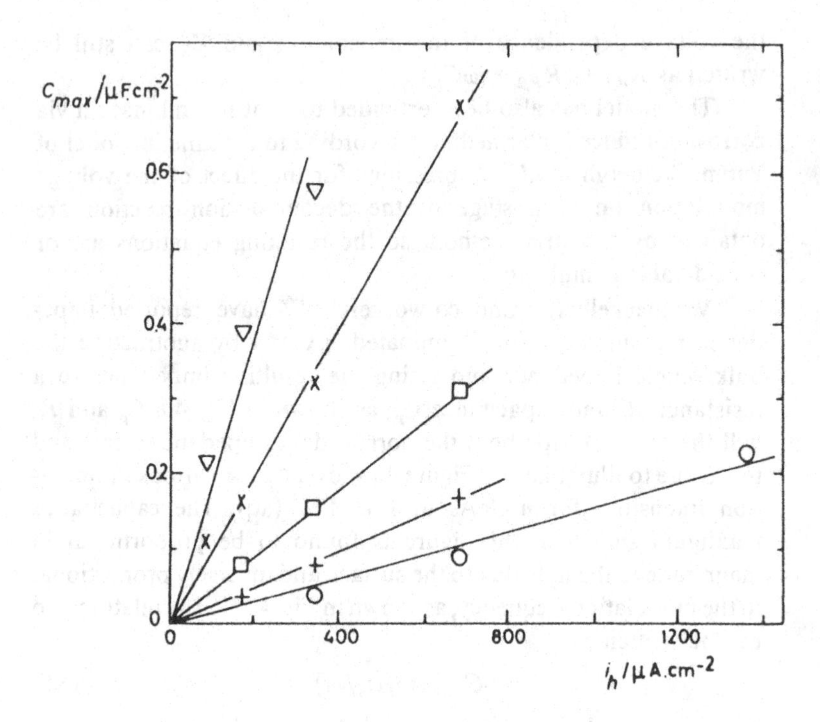

Figure 42. Variation of capacitance-peak maximum with hole flux to the surface for *n*-GaAs $\langle\bar{1}\bar{1}\bar{1}\rangle$, with electrolyte as in Fig. 40. Measurement frequency *f*: ▽, 0.5; ×, 1.0; □, 2.5; +, 5.0; ○, 10 kHz.[199]

hole current, and ω; from the experimental value of $\gamma = 6\ V^{-1}$, we obtain $\alpha \approx 2\beta_p r$. However, experimentally the capacitance peak is observed at a potential where the photocurrent is negligible, so that the rate of recombination must far exceed that of hole transfer to redox couple/corrosion reaction and, therefore, $\alpha \ll \beta_p r$. Further difficulties with both this model and that including recombination via decomposition reactions are the failure to reproduce the positive shift of C_{max} as the frequency is raised and the considerable underestimation of the peak width. The latter may be associated with the shift in flat-band potential under illumination, and with surface inhomogeneity, leading to a distribution of surface potentials. Relatively little information on the surface recombination centers appears to emerge from this model owing to the kinetic scheme used. Unlike the theories of dark impedance, there is no dependence on the surface-state energy as thermal emission is neglected, and the fit is not sufficiently good to be able to deduce anything about the state density either.

Results were also interpreted with the *series* resistor–capacitor model in parallel with the depletion layer to ascertain the usefulness of the recombination resistance (R_R) concept.[200] In accordance with the work of van den Meerakker *et al.*,[197] over a considerable potential region the recombination resistance was observed to be insensitive to electrode potential, as shown in Fig. 43, as well as the mechanism of hole creation and the electrolyte composition, but inversely proportional to the photogenerated hole current. In terms of the model for impedance due to illumination given in Eq. (43), the generation–recombination resistance should be

$$R_r = \frac{k_B T}{e_0 i_r}\left[1 + \frac{\beta_p p_s \beta_n n_s}{\omega^2 + (\beta_p p_s)^2}\right] \tag{58}$$

where i_r is the recombination current. Generally, R_r might be expected to be frequency-dependent from Eq. (58), but there will be two cases where R_r will be independent of frequency:

(a) for $\omega \ll \beta_p p_s$

$$R_r \approx (k_B T/e_0 i_r)(1 + \beta_n n_s/\beta_p p_s) \tag{59}$$

(b) for $\omega \gg \beta_p p_s$ and $\omega^2 \gg \beta_p p_s \beta_n n_s$

$$R_r \approx (k_B T/e_0 i_r) \tag{60}$$

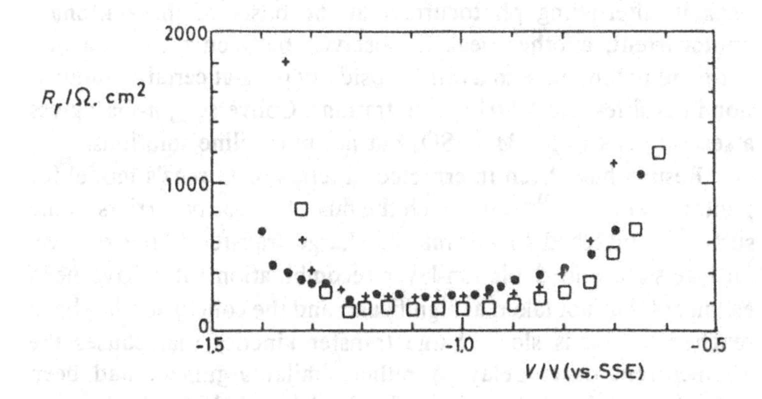

Figure 43. Recombination resistance, R_r, as a function of electrode potential at an *n*-GaAs $\langle\bar{1}\bar{1}\bar{1}\rangle$ electrode in the same electrolyte as in Fig. 40: ●, f = 1 kHz; +, f = 10 kHz; □, f = 50 kHz.[200]

In the region between flat-band potential and photocurrent onset, recombination is essentially complete, and the surface states are likely to be entirely populated by electrons, so $\beta_n n_s \gg \beta_p p_s$. The recombination current will then be given by

$$i_r \approx e_0 \beta_p \cdot r p_s \quad (61)$$

so that p_s should vary little with applied potential in this region. As a result, for case (a), R_r will be proportional to n_s and vary exponentially with applied potential. The fact that this is *not* observed in the region of complete recombination suggests that case (b) applies. In the potential region where recombination decreases and photocurrent increases, R_r is observed to increase, as predicted by Eq. (61). Where recombination is complete, the constant of proportionality between R_r^{-1} and i_r is measured as 29 V^{-1}, which is in quite good agreement with the predicted value of 38 V^{-1}. Vanmaekelbergh *et al.*[200] were, therefore, able to show that a recombination resistance independent of surface pretreatment was consistent with a surface recombination mechanism and that it was not necessary to assign the delay of photocurrent onset to depletion-layer recombination as suggested by van den Meerakker *et al.*[197]

Lorenz and co-workers have investigated the photocurrent response to voltage modulation at *n*-GaAs and *n*-GaP electrodes for a variety of electrolytes.[188,201,202] The results are particularly interesting for *n*-GaAs in alkaline electrolyte, because as well as a peak in alternating photocurrent at the onset of the stationary photocurrent, another peak is observed between this onset and flat-band potential, with a third subsidiary peak at certain illumination intensities and KOH concentrations. Conversely, *n*-GaP gives a second peak in 0.5 *M* H_2SO_4 but *not* in alkaline solutions.

Results have been interpreted in terms of Lorenz's model for photocurrents,[201,203,204] in which the flux of minority carriers to the surface is matched to the rate of charge transfer at the surface. Surface-state and depletion-layer recombination rates have been estimated, but not taken as significant, and the conclusion has been reached that it is slow charge transfer kinetics that causes the photocurrent onset delay. A rather similar argument had been suggested previously to account for the delayed photocurrent onset in *n*-Fe_2O_3,[205] though in this case there were cogent chemical reasons for believing that strongly relaxed Fe(IV) states might well be very slow in oxidizing water. Interestingly, despite the fairly large limiting photocurrents observed under the light intensity used, Lorenz and co-workers were unable to distinguish a flat-band shift for *n*-GaAs on illumination.

Figure 44 show values of the alternating photocurrent on a negative voltage sweep, starting at −0.5 V versus SCE. Considerable hysteresis is evident for more negative turning potentials and higher illumination intensities, indicating a large amount of surface change. The first peak on the voltage sweep is only weakly dependent on the frequency of modulation, whereas the two peaks closer to flat-band potential vary much more strongly with frequency, although at quite different rates. The first peak was assigned to the photocorrosion of the electrode, but the two frequency-dependent

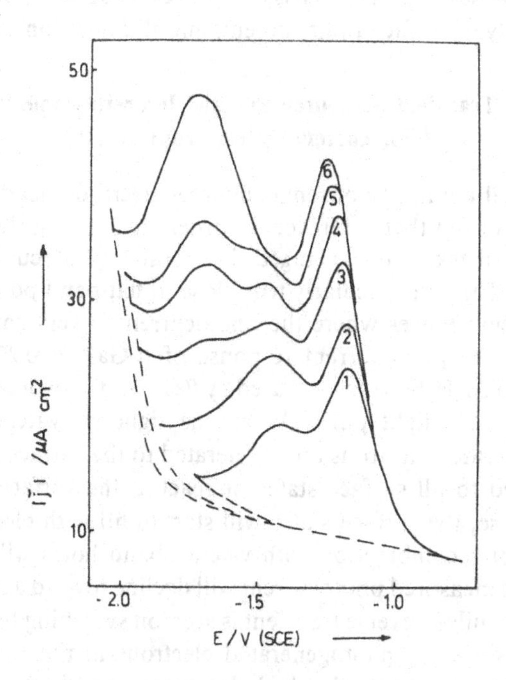

Figure 44. pH dependence of the potential-modulated photocurrent for ⟨100⟩ *n*-GaAs in aq. NaOH at 3 kHz. Amplitude of voltage oscillation, 0.5 mV. Stationary limiting photocurrent, 3 mA cm^{-2}. [OH^-]: (1) 0.1, (2) 0.25, (3) 0.75, (4) 1, (5) 1.33, (6) 2.0 *M*.[202]

peaks are believed to result from chemisorption processes on the surface. It would be interesting to see how much the results would be affected by the presence of a strongly stabilizing electrolyte such as 1 M Se^{2-}/Se_2^{2-} in 1 M KOH.

The rejection of surface-state arguments by Lorenz and co-workers undoubtedly places them out of the mainstream of investigation on GaAs and GaP, and their models, while contentious, have not really been tested by direct spectroelectrochemical techniques to the extent they warrant. However, even the theories of Vanmaekelbergh and co-workers that were reviewed above are very approximate, and the fundamental limitation of potential-modulated photocurrent remains that of relating the results obtained to a convincing molecular model of the surface: this is particularly true of the surface of n-GaAs, where response is highly sensitive to electrolyte, surface pretreatment, and illumination intensity.

3. Transient Measurements and Intensity-Modulated Photocurrent Spectroscopy (IMPS)

Transient illumination of semiconductor electrodes leads to a pulse of photocurrent that can reveal information about surface kinetics which is masked or averaged in normal photocurrent sweep methods. This is particularly true close to flat-band potential or in other circumstances where the photocurrent is very small. Figure 45 shows the photocurrent response of p-GaP at 0.225 V versus SCE in 0.5 M H_2SO_4, as measured by Peter and co-workers[127] under a square-wave light pulse. When the light is switched on, the photogenerated electrons are accelerated to the surface and can be considered to fill surface states or react at the surface. After the initial pulse, the surface states will start to fill with electrons, and the rate of recombination with valence-band holes will rise; as a result, the measured photocurrent will decline toward a steady-state value. A similar, reverse transient is seen on switching the illumination off, since the photogenerated electrons in the surface states then migrate back into the depletion region and bulk of the semiconductor until the steady-state dark surface-state population is reattained.

If only recombination is considered, then the decay of the *excess* current, $i(t) - i(\infty)$, should be first order in surface-state

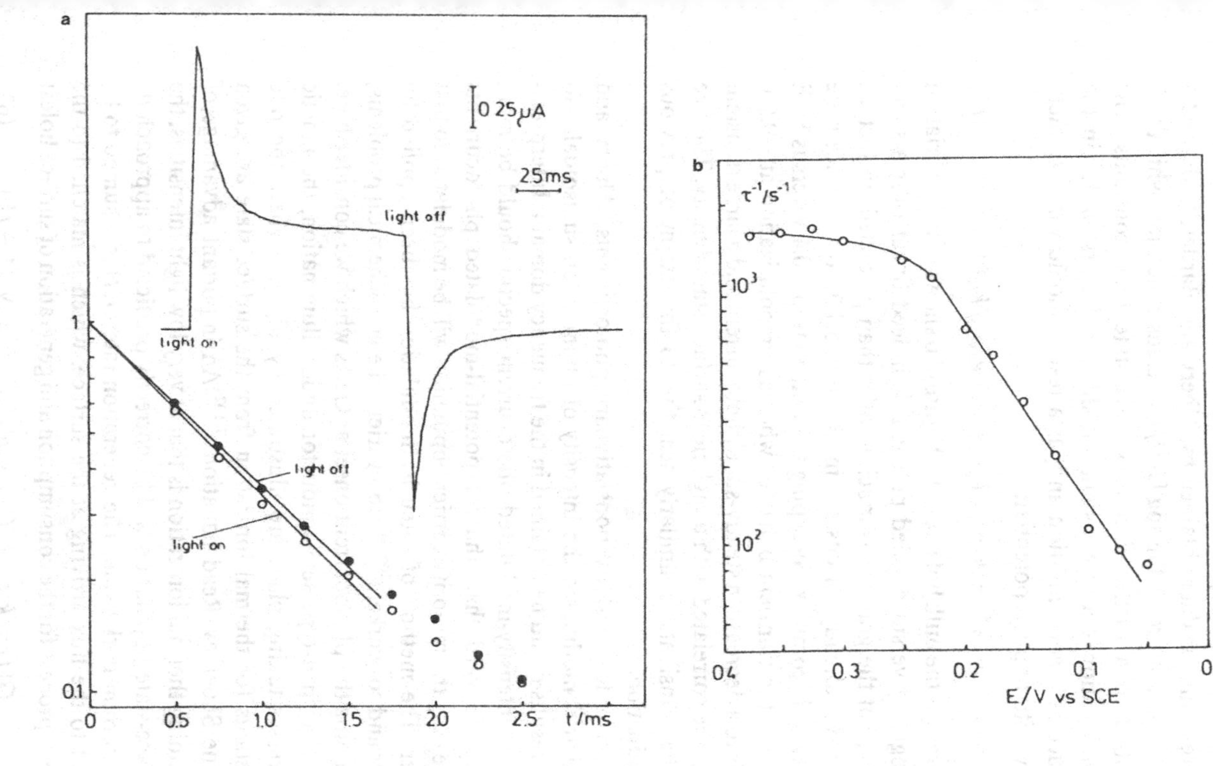

Figure 45. (a) Photocurrent transient behavior of *p*-GaP in 0.5 *M* H_2SO_4 at 0.225 V. The transients for both light on and light off are seen to be close to exponential in their decay; (b) logarithm of the inverse decay time, τ^{-1}, as a function of applied potential.[127]

population and therefore exponential. Figure 45a shows that the decay is close to exponential in the data of Peter and co-workers whereas similar data obtained by Albery and Bartlett[114] showed a distinct curvature in the $\log_e[i(t)/i_0]$ versus time plot, which was ascribed to a distribution of surface-state time constants. The recombination decay constant should also be first order in the surface hole concentration and, as a result, should vary exponentially with applied potential:

$$\tau^{-1} = k_p p_s = k_p p_0 \exp(-e_0\phi_{sc}/k_B T) \tag{62}$$

where p_0 is the bulk hole density. Figure 45b shows that the variation of $\log(\tau^{-1})$ versus applied potential is indeed linear over a small range, but the gradient is much smaller than predicted. In fact, as we have already discussed in Section 5.1.2, the surface-state behavior is complex. Sweeping the potential to +0.4 V versus SCE removes the transient behavior whereas prepolarization at −0.5 V leads to a decrease in steady-state photocurrent and a large transient response correlated with a high near-surface state density. The size of the transients is similarly sensitive to the extent of previous illumination.

In order to study nonstationary photocurrents, Albery and Bartlett[126] modulated the intensity of illumination sinusoidally so that results could be obtained in the frequency domain. Interpretation of intensity-modulated photocurrent spectra should be more straightforward than that of potential-modulated photocurrents since only the minority carrier population will be modulated to first order. The method of intensity modulation has been applied by Peter and co-workers[206–209] to a variety of electrochemical problems, including the photocurrent onset at GaAs which we consider here.

For an n-type semiconductor under illumination, the kinetic processes taking place are illustrated by Fig. 46, where the rate constants for thermal ionization from the surface state are given by the Shockley–Read equations.[7,102] An important advantage of including thermal ionization is that at very low light intensities, the surface-state population will be correctly predicted to approach its equilibrium dark value. The generation rate $G(t)$ is assumed to be equal to the flux arriving at the surface, which is matched to the kinetic process for the consumption and generation of surface holes:

$$G(t) = k_p p'_s N_{ss} f'_{ss} + k_v p'_s c'_{red} - k_p p_1 N_{ss}(1 - f'_{ss}) \tag{63}$$

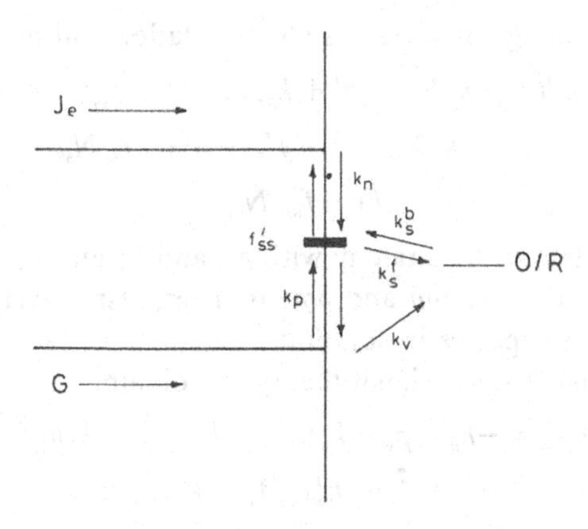

Figure 46. Kinetic scheme for processes at an illuminated *n*-type semiconductor electrode with a single set of surface states.[206]

where f'_{ss} is the time-dependent surface-state population, c'_{red} is the time-dependent concentration of oxidizable species at the semiconductor surface, and the prime represents a time-dependent variable of the form $x' = x + \tilde{x}\exp(i\omega t)$. The factor $k_p p_1$ represents the thermal ionization rate constant within the Shockley–Read theory for hole thermal ionization from the surface state, where

$$p_1 = N_v \exp[(E_v - E_{ss})/k_B T] \tag{64}$$

as defined above. Using the small-signal approximation:

$$\tilde{g} = k_p N_{ss} f_{ss} \tilde{p}_s + k_p N_{ss} \tilde{f}_{ss} p_s + k_v \tilde{p}_s c_{red} + k_p p_1 N_{ss} \tilde{f}_{ss} \tag{65}$$

if the time variation of the redox concentration, $\tilde{c}_{red}$, can be neglected. The dc quantities, f_{ss} and p_s, will vary with applied potential and illumination in a manner which can be calculated by a steady-state analysis.

If a surface hole lifetime τ_v is defined:

$$\tau_v^{-1} = (k_v c_{red} + k_p f_{ss} N_{ss}) \tag{66}$$

then

$$\tilde{p}_s = [\tilde{g} - k_p N_{ss}(p_s + p_1)\tilde{f}_{ss}]\tau_v \tag{67}$$

and, to first order,

$$\tilde{p}_s = \tilde{g}\tau_v \tag{68}$$

The rate of change of surface-state population will be

$$N_{ss}\cdot df_{ss}/dt = -k_pN_{ss}f'_{ss}p'_s + k_pN_{ss}p_1(1-f'_{ss}) - k_nN_{ss}f'_{ss}n_1 + k_nN_{ss}n_s(1-f'_{ss}) - k_s^f c_{ox}f'_{ss}N_{ss} + k_s^b c_{red}(1-f'_{ss})N_{ss} \quad (69)$$

where n_1 is defined by analogy with p_1, and k_s^f and k_s^b are the rate constants for the *electron* and *hole* transfer, respectively, from the surface state to species in solution.

Again using small-signal theory, we obtain:

$$i\omega\tilde{f}_{ss} = -k_p\tilde{f}_{ss}p_s - k_pf_{ss}\tilde{p}_s - k_pp_1\tilde{f}_{ss} - k_nn_1\tilde{f}_{ss} - k_nn_s\tilde{f}_{ss} - k_s^f c_{ox}\tilde{f}_{ss} - k_s^b c_{red}\tilde{f}_{ss} \quad (70)$$

so that, if we define

$$\tau_s^{-1} = k_n(n_s + n_1) + k_s^f c_{ox} + k_s^b c_{red} \quad (71)$$

and substitute for $\tilde{p}_s$ from the first-order expression, Eq. (68),

$$\tilde{f}_{ss} = \frac{-k_pf_{ss}\tau_v\tilde{g}}{i\omega + \tau_s^{-1} + k_p(p_s + p_1)} \quad (72)$$

The majority carrier density is assumed to remain at equilibrium with the bulk of semiconductor so the electron flux to the surface state can be calculated:

$$j_e(t) = N_{ss}[k_nn_s(1-f'_{ss}) - k_nn_1f'_{ss}] \quad (73)$$

whence, from small-signal theory,

$$\tilde{j}_e = -N_{ss}k_n(n_s + n_1)\tilde{f}_{ss} \quad (74)$$

$$= \frac{N_{ss}k_pk_nf_{ss}\tau_v(n_s + n_1)\tilde{g}}{i\omega + \tau_s^{-1} + k_p(p_s + p_1)} \quad (75)$$

Subtraction of the modulated recombination electron flux from the hole flux to the surface, $\tilde{g}$, gives the total ac current, which can be normalized to

$$\frac{\tilde{j}_{photo}}{\tilde{g}} = 1 - \frac{N_{ss}k_pk_nf_{ss}\tau_v(n_s + n_1)}{i\omega + \tau_s^{-1}} \quad (76)$$

provided relaxation is more rapid via redox species and the conduction band rather than equilibration with surface-hole density. The real and imaginary parts can then be written in the form:

$$\mathrm{Re}(\tilde{j}_{photo}/\tilde{g}) = 1 - \beta\tau_s/(1 + \omega^2\tau_s^2) \quad (77)$$

$$\text{Im}(\tilde{j}_{\text{photo}}/\tilde{g}) = \beta\omega\tau_s^2/(1 + \omega^2\tau_s^2) \tag{78}$$

These equations predict that $(\tilde{j}_{\text{photo}}/\tilde{g})$ will describe a semicircle in the upper complex plane, with a high-frequency intercept on the real axis at unity. The zero-frequency intercept will be at $1 - \beta\tau_s$ on the real axis, and the top of the semicircle, corresponding to the maximum imaginary value, will be at $\omega = \tau_s^{-1}$, so that the surface-state relaxation time should be readily available.

Figure 47 shows theoretical plots of the normalized ac photocurrent, where the response is seen to depend strongly on the band bending, through the dependence of both β and τ_s on the majority carrier density. Closer to flat-band potential, the surface electron density is higher and recombination correspondingly rapid, so that at low frequencies the photocurrent tends to zero. The time constant τ_s is smaller owing to the more rapid recombination rate so that higher frequencies are needed to obtain the full semicircle response.

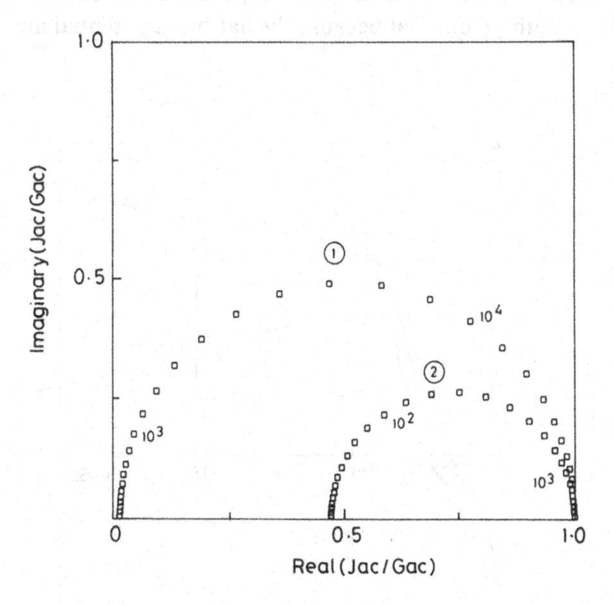

Figure 47. Modulated photocurrent response calculated from a set of 10^{13} cm^{-2} surface states, 0.5 eV below the CB, with frequencies as marked: (1) $\phi_{sc} = 0.2$ V, $\beta = 3.36 \times 10^4$ s^{-1}, $\tau_s = 2.93 \times 10^{-5}$ s; (2) $\phi_{sc} = 0.3$ V, $\beta = 6.25 \times 10^2$ s^{-1}, $\tau_s = 8.47 \times 10^{-4}$ s.[207]

For larger depletion-layer voltages, n_s, and therefore the recombination rate, is lower, so that the zero-frequency intercept moves away from the origin and the time constant τ_s is larger.

Similarly, an increase in reducing agent concentration will decrease τ_s and move the zero-frequency intercept away from the origin. An increase of surface-state density will increase recombination rate, and therefore move the zero-frequency intercept toward the origin, but it will not alter the time constant.

Figure 48a shows a photocurrent–voltage plot for (100) *n*-GaAs in 0.1 *M* KOH/0.1 *M* Na_2S/0.1 *M* S_2^{2-} and correspondingly intensity-modulated photocurrent (IMP) spectra, as obtained by Li and Peter.[207] Close to flat-band potential, the zero-frequency intercept is near the origin owing to the high recombination rate, as predicted. However, the shape of the response deviates considerably from theory, and the semicircles appear flattened, a result consistent with a distribution of surface rate constants. The result at −1.32 V versus SCE could even be modeled by two separate time constants, τ_s. Li and Peter pointed out that because the flat-band potential measured

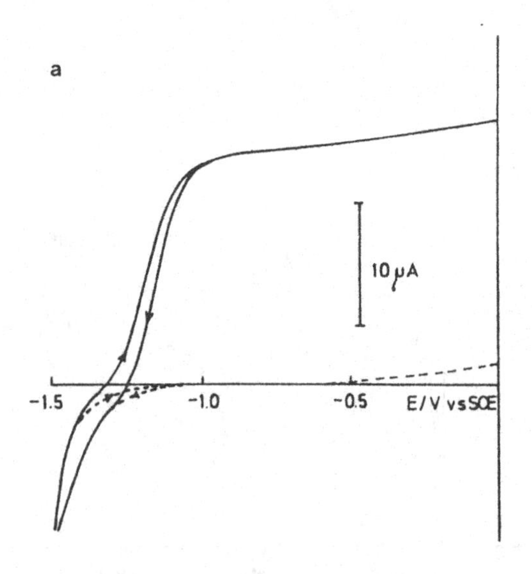

Figure 48. (a) Photocurrent–voltage curve of *n*-GaAs in 0.1 *M* KOH/0.1 *M* S^{2-}/0.1 *M* S_2^{2-}; (b) IMPS plots for the same system at different applied potentials.[207]

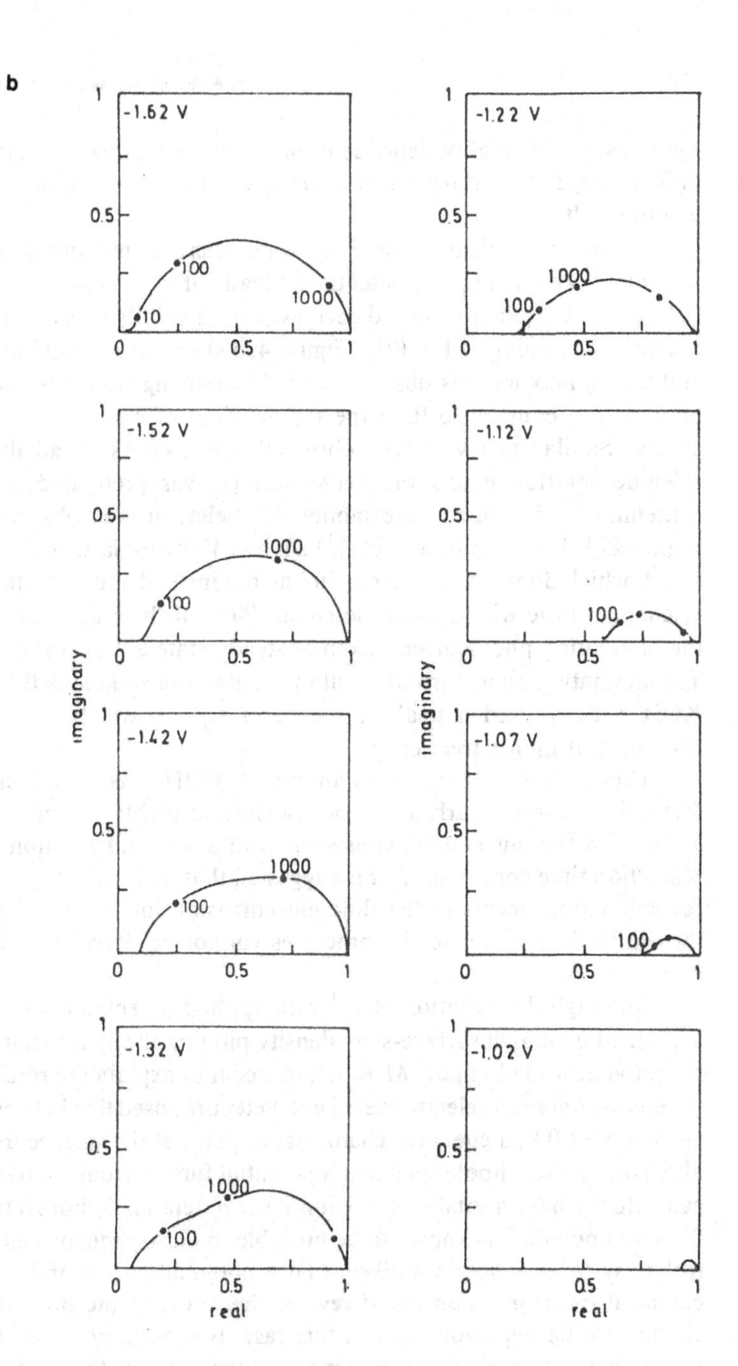

Figure 48. (*Continued*)

by impedance is highly dependent on electrode pretreatment, two different types of surface might coexist, and two time constants τ_s would result.

A further problem is that τ_s does not increase monotonically with more positive applied potentials. Ideally, if $n_s \gg n_1$ and $k_n n_s \gg (k_s^f c_{ox} + k_s^b c_{red})$, then τ_s should decrease exponentially with applied potential according to Eq. (71). Figure 49b shows that a minimum and then a maximum is observed in τ_s^{-1} in moving from flat-band potential, in contrast to the expected gradient of one decade per 59 mV. Similar behavior was observed for n-GaAs in alkaline selenide solution unless the GaAs surface was pretreated with ruthenium,[209] in which case monotonic behavior was observed. Figure 49d shows a plot of $\log(\tau_s^{-1})$ against V for n-GaAs in 0.1 M KOH, which does not show a maximum, but instead a region where τ_s changes little with applied potential. Plots of the magnitude of the alternating photocurrent taken at steady state are also shown. The alternating photocurrent in sulfide-containing and blank 0.1 M KOH both showed a peak in the onset region, which did not disappear at higher frequency.

The peak in $|\tilde{j}|$ observed in 0.1 M KOH was broad and featureless, and similarly IMP spectra showed highly compressed semicircles for this system, consistent with a wide distribution of relaxation time constants τ_s. It is suggested that this arises because recombination occurs in the different corrosion intermediates on the electrode surface, so the processes cannot be described with one τ_s.

Although the variation of τ_s^{-1} with applied potential could be explained by a high surface-state density pinning the space-charge potential of n-GaAs in 0.1 M KOH, this cannot explain the results in sulfide-containing electrolyte. Li and Peter proposed that between −1.5 and −1.0 V, a chemical change takes place at the surface that alters the surface dipole such that a potential further from flat band can actually have a smaller depletion-layer potential. Although the flat-band potential is known to be unstable in GaAs/aqueous electrolyte systems, it seems unlikely that a potential-change-induced chemical transformation could reverse the effect of the potential change on the depletion layer in this case. It should, however, be noted that just such a phenomenon is observed in the electroreflectance of n-GaAs in neutral electrolyte at anodizing potential, which we believe to result from the formation of an oxide layer. If the value of τ_s^{-1} is believed to be controlled entirely by the surface potential, then the peak in Fig. 49b will result from a reduction in ϕ_{sc} of less than 50 mV as spectra are measured at progressively higher potentials. Such a shift will be difficult to detect with available techniques.

However, the implication is also that the steady-state space-charge voltage changes little over a potential range of 500 mV between the flat-band potential and the photocurrent onset. Peat and Peter[209] made a similar assertion for *p*-GaAs in 0.5 *M* H_2SO_4 from IMPS measurements, and Meissner and Memming[210] even went so far as to suggest that the electrode may be pinned from flat-band potential to photocurrent onset in this system. Clearly, an independent method is needed to measure band bending in the steady state, and we discuss this further in Section VII.6 on electroreflectance measurements. The links between τ_s, applied potential, and surface structure remain confused at the moment, and further theoretical progress is only likely once the depletion-layer voltage and interfacial chemistry are better characterized.

Peter and co-workers have used IMPS to study the kinetics of photocurrent doubling in oxygen reduction on *p*-GaAs in 0.5 *M* H_2SO_4[209] and photocurrent quadrupling on *n*-Si in NH_4F solution.[211] A recent development has been to use a modulated beam of very small cross section ($<10\ \mu m$) that can be rastered across the electrode surface to give a spatially resolved photocurrent response.[212] In the potential regime of limiting photocurrent, the response across the *n*-GaAs surface in 0.1 *M* KOH/Na_2S/S was fairly even, and little contrast was observed over the scan. However, close to the flat-band potential, regions of high photocurrent were distinguished from dark patches of very low photocurrent, which could be correlated with scratches and areas of surface damage visible with an optical microscope. Even below the flat-band potential, patches could still be found that gave a photoresponse, and these areas were assumed to be pinned positive of the flat-band potential.

Intensity-modulated photocurrent measurements have clearly developed into a powerful new probe of surface-state kinetics. As with all kinetic studies, it must be emphasized that unambiguous interpretation of the data can only be provided once the surface

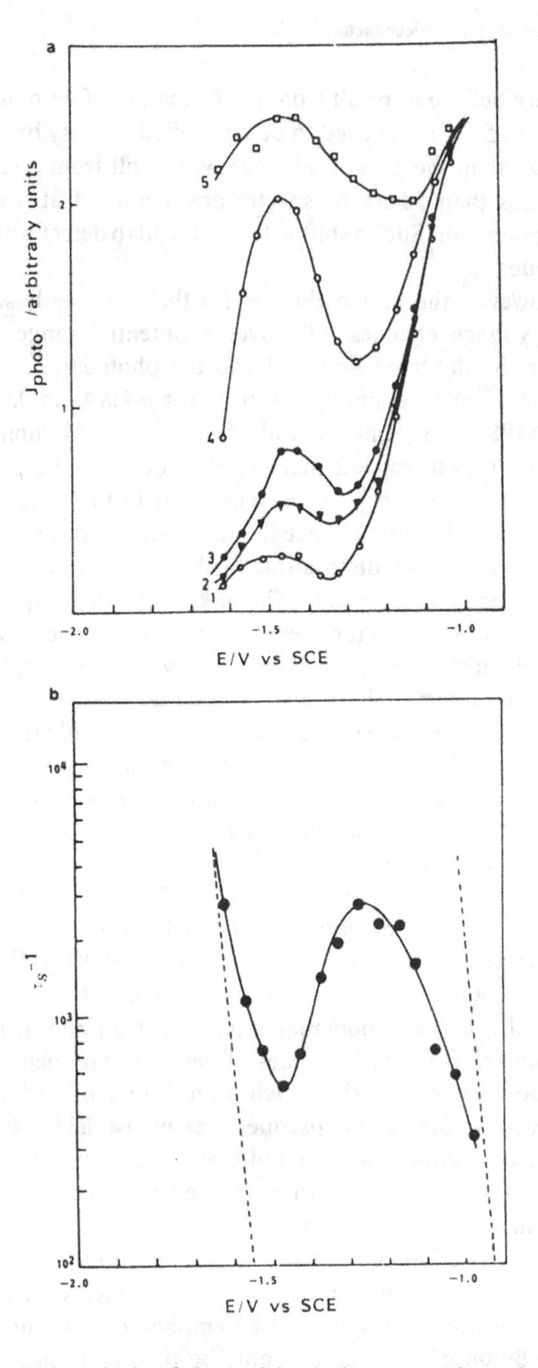

Figure 49. Magnitude of the modulated photocurrent for various frequencies as a function of applied potential for *n*-GaAs in 0.1 *M* KOH/0.1 *M* Na_2S/0.1 *M* S_2^{2-} (a) and 0.1 *M* KOH (c) at frequencies of (1) 1 Hz, (2) 10 Hz, (3) 100 Hz, (4) 1 kHz, and (5) 10 kHz, and

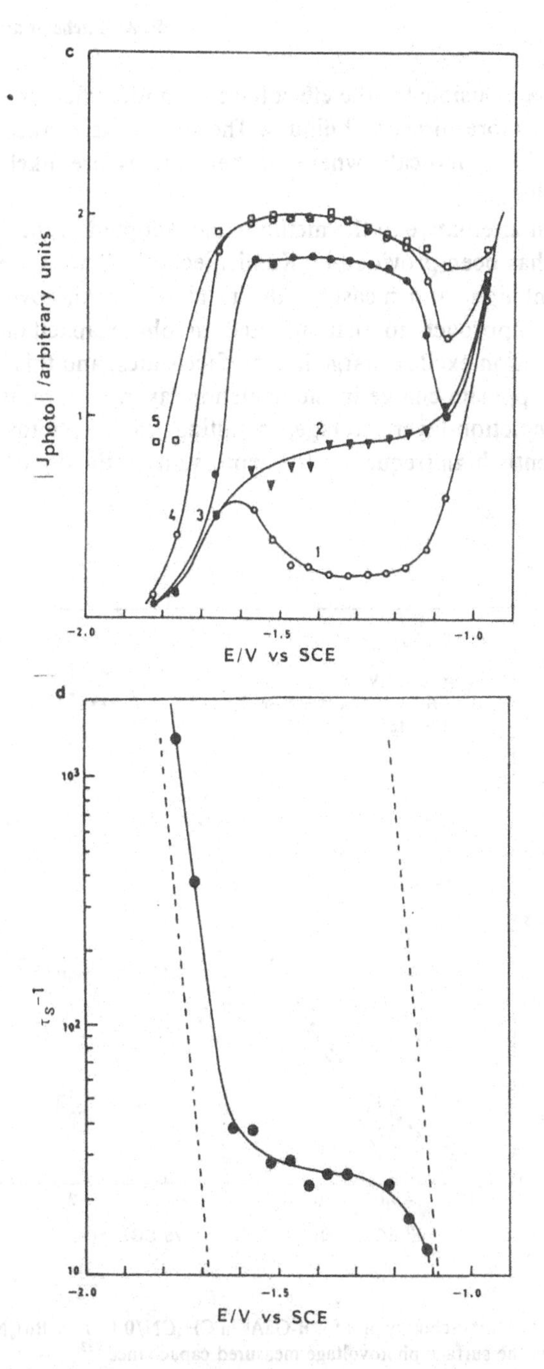

Figure 49. (*Continued*)
plots of τ_s^{-1} as a function of applied potential for sulfide electrolyte (b) and 0.1 *M* KOH (d). The magnitude of τ_s^{-1} is taken as the frequency at which $(\tilde{j}/\tilde{g})$ has its maximum imaginary value.[207]

states responsible for the effect have been identified and characterized by more direct techniques. The real value of these measurements is to indicate where further studies are likely to prove rewarding.

An alternative to the methodology adopted in the IMPS technique has been provided by Kamieniecki,[213,214] who uses chopped incident light and measures the resultant ac photovoltage. In a similar approach to that adopted in photocapacitance studies, illumination excites charge into surface states, and this is balanced by an opposite charge in the depletion layer. In turn, this reduces the depletion-layer voltage, resulting in a photovoltage. At sufficiently high frequency, the expression for the photovoltage can

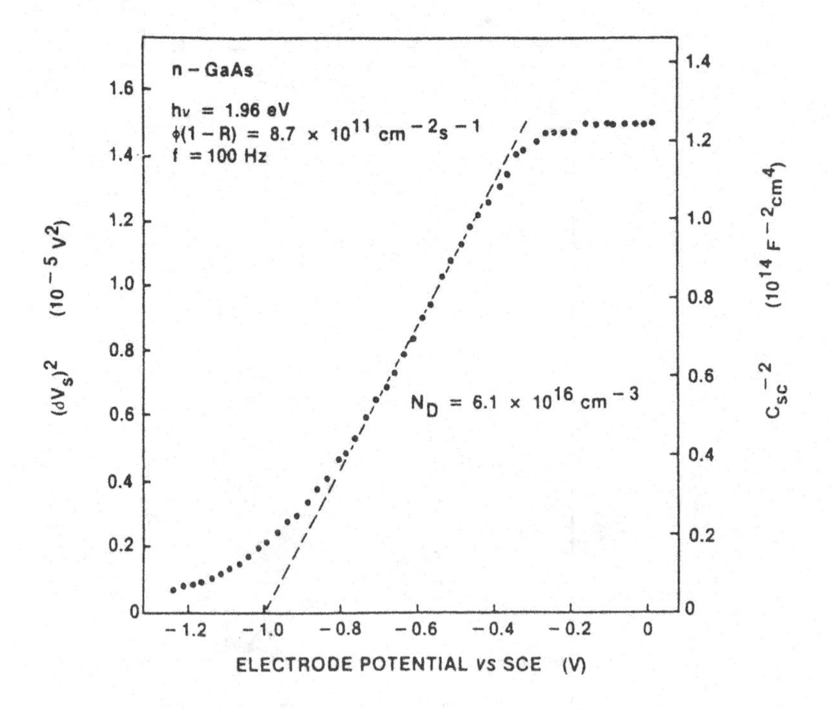

Figure 50. Mott-Schottky plot for n-GaAs in CH_3CN/0.1 M $(n\text{-Bu})_4N^+BF_4^-$ determined by the surface photovoltage measured capacitance.[213]

be approximated by

$$\delta V_s = \phi(1 - R)e_0 C_{sc}^{-1}/4\nu \tag{79}$$

where ϕ is the incident light intensity, R is the loss due to reflection, and ν is the frequency of the intermittent light. This equation allows us to calculate Mott–Schottky plots such as that shown in Fig. 50 for n-GaAs in a nonaqueous electrolyte. The major advantage of this approach is that it can be applied even in systems of very high impedance, such as those encountered in nonaqueous solvents, where extraction of C_{sc} from conventional ac studies can be very difficult.

4. Subbandgap Photocurrents

The spectroscopic study of electronic transitions to and from surface states has several inherent difficulties. The previous discussion has shown that surface-state densities below 10^{13} cm^{-2} may still have profound effects on the properties of the interface. However, the absorption cross section for such a small density of states will be very small indeed and will be obscured by bandgap transitions unless light of energy higher than the bandgap is rigorously eliminated. In addition, subbandgap light may be absorbed by defect bulk states, which will vary in density and energy from sample to sample.

Butler and co-workers have carried out an extensive series of studies on p-GaP.[215–217] Figure 51 shows photocurrent spectra normalized with respect to illumination for p-GaP in 0.1 M $HClO_4$. Etched samples showed little subbandgap photocurrent (curve a in Fig. 51), indicating a low surface density of states, but mechanical polishing with 0.25-μm diamond paste gave rise to the response shown as curve b in Fig. 51, with a well-defined photocurrent maximum below 2.0 eV. Curve c in Fig. 51 shows the photocurrent spectrum of a p-GaP electrode that had been aged by passing large cathodic currents; in this case there is a large subbandgap photocurrent and the growth of a maximum in the response. For both the mechanically damaged and cathodically aged samples, the subbandgap photocurrent could be removed by chemical or anodic

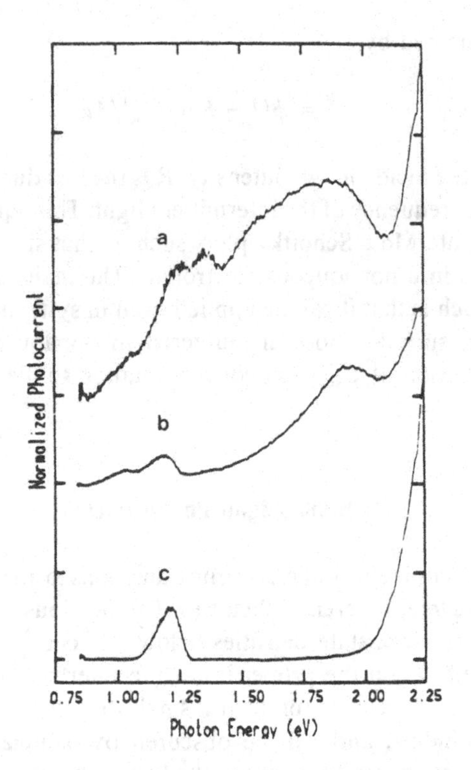

Figure 51. Subbandgap photocurrents for *p*-type GaP in 0.1 *M* H_2SO_4 at −2.0 V vs. SCE with various surface pretreatments. The peak at 1.2 eV results from a second-order reflection of the monochromator grating: (a) Electrode continually aged for one hour; (b) surface mechanically polished with 0.25-μm diamond paste; (c) polished surface etched at 4 μm.[215]

etching. Butler and Ginley[217] suggested that the defect states involved could be related to Ga vacancies.

Treatment of etched *p*-GaP with 0.07 *M* aqueous Ru(III) solutions leads to a large increase in subbandgap photoresponse,[125] even in the presence of a mechanically damaged surface.[217] This conflicts with the model of Heller,[86] which attributed the decrease in recombination velocity due to ruthenium treatment of GaP to an interaction between the ruthenium and the surface states already present, moving the latter out of the bandgap. In fact, treatment with ruthenium solutions appears to *increase* the surface-state density on GaP, and it was proposed by Dare-Edwards *et al.*[125] and Butler and Ginley[217] that rapid charge transfer through these new surface states could account for the improved photoelectrochemical response of the GaP interface.

In the very complex case of TiO_2, measurement of the subbandgap photoresponse of a crystal of TiO_2 in phosphate buffer shows a small maximum at 550 nm (ca. 2.2 eV), which can be assigned to transitions from the top of the valence band to the unoccupied surface states.[218]

The time dependence of the photocurrent shows that the response is faradaic rather than purely transient, and the potential behavior is shown in Fig. 52. It can be seen that the photoresponse at 550 nm actually passes through a *maximum* as the potential is moved away from flat band. This is quite impossible to understand in terms of a fixed intrinsic set of surface states, if the transition were due to electronic excitation from the VB. The reason is that we would expect to see a plateau in the response as the surface states were emptied with increasing potential, and not a maximum. In fact, the effects seen can most reasonably be ascribed to a surface state that is *formed in the dark by a cathodic reaction.* Siripala and Tomkiewicz[218] ascribed this state to an intermediate in the hydrogen evolution reaction that they write as TiO_2—H, and their mechanism

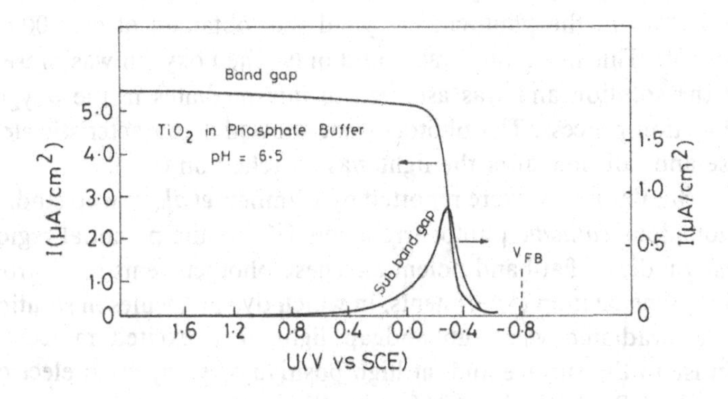

Figure 52. Photocurrent-potential behavior of the band-to-band and subbandgap photocurrents for TiO_2 in phosphate buffer.[218]

may be formally written as:

$$e_d^- + TiO_2 + H_{aq}^+ \rightarrow TiO_2{-}H$$

$$TiO_2{-}H + H_{aq}^+ + e_d^- \rightarrow TiO_2 + H_2$$

$$TiO_2{-}H + H_{aq}^+ + (h\nu)_{subband} \rightarrow p^+ + TiO_2 + H_2$$

$$e_d^- + p^+ \rightarrow \text{recombination}$$

where e_d^- is the concentration of electrons at the surface in the *dark*, and p^+ is a hole in the valence band at the surface. At positive potentials, the coverage of the surface by this intermediate, which is formed by cathodic reduction in the dark, becomes very small, and, at negative potentials, recombination will prevent the holes formed at the surface from oxidizing water. This model is very similar to one suggested by Dare-Edwards *et al.* to account for the photocurrent in *p*-GaP.[125] An important observation made by Siripala and Tomkiewicz is that the total subbandgap photocurrent actually decreased with oxygen bubbling, though it is not clear from the paper whether this was primarily due to quenching of the 550-nm peak.

Other workers have also reported anomalous subbandgap photocurrents in the cathodic region, particularly when a steady-state cathodic dark current is being sustained by the material. A typical early report by Morisaki *et al.*[219] showed that a local maximum in the photocurrent yield was obtained at ca. 800 nm (1.7 eV). This maximum was found only when oxygen was present in the solution and was ascribed to intermediates in the oxygen reduction process. The photocurrent showed a characteristic slow rise and fall time after the light was switched on or off.

Similar results were reported by Hamnett *et al.*,[220] who studied anomalous *cathodic* photocurrents on TiO_2 in the potential region just anodic of flat-band potential. These photocurrents arose from photosensitization experiments, in which dye molecules in solution were irradiated with subbandgap light. The excited molecules diffuse to the surface and, at high positive bias, inject an electron into the CB, giving rise to a fast anodic photocurrent. However, at less positive bias, a slow-rise-time cathodic photocurrent was also observed, which was traced to Förster-type energy transfer between excited dye and a surface electronic state. This is the *inverse* of the process seen by Morisaki *et al.*[219] and the model put forward by Hamnett *et al.*[220] involved the partial depopulation of surface states created by the ongoing cathodic dark current; this depopulation led to a reduction in the band bending in the semiconductor and an enhancement of the cathodic process already taking place in the dark. As with the data of Morisaki *et al.*, the results of Hamnett *et al.* also depended on the presence of oxygen in solution. Clearly, the two experiments are complementary, and both point to the existence of a surface state in the midgap region. Hamnett *et al.* suggested, on the basis of indirect evidence, that the most likely surface species was O_2^-.

Although these papers all differ in detail, the central thesis is that an ongoing reduction current can lead to the formation of intermediates at the surface that can act as surface recombination states. Since oxygen reduction has clearly been implicated in several of these papers in the generation of intermediates, the mechanism of oxygen reduction has attracted some attention. RRDE studies have been carried out on n-TiO_2 by Parkinson *et al.*,[221] Salvador and co-workers,[222-226] and Nogami *et al.*[227] Salvador and co-workers, on the basis of the variation of O_2 reduction with pH, have tentatively identified *two* types of intrinsic surface state associated with OH^-_{surf} species. The first, essentially a $Ti^{3+/4+}$—OH^- couple, lies near the CB, and the second, associated with the Ti^{4+}—$OH^{O/-}$ couple, lies above the VB. In addition, Salvador and co-workers have suggested that reduction of oxygen may lead to the formation of a strongly adsorbed form of H_2O_2, and this will provide a possible further source of midgap states. The main evidence for the first of these states, the $Ti^{3+/4+}$—OH^- state near the CB, is kinetic in origin. If there are surface states that act as intermediates in the reduction reaction, then we can imagine a two-step process: in the first step, electrons in the conduction band are captured inelastically by the surface state, and in the second step, electrons from the surface state are captured by the reductant in solution.

$$e^-_{cb}(\text{surf}) + \text{s.s.} \rightarrow \text{s.s.}^-; \qquad \text{rate constant } k_1$$

$$\text{s.s.}^- + O_2 \rightarrow \text{s.s.} + O_2^-; \qquad \text{rate constant } k_2$$

If the total number density of surface states is N_{ss}, and the number concentration of electrons in the conduction band at the surface is n_s, then the total faradaic current due to electron capture by O_2 is given, in the steady-state approximation, by

$$i^- = i^-_{sat} C_{ox}/(K + C_{ox}) \tag{80}$$

where C_{ox} is the concentration of O_2, and $K = k_1 n_s / k_2$. The value of i^-_{sat} is given by $-e_0 N_{ss} k_1 n_s$ and represents the maximum rate at which the overall electron transfer reaction can occur at limitingly high rate for the concentration of the solution species. Clearly, i^-_{sat} can be obtained from a plot of $1/i^-$ versus $1/C_{ox}$, and its variation with applied potential will reflect the variation of n_s with potential. In the absence of any pinning, this would lead to an expected Tafel slope of 60 mV, but in practice Talfalla and Salvador[224] found a value close to 100 mV, suggesting that partial pinning is taking place.

The amount of H_2O_2 detected at the ring in studies of this sort seems to depend both on the potential, with respect to the flat-band potential, and on the pH. Parkinson *et al.*[221] and Talfalla and Salvador[224] both report that, negative of the flat-band potential, very little H_2O_2 is produced, presumably because the very high concentration of electrons at the degenerate surface leads to rapid reduction of all physisorbed or chemisorbed H_2O_2. However, in alkaline solution, at potentials just positive of the flat-band potential, and appreciable amount of HO_2^- is detected at the ring; this may reflect the fact that reduction of H_2O_2 itself cannot be efficiently mediated by the $Ti^{3+/4+}$—OH^- state in alkali, owing to poor overlap.

Subbandgap photocurrents on n-TiO_2 have also been observed at more positive potentials, and Laser and Gottesfeld[228] investigated these on a polycrystalline Ti-oxide-film electrode using a dual-beam experiment. They found an enhancement of the subbandgap photocurrent in the presence of bandgap illumination, which appeared to suggest, as a possible origin of the former, excitation from photoinduced surface states lying just above the VB. There are, however, difficulties with this interpretation: the potential dependence of the enhanced and unenhanced subbandgap photocurrents is rather different, and Butler *et al.*[229] have shown that for a TiO_2 single-crystal electrode, the subbandgap photocurrent $J_\phi \propto (V - V_{fb})^{1/2}$ and $J_\phi^{2/3} \propto (E_g - h\nu)$, with E_g the bandgap energy.

Both of these relationships are difficult to rationalize as arising from surface-state effects, but can easily be derived assuming that the subbandgap photocurrent at positive potentials arises from quasi-bulk processes taking place in the depletion layer. This has been challenged by Liu *et al.*,[230] who, on the basis of systematic measurements of the dependence of subbandgap photocurrent on wavelength and bias voltage, assigned the behavior to excitation from surface states lying ca. 1.8 eV above the VB. A synthesis of the various views has been given by Salvador,[231] who has clearly shown that the subbandgap photocurrent at anodic potentials has two components, an anodic transient, $i_{ph,tr}$, and a steady-state current, $i_{ph,st}$. The former arises from the excitation of bandgap states in the depletion layer, which gives rise to a redistribution of potential at the interface and a concomitant displacement current. The latter arises from a two-photon process in which one photon can excite an electron in a bandgap state to the CB, and a second can excite an electron from the VB to the unoccupied intergap state. The hole in the VB can migrate to the surface, where formation of photogenerated intermediates such as OH· can take place. The bulk intergap states postulated by Salvador and Butler are now believed to be associated with either oxygen vacancies or Ti^{3+} interstitials and may lie up to ca. 1 eV below the CB.

5. Photocapacitance Spectroscopy

Subbandgap illumination changes the population of surface states lying within the bandgap as well as generating a photocurrent, and this leads to a further method of surface-state characterization. Small changes in the charge held either by surface states or bulk states in the depletion layer can have a large effect on the potential distribution, as we have seen many times already. The change in potential distribution can be monitored by high-frequency impedance measurements of the electrode capacitance, from which the depletion-layer width can be calculated. By systematically altering the wavelength of the monochromatic incident light, this change can be obtained as a function of photon energy. This method has been applied to the characterization of a variety of energy levels in semiconductors[232-234] and has been introduced to electrochemistry through the work of Tench and co-workers.[235-237]

Subbandgap illumination will be able to excite valence-band electrons into either empty bulk trap or surface states or to excite electrons from these types of states to the conduction band. The effect of a change in *surface* charge can be calculated from the capacitance of the Helmholtz layer, C_H. The shift in flat-band potential will be $\Delta V_{fb} = \Delta Q_{ss}/C_H$, and the associated change in the depletion-layer capacitance due to this shift is given, from the Mott–Schottky equation, by

$$\Delta C_s = C_{sc} \cdot \Delta Q_{ss}/2C_H\phi_{sc} \qquad (81)$$

for an *n*-type semiconductor. For a *p*-type semiconductor, an increase in surface *negative* charge will decrease the depletion-layer potential and hence increase the interfacial capacitance, so the equation above should be prefixed with a minus sign.

If bulk defect states have their population changed by a uniform amount across the depletion layer, then the change in capacitance due to a change $e_0\Delta N_t$ in total charge due to bulk states is

$$\Delta C_b = -C_{sc} \cdot \Delta N_t/2N_D \qquad (82)$$

and for a *p*-type semiconductor, the sign must be changed as before. Unfortunately, the change in trap population is unlikely to be uniform across the depletion layer, since both majority and minority carrier densities are strong functions of position, and we must utilize a more general expression of the form

$$\Delta C_b = -(C_{sc}/2) \cdot (N_t/N_D) \cdot (1/W) \int_0^W \delta f(x)\, dx \qquad (83)$$

where $\delta f(x)$ is the change in the trap population, and W is the space-charge width.

This technique has been applied to *n*-ZnO,[235,238] thin-film *p*-Zn_3P_2,[237] *p*-InP,[239] and both single-crystal and thin-film CdSe.[240–242] Porous thin-film electrodes present a particular problem to the electrochemist since their capacitance is usually very frequency-dependent at all but the lowest frequencies, and any leakage current to the back of the electrode will obscure the small capacitance change which takes place on illumination. For single-crystal electrodes, by way of contrast, measurements are carried out at high frequency to eliminate interference from surface states with moderate time constants.

Allongue and Cachet[243-247] have made a series of studies on the photocapacitance of *n*- and *p*-type GaAs, which can be related to their earlier work discussed above. Figure 53 shows the photocapacitance spectrum of *n*-GaAs in 1 *M* KOH. In common with spectra obtained with other electrolytes, this spectrum is dominated by a negative-going peak at 0.85 eV. The increase in ΔC starting at this energy can be assigned to excitation from the bulk trap commonly represented by EL2. Allongue and Cachet proposed that a negative-going change in ΔC should occur close to 0.6 eV, due to the excitation from the VB to EL2, which could account for the drop in ΔC on approach to 0.85 eV. The size of the EL2 peak is highly dependent on the applied potential, with the result that the variation of the other features with potential is obscured. In order to separate the effect of interface changes from changes in ϕ_{sc}, spectra were taken with the applied potential adjusted to ϕ_{sc} = 0.95 eV for all electrolytes. The size of the EL2 peak is little changed by the variation in electrolyte, provided the space-charge potential remains the same, a result consistent with the assignment of this peak to bulk traps.

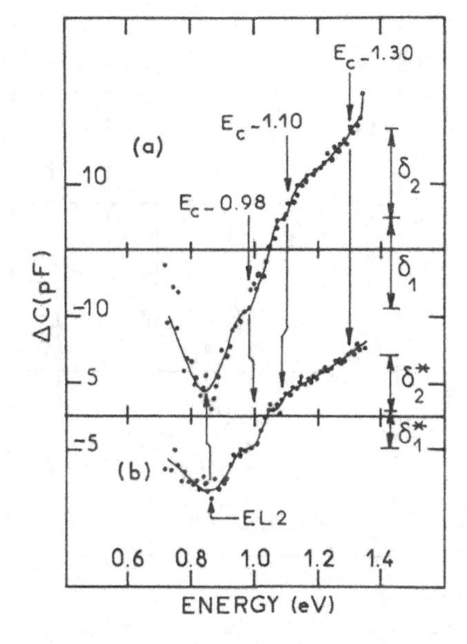

Figure 53. Normalized photocapacitance spectra for $\langle\overline{1}\overline{0}\overline{0}\rangle$ *n*-GaAs in 1 *M* KOH; V_{sc} = 0.95 V and incident photon flux = 10^{16} cm^{-2} s^{-1}. (a) Single-beam experiment; (b) also illuminated by suprabandgap light to give a photocurrent of 2 μA cm^{-2}.[245]

Figure 53 shows the photocapacitance spectrum of ⟨100⟩ *n*-GaAs at 25 kHz ($N_D = 2 \times 10^{16}\ \text{cm}^{-3}$) in 1 *M* KOH obtained upon illumination from the *back* face of the crystal by subbandgap light. At higher energy than the EL2 peak, two positive steps in ΔC are observed close to 0.98 eV and 1.10 eV, and a further rise in ΔC is assigned at 1.30 eV. The positive features imply excitation to the CB and are observed for all the interfaces studied. Allongue and Cachet assigned these steps to surface corrosion states because of their dependence on surface conditions, and their sizes are denoted by δ_1 and δ_2 in the spectra illustrated.

The size and shape of these features change in stabilizing electrolytes, as can be seen in Fig. 54, where the electrode is immersed in a polysulfide solution. Figures 53 and 54 show the effect of illuminating the front fact of the electrode with He-Ne light at 633 nm to give a photocurrent of 2 μA cm^{-2}. The suprabandgap light causes the photocapacitance signal size to *drop*, as the population of the interband states is altered. As the stabilizing ability of the electrolyte increases, the ratio of the step sizes in the presence and absence of suprabandgap illumination, δ_1^*/δ_1 and

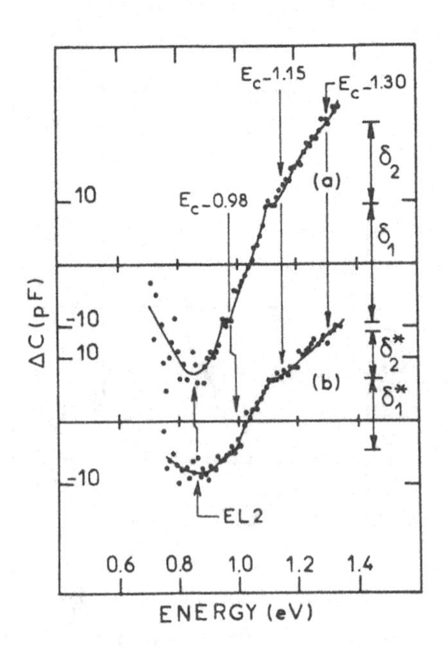

Figure 54. Normalized photocapacitance spectra for *n*-GaAs in 1 *M* KOH/0.1 *M* K_2S/0.1 *M* K_2S_2; $V_{sc} = 0.95$ V and incident photon flux = 10^{16} cm^{-2} s^{-1}. (a) Single-beam experiment; (b) also illuminated by suprabandgap light to give a photocurrent of 2 μA cm^{-2}.[245]

δ_2^*/δ_2, also increases. More stabilizing electrolytes are more effective at refilling corrosion states that have undergone hole capture and therefore decrease the corrosion state population change that occurs on illumination.

From Eq. (81) it should be possible to estimate the density of surface-state population change taking place. In Fig. 53a, $\delta_1 = 13.5\,\text{pF}$, $\phi_{sc} = 0.95\,\text{V}$, and $C_{sc} = 14\,\text{nF}$, and if the Helmholtz capacitance is taken as $C_H = 20\,\mu\text{F cm}^{-2}$, then the population change can be estimated as $5 \times 10^{10}\,\text{cm}^{-2}\,\text{eV}^{-1}$. This is very low compared to the common estimate of surface density of $N_{ss} \approx 10^{13}\,\text{cm}^{-2}$ and implies that the spectra observed may result from rather small changes in the total surface-state population.

Measurements were also carried out on *n*-GaAs coated with submonolayers of platinum or ruthenium, such that the metal forms small clusters on the surface, which act to stabilize the electrode. The presence of the metals resulted in two new features in the photocapacitance in both cases, which were attributed to metal-induced surface states. The features observed are summarized in Fig. 55. Platinum coating the surface seems to decrease the feature at 1.14 eV. A slightly different picture is then advanced for the stabilization of *n*-GaAs by metal coatings compared to the earlier model suggested by Heller.[86] The metal coating does appear partially to passivate the corrosion states, but additional states are also introduced close to the band edges, which can act as pathways for efficient charge transfer to the electrolyte. This is similar to the

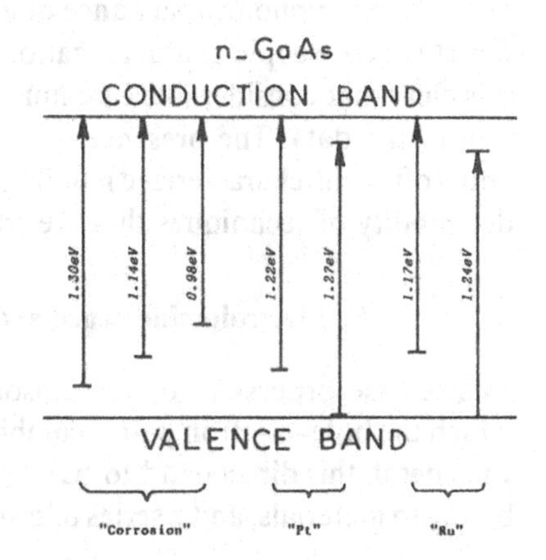

Figure 55. States assigned to features in the *n*-GaAs photocapacitance spectra.[245]

earlier conclusions of Dare-Edwards *et al.*[125] and Butler and Ginley[217] discussed above, which were based on ac impedance and subbandgap photocurrent data.

Goodman *et al.*[239] reported that electrodeposition of cobalt followed by a platinum treatment on *p*-InP etched in HCl (aq) dramatically changed the photocapacitance. Initially, a single positive-going feature was observed, which considerably diminished on Co–Pt treatment, while a new feature appeared, of about three times the intensity of the initial signal. Again, it appears that original surface states are partially passivated, but that a new set of surface states are introduced by the metal treatment.

The spectra illustrated above show that for GaAs in aqueous electrolyte, the technique is close to its noise-imposed limit. Haak and Tench have measured photocapacitance spectra for *n*-GaAs in 0.1 *M* tetraethylammonium perchlorate/acetonitrile electrolyte, which do show the same features as Figs. 53 and 54, though the size is much greater. However, the assignment of Haak and Tench[237] differs radically from that put forward by the French group. They suggested that most of the features observed are, in fact, due to bulk states in their Bridgeman grown crystals. The decrease in capacitance above 0.7 eV is assigned to a state independent of EL2, whereas the increase above 0.85 eV is assigned to the EL2 trap. The step at 0.92 eV is assigned to the presence of chromium in the crystal, while the state at 1.09 eV remains unidentified. The total trap concentration estimated from photocapacitance measurements is ca. $10^{14}\ cm^{-3}$, a very plausible value.

Although photocapacitance might appear to offer an attractive direct spectroscopic characterization of surface-state distributions, it is clear that considerable care must be exercised in the interpretation of the data. The presence of contributions from both surface and (often ill-characterized) bulk states serves to underline the desirability of techniques that are truly surface sensitive.

6. Electroluminescence and Photoluminescence

The reverse process to optical absorption of light is emission, in which the hole–electron pair recombines with emission of a photon. In general, this direct band-to-band process is not favored for wider bandgap materials, and a series of transitions involving intermediate electronic states within the bandgap is the usual relaxation mode. Relaxation to bandgap states may be accompanied either by multiple-phonon emission or emission of a single photon, the latter process being termed luminescence. Clearly, the energy of a luminescent photon will yield information about intermediate levels in semiconductors, and the theory and practice of this technique have been reviewed by Bebb and Williams.[248,249]

In the area of semiconductor electrochemistry, two methods exist for producing the electron and hole population inversion necessary for luminescence to take place. In the technique of *electro*luminescence, the semiconductor is biased close to the flat-band potential or into the accumulation region, so that the majority carrier density at the surface is high, and luminescence is induced by injection of minority carriers from a redox couple in solution. The alternative technique of photoluminescence uses suprabandgap light to excite electron-hole pairs and to follow luminescence via surface or bulk states of intermediate energy. Both techniques have been extensively used.[250-258] Of the materials investigated, both *n*-CdS and *n*-ZnO show features that appear to correspond to the bandgap transition, in addition to transitions that originate from intergap states, which are shown by all semiconductors studied. The differences sometimes observed between photoluminescence and electroluminescence spectra show that the states sampled by these two techniques can be quite distinct, as illustrated in Fig. 56, and Nakato *et al.*[257,258] have assigned the electroluminescence features seen in a variety of oxide and sulfide semiconductors entirely to surface states, but specific assignment of the subbandgap transitions is plagued by uncertainty over whether they are CB $\rightarrow$ defect state or defect state $\rightarrow$ VB in origin.

Electroluminescence studies are limited to electrolyte systems where there is a redox couple capable of injecting minority carriers into the electrode, and it is also limited to potentials close to flat-band potential. Photoluminescence, however, allows studies over a wide range of potentials, and direct comparison can be made with measurements made in blank electrolyte or even in air. Decker *et al.*[259] have compared the two forms of luminescence in a series of electrolytes for *n*-GaAs. Electroluminescence spectra were obtained by pulsing the GaAs to cathodic potentials in order to avoid electrode corrosion and are shown in Fig. 57 before normaliz-

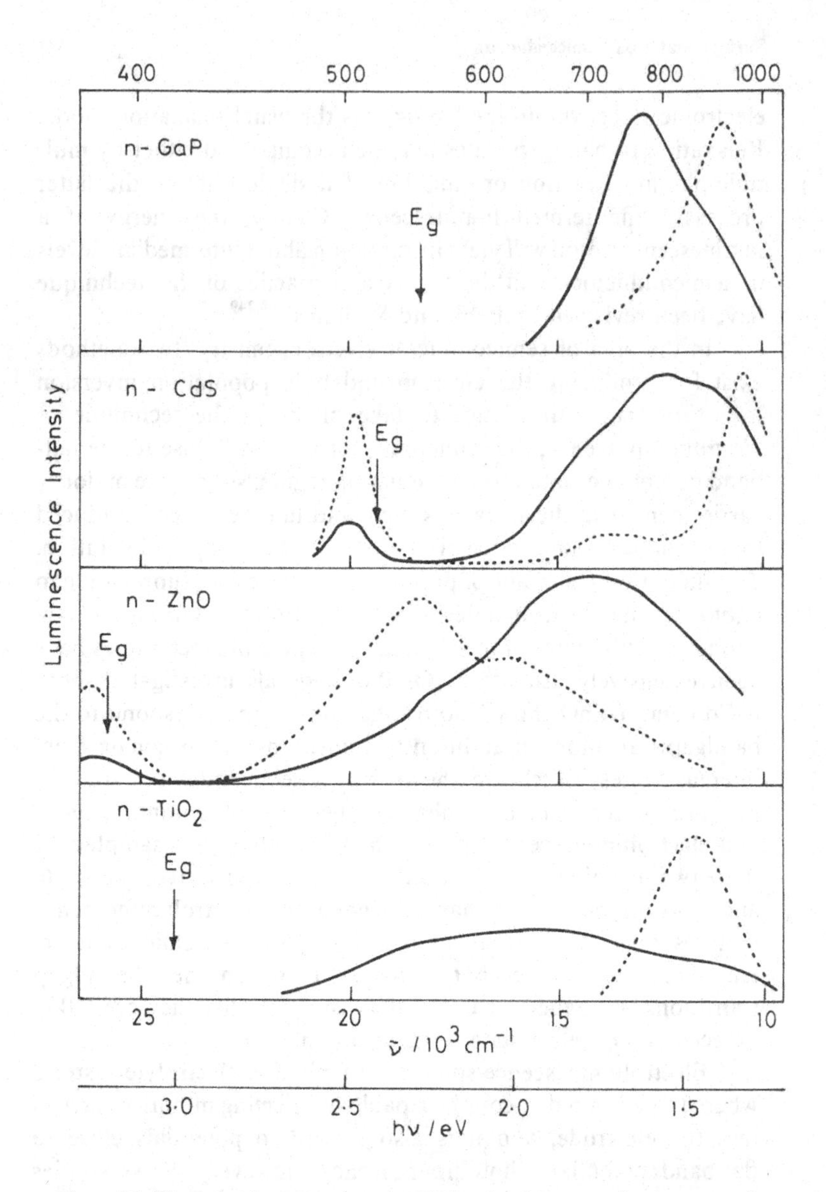

Figure 56. Electroluminescence (——) and photoluminescence, (· · ·) spectra of *n*-type semiconductors, the PL spectra being measured close to flat-band potential but in the absence of redox couples in solution. The EL spectra were obtained under the following conditions: (a) ⟨111⟩ GaP, $[Fe(CN)_6]^{3-}$ at pH 9, −2.0 V vs. SCE; (b) ⟨1000⟩ *n*-CdS, $S_2O_8^{2-}$ at pH 12.6, −1.1 V vs. SCE; (c) *n*-ZnO, $S_2O_8^{2-}$ at pH 1.0, −2.5 to +0.5 V vs. SCE, and (d) ⟨001⟩ *n*-TiO_2, $S_2O_8^{2-}$ at pH 5.0, −4.0 V vs. SCE.[257]

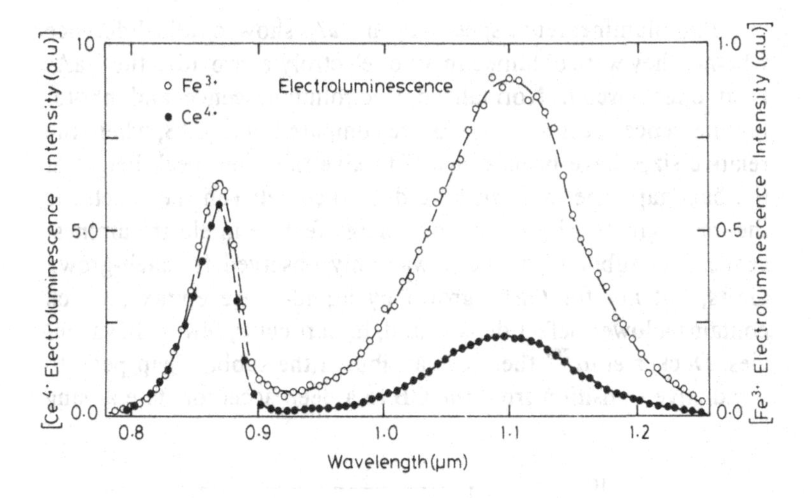

Figure 57. Non-normalized EL spectra for *n*-GaAs in 0.01 *M* $Fe_2(SO_4)_3$/1.0 *M* H_2SO_4 and in 0.03 *M* $Ce(SO_4)_2$/1.0 *M* H_2SO_4. The electrode was continuously pulsed between 0 (20 s) and −1.75 V (2 s) vs. SCE. Note the tenfold increase in sensitivity for the Fe^{3+} (aq) spectrum.[259]

ation of the photomultiplier tube response. From the results it must be concluded that both Ce^{4+} and Fe^{3+} are capable of injecting holes into the VB, but the process must be considerably more efficient for Ce^{4+} because the luminescence intensity at the bandgap is ten times that found for Fe^{3+}, as is clearly shown in Fig. 57. This is unsurprising in that the redox potential of the $Ce^{4+/3+}$ couple lies well below the commonly assumed position of the VB edge, whereas the redox potential of $Fe^{3+/2+}$ lies far closer to this edge. If the $[Fe(CN)_6]^{3-/4-}$ couple is used as an oxidant, luminescence again increases since the ferricyanide couple lies below the VB edge at the pH used.

Even though the electroluminescence spectra of Fig. 57 are not normalized, and the low-energy peak is, therefore, underestimated, there is no doubt that this peak is relatively much more intense when the redox couple is $Fe^{3+/2+}$ rather than $Ce^{4+/3+}$. Presumably, this state, of energy 0.87 eV, corresponds to transitions from the CB to a surface state depopulated by electron transfer to Fe^{3+}, a result consistent with the relatively high energy of the redox potential.

Photoluminescence spectra from GaAs showed little difference whether they were obtained in air or electrolyte, provided the GaAs is at open circuit. Normalized electroluminescence and photoluminescence spectra for GaAs are compared in Fig. 58, where the relative sizes have been adjusted to give the same peak height at the bandgap. The most striking difference between the spectra is the much greater size of the low-energy feature in electroluminescence. The subbandgap peak was only observed for melt-grown GaAs, but not for GaAs grown by liquid-phase epitaxy, which contains a lower defect density and, in particular, fewer Ga vacancies. Decker *et al.*[259] therefore attributed the subbandgap peak to a radiative transition from the CB to a deep acceptor state arising

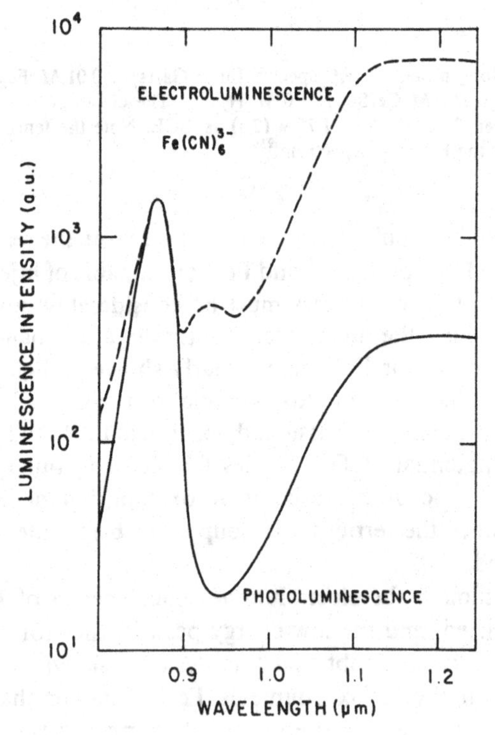

Figure 58. Normalized PL (——) and EL(- - -) spectra of *n*-GaAs in 0.03 *M* $[Fe(CN)_6]^{3-}$/0.1 *M* KOH. PL spectrum was measured at open circuit, and EL spectrum at −1.75 V vs. SCE.[259]

from the presence of Ga vacancies. The much greater intensity of the subbandgap peak in electroluminescence as compared to photoluminescence suggests that there may be a much higher concentration of such defects in the near-surface region, perhaps as a result of corrosion damage or perhaps simply because only at the surface can serious depopulation of such defects occur. Alternatively, the main CB → VB transition may be reduced in intensity at the surface through nonradiative processes, which would have the same effect, given the way in which normalization is carried out.

A further feature apparent from Fig. 58 is that the bandgap luminescence is commonly rather broader on the high-energy side for electroluminescence. In photoluminescence, the same spectrum of energy is believed to be emitted as in electroluminescence, but because the process takes place within the semiconductor, the higher light energies will tend to be reabsorbed before reaching the surface, and the resultant spectrum is narrower.

Photoluminescence and electroluminescence have also been used to probe the identity of the midgap state of TiO_2. Unambiguous evidence for this state was first obtained from electroluminescence experiments carried out by Noufi *et al.*[260] Reduction of $S_2O_8^{2-}$ on TiO_2 was found to lead to a luminescence signal; this is typical of many semiconductors and arises because the first one-electron transfer takes place as

$$S_2O_8^{2-} + e_{cb}^- \rightarrow SO_4^{2-} + SO_4^{\cdot -}$$

and the $SO_4^{\cdot -}$ generated is so powerful an oxidizing agent as to be able to inject a hole directly into the VB of most semiconductors:

$$SO_4^{\cdot -} \rightarrow SO_4^{2-} + p_{vb}^+$$

Electron–hole recombination then leads to luminescence. However, on TiO_2, this luminescence takes place primarily not with light of energy corresponding to the bandgap, but with light in the energy range 1.24–1.5 eV. It would appear that the injected holes are captured by surface states on the TiO_2, which then capture electrons with emission of midbandgap energy.

A more thoroughgoing study by Nakato *et al.*[261] confirmed this suggestion. Photoluminescence spectra of TiO_2 showed a sharp peak at 1.47 eV, with a peak width of ca. 0.3 eV. The photoluminescence also showed a sharp maximum as a function of potential, attaining this maximum at a potential just anodic of the first appearance of photocurrent. The photoluminescence was strongly quenched if a sacrifical redox species such as hydroquinone was added to the solution, confirming that these phenomena are essentially surface in origin. All this can be explained satisfactorily if we assume that a surface state at 1.47 eV below the CB edge is formed during the initial stages of the photooxidation of water. This state is comparatively long-lived, since if the potential is swept to very anodic values, where the photoluminescence disappears, and then back to cathodic values after the light is switched off, a transient dark current flows (see above) *accompanied by a transient photoluminescence.*

Although Nakato *et al.* assigned this 1.47-eV state to surface hydroxyl radicals, Salvador and Gutiérrez[262] have drawn attention to the fact that the estimated lifetime of this state (ca. 1 h) is far too long for so active a species as $OH^{\bullet}_{ads}$. They have suggested that a form of adsorbed H_2O_2 is a more likely candidate. Some support for this does indeed come from the paper of Nakato *et al.* The electroluminescence spectrum for $S_2O_8^{2-}$ reduction is quite broad and rather different from the photoluminescence spectrum, suggesting that the surface state is substantially modified in the presence of the $SO_4^{\bullet-}$ ion. However, the electroluminescence spectrum associated with H_2O_2 is much stronger and shows two peaks, at 1.47 eV and near 2.3 eV. The latter can certainly be assigned to the surface adsorbed $OH^{\bullet}_{ads}$ species described above, and the 1.47-eV peak to H_2O_2 adsorbed onto the surface.

In order to study electrochemical reactions in more depth, the variation of photoluminescence with applied potential has been studied by Ellis and co-workers[263-265] and Uosaki and co-workers[266-270] for GaAs in aqueous electrolyte. Results were interpreted initially on the basis of the "dead layer" model,[271,272] which assumes a surface layer in the semiconductor where the space charge prevents significant recombination, and hence luminescence. Illumination must, therefore, penetrate through this dead layer to photoexcite carriers in the region behind the dead layer for photoluminescence. The emitted light must then pass out through the dead layer, where further attenuation will occur due to self-absorption of the luminescence by the semiconductor. If the luminescence efficiency of the region behind the dead layer is unchanged by the dead layer size, then the quantum efficiency of luminescence, Φ_r, compared to that at the flat-band potential with no dead layer present, Φ_{FB}, will be given by

$$\Phi_r/\Phi_{FB} = \exp(-\alpha' D) \tag{84}$$

where D is the thickness of the dead layer, and $\alpha' = \alpha + \beta$, with α the absorption coefficient of the *exciting* radiation, and β the absorption coefficient for the *emitted* light which undergoes self-absorption. If the dead layer is assumed to vary with potential in a manner similar to the depletion-layer width, then Eq. (84) could be written

$$\Phi_r/\Phi_{FB} = \exp[-\alpha' C(V - V_{FB})^{1/2}] \tag{85}$$

with C a constant to be determined. Figure 59 shows the variation of photoluminescence intensity with constant potential for n-GaAs in a *di*telluride electrolyte for three excitation wavelengths. As predicted by Eq. (85), luminescence was more sensitive to applied potential when the exciting radiation was of lower wavelength, and consequently more strongly absorbed in the dead layer. The dashed

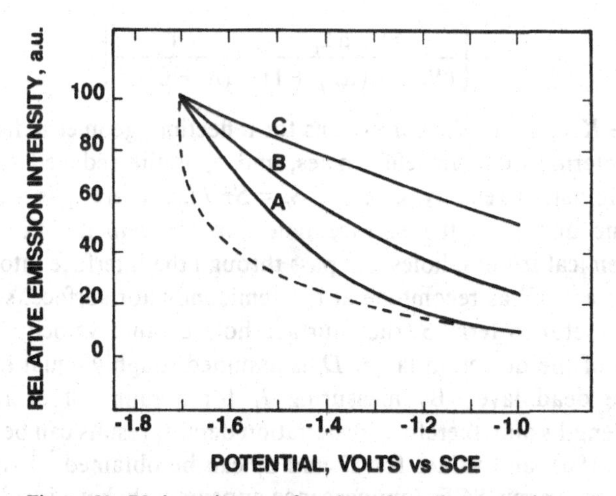

Figure 59. PL intensity as a function of applied potential and excitation wavelength for n-GaAs ($N_D = 4 \times 10^{16}$ cm^{-3}) in 7.5 M KOH/0.2 M Te^{2-}/0.001-0.006 M Te_2^{2-}. The potential was swept at 10 mV s^{-1}. Dashed line gives the variation calculated from Eq. (85) for 457.9 cm. (A) 457.9 nm, (B) 514.5 nm, (C)632.8 nm.[263]

line shows the variation calculated from Eq. (85) for radiation at 457.9 nm, and the rate of photoluminescence change is clearly overestimated close to the flat-band potential. Hobson and Ellis suggested that only ca. 50 mV of the first 200 mV applied to the electrode was actually being dropped within the space-charge region, and the rest was appearing across the Helmholtz layer.[263] If such a deduction were valid, it would make photoluminescence an extremely valuable method of characterizing the depletion-layer voltage with only one datum point. The effect of rapid surface changes and voltage sweeps on the potential distribution could then be followed, at least in a qualitative manner.

The dead layer model neglects the flux of photoexcited carriers from the bulk into the dead layer and the nonradiative loss of carriers at the surface. Mettler[273] used Eq. (19) to derive an expression for the hole flux into the dead layer, which can be matched to the hole recombination flux at the GaAs/air interface. For $\alpha^{-1} < L_p$, the hole diffusion length, the photoluminescence should be given by

$$I_L = K \exp[-(\alpha + \beta)D]\frac{\alpha L_p}{(\alpha L_p)^2 - 1} \times \left[\frac{S_r + \alpha L_p}{(S_r + 1)(\alpha L_p + 1)} - \frac{1}{(\alpha + \beta)L_p}\right] \tag{86}$$

where K contains the corrections for reflection, geometric factors, and internal quantum efficiencies, and S_r is the reduced surface recombination velocity, given by $S_r = S\tau_p/L_p$, where τ_p is the hole lifetime and S is the surface hole capture velocity. For electrochemical studies, holes can pass through the interface into electrolyte as well as recombine at the semiconductor surface, so it is more useful to term S_r the "surface hole capture velocity." The width of the depletion layer, D, is assumed roughly equal to that of the dead layer. By measuring I_L for a range of excitation wavelengths and, therefore, penetration depths, results can be fitted to Eq. (86), and values for D and S_r can be obtained.[264] Anodic of −1.5 V versus SCE, luminescence appears to behave ideally for n-GaAs in *di*telluride electrolyte, S_r has a large constant value, and Eq. (86) effectively reduces to Eq. (85). Furthermore, the value of S_r calculated for the electrolyte is the same as that obtained for the *n*-GaAs/air interface, which implies a common rate-limiting step, despite the fact that holes can undergo either recombination or charge transfer in the electrochemical case. Burk *et al.*[264] suggested that hole capture by surface states is rate limiting and that subsequent charge transfer is through the surface states, in accord with the conclusion of Allongue and Cachet in a similar system.[147]

Similar studies have been carried out on the hydrogen evolution reaction on *p*-GaAs in H_2SO_4.[265,270] The photoluminescence shows considerable hysteresis in the region between the impedance-measured flat-band potential and the photocurrent onset. Analysis with Eq. (86) shows that S_r varies considerably with potential anodic of photocurrent onset and shows hysteresis consistent with significant changes in surface chemistry in that region, which may be due to the adsorption and desorption of hydrogen atoms.

When the effect of surface recombination has been included in the interpretation, it becomes apparent that it is difficult to determine whether or not the potential distribution does or does not vary ideally. In common with several of the other techniques we have discussed, photoluminescence becomes steadily less clear-cut in its interpretation as the potential approaches the flat-band situation.

7. Electroreflectance

The electric field at the surface of a semiconductor electrode will perturb the optical properties of the semiconductor and change the electrode reflectivity. This phenomenon is most pronounced for radiation of energy close to that of transitions between critical points in the semiconductor band structure, for example, the direct bandgap transition. The effect can be measured as a function of wavelength if the potential applied to the electrode is modulated, so that the resulting very small modulation of the intensity of the reflected light can be detected with phase-sensitive techniques. Electroreflectance spectroscopy is particularly attractive as a possible method of characterizing the depletion-layer voltage since it is independent of the electrochemical processes taking place at the electrode surface, in contrast to the impedance and photoluminescence techniques discussed earlier. The principal difficulty with electroreflectance has, however, been the interpretation of the frequently complicated variations in lineshape, although considerable progress has been made in recent years toward an understanding of the effects involved.[274–276]

Electroreflectance spectroscopy was initially developed as a probe of semiconductor band structure,[277] and it is now routinely used to characterize semiconductor materials. It has recently become popular in the guise of photoreflectance, where the surface electric field is modulated by a chopped laser beam, so that bulk or heterostructure semiconductors can be characterized without the need for a surface contact.[278] As a result, there exists now a considerable body of experimental electroreflectance data in the literature for many common semiconductors to which spectra obtained in electrolyte may be compared.

Figure 60 shows a typical experimental arrangement. Monochromatic light is focused onto the electrode and reflected into a photodiode detector. A dc potential is applied to the electrode through a potentiostat, and an ac potential superimposed. As indicated above, this gives rise to an ac modulation on the intensity of the reflected light beam which is detected at the photodiode using a lock-in amplifier. The photodiode output is measured by an analog-to-digital converter, and the microcomputer ratios the rms ac signal to the dc signal to obtain the normalized electroreflectance value for each wavelength.

Expressions for the effect of an electric field on the semiconductor optical properties for a variety of critical points were evaluated by Aspnes[279,280] shortly after the earliest electroreflectance experiments. For the case of a direct allowed transition between bands that can be treated by a three-dimensional isotropic effective mass approximation, the field-induced change in the dielectric function will be given by

$$\begin{aligned}\Delta\varepsilon_2(\omega, E) = {} & B\theta^{1/2}\pi[Ai'^2(x) - xAi^2(x)]/\omega^2 \\ & - B\theta^{1/2}(-x)^{1/2}\cdot u(-x)/\omega^2\end{aligned} \tag{87}$$

$$\begin{aligned}\Delta\varepsilon_1(\omega, E) = {} & B\theta^{1/2}\pi[Ai'(x)Bi'(x) - xAi(x)Bi(x)]/\omega^2 \\ & + B\theta^{1/2}x^{1/2}u(x)/\omega^2\end{aligned} \tag{88}$$

where $x = (\omega_g - \omega)/\theta$, and $\theta = (E^2e_0^2/2\mu\hbar)^{1/3}$, with μ the reduced effective mass of valence-band hole and conduction-band electron.

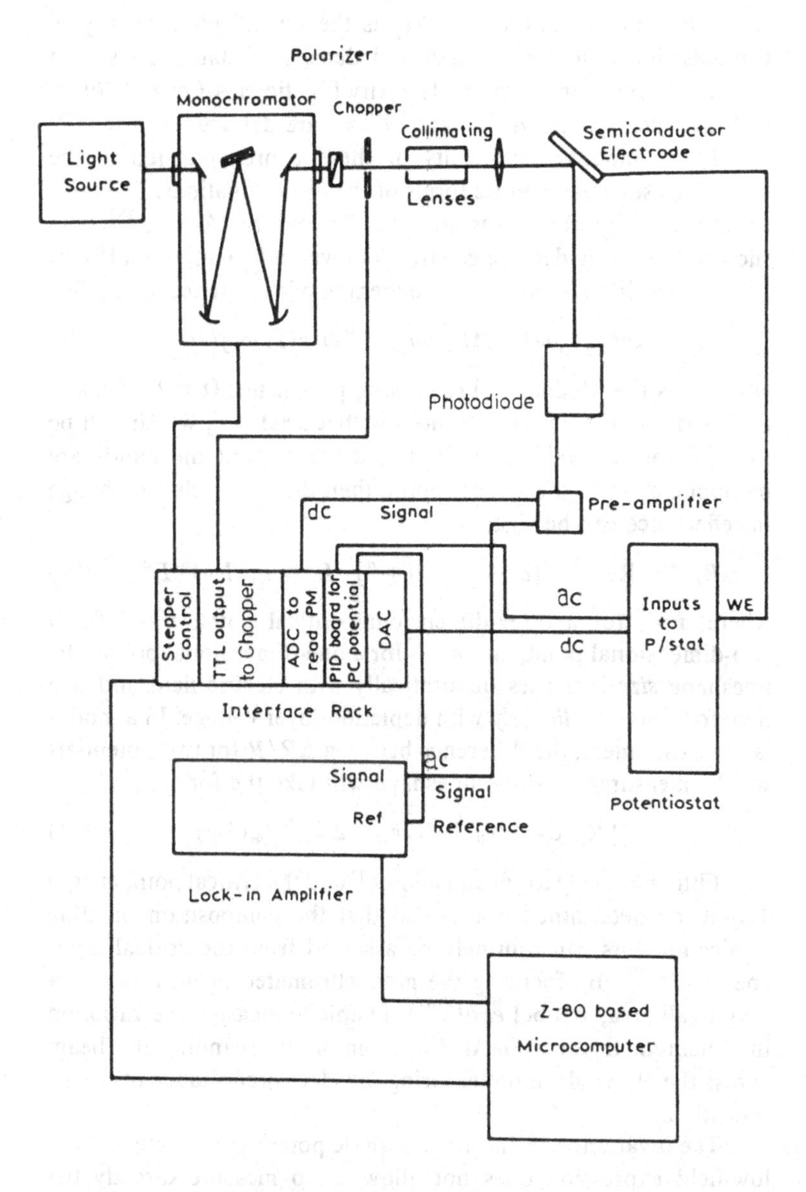

Figure 60. Typical experimental arrangement for electroreflectance (ER).

The electric field is denoted by E, $\hbar\omega$ is the energy of the monochromatic incident radiation, and $\hbar\omega_g$ is the critical point energy. B contains the oscillator strength and density of states terms, and $u(x)$ is the unit step function. The Airy functions $Ai(x)$ and $Bi(x)$, and their derivatives, $Ai'(x)$ and $Bi'(x)$, are defined elsewhere.[281]

The numerical complexity of these expressions led to the extensive use of asymptotic forms of the basic equations. The most commonly used form was initially derived by Aspnes,[282] who showed that, provided the electric field was sufficiently small compared to the lifetime-induced broadening of the excited state, then

$$\Delta\varepsilon(E, \Gamma, \omega) = (\Omega^3/3\omega^2) \cdot \partial^3[\omega^2\varepsilon(\Gamma, \omega)]/\partial\omega^3 \tag{89}$$

where Γ is the lifetime or broadening parameter, $\Omega \equiv 2^{-2/3}\theta$, and the condition for Eq. (89) to hold is that $3\hbar\Omega < \Gamma$, which will be true for the *low-field* case. If it is assumed that the bands are parabolic close to the critical point, then the field-induced change in reflectance can be written

$$\Delta R/R = \mathrm{Re}[C e^{i\theta}(\omega - \omega_g + i\Gamma)^{-n}] \cdot E^2 \equiv L_n(\hbar\omega) \cdot E^2 \tag{90}$$

where $n = \frac{5}{2}$ for a three-dimensional critical point, $n = 3$ for a two-dimensional point, and $n = \frac{7}{2}$ for a one-dimensional point. The lineshape *size* increases quadratically with electric field and will therefore increase *linearly* with depletion-layer voltage. In a modulation experiment, the difference between $\Delta R/R$ for two potentials will be measured, and the lineshape will take the form

$$\Delta R/R = (2e_0N_D/\varepsilon_0\varepsilon_{sc}) \cdot \Delta V_{sc} \cdot L_n(\hbar\omega) \tag{91}$$

Fitting Eq. (91) to spectra allows Γ and the critical point energy $\hbar\omega_g$ to be determined readily, so that the composition of alloy semiconductors can routinely be assessed from the critical point energies.[283-285] By focusing the monochromated beam down to a very small area, Wrobel *et al.*[286] were able to measure the variation in stoichiometry of a $GaAl_xAs_{1-x}$ sample by scanning the beam across the electrode and measuring the electroreflectance of the E_1 transition.

The invariance of lineshape with dc potential predicted by the low-field expression does not allow us to measure directly the space-charge voltage. However, if only a portion of the applied voltage modulation is dropped across the space-charge layer, then the signal size will be proportionately reduced in accord with Eq. (91). By using the concept of Helmholtz capacitance, Tomkiewicz and co-workers calculated the fraction of potential dropped across the Helmholtz layer by surface charging, and hence the effect on the signal size of such a drop[287,288]:

$$\Delta R/R = (2e_0N_D/\varepsilon_0\varepsilon_{sc}) \cdot L_n(\hbar\omega) \cdot [1 - (e_0\Delta N_{ss}/C_H\Delta V)] \cdot \Delta V \tag{92}$$

where ΔN_{ss} is the change in surface-state occupancy induced by the potential change ΔV. Figure 61 shows the variation of $\Delta R/R$ with applied dc potential with a modulation of 100 $mV_{p\text{-}p}$ for a single crystal and a polycrystalline sample of CdSe. The actual lineshape does not change with the dc potential, so it is assumed that low-field conditions apply. The decrease in $\Delta R/R$ for single-crystal *n*-CdSe close to flat-band potential and the drop in $\Delta R/R$ for the polycrystalline sample at more anodic potentials can then be assigned to the presence of densities of surface states which are

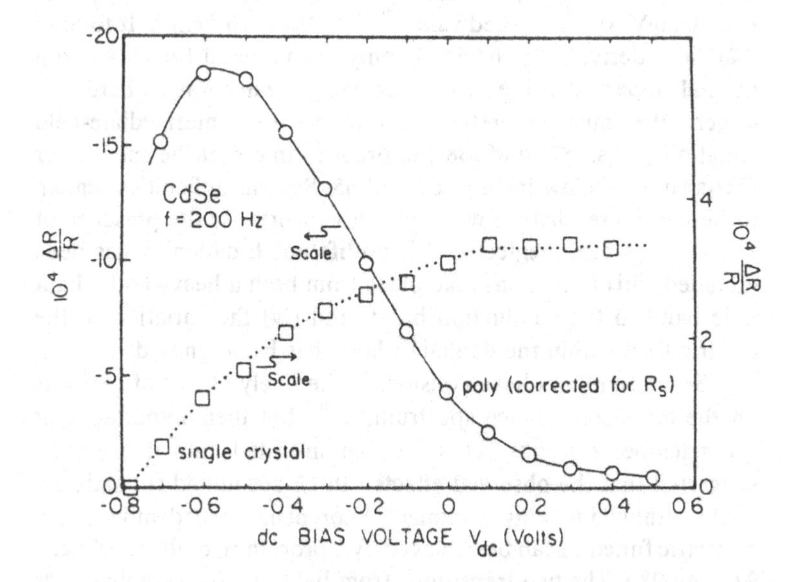

Figure 61. ER signal amplitude for *n*-CdSE electrodes in 1:1 1 *M* $Na_2S/S_2^{2-}/NaOH$ measured with 100-mV p-p modulation at 200 Hz. The signal for the polycrystalline electrode is corrected for the ac voltage dropped across a large bulk resistance.[287]

sufficient to pin the Fermi level, at least partially. Similar analyses have been attempted for CdS,[289] GaAs,[290] TiO_2,[291] and Si.[292]

The rate at which surface-state populations are able to relax can be studied by varying the frequency of modulation and observing the variation of $\Delta R/R$. If both the in-phase and quadrature values of $\Delta R/R$ are measured, then any phase change in the depletion-layer voltage can be identified.[293] This measurement can be carried out readily even in systems that are difficult to study by other methods.

However, the third-derivative lineshape frequently does *not* give satisfactory fits to measured spectra, and Raccah and coworkers have extended the model within the low-field assumption to include the effect of defects and electrostriction on the electroreflectance lineshape.[294,295] The resultant improvement on lineshape is, however, at the expense of additional parameters to include defect-induced changes in the local crystal field and critical point energy.

The low-field assumptions that underlie the work described above are very restrictive: for GaAs and InP, for example, with $\Gamma \leq 10$ meV (the measured value[278,283]), $N_D \ll 10^{15}$ cm^{-3}. It follows that third-derivative behavior is only encountered for GaAs with normal dopant densities at higher energy transitions, where Γ is larger. We must, therefore, return to the intermediate-field equations, Eqs. (87) and (88), in order to interpret the spectra for GaAs shown below in Figs. 62 and 65. Several difficulties remain to be considered before we apply this theory: (i) the presence of excitons has been neglected; (ii) no lifetime broadening has been included; (iii) transitions take place from both a heavy and a light hole band to the conduction band; and (iv) the variation of the electric field within the depletion layer has been ignored.

Several authors have considered the likely effects of excitons on the electroreflectance spectrum,[296-300] but their importance at room temperature has yet to be demonstrated, and it has been suggested that the observed effects can be accounted for satisfactorily without invoking excitons.[276] Lorentzian broadening of the dielectric function can be achieved by a procedure outlined in Refs. 275 and 280. The two transitions from light and heavy hole bands can be treated independently, and their relative strengths have been calculated.[301]

The fourth problem above is the most serious. Light reflected at the surface of the electrode will penetrate through the depletion layer and sample a range of electric fields, an effect that is particularly important at the bandgap transition, where the penetration of the light is high. Aspnes and Frova[302] have treated the problem with a WKB-type solution for Ge electroreflectance, but Hamnett and co-workers[274-276] have used a multilayer reflectance routine[303] so that electroreflectance spectra can be calculated for any angle of incidence and in the presence of surface layers.

Figure 62 shows spectra for *p*-GaAs in 0.5 *M* H_2SO_4 at a series of applied potentials.[304] The lineshape is observed to broaden as the electrode is biased away from the flat-band potential, and a shoulder can be seen to develop on the main peak. Theoretical spectra were calculated with the experimentally determined dopant density, so that the only parameter that needs to be determined by fitting to the experimental spectra is the variation of the lifetime broadening with energy. The theoretical spectra are illustrated in Fig. 63 and appear to model the experimental broadening and lineshape changes very successfully. The variation of peak positions with applied potential in the experimental and theoretical spectra is compared in Fig. 64. Cathodic of −0.6 V versus SCE, spectra vary as predicted for a flat-band potential of −0.2 V versus SCE, which demonstrates a classical variation of the space-charge voltage with the applied potential. However, over the potential range −0.4 to 0.0 V versus SCE, the electroreflectance spectrum changes very little in contrast to the rapid narrowing expected from theory on approach to the flat-band potential. The implication is that the dc space-charge potential is *pinned* at 0.2 V over this potential range, a deduction that can be made from the *shape* rather than the *size* of the measured spectrum. The onset of Fermi level pinning coincides with the photocurrent onset potential of −0.5 V versus SCE for *p*-GaAs in 0.5 *M* H_2SO_4[98] and confirms the suggestion made by Peat and Peter[209] that the electrode is substantially pinned between the flat-band potential and the photocurrent onset. However, contrary to the suggestion of Meissner and Memming,[210] the space-charge voltage is pinned close to 0.2 V anodic of the photocurrent onset and not to the flat-band potential itself.

Electroreflectance can also be used to characterize electrodes that are very far from ideal. Figure 65 shows spectra measured on

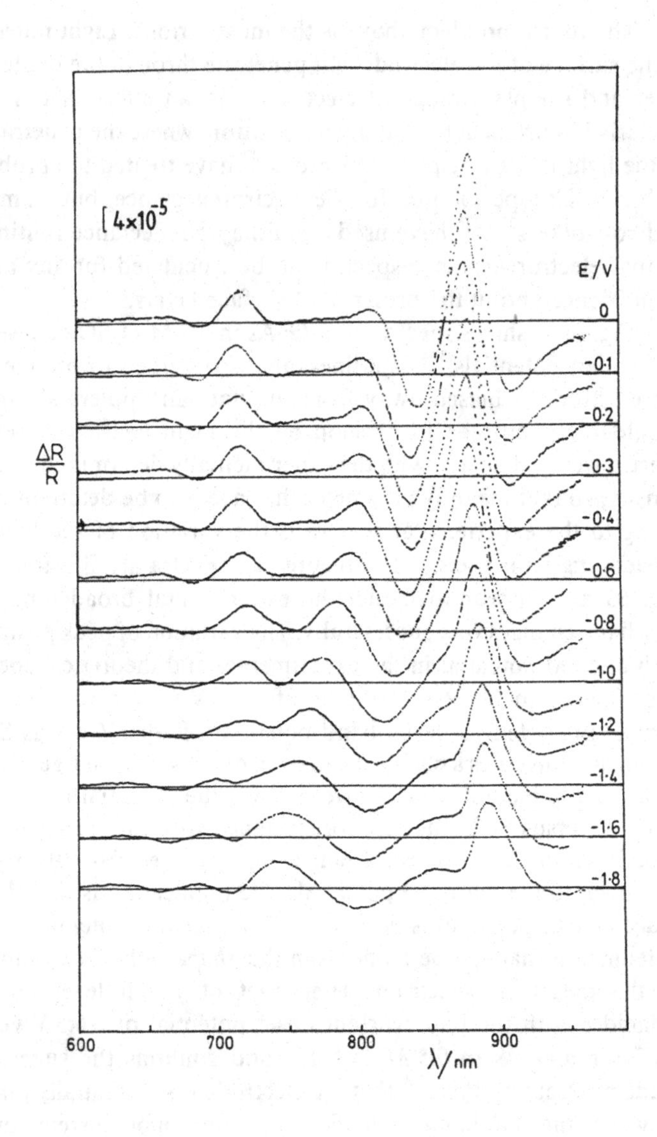

Figure 62. ER spectra of p-GaAs (etched in Br_2/methanol) in 0.5 M H_2SO_4 with 16-mV square-wave modulation at 270 Hz; p-polarized light.

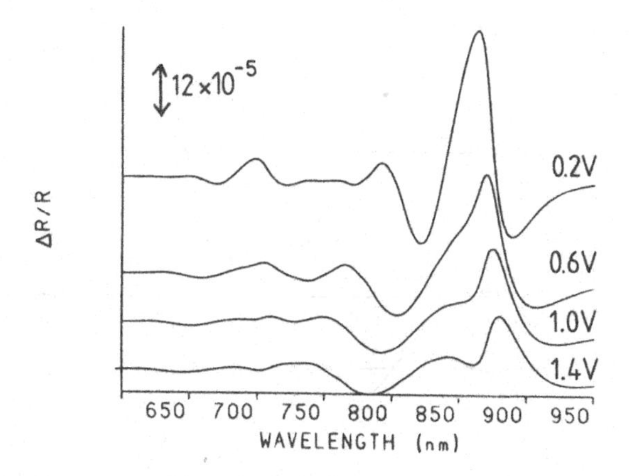

Figure 63. Theoretical ER spectra for p-GaAs calculated under the conditions of Fig. 62.

an n-GaAs electrode, of similar dopant density to the previous p-GaAs sample, where the electrolyte is Na_2SO_4 at pH 5. The spectra appear inverted compared to those of the p-type electrode because the effect of a voltage modulation on the magnitude of the depletion-layer voltage is reversed on changing the carrier type. Substantially the same lineshapes are observed as in Fig. 62, but the shape changes

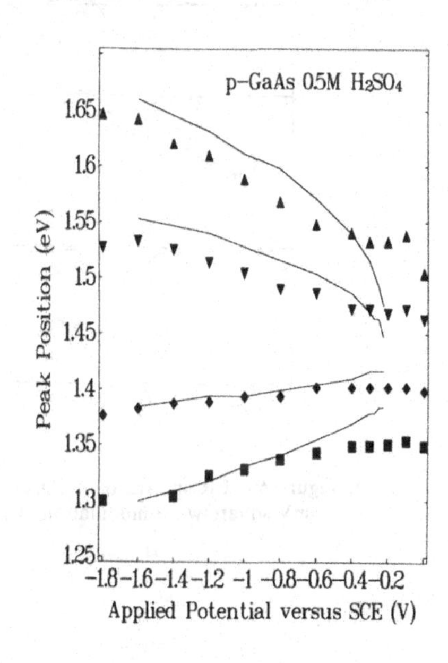

Figure 64. Plot of peak positions versus applied potential for 0.5 M H_2SO_4/p-GaAs: (■, ◆, ▼, ▲) experimental results; ——, theoretical results for $E_g = 1.41$ eV.

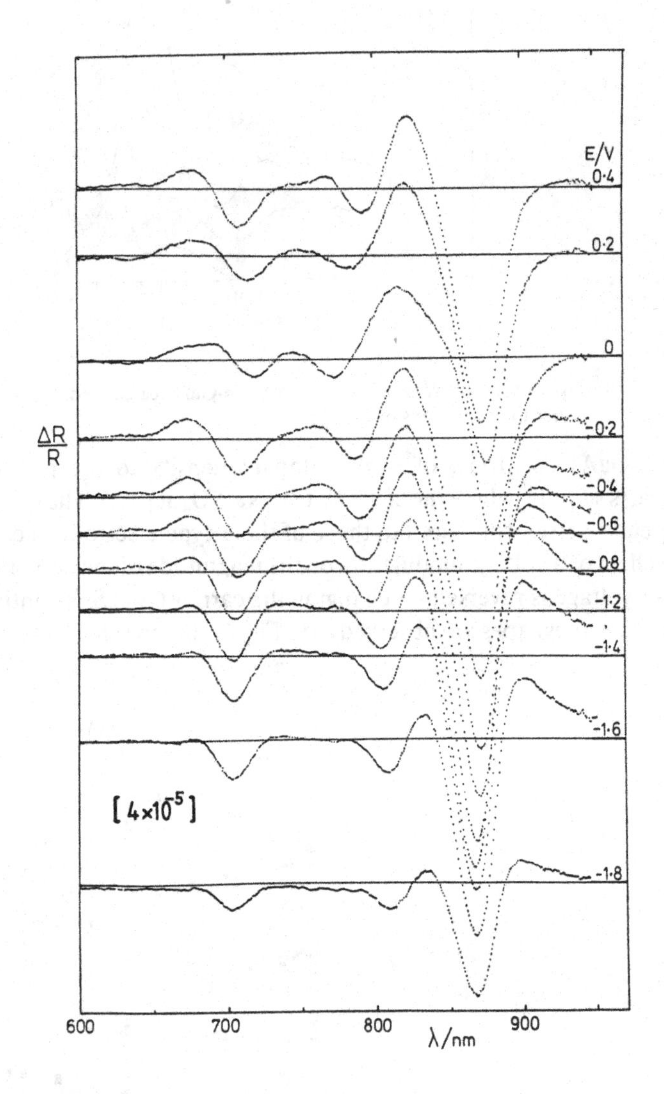

Figure 65. ER spectra of n-GaAs in 0.1 M Na_2SO_4 at pH 5 with 16-mV square-wave modulation at 270 Hz; p-polarized light.

with applied potential much more slowly for *n*-GaAs in this near-neutral electrolyte. This can most clearly be seen if peak position is plotted against voltage as in Fig. 66, in which theory and experiment can be compared directly. The evolution of lineshape with voltage is considerably more rapid for the *same* electrode in 0.1 *M* KOH,[305] 0.1 *M* KOH/0.005 *M* Na_2Se/Se_2^{2-},[304] and 0.1 *M* KOH/0.01 *M* Na_2S/S_2^{2-}.[304] From Fig. 66 we can deduce that in neutral solution, only a minority of any change in applied potential is dropped across the space-charge layer in the potential range −1.6 to −0.2 V versus SCE. Gallium arsenide will have a surface oxide layer that will not be removed by the electrolyte at pH 5, and it might therefore be expected that a considerable fraction of the applied potential is, in fact, dropped across this oxide film, or across surface states associated with the film. However, the three spectra at most anodic potentials show an unusual reversal of the broadening trend with increasing reverse bias. At 0 V versus SCE, the shoulder on the main peak has started to develop, but at 0.2 and 0.4 V versus SCE, the shoulder has collapsed back. In this experiment, it appears that increasing the potential above 0.0 V versus SCE can actually lead to a drop in the space-charge voltage,

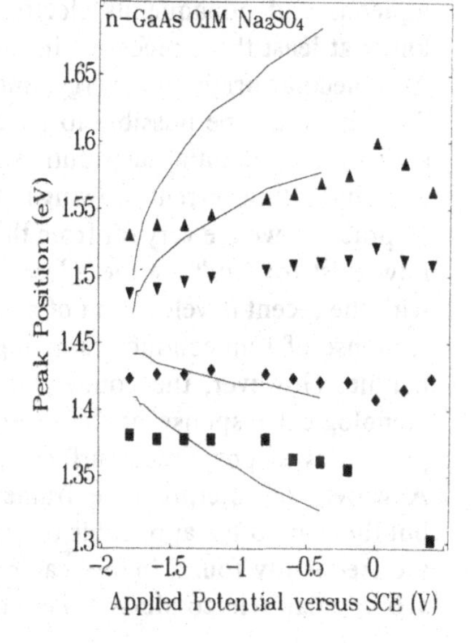

Figure 66. Plot of peak positions versus applied potential for *n*-GaAs in 0.1 *M* H_2SO_4 at pH 5; (■, ◆, ▼, ▲) experimental results; ——, theoretical results assuming $V_{fb} = -1.81$ V vs. SCE and conditions as in Fig. 63.

which is presumably caused by potential-induced degradation of the electrode surface.

Subbandgap states have also been investigated with electroreflectance on TiO_2,[306,307] and the results have been interpreted as due to surface states owing to their sensitivity to electrolyte at the interface. Care must, however, be taken in assigning extra peaks close to the bandgap to impurities[308,309] or surface-state effects owing to the likelihood that they may arise from interference effects within the depletion layer.[275,276]

Electrolyte electroreflectance is emerging as an important tool for the study of the space-charge potential in the presence of high surface-state densities which can induce pinning. Recent developments have allowed spectra to be successfully modeled for the first time, and the results have indicated that the potential distribution may be very far from ideal. It is increasingly evident that the classical depletion-layer model must be treated, even in electrolyte, with considerable skepticism.

VIII. CONCLUSIONS

It is evident from the discussion above that we are still some way from understanding the surface chemistry of semiconductors in aqueous and nonaqueous electrolytes. Ideally, we would like to know at least three pieces of information about any surface state: its molecular origin, its energy, and its surface density. From these data, it should be possible to predict the influence of the surface state on the potential distribution at the interface and the faradaic and alternating currents flowing in the dark and under illumination. In practice, we are very far from this ideal for any surface: methods now exist for studying the *effects* of surface states in some detail, with the recent development of good models for the electro-optical response of semiconductors giving us a most valuable additional handle. However, the connection between the observed phenomenological response of the semiconductor and the microscopic properties of postulated surface states is not yet well established. A variety of interpretative frameworks have been put forward, but these are often applicable only to one type of measurement and are frequently found inapplicable to a whole series of studies on a particular semiconductor. Perhaps the most difficult task is the identification of the molecular origin of the surface states identified by macroscopic measurements. The effectiveness of such recently developed techniques as *in situ* Fourier transform infrared (FTIR) reflectance, which has been so successful in metal-electrode studies, is to some extent vitiated by the very low coverage of surface states required for them to have a major effect on the semiconductor properties. It is possible that scanning tunneling microscopy (STM) may have a major role to play here, and certainly the identification of defects at the atomic or molecular level is now feasible with this technique, even in aqueous electrolyte.

In this chapter, we have emphasized the nature of the surface-state problem and the ways in which electrochemists have attempted to cope with the difficulties encountered in an unfamiliar area. The strategic rewards of success in this area are considerable: the better understanding of photoelectrochemical efficiencies and of semiconductor etching processes is fundamental to applications of semiconductor electrochemistry in a wide forum. It must be a primary goal of electrochemistry in the next decade to unravel more of the problems encountered at the semiconductor/electrolyte interface.

REFERENCES

[1] W. H. Brattain and C. G. B. Garrett, *Bell Syst. Tech. J.* **34** (1955) 129.
[2] J. Bardeen, *Phys. Rev.* **71** (1947) 717.
[3] H. O. Finklea, Ed., *Semiconductor Electrodes*, Elsevier, Amsterdam, 1989.
[4] V. A. Myamlin and Yu. V. Pleskov, *Semiconductor Electrochemistry*, Plenum Press, New York, 1967.
[5] A. Hamnett, in *Comprehensive Chemical Kinetics*, Vol. 27, Ed. by R. G. Compton and A. Hamnett, Elsevier, Amsterdam, 1987, p. 61.
[6] I. E. Tamm, *Z. Phys.* **76** (1932) 849; *Phys. Z. Sowjetunion* **1** (1932) 733.
[7] W. Shockley, *Phys. Rev.* **56** (1939) 317.
[8] A. Many, Y. Goldstein, and N. B. Grover, *Semiconductor Surfaces*, North-Holland Press, Amsterdam, 1965.
[9] S. G. Davison and J. D. Levine, *Solid State Phys.* **25** (1970) 1.
[10] J. Bardeen, *Phys. Rev.* **71** (1947) 717.
[11] I. Lindau and T. Kendelwicz, *CRC Crit. Rev. Solid State Mater. Sci.* **13** (1986) 27.
[12] A. Zur, T. C. McGill, and D. L. Smith, *Phys. Rev. B* **28** (1983) 2060.
[13] J. M. Palau, A. Ismail, and L. Lassabatere, *Solid-State Electron.* **28** (1985) 499.
[14] L. J. Brillson, *Surf. Sci. Rep.* **2** (1982) 127.
[15] C. B. Duke and A. Paton, *Surf. Sci.* **164** (1985) L797.
[16] C. B. Duke, *Appl. Surf. Sci.* **11/12** (1982) 1.
[17] Z. Zunger, *Phys. Rev. B* **24** (1981) 4372.
[18] D. J. Chadi, *Phys. Rev. B* **18** (1978) 1800.
[19] W. Mönch, *Surf. Sci.* **132** (1983) 92.

[20] K. Stiles, D. Mao, and A. Kahn, *J. Vac. Sci. Technol.*, *B* **64** (1988) 1170.
[21] S. M. Sze, *Physics of Semiconductor Devices*, 2nd ed., Wiley-Interscience, 1981.
[22] V. Heine, *Phys. Rev.* **138** (1965) A1689.
[23] J. Tersoff, *Phys. Rev. Lett.* **52** (1984) 465.
[24] R. Ludeke, *Phys. Rev. B* **40** (1989) 1947.
[25] W. Mönch, *J. Vac. Sci. Technol.*, *B* **6** (1988) 1270.
[26] W. E. Spicer, P. W. Chye, P. R. Skeath, C. Y. Su, and I. Landau, *J. Vac. Sci. Technol.* **16** (1979) 1422.
[27] E. R. Weber, H. Ennen, V. Kaufmann, J. Windscheiff, J. Schneider, and T. Wosinski, *J. Appl. Phys.* **53** (1982) 6140.
[28] W. E. Spicer, Z. Liliental-Weber, E. Weber, N. Newman, T. Kendelwicz, R. Cao, C. McCants, P. Mahowald, K. Miyano, and I. Lindau, *J. Vac. Sci. Technol.*, *B* **6** (1988) 1245.
[29] J. M. Woodall and J. L. Freeouf, *J. Vac. Sci. Technol.* **19** (1981) 794.
[30] J. M. Woodall and J. L. Freeouf, *J. Vac. Sci. Technol.*, *B* **2** (1984) 510.
[31] H. Hasegawa and H. Ohno, *J. Vac. Sci. Techol.*, *B* **4** (1986) 1130.
[32] H. Hasegawa, L. He, H. Ohno, T. Sawada, T. Hada, Y. Abe, and H. Takahasi, *J. Vac. Sci. Technol.*, *B* **5** (1987) 1097.
[33] L. J. Brillson, R. E. Viturro, C. Mailhiot, J. L. Shaw, N. Tache, J. McKinley, G. Margaritondo, J. M. Woodall, P. D. Kirchner, G. D. Petit, and S. L. Wright, *J. Vac. Sci. Technol.*, *B* **6** (1988) 1263.
[34] R. E. Viturro, S. Chang, J. L. Shaw, C. Mailhiot, L. J. Brillson, A. Terrassi, Y. Hwu, G. Margaritondo, P. D. Kirchner, and J. M. Woodall, *J. Vac. Sci. Technol.*, *B* **7** (1989) 1007.
[35] C. J. Sandroff, R. N. Nottenburg, J. C. Bischoff, and R. Bhat, *Appl. Phys. Lett.* **51** (1987) 33.
[36] C. J. Sandroff, M. S. Hegde, L. A. Farrow, C. C. Chang, and J. P. Harbison, *Appl. Phys. Lett.* **54** (1989) 362.
[37] C. J. Sandroff, M. S. Hegde, and C. C. Chang, *J. Vac. Sci. Technol.*, *B* **7** (1989) 841.
[38] T. Tiedje, P. C. Wong, K. A. R. Mitchell, W. Eberhardt, Z. Fu, and D. Sondericker, *Solid State Commun.* **70** (1989) 355.
[39] C. J. Spindt, R. S. Besser, R. Cao, K. Miyano, C. R. Helms, and W. E. Spicer, *Appl. Phys. Lett.* **54** (1989) 1148.
[40] R. S. Besser and C. R. Helms, *Appl. Phys. Lett.* **52** (1988) 20.
[41] H. Hasegawa, H. Ishii, T. Sawada, T. Saitoh, S. Konishi, Y. Liu, and H. Ohno, *J. Vac. Sci. Technol.*, *B* **6** (1988) 1184.
[42] O. K. Gaskill, N. Bottka, and R. S. Sillman, *J. Vac. Sci. Technol.*, *B* **6** (1988) 1497.
[43] J. F. DeWald, in *Semiconductors*, Ed. by N. B. Hannay, ACS Monograph 140, Reinhold, New York, 1959, p. 727.
[44] M. Green, in *Modern Aspects of Electrochemistry*, No. 2, Ed. by J. O'M. Bockris, Butterworths, London, 1959, p. 343.
[45] P. J. Holmes, *The Electrochemistry of Semiconductors*, Academic Press, New York, 1961.
[46] V. A. Myamlin and Yu. V. Pleskov, *Electrochemistry of Semiconductors*, Plenum Press, New York, 1967.
[47] H. Gerischer, in *Physical Chemistry: An Advanced Treatise*, Vol. IXA, Ed. by H. Eyring, Academic Press, New York, 1970, p. 463.
[48] S. R. Morrison, *Prog. Surf. Sci.* **1** (1971) 105.
[49] S. R. Morrison, *Electrochemistry at Semiconductor and Oxidized Metal Electrodes*, Plenum Press, New York, 1980.
[50] R. Memming, in *Electroanalytical Chemistry*, Vol. 11, Ed. by A. J. Bard, Marcel Dekker, New York, 1979.

[51] R. Memming, in *Comprehensive Treatise of Electrochemistry*, Vol. 7, Ed. by B. E. Conway, J. O'M. Bockris, E. Yeager, S. U. M. Khan, and R. E. White, Plenum Press, New York, 1983, p. 529.
[52] R. H. Wilson, *CRC Crit. Rev. Solid State Mater. Sci.* **10** (1980) 2.
[53] W. P. Gomes and F. Cardon, *Prog. Surf. Sci.* **12** (1982) 155.
[54] L. M. Peter, in *RSC Specialist Periodical Reports: Electrochemistry*, Vol. 9, Ed. by D. Pletcher, Royal Society of Chemistry, London, 1984, p. 66.
[55] Yu. V. Pleskov and Yu. Ya. Gurevich, in *Modern Aspects of Electrochemistry*, Vol. 16, Ed. by J. O'M. Bockris, Plenum Press, New York, 1986, p. 189.
[56] Yu. V. Pleskov and Yu. Ya. Gurevich, *Semiconductor Photoelectrochemistry*, Plenum Press, New York, 1986.
[57] R. Memming, *Ber. Bunsenges. Phys. Chem.* **91** (1987) 353.
[58] R. A. Marcus, *J. Chem. Phys.* **43** (1965) 679.
[59] H. Gerischer, *Surf. Sci.* **18** (1969) 97.
[60] S. Trassatti, in *Comprehensive Treatise of Electrochemistry*, Vol. 1, Ed. by J. O'M. Bockris, B. E. Conway, and E. Yeager, Plenum Press, New York, 1980, p. 45.
[61] R. Memming and G. Schwandt, *Electrochim. Acta* **13** (1968) 1299.
[62] V. A. Tyagai and G. Ya. Kolbasov, *Surf. Sci.* **28** (1971) 423.
[63] R. Memming, *J. Electrochem. Soc.* **125** (1978) 117.
[64] D. Meissner, I. Lauermann, R. Memming, and B. Kastening, *J. Phys. Chem.* **92** (1988) 3484.
[65] D. Meissner, C. Sinn, R. Memming, P. H. L. Notten, and J. J. Kelly, in *Homogeneous and Heterogeneous Photocatalysis*, NATO ASI Ser., Ser. C, Vol. 174, D. Reidel, Dordrecht, 1986, p. 317.
[66] K. Tubbesing, D. Meissner, R. Memming, and B. Kastening, *J. Electroanal. Chem.* **214** (1986) 685.
[67] J. W. Conely, C. B. Duke, G. D. Mahan, and J. J. Tiemann, *Phys. Rev.* **150** (1966) 466.
[68] L. B. Freeman and W. B. Dahlke, *Solid-State Electron.* **13** (1970) 1483.
[69] B. Pettinger, H. Schöppel, and H. Gerischer, *Ber. Bunsenges. Phys. Chem.* **77** (1973) 960.
[70] J. Ulstrup, *J. Chem. Phys.* **63** (1975) 4358.
[71] W. Schmickler, *Ber. Bunsenges. Phys. Chem.* **82** (1978) 477.
[72] K. Kobayashi, X. Aikawa, and M. Sukigara, *J. Electroanal. Chem.* **134** (1982) 11.
[73] J. C. Tranchart, L. Hollan, and R. Memming, *J. Electrochem. Soc.* **125** (1978) 1185.
[74] H. Gerischer, *Ber. Bunsenges. Phys. Chem.* **69** (1965) 578.
[75] H. Gerischer and W. Mindt, *Electrochim. Acta* **13** (1968) 1329.
[76] H. Gerischer, *J. Vac. Sci. Technol.* **15** (1978) 1422.
[77] J. J. Kelly, J. E. A. M. van den Meerakker, P. H. L. Notten, and R. P. Tijburg, *Philips Tech. Rev.* **44** (1988) 61.
[78] P. H. L. Notten, *J. Electroanal. Chem.* **224** (1987) 211.
[79] L. Hollan, J. C. Tranchant, and R. Memming, *J. Electrochem. Soc.* **126** (1979) 855.
[80] W. J. Albery and M. L. Hitchman, *Ring-Disc Electrodes*, Clarendon Press, Oxford, 1973.
[81] J. J. Kelly and P. H. L. Notten, *Electrochim. Acta* **29** (1984) 589.
[82] F. Decker, B. Pettinger, and H. Gerischer, *J. Electrochem. Soc.* **130** (1983) 1335.
[83] H. Gerischer and I. Wallem-Mattes, *Z. Phys. Chem. Neue Folge* **64** (1969) 187.
[84] D. Vanmaekelbergh, J. J. Kelly, S. Lingier, and W. P. Gomes, *Ber Bunsenges. Phys. Chem.* **92** (1988) 1068.
[85] D. Vanmaekelbergh, W. P. Gomes, and F. Cardon, *Ber. Bunsenges. Phys. Chem.* **89** (1985) 987.

[86] A. Heller, in *Photoeffects at Semiconductor-Electrolyte Interfaces*, Ed. by A. J. Nozik, ACS Symposium Series, 146, American Chemical Society, Washington, D.C., 1981, p. 57.
[87] T. Solomun, W. Richtering, and H. Gerischer, *Ber. Bunsenges. Phys. Chem.* **91** (1987) 412.
[88] D. Vanmaekelbergh and J. J. Kelly, *J. Electrochem. Soc.* **136** (1989) 108.
[89] K. H. Beckmann and R. Memming, *J. Electrochem. Soc.* **116** (1969) 368.
[90] Y. Nakato, A. Tsumura, and H. Tsubomura, *Bull. Chem. Soc. Jpn.* **55** (1982) 3390.
[91] Y. Nakato, K. Morita, and H. Tsubomura, *J. Phys. Chem.* **90** (1986) 2718.
[92] W. W. Gärtner, *Phys. Rev.* **116** (1959) 84.
[93] P. Lemasson, A. Etchebery, and J. Gautron, *Electrochim. Acta* **27** (1982) 607.
[94] A. Etchebery, M. Etman, B. Fotouhi, J. Gautron, S. Sculfort, and P. Lemasson, *J. Appl. Phys.* **53** (1982) 8867.
[95] V. A. Tyagai, *Russ. J. Phys. Chem.* **38** (1965) 1335.
[96] R. H. Wilson, *J. Appl. Phys.* **48** (1977) 4292.
[97] M. P. Dare-Edwards and A. Hamnett, *J. Electroanal. Chem.* **105** (1979) 283.
[98] J. J. Kelly and R. Memming, *J. Electrochem. Soc.* **129** (1982) 730.
[99] K. Schröder and R. Memming, *Ber. Bunsenges. Phys. Chem.* **89** (1985) 385.
[100] M. A. Butler and D. S. Ginley, *J. Electrochem. Soc.* **127** (1980) 1273.
[101] R. L. Wheeler and N. Hackerman, *J. Phys. Chem.* **92** (1988) 1601.
[102] W. Shockley and W. T. Read, *Phys. Rev.* **87** (1952) 835.
[103] C. Sah, R. Noyce, and W. Shockley, *Proc. IRE* **45** (1957) 1228.
[104] J. Reichman, *Appl. Phys. Lett.* **36** (1980) 574.
[105] J. Reichman and M. A. Russak, in *Photoeffects at Semiconductor-Electrolyte Interfaces*, Ed. by A. J. Nozik, ACS Symposium Series 146, American Chemical Society, Washington, D.C., 1981, p. 359.
[106] H. Gerischer and J. Gobrecht, *Ber. Bunsenges. Phys. Chem.* **82** (1978) 520.
[107] D. Cahen, B. Vainas, and J. M. Vandenberg, *J. Electrochem. Soc.* **128** (1981) 1484.
[108] R. D. Rauh, in *Semiconductor Electrodes*, Ed. by H. O. Finklea, Elsevier, Amsterdam, p. 277.
[109] G. Horowitz, *Appl. Phys. Lett.* **40** (1982) 409.
[110] F. El Guibaly, K. Kolbow, and B. L. Funt, *J. Appl. Phys.* **52** (1981) 3480.
[111] P. Panayotatos and H. C. Card, *Solid-State Electron.* **23** (1980) 41.
[112] D. Haneman and J. F. McCann, *Phys. Rev. B* **25** (1982) 1241.
[113] W. J. Albery, P. N. Bartlett, M. P. Dare-Edwards, and A. Hamnett, *J. Electrochem. Soc.* **128** (1981) 1492.
[114] W. J. Albery and P. N. Bartlett, *J. Electrochem. Soc.* **130** (1983) 1699.
[115] R. L. Van Meirhaeghe, F. Cardon, and W. P. Gomes, *Ber. Bunsenges. Phys. Chem.* **83** (1979) 236.
[116] A. Heller, K. C. Chang, and B. Miller, *J. Electrochem. Soc.* **124** (1977) 697.
[117] H. Gerischer, *J. Electroanal. Chem.* **150** (1983) 553.
[118] R. H. Wilson, *J. Electrochem. Soc.* **127** (1980) 228.
[119] L. M. Peter, J. Li, and R. Peat, *J. Electroanal. Chem.* **165** (1984) 29.
[120] H. Gobrecht and R. Blasser, *Electrochim. Acta* **13** (1968) 1285.
[121] K. Rajeshwar, *J. Electrochem. Soc.* **129** (1982) 1003.
[122] Y. Nakato, S. Tonomura, and H. Tsubomura, *Ber. Bunsenges. Phys. Chem.* **80** (1976) 1289.
[123] M. A. Butler and D. S. Ginley, *J. Electrochem. Soc.* **127** (1980) 1273.
[124] K. Uosaki and H. Kita, *J. Electrochem. Soc.* **128** (1981) 2153.
[125] M. P. Dare-Edwards, A. Hamnett, and J. B. Goodenough, *J. Electroanal. Chem.* **118** (1981) 109.
[126] W. J. Albery and P. N. Bartlett, *J. Electrochem. Soc.* **129** (1982) 2254.
[127] J. Li, R. Peat, and L. M. Peter, *J. Electroanal. Chem.* **165** (1984) 41.

[128] M. A. Butler and D. S. Ginley, *Appl. Phys. Lett.* **36** (1980) 845.
[129] F. Cardon, W. P. Gomes, F. Van den Kerchove, D. Vanmaekelbergh, and F. Van Overmeire, *Faraday Disc. Chem. Soc.* **70** (1980) 153.
[130] W. P. Gomes, F. Van Overmeire, D. Vanmaekelbergh, F. Van den Kerchove, and F. Cardon, in *Photoeffects at Semiconductor-Electrolyte Interfaces*, Ed. by A. J. Nozik, ACS Symposium Series 146, American Chemical Society, Washington, D.C., 1981, p. 119.
[131] D. Vanmaekelbergh, W. P. Gomes, and F. Cardon, *J. Electrochem. Soc.* **129** (1982) 546.
[132] D. Vanmaekelbergh, W. P. Gomes, and F. Cardon, *J. Chem. Soc., Faraday Trans. 1* **79** (1983) 1391.
[133] D. Vanmaekelbergh, W. Rigole, and W. P. Gomes, *J. Chem. Soc., Faraday Trans. 1* **79** (1983) 2820.
[134] S. Lingier, D. Vanmaekelbergh, and W. P. Gomes, *J. Electroanal. Chem.* **228** (1987) 77.
[135] H. Gerischer and H. Lübke, *Ber. Bunsenges. Phys. Chem.* **87** (1983) 123.
[136] K. W. Frese, M. J. Madou, and S. R. Morrison, *J. Phys. Chem.* **84** (1980) 3172.
[137] K. W. Frese, M. J. Madou, and S. R. Morrison, *J. Electrochem. Soc.* **128** (1981) 1527.
[138] M. Green, *J. Chem. Phys.* **31** (1959) 200.
[139] A. J. Bard, A. B. Bocarsley, F.-R. F. Fan, E. G. Walton, and M. S. Wrighton, *J. Am. Chem. Soc.* **102** (1980) 3671.
[140] F.-R. Fan and A. J. Bard, *J. Am. Chem. Soc.* **102** (1980) 3677.
[141] F. Di Quarto and A. J. Bard, *J. Electroanal. Chem.* **127** (1981) 43.
[142] G. Nagasubramanian, B. L. Wheeler, and A. J. Bard, *J. Electrochem. Soc.* **130** (1983) 1680.
[143] M. A. Butler and D. S. Ginley, *J. Electrochem. Soc.* **125** (1978) 228.
[144] M. L. Rosenbluth and N. S. Lewis, *J. Phys. Chem.* **93** (1989) 3735.
[145] G. Horowitz, P. Allongue, and H. Cachet, *J. Electrochem. Soc.* **131** (1984) 2563.
[146] C. D. Jaeger, H. Gerischer, and W. Kautek, *Ber. Bunsenges. Phys. Chem.* **86** (1982) 20.
[147] P. Allongue and H. Cachet, *Solid State Commun.* **55** (1985) 49.
[148] A. C. Brown, Thesis, Oxford, 1988, unpublished.
[149] S. J. Higgins, R. Revill, and A. Hamnett, unpublished work.
[150] P. H. L. Notten, *J. Electroanal. Chem.* **224** (1987) 211.
[151] P. Allongue, H. Cachet, and G. Horowitz, *J. Electrochem. Soc.* **130** (1983) 2352.
[152] K. Tubesing, D. Meissner, R. Memming, and B. Kastening, *J. Electroanal. Chem.* **214** (1986) 685.
[153] Y. Nakato, A. Tsumaro, and H. Tsubomura, *J. Electrochem. Soc.* **127** (1980) 1502.
[154] D. Meissner, R. Memming, and B. Kastening, *J. Phys. Chem.* **92** (1988) 3476.
[155] D. Meissner, I. Lauermann, R. Memming, and B. Kastening, *J. Phys. Chem.* **92** (1988) 3484.
[156] M. P. Dare-Edwards, A. Hamnett, and P. R. Trevellick, *J. Chem. Soc., Faraday Trans. 1* **79** (1983) 2111.
[157] M. P. Dare-Edwards and A. Hamnett, *J. Electroanal. Chem.* **105** (1979) 283.
[158] H. O. Finklea, *J. Electrochem. Soc.* **129** (1982) 2003.
[159] G. Cooper, J. A. Turner, and A. J. Nozik, *J. Electrochem. Soc.* **129** (1982) 1973.
[160] J. Schoonman, K. Vos, and G. Blasse, *J. Electrochem. Soc.* **128** (1981) 1154.
[161] G. Nogami, *J. Electrochem. Soc.* **129** (1982) 2219.
[162] H. Morisaki, T. Baba, and K. Yazawi, *Phys. Rev. B* **21** (1980) 837.
[163] G. Nogami, *J. Electrochem. Soc.* **132** (1985) 76.
[164] G. Nogami, R. Shiratsuchi, H. Nakamura, and H. Taniguchi, *J. Electrochem. Soc.* **132** (1985) 1663.

[165] G. Nogami, *J. Electrochem. Soc.* **133** (1986) 525.
[166] M. Tomkiewicz, *J. Electrochem. Soc.* **126** (1979) 1505.
[167] M. Tomkiewicz, *J. Electrochem. Soc.* **126** (1979) 2220.
[168] W. Siripala and M. Tomkiewicz, *J. Electrochem. Soc.* **128** (1981) 2491.
[169] K. Kobayashi, Y. Aikawa, and M. Sukigara, *J. Appl. Phys.* **54** (1983) 2526.
[170] P. Zoltowski, *J. Electroanal. Chem.* **178** (1984) 11.
[171] R. Memming and G. Schwandt, *Surf. Sci.* **5** (1966) 97.
[172] M. Wolovelsky, J. Levy, Y. Goldstein, A. Many, S. Z. Weisz, and O. Resto, *Surf. Sci.* **171** (1986) 442.
[173] D. R. Turner, *J. Electrochem. Soc.* **105** (1958) 402.
[174] D. R. Turner, in *The Electrochemistry of Semiconductors*, Ed. by P. J. Holmes, Academic Press, London, 1962, p. 155.
[175] R. Memming and G. Schwandt, *Surf. Sci.* **4** (1966) 109.
[176] M. I. J. Beale, N. G. Chew, M. J. Uren, A. G. Cullis, and J. D. Benjamin, *Appl. Phys. Lett.* **46** (1985) 86.
[177] M. I. J. Beale, J. D. Benjamin, M. J. Uren, N. G. Chew, and A. G. Collis, *J. Cryst. Growth* **73** (1985) 622.
[178] R. Herino, G. Bomchil, K. Barla, G. Bertrand, and J. L. Ginoux, *J. Electrochem. Soc.* **134** (1987) 1994.
[179] J. D. Lecuyer, Ph.D. thesis, University of Birmingham, U.K., 1990.
[180] K. Barla, G. Bomchil, R. Herino, J. G. Pfister, and J. Baruchal, *J. Cryst. Growth* **68** (1984) 721.
[181] M. Matsumura and S. R. Morrison, *J. Electroanal. Chem.* **147** (1983) 157.
[182] J. P. G. Farr and I. Sturland, unpublished work.
[183] K. H. Beckmann, *Surf. Sci.* **3** (1965) 314.
[184] L. G. Earwaker, J. P. G. Farr, P. E. Grzeszczyk, I. M. Sturland, J. M. Keen, *Nucl. Instrum. Methods B* **9** (1985) 317.
[185] C. M. Braun, A. Fujishima, and K. Honda, *Chem. Lett.* **1985** 1763.
[186] C. M. Braun, A. Fujishima, and K. Honda, *J. Electroanal. Chem.* **205** (1986) 291.
[187] J. F. McCann and S. P. S. Badwal, *J. Electrochem. Soc.* **129** (1982) 551.
[188] B. Wolf and W. Lorenz, *Electrochim. Acta* **28** (1983) 699.
[189] K. Kobayashi, M. Takata, S. Okamoto, Y. Aikawa, and M. Sukigara, *Chem. Phys. Lett.* **96** (1983) 366.
[190] P. Allongue and H. Cachet, *J. Electroanal. Chem.* **119** (1981) 371.
[191] P. Allongue and H. Cachet, *J. Electrochem. Soc.* **132** (1985) 45.
[192] D. Vanmaekelbergh, W. P. Gomes, and F. Cardon, *Ber. Bunsenges. Phys. Chem.* **89** (1985) 994.
[193] K. Chandrasekaran, M. Weichold, F. Gutmann, and J. O'M. Bockris, *Electrochim. Acta* **30** (1985) 961.
[194] R. F. Pierret and C. T. Sah, *Solid-State Electron.* **13** (1970) 269.
[195] R. F. Pierret and C. T. Sah, *Solid-State Electron.* **13** (1970) 289.
[196] J. J. Kelly and P. H. L. Notten, *J. Electrochem. Soc.* **130** (1983) 2452.
[197] J. E. A. M. van den Meerakker, J. J. Kelly, and P. H. L. Notten, *J. Electrochem. Soc.* **132** (1985) 638.
[198] D. Vanmaekelbergh and F. Cardon, *J. Phys. D* **19** (1986) 643.
[199] D. Vanmaekelbergh, W. P. Gomes, and F. Cardon, *J. Electrochem. Soc.* **134** (1987) 891.
[200] D. Vanmaekelbergh, W. P. Gomes, and F. Cardon, *Ber. Bunsenges. Phys. Chem.* **90** (1986) 431.
[201] W. Lorenz, H. Handschuh, C. Aegerter, and H. Herrnberger, *J. Electroanal. Chem.* **184** (1985) 61.
[202] R. Sourisseau and W. Lorenz, *J. Electroanal. Chem.* **274** (1989) 123.
[203] W. Lorenz and M. Handschuh, *J. Electroanal. Chem.* **178** (1984) 197.

[204] M. Handschuh, *J. Electroanal. Chem.* **221** (1987) 23.
[205] P. R. Trevellick, M. P. Dare-Edwards, and A. Hamnett, *J. Chem. Soc., Faraday Trans. 1* **79** (1983) 2027.
[206] J. Li and L. M. Peter, *J. Electroanal. Chem.* **193** (1985) 27.
[207] J. Li and L. M. Peter, *J. Electroanal. Chem.* **199** (1986) 1.
[208] R. Peat, L. M. Peter, and L. M. Abrantes, *Ber. Bunsenges. Phys. Chem.* **91** (1987) 381.
[209] R. Peat and L. M. Peter, *J. Electroanal. Chem.* **209** (1986) 307.
[210] D. Meissner and R. Memming, Photocatalytic Production of Energy-Rich Compounds, Ed. by G. Grassi and D. O. Hall, CEC Report: EUR 11371, 1988, p. 138.
[211] H. J. Lewerenz, J. Stumper, and L. M. Peter, *Phys. Rev. Lett.* **61** (1988) 1989.
[212] R. Peat, A. Riley, D. E. Williams, and L. M. Peter, *J. Electrochem. Soc.* **136** (1989) 3352.
[213] E. Kamieniecki, *J. Vac. Sci. Technol.* **20** (1982) 811.
[214] E. Kamieniecki, *J. Appl. Phys.* **54** (1983) 6481.
[215] M. A. Butler and D. S. Ginley, *J. Electrochem. Soc.* **128** (1981) 712.
[216] M. A. Butler, G. W. Arnold, and D. K. Brice, *J. Electrochem. Soc.* **129** (1982) 2735.
[217] M. A. Butler and D. S. Ginley, *Appl. Phys. Lett.* **42** (1983) 582.
[218] W. Siripala and M. Tomkiewicz, *J. Electrochem. Soc.* **129** (1982) 1240.
[219] H. Morisaki, M. Hariya, and K. Yazawa, *Appl. Phys. Lett.* **30** (1977) 7.
[220] A. Hamnett, M. P. Dare-Edwards, R. D. Wright, K. R. Seddon, and J. B. Goodenough, *J. Phys. Chem.* **83** (1979) 3280.
[221] B. Parkinson, F. Decker, J. F. Julião, M. Abramovitch, and H. C. Chagas, *Electrochim. Acta* **25** (1980) 521.
[222] P. Salvador and C. Gutierrez, *Chem. Phys. Lett.* **86** (1982) 131.
[223] C. Gutierrez and P. Salvador, *J. Electroanal. Chem.* **138** (1982) 457.
[224] D. Talfalla and P. Salvador, *Ber. Bunsenges. Phys. Chem.* **91** (1987) 479.
[225] D. Talfalla and P. Salvador, *J. Electroanal. Chem.* **237** (1987) 225.
[226] P. Salvador and C. Gutierrez, *J. Electrochem. Soc.* **131** (1984) 326.
[227] G. Nogami, K. Ohno, and R. Shiratsuchi, *J. Electrochem. Soc.* **136** (1989) 3724.
[228] D. Laser and S. Gottesfeld, *J. Electrochem. Soc.* **126** (1979) 475.
[229] M. A. Butler, M. Abramovich, F. Decker, and J. F. Julião, *J. Electrochem. Soc.* **128** (1981) 200.
[230] C. Liu, Y. Chen, and W. Li, *Surf. Sci.* **163** (1985) 383.
[231] P. Salvador, *Surf. Sci.* **192** (1987) 36.
[232] F. D. Hughes, *Acta Electronica* **15** (1972) 43.
[233] K. Sakai and T. Ikoma, *Appl. Phys. Lett.* **5** (1974) 165.
[234] D. Bois, *J. Phys.* **35** (C-3) (1974) 241.
[235] D. M. Tench and H. Gerischer, *J. Electrochem. Soc.* **124** (1977) 1612.
[236] R. Haak, C. Ogden, and D. Tench, *J. Electrochem. Soc.* **129** (1982) 891.
[237] R. Haak and D. M. Tench, *J. Electrochem. Soc.* **131** (1984) 275.
[238] S. Nakabayashi, A. Kira, and M. Ipponmatsu, *J. Phys. Chem.* **93** (1989) 5543.
[239] C. E. Goodman, B. W. Wessels, and P. G. P. Ang, *Appl. Phys. Lett.* **45** (1986) 442.
[240] R. Haak and D. Tench, *J. Electrochem. Soc.* **131** (1984) 1442.
[241] R. Haak, D. Tench, and M. Russak, *J. Electrochem. Soc.* **131** (1984) 2709.
[242] M. Cocivera, W. M. Sears, and S. R. Morrison, *J. Electroanal. Chem.* **216** (1987) 41.
[243] P. Allongue and H. Cachet, *Ber. Bunsenges. Phys. Chem.* **91** (1987) 386.
[244] P. Allongue and H. Cachet, *Ber Bunsenges. Phys. Chem.* **92** (1988) 566.
[245] P. Allongue, *Ber. Bunsenges. Phys. Chem.* **92** (1988) 895.
[246] P. Allongue and H. Cachet, *Electrochim. Acta* **33** (1988) 79.
[247] P. Allongue, E. Souteyrand, and L. Allemand, *J. Electrochem. Soc.* **136** (1989) 1027.
[248] H. B. Bebb and E. W. Williams, in *Semiconductors and Semimetals*, Vol. 8, Ed. by R. K. Willardson and A. C. Beer, Academic Press, New York, 1972, p. 181.

[249] E. W. Williams and H. B. Bebb, in *Semiconductors and Semimetals*, Vol. 8, Ed. by R. K. Willardson and A. C. Beer, Academic Press, New York, 1972, p. 321.
[250] K. H. Beckmann and R. Memming, *J. Electrochem. Soc.* **116** (1969) 368.
[251] R. Pettinger, H. R. Schöppel, and H. Gerischer, *Ber. Bunsenges. Phys. Chem.* **80** (1976) 849.
[252] H. H. Streckert, J. Tong, and A. B. Ellis, *J. Am. Chem. Soc.* **104** (1982) 581.
[253] H. H. Streckert, J. Tong, M. K. Carpenter, and A. B. Ellis, *J. Electrochem. Soc.* **129** (1982) 772.
[254] A. Etcheberry, J. Gautron, and J. L. Sculfort, *Appl. Phys. Lett.* **46** (1985) 744.
[255] I. J. Ferrer and P. Salvador, *J. Appl. Phys.* **66** (1989) 2568.
[256] R. N. Noufi, P. A. Kohl, S. N. Frank, and A. J. Bard, *J. Electrochem. Soc.* **125** (1977) 246.
[257] Y. Nakato, A. Tamura, and H. Tsubomura, *Chem. Phys. Lett.* **85** (1982) 387.
[258] Y. Nakato, A. Tamura, and H. Tsubomura, *J. Phys. Chem.* **87** (1983) 2402.
[259] F. Decker, M. Abramovitch, and P. Motisuke, *J. Electrochem. Soc.* **131** (1984) 1173.
[260] R. N. Noufi, P. A. Kohl, S. N. Frank, and A. J. Bard, *J. Electrochem. Soc.* **125** (1977) 246.
[261] Y. Nakato, A. Tamura, and H. Tsubomura, *J. Phys. Chem.* **87** (1983) 2402.
[262] P. Salvador and C. Gutiérrez, *J. Phys. Chem.* **88** (1984) 3696.
[263] W. S. Hobson and A. B. Ellis, *J. Appl. Phys.* **54** (1983) 5956.
[264] A. A. Burk, P. B. Johnson, W. B. Hobson, and A. B. Ellis, *J. Appl. Phys.* **59** (1986) 1621.
[265] P. B. Johnson, C. S. McMillan, A. B. Ellis, and W. S. Hobson, *J. Appl. Phys.* **62** (1987) 4903.
[266] S. Kaneko, K. Uosaki, and H. Kita, *Chem. Lett.* **1986** 1951.
[267] S. Kaneko, K. Uosaki, and H. Kita, *J. Phys. Chem.* **90** (1986) 6654.
[268] K. Uosaki and H. Kita, *Ber. Bunsenges. Phys. Chem.* **91** (1987) 447.
[269] K. Uosaki, Y. Shigematsu, and H. Kita, *Chem. Lett.* **1988** 1815.
[270] K. Uosaki, Y. Shigematsu, S. Kaneko, and H. Kita, *J. Phys. Chem.* **93** (1989) 6521.
[271] D. B. Wittry and D. F. Kyser, *J. Appl. Phys.* **38** (1967) 375.
[272] R. E. Hollingworth and R. J. Sites, *J. Appl. Phys.* **57** (1982) 5357.
[273] K. Mettler, *Appl. Phys.* **12** (1977) 75.
[274] A. Hamnett, R. L. Lane, S. Denison, and P. R. Trevellick, in *Comprehensive Chemical Kinetics*, Vol. 29, Ed. by R. G. Compton and A. Hamnett, Elsevier, Amsterdam, 1989, p. 385.
[275] R. A. Batchelor, A. C. Brown, and A. Hamnett, *Phys. Rev. B* **41** (1990) 1401.
[276] R. A. Batchelor and A. Hamnett, in *Int. Conf. Modulation Spectr.*, F. H. Pollak, M. Cardona, and D. E. Aspnes, Eds., Bellingham, Washington, *Proc. SPIE* 1286, p. 175 (1990).
[277] K. L. Shaklee, F.H Pollak, and M. Cardona, *Phys. Rev. Lett.* **15** (1965) 883.
[278] N. Bottka, D. K. Gaskill, R. S. Sillmon, R. Henry, and R. Glosser, *J. Electron. Mater.* **17** (1988) 161.
[279] D. E. Aspnes, *Phys. Rev.* **147** (1966) 554.
[280] D. E. Aspnes, *Phys. Rev.* **153** (1966) 972.
[281] M. Abramovitch and I. A. Stegun, Eds., *Handbook of Mathematical Functions*, Dover Publications, New York, 1970.
[282] D. E. Aspnes, *Surf. Sci.* **37** (1973) 418.
[283] N. Neff, P. Lange, M. L. Fearheiley, and K. J. Bachmann, *Appl. Phys. Lett.* **47** (1985) 1069.
[284] D. Huang, G. Ji, U. K. Reddy, H. Morkoç, F. Xiong, and T. A. Tombrello, *J. Appl. Phys.* **63** (1988) 5447.
[285] J. M. Rodriguez, G. Armelles, and P. Salvador, *J. Appl. Phys.* **66** (1989) 3929.

[286] J. M. Wrobel, L. C. Bassett, J. L. Aubel, S. Sundaram, J. L. Davis, and J. Comas, *J. Vac. Sci. Technol., A* **5** (1987) 1464.
[287] R. P. Silberstein, F. H. Pollak, J. K. Lyden, and M. Tomkiewicz, *Phys. Rev. B* **24** (1981) 7397.
[288] M. Tomkiewicz, W. Siripala, and R. Tenne, *J. Electrochem. Soc.* **131** (1984) 736.
[289] I. J. Ferrer and P. Salvador, *Ber. Bunsenges. Phys. Chem.* **91** (1987) 374.
[290] P. Lemasson and C. N. Van Huong, *J. Electrochem. Soc.* **135** (1988) 2080.
[291] I. J. Ferrer, H. Muraki, and P. Salvador, *J. Phys. Chem.* **90** (1986) 2805.
[292] M. C. A. Fantini, W. M. Shew, M. Tomkiewicz, and J. P. Gambino, *J. Appl. Phys.* **65** (1989) 4884.
[293] R. P. Silberstein, J. K. Lyden, M. Tomkiewicz, and F. H. Pollak, *J. Vac. Sci. Technol.* **19** (1981) 406.
[294] P. M. Raccah, J. W. Garland, Z. Zhang, U. Lee, D. Z. Xue, L. L. Abels, S. Ugar, and W. Wilinsky, *Phys. Rev. Lett.* **54** (1984) 1958.
[295] P. M. Raccah, J. W. Garland, Z. Zhang, L. L. Abels, S. Ugar, S. Mioc, and M. Brown, *Phys. Rev. Lett.* **55** (1985) 1323.
[296] C. B. Duke and M. E. Alferieff, *Phys. Rev.* **145** (1966) 583.
[297] H. I. Ralph, *J. Phys. C.* **1** (1968) 378.
[298] J. D. Dow and D. Redfield, *Phys. Rev. B* **1** (1970) 3358.
[299] D. F. Blossey, *Phys. Rev. B* **2** (1970) 3976.
[300] D. F. Blossey, *Phys. Rev. B* **3** (1971) 1382.
[301] G. M. Wysin, D. L. Smith, and A. Redondo, *Phys. Rev. B* **38** (1988) 12514.
[302] D. E. Aspnes and A. Frova, *Solid State Commun.* **7** (1969) 155.
[303] W. K. Paik, in *MTP Review of Science, Series One*, Vol. 6, Ed. by J. O'M. Bockris, Butterworths, London, 1973, p. 239.
[304] R. A. Batchelor, A. Hamnett, R. Peat, and L. M. Peter, *J. Appl. Phys.* **70** (1991) 266.
[305] L. M. Abrantes, R. Peat, L. M. Peter, and A. Hamnett, *Ber. Bunsenges. Phys. Chem.* **91** (1987) 369.
[306] S. S. M. Lu, F. H. Pollak, and P. M. Raccah, *Electrocatalysis on Non-metallic Surfaces*, NBS Spec. Publ. 455, Ed. by A. D. Franklin, U.S. Government Printing Office, Washington, D.C., 1976, p. 125.
[307] W. Siripala and M. Tomkiewicz, *Phys. Rev. Lett.* **50** (1983) 443.
[308] E. W. Williams and V. Rehn, *Phys. Rev.* **172** (1968) 798.
[309] E. W. Williams, *Solid State Commun.* **7** (1969) 511.

4

Electrodeposition of Nickel–Iron Alloys

Stojan S. Djokić

Department of Chemistry, University of Ottawa, Ottawa, Ontario K1N 9B4, Canada; present address: *Sheritt Gordon Limited, Fort Saskatchewan, Alberta T8L 3W4, Canada*

Miodrag D. Maksimović

Faculty of Technology and Metallurgy, University of Belgrade, Belgrade, Yugoslavia

I. INTRODUCTION

Electrodeposition of thin Ni–Fe alloy films is very important because of their diverse applications in many different industrial fields. Applications include high-speed memory units for computers and magnetic shielding, as well as electronic wire condensers, as structures for bubble memory systems in the microelectronics industry, and in rocketry and space technology.[1-10] Further, Ni–Fe coatings can be used as decorative and protective materials.

Ni–Fe alloy electrodeposition has been utilized to produce high-speed tools, surgical instruments, and printed circuits, which are of major commercial importance. Such applications require uniform and invariant physicochemical properties of the alloy deposits.

Modern Aspects of Electrochemistry, Number 22, edited by John O'M. Bockris *et al.* Plenum Press, New York 1992.

The properties of Ni–Fe alloys are affected dramatically by composition, roughness of the substrate, crystal size, and structure and internal stress in the deposit. In turn, these parameters very much depend on the current density, agitation, and other plating variables.

The aim of this chapter is to present information on the process of electrodeposition of Ni–Fe alloys under dc and periodically changing current conditions and on the properties of these materials.

II. PHENOMENOLOGY OF Ni–Fe ALLOY ELECTRODEPOSITION

Simultaneous electrodeposition of Ni and Fe is a complex process. The thermodynamic as well as kinetic aspects of this process are very complicated and not yet well understood. The standard electrode potentials are −0.25 V for nickel and −0.44 V for iron. The corresponding values of the overpotentials depend upon various parameters, including, of course, the current density. In a 0.5 *M* sulfate solution at pH 3.0–3.5 at 25°C and at a current density of 1.5 A dm^{-2}, the values of the overpotential are 0.613 V for nickel and 0.161 V for iron.[11]

In the electrodeposition of alloys of the iron group metals from simple- or complex-salt baths, the less noble metal is preferentially reduced, and consequently its relative content in the deposit is higher than that in the bath. Such a phenomenon is known as *anomalous codeposition.*[12] Anomalous codeposition of Ni–Fe alloys is probably the main reason leading to certain disagreements among the results of several investigators, as reported in the literature.

The phenomenon of anomalous codeposition of Ni–Fe alloys has been extensively studied. However, it is important to note that the mechanism of this phenomenon is not yet completely understood. Glasstone and Symes assumed that in the process of codeposition of Ni and Fe, Fe suppresses the transition of Ni from a so-called metastable to a stable state.[13,14] However, further investigations have not provided evidence for the existence of a metastable state of nickel, and also it is not clear how iron disturbs the electron transfer involving nickel.[15]

A very important factor that affects not only the composition but also the properties of the electrodeposited films is simultaneous hydrogen evolution during electrodeposition of Ni–Fe alloys.

Most of the hydrogen atoms that are produced during electrodeposition of the metals form molecular hydrogen which, as bubbles, is removed from the cathode surface. Another small fraction of hydrogen becomes adsorbed on the cathode surface and, with the growth of the metal, becomes incorporated in the crystal lattice of the electrodeposited metals. Some H atoms can become directly sorbed into the lattice of the metals, when the latter are transition metals, especially of group VIII. The quantity of hydrogen occluded in the deposit of ferromagnetic metals is approximately 0.45 at.%.[6]

Frumkin[6] analyzed possible ways in which hydrogen can be incorporated in electrodeposits in relation to the nature of the metal and the conditions of electrodeposition. During electrolysis, hydrogen can be incorporated into deposits in the form of solid solutions with metals or as hydride phases. With thermal treatment of these electrodeposits, the incorporated hydrogen can be removed, which leads to the deformation of the crystal lattice and ultimately causes changes in the properties of the metals electrodeposited.

In the electrodeposition of Ni–Fe alloys, it was shown that the current efficiency for simultaneous evolution of hydrogen depends on the applied current density.[7] In the current density range 0.025–0.3 A dm^{-2}, the current efficiency for hydrogen evolution passes through a minimum at ca. 0.1 A dm^{-2}. At the same current density, the current efficiencies for nickel or iron deposition (in other words, the current efficiency for Ni–Fe alloy electrodeposition) pass through a corresponding maximum (Fig. 1).

In terms of electrocatalytic effects in the H_2 evolution reaction, the changing composition of the surface of electrodeposited alloys, for example, Ni–Fe, *during* their deposition can lead to a *time-dependent* current efficiency for metal deposition relative to the H_2 evolution rate.

Vagramyan and Fatueva[16] have investigated Ni–Fe alloy electrodeposition from simple sulfate baths. They assumed that during simultaneous discharge of metal ions a mutual influence arises due to the changes in the state of the ions in solution, changes of concentration of the ions discharged, and changes in the nature of

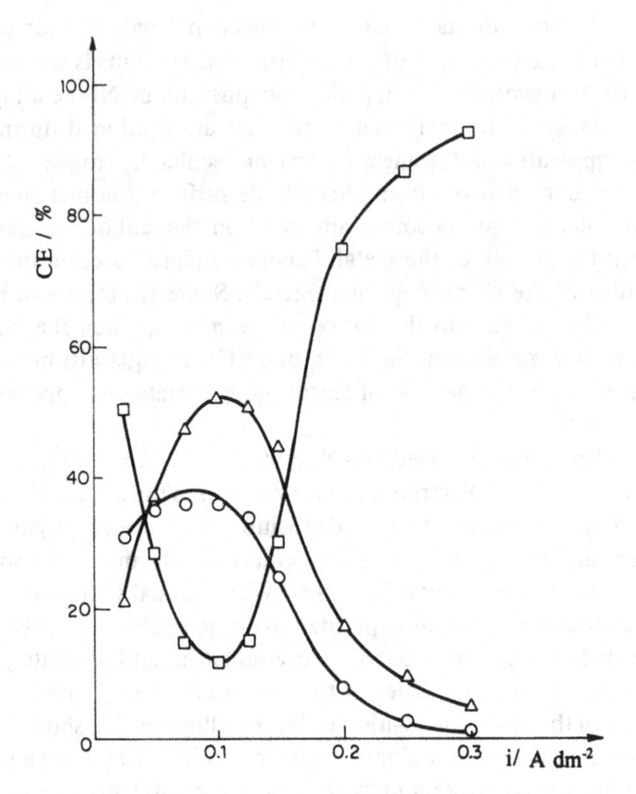

Figure 1. The dependence of current efficiencies of Fe (○) and Ni (△) deposition and H_2 evolution (□) on the total current density in the electrodeposition of Ni–Fe alloy under dc conditions.[7]

the substrate and the state of the electrode surface. In the case of electrodeposition leading to codeposition of Fe with Ni, the reduction rate of Fe^{2+} becomes three times higher, as a result of these changes, and the discharge rate of Ni^{2+} ions is approximately eight times lower than in the reduction of these ions separately. Based on the results obtained, Vagramyan and Fatueva considered that anomalous codeposition can be eliminated when electrolysis is carried out at a low cathodic potential and high temperature. Under these conditions, the Ni–Fe alloy obtained contained more nickel than iron.

Working with a dropping-mercury electrode, Dahms[17] clearly demonstrated the absence of any influence due to electrode material on the anomalous codeposition effect. In this work, it was shown that an increase of pH at the cathode surface, with consequent hydrolysis effects, causes anomalies in the deposition of iron group metals which are analogous to those observed in the plating of Ni–Fe alloys. On the basis of the assumption that the diffusion-layer thickness, δ, is identical for all species (molecular or ionic) in solution, the following equation has been obtained[18]:

$$\frac{i_H}{F}\delta = D(H^+)[C(H^+) - C_s(H^+)] + D(OH^-)K_w\frac{C(H^+) - C_s(H^+)}{C(H^+)C_s(H^+)} + nD(H_nA)\frac{C^n(H^+)}{[K_d + C^n(H^+)]} \quad (1)$$

where i_H is the current density, F is the Faraday constant, D is the diffusion coefficient, C is the concentration in solution, C_s is the surface concentration, K_w is the water dissociation constant, n is the number of electrons, and K_d is the dissociation constant of the species H_nA.

Because of the increase of electrode potential, thus leading to an increase of pH at the cathode surface (Fig. 2), Dahms and Croll[18] considered that, under these conditions, the ferrous hydroxide would be formed, with the subsequent suppression of Ni^{2+} discharge. Based on these results, they proposed the following type of mechanism of Ni–Fe alloy electrodeposition:

$$2H_2O + 2e^- = H_2 + 2OH^- \quad (2)$$

$$Fe^{2+} + 2OH^- = Fe(OH)_{2,ads} \qquad (\theta_{Fe(OH)_2} \rightarrow 1) \quad (3)$$

At θ:

$$Fe(OH)_{2,ads} + Fe(OH)_n^{(2-n)+} + 2e^- = Fe + Fe(OH)_{2,ads} + nOH^- \quad (4)$$

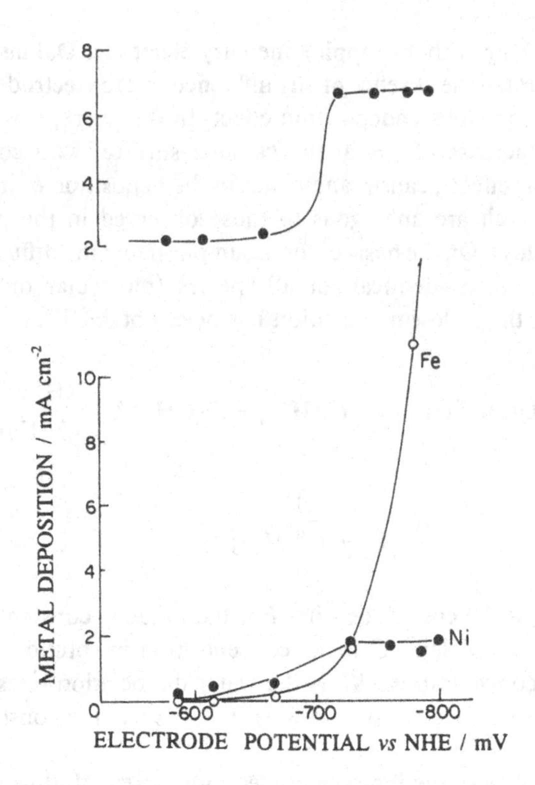

Figure 2. Current-voltage curves of iron and nickel deposition from 0.5 *M* $NiSO_4$, 0.5 *M* $FeSO_4$, 0.01 *M* H_2SO_4 solution (*bottom*) and pH at electrode surface calculated from partial current of hydrogen evolution (*top*).[18]

At $1 - \theta$:

$$Ni(OH)_n^{(2-n)+} + 2e^- = Ni + nOH^- \quad (5)$$

where θ is the surface coverage with ferrous hydroxide.

Giuliani and Lazzari[19] assumed that the percentage of Fe in the deposit will be low at the beginning of electrolysis, until a sufficient concentration of OH^- is generated at the electrode surface so that hemihydroxide compounds can be formed. At this point, Ni is discharged at the limiting current as the percentage of Fe increases. As soon as the cathode polarization is sufficiently high to discharge $NiOH^+$, the iron content in the deposit decreases.

Beltowska-Lehman and Riesenkampf[20] determined total and partial polarization curves for the deposition of Ni and Fe separately as well as for their codeposition from acidic sulfate baths, by using a system with a rotating disk electrode. It was found that, for the reduction of both Ni^{2+} and Fe^{2+}, the reaction kinetics are mixed, with the activation kinetics being dominant. In both cases, a significant role of pH can be attributed to the alkalization of the cathode boundary layer. The codeposition of thin permalloy films showed that the mechanism of the electrode reactions involved was consistent with the model suggested by Dahms and Croll.[18]

Nakamura and Hayashi[8] investigated the temperature and the pH effects on the anomalous codeposition of Ni–Fe alloys. They showed that the polarization curves for Ni–Fe alloy deposition shifted toward a more negative value with an increase in Fe^{2+}-ion concentration in the electrolyte. The effect of pH on the polarization curves in the range of current density from 1 to 8 A dm^{-2} was insignificant. A decrease in electrolyte temperature from 70 to 15°C shifted the polarization curves for Ni–Fe alloy deposition toward more negative values by 0.30 V in the current density range of 0.1 to 6 A dm^{-2}. Ni–Fe alloy deposition seemed to be hindered by a decrease in temperature.

Lieder and Biallozor[21] investigated the codeposition of Ni and Fe from chloride solutions containing organic additives such as D-mannitol, citric acid, or ascorbic acid. They showed that the ratio of Fe to Ni content in the deposit was always higher than the ratio of their ion concentrations in the solutions. Based on these results, they suggested that Ni–Fe alloy codeposition might proceed in the following way:

1. The first cathodic process is discharge of Ni^{2+}_{aq}.
2. When the cathode surface is covered by a very thin film of Ni (perhaps a monolayer), the H_2O molecules are chemisorbed on it with the creation of $Ni(OH)^{+}_{ads}$ species.
3. As a result of competition between Ni^{2+} and Fe^{2+} ions to occupy active sites, the preferential deposition of Fe and suppression of Ni^{2+} discharge are observed.

The explanation of the mechanism of Ni–Fe codeposition suggested by Lieder and Biallozor[21] did not confirm the theory of Dahms and Croll.[18] This explanation is probably in accordance with the model of Ni^{2+} discharge given in Biallozor and Lieder's publication,[22] but there is not strong evidence for the proposed mechanism of Fe^{2+} discharge.

A mathematical model for anomalous codeposition of Ni-Fe alloy on a rotating disk electrode (RDE) has been developed by Hessami and Tobias.[23] They assumed that the kinetic parameters of the alloy deposition were the same as those for the deposition of the individual metals (for both metal and metal hydroxide ions) and that the discharge of metal hydroxide ions takes place only at a fraction of the surface sites, proportional to their interfacial concentration. The authors compared the predictions of this mathematical model with the experimental behavior reported by Andricacos *et al.*,[24] and the results were in qualitative agreement. The following important characteristics of Ni-Fe anomalous deposition at an RDE system are successfully predicted by this model:

1. The ratio of Fe to Ni is much higher in the alloy than in the electrolyte.
2. The dependence of Fe content in the alloy on applied potential passes through a maximum.
3. The presence of Fe inhibits the discharge of Ni.
4. Current efficiency increases with increase of applied potential.

In spite of the fact that anomalous deposition of Ni-Fe alloys has been studied by several investigators, this process is not yet fully explained. Obviously, extensive experimental studies are needed.

III. THE EFFECT OF THE ELECTRODEPOSITION CONDITIONS ON THE COMPOSITION OF Ni–Fe ALLOYS

1. Electrolytes for Ni–Fe Alloy Electrodeposition

In the electrodeposition of ferromagnetic metals and their alloys, various electrolytes can be used. However, of practical importance for Ni-Fe electrodeposition are sulfate, chloride, and sulfamate and also mixed solutions (sulfate-chloride, citrate-ammonium, fluoride-chloride, etc.). The role of anions is sometimes decisive for the production of desired properties of Ni-Fe alloys.

(i) *Electrodeposition from Simple Salt Solutions*

Metals such as nickel and iron can be electrodeposited from aqueous solutions of their simple salts. In such solutions, the Ni^{2+} ions are sufficiently stable, but Fe^{2+} ions in the presence of oxygen can, of course, be oxidized to Fe^{3+} ions.[25] It is necessary to take this into account in the preparation of sulfate and/or chloride solutions. Thus, first it is necessary to prepare a solution of Ni^{2+} and all the other components without Fe^{2+}. Just before electrolysis, the Fe^{2+}-containing compounds should be added.

Ni-Fe alloys with the same composition can be obtained from solutions with different metal concentration ratios. For example, in the electrodeposition of permalloy films, electrolytes having a $[Ni^{2+}]/[Fe^{2+}]$ ratio of 50[26] and of about 4.7[27–29] were used. Although there is some disagreement among the published results, in most cases the increase of total metal-ion concentration in solution at a constant $[Ni^{2+}]/[Fe^{2+}]$ ratio leads to an increase in the iron content of the deposit.[6,30,31] On the other hand, there is difficulty in reaching a general conclusion about the effect of the $[Ni^{2+}]/[Fe^{2+}]$ ratio on the electrodeposition of Ni-Fe alloys. This is a very complicated process depending on other parameters such as the presence of various additives, temperature, pH, and current density. Under most conditions, Fe deposits preferentially to Ni, although it is less noble, thus making the codeposition anomalous. This type of deposition mainly depends on temperature, current density, and probably other electrolysis parameters.

First of all, in practice, sulfate solutions are used; the advantage of using these solutions is their stability: SO_4^{2-} ions cannot be reduced at the cathode and are not oxidized at the anode. Sulfate solutions have a good conductivity, and they are not very aggressive or toxic in comparison to chloride and sulfamate solutions. The deposits obtained from sulfate solutions, however, contain strongly bonded H, more than those obtained from chloride solutions.[32]

In some cases, chloride solutions have been used in the electrodeposition of Ni-Fe alloys.[31,33] Klok and White[33] used chloride solutions of Ni^{2+} and Fe^{2+} and showed that the $[Ni^{2+}]/[Fe^{2+}]$ ratio in the solution is the same as that in the alloy deposit, amounting to 3:97. On varying the $[Ni^{2+}]/[Fe^{2+}]$ ratio in chloride solutions, within the range 0.82-1.22, it was shown that the Ni-Fe alloy, deposited at a current density of 2 A dm^{-2}, had a constant composition.

(ii) Electrodeposition from Other Solutions

Several investigators have studied the electrodeposition of Ni-Fe alloys from baths other than sulfate and/or chloride solutions. These baths contained one of more agents complexing Ni^{2+} and/or Fe^{2+} ions. However, the role of different agents that complex Fe^{2+} and Ni^{2+} ions in the electrodeposition process has not yet been critically evaluated.

It is known that Ni can be deposited from cyanide complexes only with a very low cathodic current efficiency, but iron cannot be deposited from cyanide complexes at all. In the earlier work of Stout and Carol,[34] it was shown that Ni-Fe alloys can be deposited from a solution containing tartrate and no free cyanide. They produced Ni-Fe alloys with a current efficiency of 60%. In their opinion, the Fe is deposited from a tartrate complex. Unfortunately, other data for electrodeposition of Ni-Fe alloys from cyanide solutions are not available in the literature.

Andryushenko *et al.*[35] investigated the electrodeposition of Ni-Fe alloys from chloride solutions of Ni^{2+} and Fe^{2+} which contained sufficient potassium pyrophosphate to be considered as pyrophosphate baths. Based on the polarization curves obtained, they concluded that the Ni-Fe alloy film is formed by simultaneous discharge of complex ions on the cathode. Electrodeposition of Ni-Fe alloys from pyrophosphate solutions was reported by other workers.[36-38] Srimathi and Mayanna[39] electrodeposited Ni-Fe alloy from alkaline sulfate solutions containing pyrophosphate and ethylendiamine. With increasing pyrophosphate concentration, the Ni content in the alloy increased as well. On the other hand, with an increase of the ethylendiamine concentration, the Ni content in the alloy significantly decreased. This behavior was explained by the complexing of metal ions with a pyrophosphate and/or ethylendiamine. It was shown that, in some cases, phosphorus was incorporated in the Ni-Fe alloy deposit.[40-48] Usually, phosphorus is incorporated into a deposit from sodium hypophosphite. The content of phosphorus in the electrodeposits obtained is about 1–3%. Similarly, it was found that As can be incorporated (1–15% As) in the alloy deposit (6–20% Fe) if the plating solution contains sodium arsenate.[49] In some cases, Ni–Fe alloys are obtained which contain Se, Te, or Sb.[50] Ni–Fe alloy films obtained from solutions with thiourea as additive contained sulfur,[51] the amount of which increased linearly from 1 to 6%, with increasing thiourea concentration in the solution.

It is important to note the work on electrodeposition of Ni–Fe alloys from solfamate solutions. Several authors have used such solutions to obtain deposits of Ni–Fe alloy having various compositions and properties.[52–58] The production of thin, cylindrical Ni–Fe alloy magnetic films from sulfamate solutions was originally proposed by Long.[56] Such solutions make possible the application of very high cathodic current densities for Ni–Fe alloy electrodeposition, without decreasing the current efficiency.

The effects of sulfamic acid and sulfosalicylic acid on the electrodeposition of Ni–Fe alloys were reported by Srimathi and Mayanna.[58] Some of the observed spectral data indicate the complexation of both Fe^{2+} and Ni^{2+} ions with sulfamic and sulfosalicylic ions. Experiments showed a slight decrease in the Ni content of the alloy with increase in the concentration of either sulfamic or sulfosalicylic acid.

Barauskas and Guttensohn[59] found that basic ammoniacal citrate solution is more acceptable than acid sulfate solution in electrodeposition of Ni–Fe alloys for use in plated wire memory production.

Djokić *et al.*[60] investigated the effects of sodium potassium tartrate on the composition of Ni–Fe alloys. They found that with an increase in the tartrate concentration, the Fe content in the alloy increases as well. The sodium potassium tartrate affects the alloy composition when its concentration is lower than the total concentration of Ni^{2+} and Fe^{2+} in the solution. In the tartrate concentration range higher than total concentration of Ni^{2+} and Fe^{2+} ions, the composition of the alloy remains practically constant.

The reasons for the use of various additives in the electrodeposition of Ni–Fe alloy films are very different. Additives can be used to increase the conductivity of the bath, to complex ions and buffer the solutions, as well as to decrease deposit stresses and enable production of smooth deposits of good quality. In general, chloride and sulfate salts of sodium, potassium, calcium, magnesium, etc., are used as additives to increase the bath conductivity. In some cases, NH_4Cl was used for this purpose.[6] Although boric acid does not eliminate the anomalous codeposition effect in Ni–Fe alloy plating,[61] it is used not only in sulfate and chloride solutions, but also in sulfamate solutions. Boric acid is used extensively in Ni–Fe alloy plating systems as buffer to decrease the rise of pH at the surface. However, according to Horkans,[2] boric acid does not provide buffering during H^+ discharge in Na_2SO_4 and NaCl solutions. She found that the limiting current for H_2 evolution is higher in sulfate than in chloride or chlorate solutions at the same pH. However, when H_3BO_3 is added to sulfate solutions, the limiting current for H_2 evolution decreases; it is postulated that boric acid can be adsorbed, causing a decrease of active surface area. In NaCl solutions, boric acid does not, however, lead to an effect on limiting current. The Fe content is lower in deposits obtained from chloride than from sulfate solutions but increases upon addition of boric acid.[62]

The regulation of pH is especially important for acidic solutions because of simultaneous H_2 evolution. For this purpose, H_3BO_3, citric acid and sodium acetate as well as salts of weak acids can be used. Citric acid and its salts are good complexing agents of Fe^{2+} ions. Korovin's observations[63] indicate increasing Fe content in the deposit with increase of the citrate concentration in the electrolyte.

Ni–Fe alloy films have been deposited on a beryllium-bronze wire from a sulfate-chloride solution of Ni^{2+} and Fe^{2+}, containing sodium lauryl sulfate and thiourea.[57,64] Bielinski and Przyluski showed that thiourea acts as a depolarizer in the cathodic Ni^{2+} reduction process through the reaction:

$$(NH_2)_2CS + Ni^{2+} + 2H_{ads} = NiS + (NH_2)_2CH_2 \qquad (6)$$

In contrast, with sulfate–chloride solutions containing saccharin and thiourea, the deposition process was less sensitive to changes in pH.

Srimathi and Mayanna[65] reported that Ni–Fe alloys can be produced from acid sulfate baths of various $[Ni^{2+}]/[Fe^{2+}]$ ratios, with and without different amines present. The amines employed in this work were ethylamine (EA), diethylamine (DEA), triethylamine (TEA), butylamine (BA), and ethylenediamine (EDA). They showed that in the presence of amines, the extent of preferential deposition of Fe was reduced considerably; the corresponding increase of Ni content in the deposit was in the order EDA > EA ~ DEA ~ TA > BA. On the other hand, the values of current efficiency were in the order EDA > BA ~ TEA ~ DEA > EA.

In many cases, the organic compounds are adsorbed on the cathodic surface. Sodium lauryl sulfate, saccharin, thiourea, phenol, and other noncolloidal organic compounds have been used as surfactants in the electrodeposition of thin Ni-Fe alloy films.[6,66-69] The mechanisms of action of different additives in the plating of thin magnetic films have not yet been critically or fully evaluated. However, the presence of such agents in the bath allows better controllability of the electrodeposition process in comparison with what can be obtained using other methods for production of thin magnetic films.

2. The Effect of Plating Conditions on the Composition of Ni–Fe Alloys

(i) *The Effect of pH*

A change in the pH of the solution affects the composition, internal stress, size and orientation of crystallites, and other properties of the deposit. The composition of Ni-Fe alloys electrodeposited from plating solutions of simple salts in the pH range of about 3-5 does not show any definite trend with respect to the effect of pH. When electrodeposition is carried out at 0.25 A dm^{-2}, an increase in the pH from 3 to 5 results in a decrease in the iron content from 25 to 15%.[14] On the other hand, Ni-Fe alloys obtained at 0.5 A dm^{-2} show an increase in Fe content from 75 to 80% with an increase in the pH from 3 to 5.[13] At sufficiently high current density (4 A dm^{-2}), for the same change in pH, the Fe content in the deposit remains approximately constant (80%).[13] With an increase in the pH of sulfate solutions containing boric acid from 1 to 3, the composition essentially does not change.[70]

The composition of electrodeposited Ni–Fe alloys changes very much with change in the composition of solutions containing citrate ions[71]: with an increase in the pH, the Ni content in the bath increases. At pHs above 6, the deposits obtained contain predominantly Ni. This is explained by the change in the relative ability of citrate ion to complex Fe^{2+} and Ni^{2+} with variation of pH. For instance, at pH 3, Ni^{2+} ions should be preferentially complexed by citrate. With increasing pH, however, citrate ions become preferentially complexed with Fe^{2+}, thus causing an increase in the Ni content in the deposit.

Horkans[62] reported that a decrease in pH generally causes a decrease in current efficiency and an improvement in the appearance of the deposit obtained. With increasing pH, the Fe content is higher at low potentials but decreases sharply at higher potentials.

With an increase in the pH of sulfamate solutions, it has been shown[65] that Ni content in the deposit increases. Nakamura and Hayashi[8] found that with an increase in the pH of the solution from 2 to 4.5, the Fe content in the alloy was higher by about 10%, at all investigated current densities. On the other hand, the deposits of Ni–Fe alloy obtained from sulfate–tartrate solutions at a current density of 0.1 A dm^{-2} suffered a decrease of Fe content from about 70 to 35% with an increase in the pH of the solution from 2.5 to 5.[7] According to a mathematical model of Hessami and Tobias,[23] the concentration of metal hydroxide ions depends on both pH and the dissociation constant; higher pH increases the concentration of such ions, causing a reduction in overpotential and an increase in the partial current for metal deposition. The authors explained some quantitative disagreement between the model[23] and experimental data[24] by uncertainty in the values of the dissociation constants for $FeOH^+$ and $NiOH^+$.

(ii) *The Effect of Temperature*

The working temperature for electrodeposition of thin magnetic films is determined on the basis of the total concentration of salts in the solution. When the salt concentrations exceed the solubilities of the salts at room temperature, it is normally necessary to raise the temperature of the solution.

Sysoeva[72] showed that with increase of the solution temperature from ca. 25 to 70°C, the Fe content in the alloy increased. On the other hand, with an increase in the temperature from 25 to 100°C, the rate of the discharge of Ni^{2+} ions increases.[16] The Fe content in the deposits of Ni–Fe alloys obtained at low current densities decreases with increase of the temperature from 20 to 70°C.[73] In electrodeposition at current densities of 1.5–2.0 A dm^{-2}, the composition of the alloy did not depend on the temperature. With an increase of the temperature of a sulfamate solution from 20 to 50°C, the current efficiency increased from 20 to 50%, and the Fe content in the Ni–Fe alloy decreased from 20 to 10%.[57]

Fedoseeva and Vagramyan[74] investigated the effect of temperature on the electrodeposition of Ni–Fe alloy films under pulsating current conditions. They showed that with an increase in the temperature, the rate of Ni^{2+} discharge becomes higher than that for Fe^{2+} discharge (activation energy effect). Experimental results show that the deposit obtained at 25°C contains approximately 70% Fe, while that obtained at 130°C consists only of nickel.

Fedoseeva *et al.*[75] suggested a method for the approximate calculation of the composition of electrodeposited Ni–Fe alloys according to the following equation:

$$\frac{i_1}{i_2} = \frac{i_{01}a_1}{i_{02}a_2}\exp\left(-\frac{nF[\alpha_1(E - E_{e,1}) - \alpha_2(E - E_{e,2})]}{RT}\right)\exp\left(-\frac{nF\psi}{RT}\right) \tag{7}$$

Here, E is the deposition potential of the alloy, ψ is the deviation of the stationary potential from the equilibrium value, R is the universal gas constant, and T is the temperature, and i_i, i_{0i}, a_i, α_i, and $E_{e,i}$ are the current density, the exchange current density, the activity, the transfer cofficient, and the equilibrium potential, respectively, for Ni^{2+} ($i = 1$) and Fe^{2+} ($i = 2$). All parameters in this equation are taken with the same values as for the discharge of the individual ions.[76] Experimental results showed that changes of the parameters such as current density, concentration, and pH at a low (50°C) and at a high temperature (120°C) had little influence on the [Ni]/[Fe] ratio in the deposit.[70] According to these authors, the temperature is a basic factor that influences the ratio of discharge rates of Ni^{2+} and Fe^{2+} during their coupled electrodeposition. The deposits obtained at 50°C mainly contained Fe (90%), while those obtained at 120°C contained more than 90% Ni, which is consistent with the proposed equation (Eq. 7). The effect is evidently one arising mainly from differences in activation energy, though the latter can itself change with temperature if speciation of dissolved or complexed Ni^{2+} or Fe^{2+} depends on temperature.

In electrodeposition of Ni–Fe alloys at 1 A dm^{-2}, with decreasing temperature from 70 to 15°C, the Fe content in the deposit increases from 12 to 60%.[8] Srimathi and Mayanna[65] showed that with increase of the solution temperature, the Ni content in the alloy increased as well.

In most cases, during alloy electrodeposition, increase of the solution temperature leads to an increase in the relative amount of the more noble component in the alloy deposit. This is in accordance with Eq. (7), as also found experimentally by several investigators.[57,73–76]

The complexity of the dependence of composition on temperature demands, for all cases, a selection of a specific working temperature for the electrodeposition process. The range of the working temperatures, in general, is determined by the desired composition and properties of the alloy, the composition of the solution, and the parameters of the electrodeposition process.

(iii) *The Effect of Deposit Thickness*

The effect of the deposit thickness on the composition of Ni–Fe alloys has been investigated by several workers. In general, the composition of electrodeposited Ni–Fe alloys exhibits a significant gradient in iron concentration in a direction perpendicular to the substrate surface. The gradient of the composition of Ni–Fe alloy thin films was studied by Cocket and Spencer-Timms.[77] They obtained magnetic alloy films having thicknesses of 50–200 nm. The composition of these films depends on the thickness. Up to a thickness of 150 nm, the Fe content in the alloy was 13.5%. However, for thicknesses between 150 and 200 nm the fraction of Fe in the deposit was 24.3%.

Dahms[78] studied the effect of the thickness of Ni–Fe alloy films on their composition. In considering his experimental results and his theoretical treatment, Dahms postulated the following:

1. Several monolayers deposited in the beginning consist only of Ni.

2. On these monolayers of Ni, the Ni–Fe alloy is deposited. The Fe content in the alloy decreases during this time, approaching the value corresponding to the overall alloy composition.

The experimental results are consistent with Dahms's theoretical treatment (Fig. 3).

Doyle[26] investigated electrodeposition of Ni–Fe alloys at planar and wire electrode substrates. His results showed the variation of composition depending on the geometry of the electrode. For instance, the Fe content in the alloy electrodeposited on a wire electrode was only ca. 4%, which was significantly smaller than in the case of a Ni–Fe alloy electrodeposited on a planar electrode

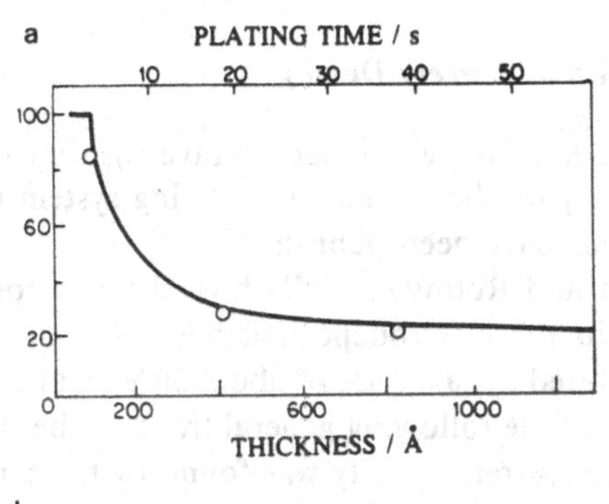

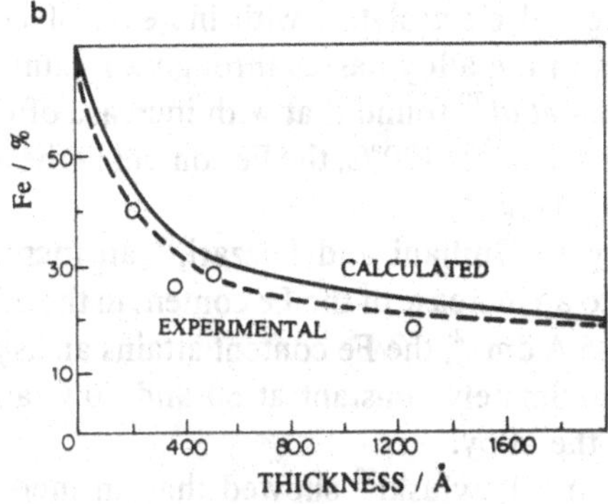

Figure 3. Comparison of galvanostatic (a) and potentiostatic (b) data with calculated curves.[78]

for the same deposit thickness (20-250 nm) and from the same solution.

Venkatasetty[4] showed that the Fe content is practically independent of the deposit thickness, within 100-1000 nm. Similar results, showing that Ni-Fe alloy composition is independent of the thickness above 100 nm, were obtained by Srimathi and Mayanna.[65]

In order to obtain a homogeneous Ni-Fe alloy composition, several authors have recommended electrodeposition at constant cathodic potential.[66,77,79] The gradient of the Ni-Fe alloy composition is also reduced by various additives, for example, saccharine.[67] A similar trend was observed when thiourea was used.[28,66]

Based on the literature data, it can be concluded that electrodeposited Ni-Fe alloys exhibit composition gradients, especially in the case of thin films with thickness of <100 nm.

(*iv*) *The Effect of Current Density*

A number of papers aimed at investigation of the effect of current density on the Ni-Fe alloy plating system under a variety of conclusions have been published.

Sysoeva and Rotinyan[73,80,81] showed that dependence of the Fe content in the electrodeposited alloy on the applied current density exhibited a maximum of about 50% Fe at a current density of 0.5 A dm^{-2}. The following general trend of the dependencies of Fe content on current density was found by the same authors,[73,81] for all investigated electrolytes: with increase of current density, the Fe content in the alloy passes through a minimum.

Vagramyan *et al.*[70] found that with increase of current density from 0.2 to 25 A dm^{-2} at 120°C, the Fe content in the alloy increased from 15 to 36% (Fig. 4).

According to Giuliani and Lazzari,[19] an increase in current density leads to an increase of the Fe content in the alloy. At current densities of 4-5 A dm^{-2}, the Fe content attains an asymptotic value which is approximately constant at 60 and 70°C and amounts to about 70% in the alloy.

Bielinski and Przyluski[57] showed that an increase in current density leads to an increase of the Fe content in the alloy when electrodeposition is carried out from a sulfate solution containing

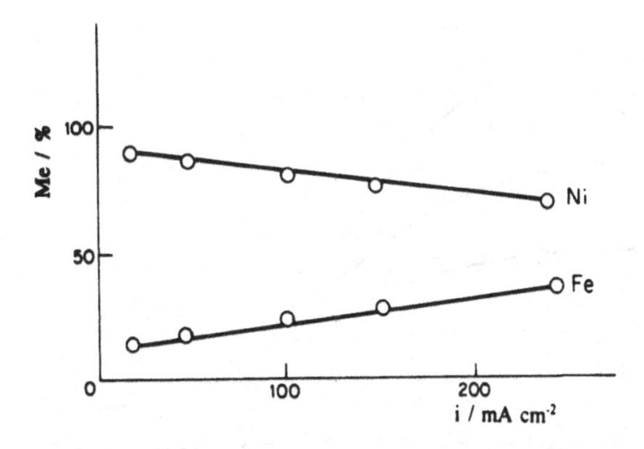

Figure 4. The dependence of Ni–Fe alloy composition on the current density (T = 120°C, pH 1.9).[70]

thiourea. With an increase in the current density from 5 to 25 A dm^{-2}, the Fe content increased from 15 to 40%. However, when electrodeposition was carried out under the same conditions but from a sulfamate solution, the Fe content in the alloy decreased from 30 to 5%.

Djokić *et al.*[60] found that with an increase in the current density, the Fe content in the alloy deposited from a solution containing sodium potassium tartrate decreases (Fig. 5).

However, Nakamura and Hayashi[8] reported that as the current density was increased from 0.5 to 10 A dm^{-2}, the Fe content in the alloy passed through a maximum at a current density of 3 A dm^{-2}. Similar trends were obtained by Horkans[62] (Fig. 6).

It is obvious that significant disagreements between published results exist; this is probably the consequence of the electrodeposition being conducted under different conditions from different solutions as well as, of course, the anomalous nature of the Ni–Fe alloy electrodeposition, which has been shown to depend quite sensitively on such conditions.

IV. THE EFFECT OF A PERIODICALLY CHANGING RATE ON THE ELECTRODEPOSITION OF NICKEL–IRON ALLOYS

The theory of the effect of electrodeposition at a periodically changing rate has been developed for pure metals,[82-84] while there

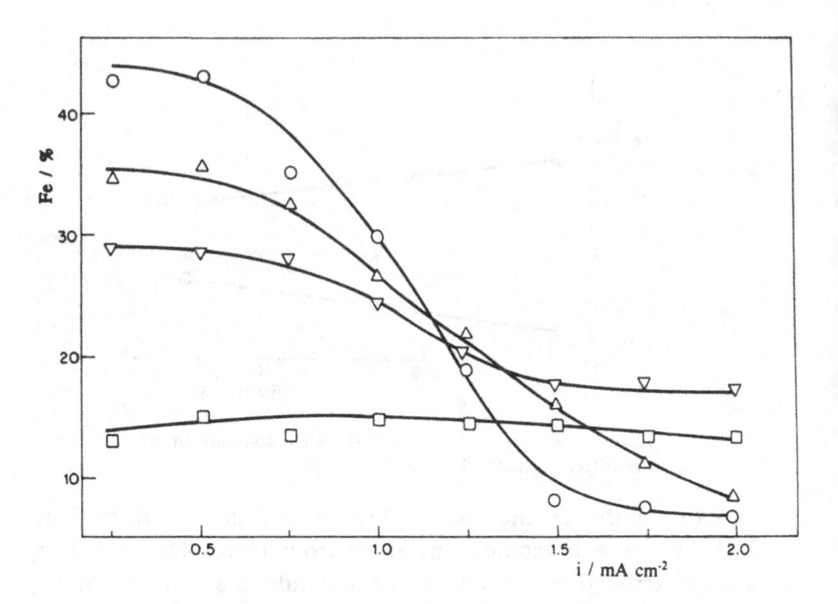

Figure 5. The dependence of Ni–Fe alloy composition on the current density for different $NaKC_4H_4O_6$ concentrations (pH 4.75): □, 0.005 *M*; ▽, 0.01 *M*; △, 0.02 *M*; ○, 0.03 *M*.[60]

are much fewer papers dealing with the corresponding problems in alloy deposition.

Cheh and co-workers[84-86] presented finite-difference models for the case of pulsating current electrodeposition of a single metal with hydrogen evolution, and for alloys. Verbrugge and Tobias[87] gave an analytical solution for the deposition of multicomponent alloys by an arbitrary current source and discussed an analogous model for controlled-potential electrolysis. Beauchamp[88] presented a mathematical model for the electrodeposition of alloys at the rotating disk electrode, including migration effects. Kovac[89] used Roseburgh and Lash Miller's solution[90] to examine the ways in which superimposed sinusoidal current influences the electrodeposition process, with and without an accompanying chemical reaction. Fedoseeva and Vagramyan[74] derived a mathematical model for Ni–Fe electrodeposition which enabled estimation of the alloy composition as a function of the pulse duration to be made.

It is important to note the papers of Sheshadri and co-workers,[91,92] who found that the composition of electrodeposited

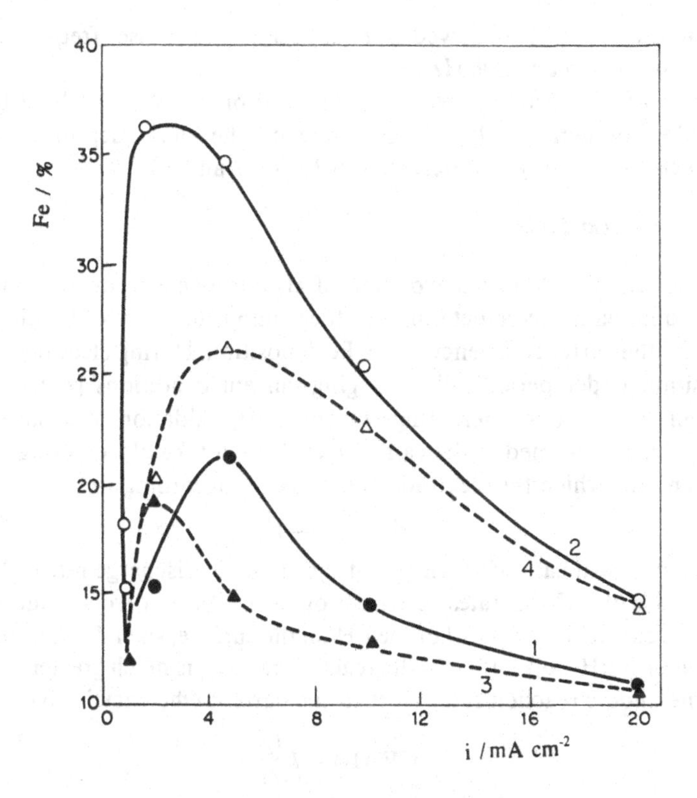

Figure 6. The dependence of Ni–Fe alloy composition on current density in various pH 3.0 solutions: 1, 0.50 *M* $NiSO_4$, 0.01 *M* $FeSO_4$; 2, 0.05 *M* $NiSO_4$, 0.01 *M* $FeSO_4$, 0.4 *M* H_3BO_3; 3, 0.50 *M* $NiCl_2$, 0.01 *M* $FeCl_2$; 4, 0.050 *M* $NiCl_2$, 0.01 *M* $FeCl_2$, 0.4 *M* H_3BO_3.[62]

Ni–Fe alloy under periodically changing current conditions depends on the ratio of the amplitude current density to the direct current density (i_p/u_{dc}), the frequency, and the waveform. Hamaev and co-workers,[93,94] investigated the electrodeposition of Ni–Fe alloys from sulfate and pyrophosphate solutions under periodically changing current. Grimmett *et al.*[95] investigated electrodeposition of Ni–Fe alloy on a rotating cylindrical electrode under pulsating current conditions. They found that during pulse-reverse plating, an increase in anodic pulse magnitude decreased the anomalous codeposition, and the greatest effect in reducing the anomalous deposition was observed for deposition at pulse frequencies between 100 and 300 Hz.

In this section, attention is focused on simple models of the effect of periodically changing rate on the electrodeposition of nickel–iron alloys as presented in Refs. 82 and 83.

(i) *Current Efficiency*

The simultaneous evolution of H_2 during electrodeposition is a dominant factor determining the composition of Ni–Fe alloys and the current efficiency of the Fe deposition. During electrodeposition under periodically changing current conditions [reversing current (rc) and alternating current (ac)], oxidation of adsorbed hydrogen, formed in the cathodic cycle, will take place during the time for which the electrode is the anode, according to

$$H^+ + e^- \rightleftarrows H \tag{8}$$

In the ideal case of balancing of the cathodic discharge rate of H^+ ions with H^+ ion rates of oxidation and diffusion from solution, one can achieve control of the pH on the surface, such that pH(surface) → pH(bulk). If cathodic reaction rate, v_c, is much greater than the anodic reaction rate, v_a, then the current efficiency is given by

$$\mathrm{CE}(H_2) = k\frac{Q_c}{Q} \tag{9}$$

where Q is the total quantity of electricity, and Q_c is the cathodic quantity of electricity. The ratio Q_c/Q is a function of the parameters of the periodically changing currents, given by

$$\frac{Q_c}{Q} = \frac{i_c t_c}{i_{av} T} \tag{10}$$

for pulsating current (pc), by

$$\frac{Q_c}{Q} = \frac{i_c t_c}{i_c t_c - i_a t_a} \tag{11}$$

for rc, and by

$$\frac{Q_c}{Q} = \frac{1}{\pi}\arcsin\left(\frac{i_{dc}}{i_p}\right) + \frac{1}{2} + \frac{\sqrt{1 - \left(\frac{i_{dc}}{i_p}\right)^2}}{\pi\frac{i_{dc}}{i_p}} \tag{12}$$

for ac.[98] Here, i_c and i_a are the cathodic and the anodic current density, respectively, i_{av} is the average current density, and t_c and t_a are the cathodic and the anodic time, respectively.

In the cases of pulsating current and any periodically changing current without an anodic component, the following equation can be applied:

$$\frac{Q_c}{Q} = 1 \tag{13}$$

Combination of Eqs. (9) and (13) gives

$$\mathrm{CE}(\mathrm{H_2}, \mathrm{pc}) = k \tag{14}$$

where k is a constant.

Figure 7 represents the calculated values of $CE(H_2)$ (full line) and the experimentally obtained values as a function of Q_c/Q for pc, rc, and ac conditions.

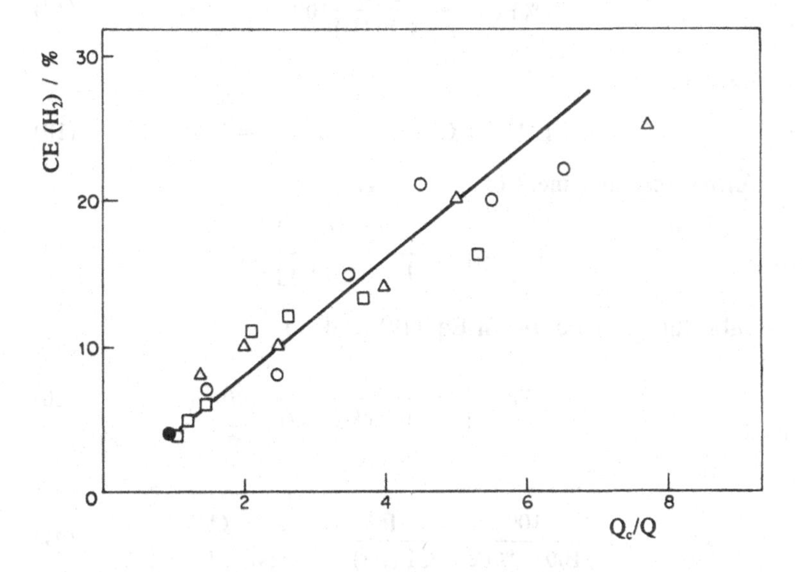

Figure 7. The dependence of $CE(H_2)$ on Q_c/Q for electrodeposition under pc (●), ac (□, $\nu = 50$ Hz), and rc conditions (△, $t_c = 10$ ms, $i_c = i_a$; ○, $t_c = t_a = 10$ ms).[96]

(ii) The Effect of the Amplitude of Periodically Changing Currents on the Alloy Composition

Assuming that the current efficiency of Ni deposition is practically constant,[95] a simple mathematical model for the dependence of alloy composition on the parameters of the periodically changing current can be derived. The iron content in the alloy is defined as follows:

$$\%\text{Fe} = \frac{m(\text{Fe})}{m(\text{Fe}) + m(\text{Ni})} 100 \tag{15}$$

where m is weight, or, assuming that the atomic weights of Fe and Ni are equal,

$$\%\text{Fe} = \frac{Q(\text{Fe})}{Q(\text{Fe}) + Q(\text{Ni})} 100 \tag{16}$$

and, expressed in terms of current efficiencies,

$$\%\text{Fe} = \frac{\text{CE(Fe)}}{1 - \text{CE}(\text{H}_2)} 100 \tag{17}$$

Because

$$\text{CE}(\text{H}_2) + \text{CE(Ni)} + \text{CE(Fe)} = 1 \tag{18}$$

further rearrangement of Eq. (17) gives

$$\%\text{Fe} = \left[1 - \frac{\text{CE(Ni)}}{1 - \text{CE}(\text{H}_2)}\right] 100 \tag{19}$$

Substitution of Eq. (9) in Eq. (19) leads to

$$\%\text{Fe} = \left[1 - \frac{\text{CE(Ni)}}{1 - (kQ_c/Q)}\right] 100 \tag{20}$$

or

$$\frac{100}{100 - \%\text{Fe}} = \frac{1}{\text{CE(Ni)}} - \frac{k}{\text{CE(Ni)}} \frac{Q_c}{Q} \tag{21}$$

Figure 8 represents experimentally obtained values of 100/(100 − %Fe) as a function of Q_c/Q for pc, rc, and ac conditions. In this

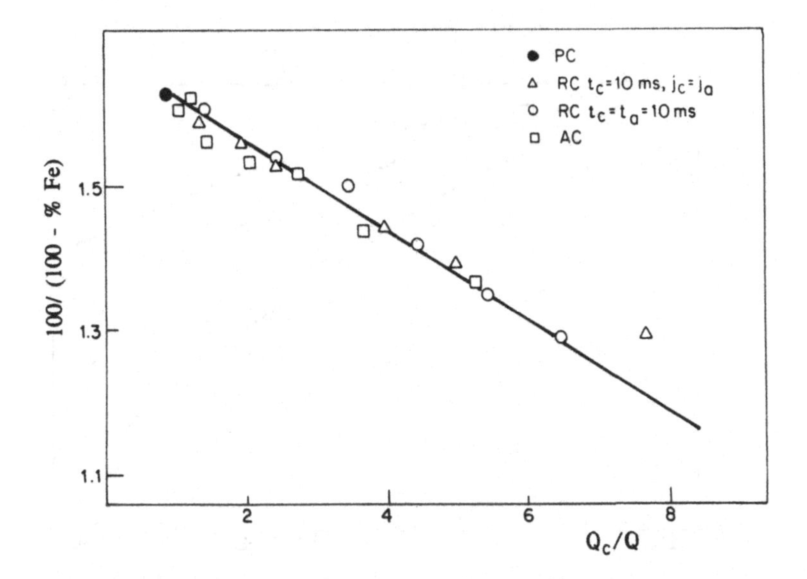

Figure 8. The effect of Q_c/Q on the Fe content in the Ni–Fe alloy for electrodeposition under pc (●), ac (□), and rc conditions (△, $t_c = 10$ ms, $i_c = i_a$; ○, $t_c = t_a = 10$ ms).[96]

way, the results obtained are in good agreement with the predictions of Eq. (21).

Using the experimental current efficiency results for Ni, *k*, and Eq. (20), the dependence of the percentage of Fe on i_p/i_{dc} for different forms of periodically changing currents can be established as shown in Fig. 9, which indicates that periodically changing currents for $i_p < i_{dc}$ do not influence alloy composition. This is in accordance with the conclusions of Srimathi *et al.*[92] For $i_p > i_{dc}$, the extents of the decrease in the Fe content for various wavefroms are in the order square > sin > triangular.

The alloy composition depends on the parameters of periodically changing currents only if $Q_c/Q > 1$ during each period of the current wave. This is fulfilled in the case of reversing current and sinusoidal alternating current superimposed on direct current for $i_p/i_{dc} > 1$. The Fe content is reduced more by rc than it is by ac. In other words, the anomalous nature of the codeposition can be reduced more under rc conditions than under ac conditions.

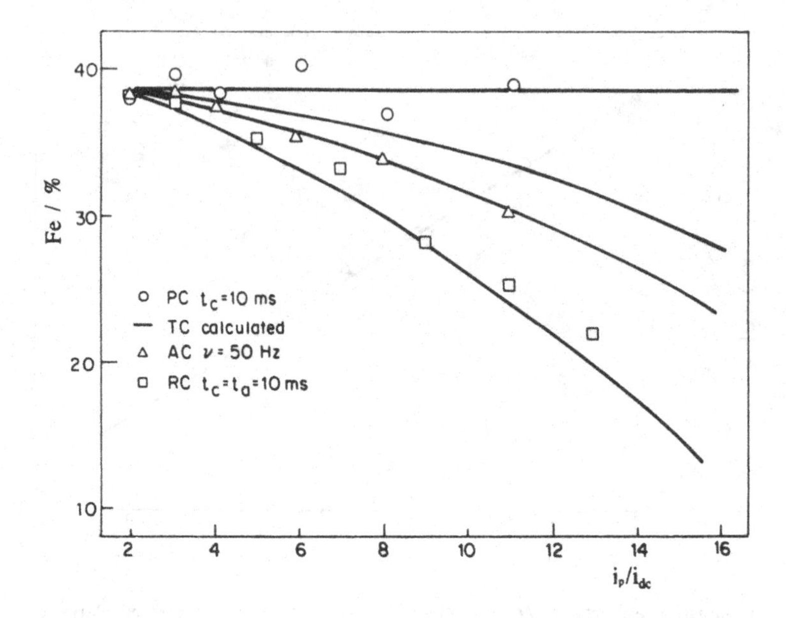

Figure 9. The effect of i_c/i_{dc} on the Fe content in the Ni–Fe alloy under pc (○, t_c = 10 ms), ac (△, ν = 50 Hz), and rc conditions (□, t_c = t_a = 10 ms). The solid curve without experimental points is the calculated curve for triangular current.[96]

(iii) *The Effect of the Frequency of Periodically Changing Currents on the Alloy Composition*

The Fe content as a function of the frequency of the periodically changing current is shown in Fig. 10. In the case of pc, there is no effect, and alloy composition remains the same as under dc conditions. The effective frequency range of ac is smaller than that of rc; that is, the composition reaches an asymptotic limiting value with increasing frequency of the periodically changing current.[99,100] Because of this, Q_c cannot be calculated from

$$Q_c = \int_{t_1}^{t_2} i(t)\,dt \tag{22}$$

where $i(t)$ is the input of periodically changing current, but from

$$Q_c = \int_{t_1}^{t_2} i_F\,dt \tag{23}$$

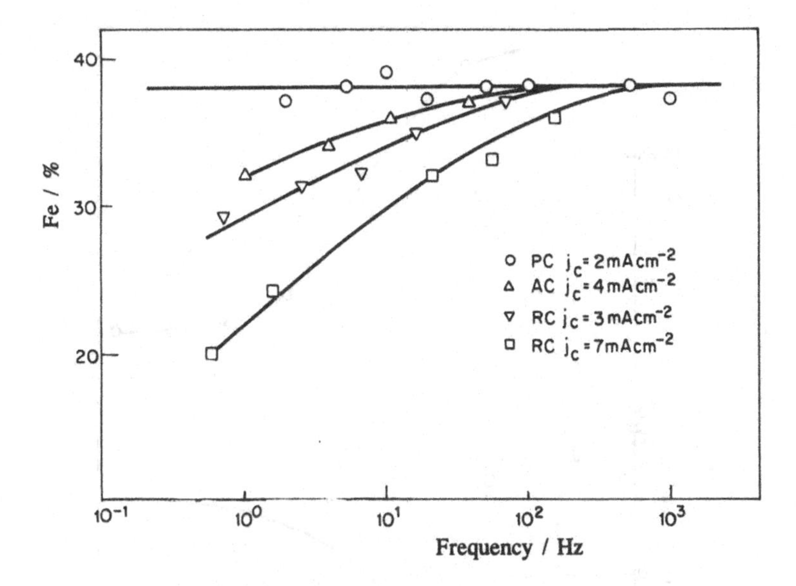

Figure 10. The effect of the frequency of periodically changing currents on the Fe content in the Ni-Fe alloy: ○, pc (i_c = 2 mA cm^{-2}); △, ac (i_c = 4 mA cm^{-2}); ▽, rc (i_c = 3 mA cm^{-2}); □, rc (i_c = 7 mA cm^{-2}).[96]

where i_F is the faradaic current. The time limits t_1, t_2, t_1', and t_2' determine the time interval for cathodic current flow. At some (sufficiently high) frequency, $t_1' = 0$ and $t_2' = T$; hence $Q_c/Q = 1$, and a further increase in frequency does not affect the alloy composition. The slower faradaic current flattening under rc than under ac conditions can be explained by faster charging of the double-layer capacity during a rectangular current pulse than during a nonrectangular one.[100]

(iv) *The Effect of Constant and Pulsating Potential on the Alloy Composition*

Figure 11 shows the stationary polarization curve for Ni-Fe electrodeposition and the Fe content in the alloy as a function of potential. It can be seen that an increase to more negative potentials leads to a reduction in the Fe content of the alloy. When the limiting current density is reached, the Fe content becomes constant, lowering the quality of the coatings.

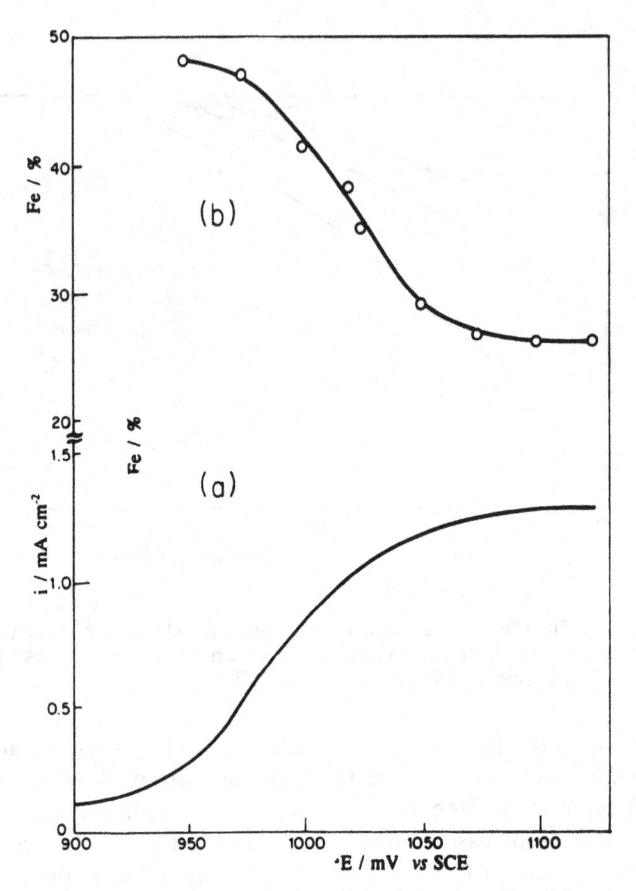

Figure 11. Polarization curve for Ni-Fe electrodeposition (a) and the effect of deposition potential on the Fe content in the Ni-Fe alloy (b).[97]

A model given by Fedoseeva *et al.*[75] can be used for the derivation of the dependence of alloy composition on potential. The partial current density for electrodeposition of the ith component is given by the following equation:

$$i_1 = i_{0i} a_i \exp\left[-\frac{nF\alpha_i}{RT}(E - E_{st,i}) \right] \tag{24}$$

and the stationary potential is

$$E_{st,i} = E_{e,i} - \psi_i \tag{25}$$

where $E_{e,i}$ is the equilibrium potential, and ψ_i is the deviation of the stationary potential from the equilibrium value. In this way, the ratio of the partial current densities, given by Eq. (7), can be simplified to

$$\frac{i_1}{i_2} = Kf \exp(-AE) \tag{26}$$

where f is a function containing all the variables that depend on potential, and K and A are constants. If we suppose that f has the form

$$f \approx \exp(-BE) \tag{27}$$

Eq. (26) becomes

$$\frac{i_1}{i_2} = K' \exp[-(A+B)E] \tag{28}$$

Since

$$\frac{\%\text{Ni}}{\%\text{Fe}} \approx \frac{i_1}{i_2} \tag{29}$$

then the alloy composition is an exponential function of potential according to:

$$\frac{\%\text{Ni}}{\%\text{Fe}} = K'' \exp[-(A+B)E] \tag{30}$$

or

$$\log\left(\frac{\%\text{Ni}}{\%\text{Fe}}\right) = \log K'' - \frac{A+B}{2.3}E \tag{31}$$

as is shown in Fig. 12.

The experimental data in Fig. 12 show that Eq. (31) is a good approximation over a potential range in which the current density is lower than the limiting value.

Experiments with pulsating potential show that the Fe content in the alloy decreases as the amplitude, E_p, increases, up to some limiting value.[97] Further increase in amplitude does not significantly affect the Fe content, and the coating quality deteriorates.

In the case of the pulsating potential, the alloy composition is mostly influenced by the more negative potential E' ($E' = E_{av} - E_p$)

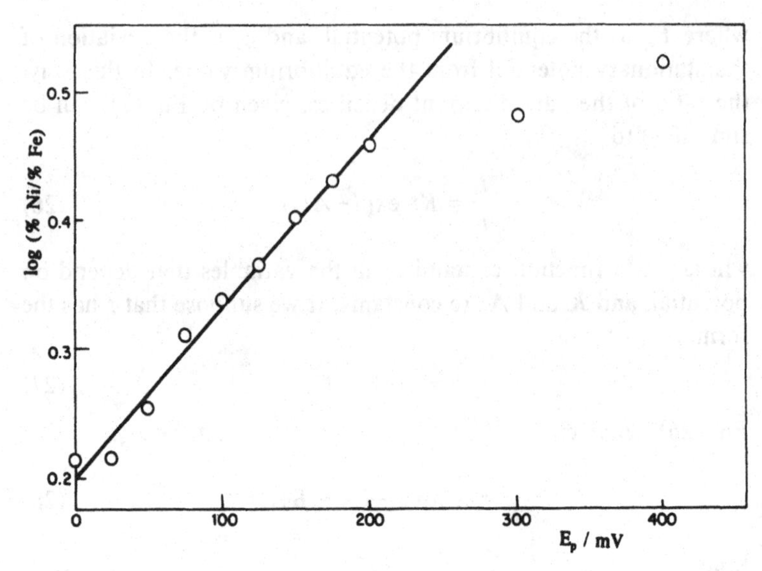

Figure 12. The effect of the deposition on log(%Ni/%Fe).[97]

rather than the less negative potential E'' ($E'' = E_{av} + E_p$), since $\exp(-AE') \gg \exp(-AE'')$ or, at least, $\exp(-AE') > \exp(-AE'')$. In this case, the partial current ratio for Ni and Fe electrodeposition is given by

$$\frac{i_1}{i_2} = Kf \exp(-AE') \tag{32}$$

Assuming that the analogy with the individual metal deposition exists,[85] then for higher frequencies and for E_{av} constant, f is constant, and thus

$$\frac{\%\mathrm{Ni}}{\%\mathrm{Fe}} = K''' \exp(-AE_p) \tag{33}$$

or

$$\log\left(\frac{\%\mathrm{Ni}}{\%\mathrm{Fe}}\right) = \log K''' - \frac{A}{2.3} E_p \tag{34}$$

as is shown in Fig. 13.

According to Eq. (34), there is a linear dependence on the potential amplitude until 200 mV, that is, over the range in which *bright* alloy coatings are produced.[97]

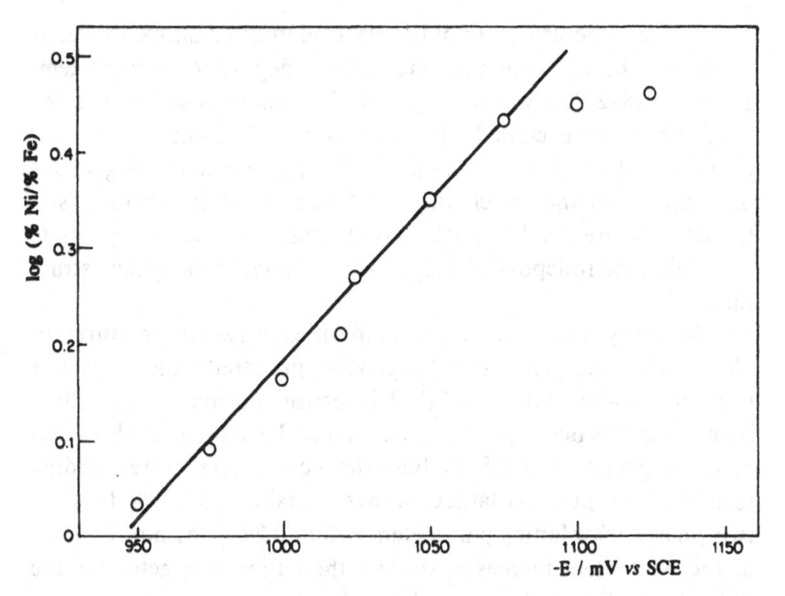

Figure 13. The effect of the pulsating potential amplitude on log(%Ni/%Fe) for $E_{av} = -1020$ mV.[97]

As in the case of periodically changing current, increase in the frequency of the pulsating potential leads to an increase in the Fe content in the alloy, until the asymptotic value (corresponding to the constant potential) is reached. In contrast to the case of periodically changing current, the asymptotic value is reached later, that is, at frequencies above 1000 Hz. This can be explained by the slower faradaic flattening effect in the case of pulsating potential, in relation to that for the case of periodically changing current.[100]

V. THE PROPERTIES OF THE ELECTRODEPOSITED Ni–Fe ALLOYS

1. Crystal Structure and Surface Morphology of the Films

The structure and morphology of the Ni–Fe alloy films are affected by the electrodeposition conditions, by the character of the substrate, and by the bath composition including the presence of special additives such as surfactants as well as of contaminants in the bath.

The specific structures of the electrodeposited alloys are determined by the fact that they are usually deposited at room temperature (18–20°C), where the possibility of mutual diffusion in the solid phase is excluded. Thus, it is very common that electrodeposited alloys consist of tiny crystals that are nonhomogeneous in composition and are characterized by marked defects of crystal lattice structure. Under specific conditions, however, it is possible to obtain electrodeposited magnetic alloys with amorphous structures.[101]

In many cases, there are similarities between structures of electrodeposited and metallurgically prepared alloys.[102] For example, Ni–Fe alloys with 31% Ni electrodeposited from a sulfate bath have the bcc type structure of the Fe lattice, while alloys containing more than 52% Ni have the fcc structure, corresponding to that of the pure Ni lattice. Alloys containing 31–52% Ni have two phases with lattice parameters within 0.29–0.357 nm. Further, as the Ni content increases, so does the lattice parameter. On the other hand, the structure of electrodeposited alloys can often be significantly different from that of the analogous metallurgical alloys. A very important factor causing the different phase composition of electrodeposited alloys is the overvoltage involved in their deposition. According to Raub,[103] the structure of an electrodeposited alloy corresponds to the structure of the corresponding metallurgical alloy only when, during the electrocrystallization, solid solutions arise already at the lower current densities.

Phase analysis of permalloy films, electrodeposited on a silver substrate, showed that permalloy has an fcc structure, with lattice parameter 0.355 nm, which corresponds to that of the solid solution of Fe in Ni (γ-phase).[104]

The structural characteristics of electrodeposited films depend very much on the conditions of electrolysis and the substrate properties. The grain size is a function of the rate of crystallization and the rate of growth of the crystallites. Increase in the current density during electrodeposition of permalloy leads to a decrease in crystallite size.[105] On the other hand, electrodeposits with small grain size can be obtained from solutions having a lower concentration of the metal ions.

Important effects on the electrodeposition rate and on the character of the cathodic processes arise when additives are present in the electrolyte. The presence of saccharin in the electrolyte leads, for example, to a decrease in crystallite size in electrodeposited permalloy films.[67] Without saccharin, an increase in the pH results in an increase in the crystallite size. However, with addition of saccharin to the bath, the crystallite size is independent of pH within the range 2.1–3.1. The Ni–Fe alloy films obtained by Srimathi and Mayanna[65] appeared bright and uniform as the Ni content of the deposit increased gradually either due to increase of the temperature or with increasing thickness of the deposit.

With increasing thickness of the deposit from 30 to 120 nm, a decrease in the crystallite size and a change in the surface morphology has been observed.[106] Thinner films (about 30 nm) have an appearance similar to that of the substrate. For thicknesses above 60 nm, the film consisted of small grains uniformly distributed on the surface. Electrodeposited Ni–Fe alloys from sulfate-chloride electrolyte had laminar microstructure which became granular.[107]

Deposits obtained from unstirred solutions exhibited considerable porosity. Those obtained from stirred solutions, however, had some porosity, while those agitated by ultrasound had no visible pores.[1] Generally, unltrasound at 24.8 kHz gave better deposits than at 37.8 kHz, where alloy deposits tended to peel off due to high internal stress.

It has been shown that with Ni–Fe alloy plating solutions containing saccharin, the Fe component exerts a *leveling* effect.[108,109] The initial distribution of iron in microprofile is nonuniform, the amount on the peaks being higher. With a decrease in the initial roughness of the cathode surface, the nonuniform distribution of iron becomes less pronounced.

It has been found[7] that electrodeposited Ni–Fe alloy films with 36% Fe and having a thickness of about 300 nm exhibit epitaxial growth. The results of X-ray diffraction experiments showed that electrodeposited Ni–Fe alloy had fcc structure with a lattice parameter $a = 0.356$ nm.

2. Magnetic Properties of Thin Ni–Fe Alloy Films

Magnetic parameters of magnetic materials are well known (Fig. 14). The magnetic properties of electrodeposited Ni–Fe alloys depend on the bath composition, electrolysis conditions, character

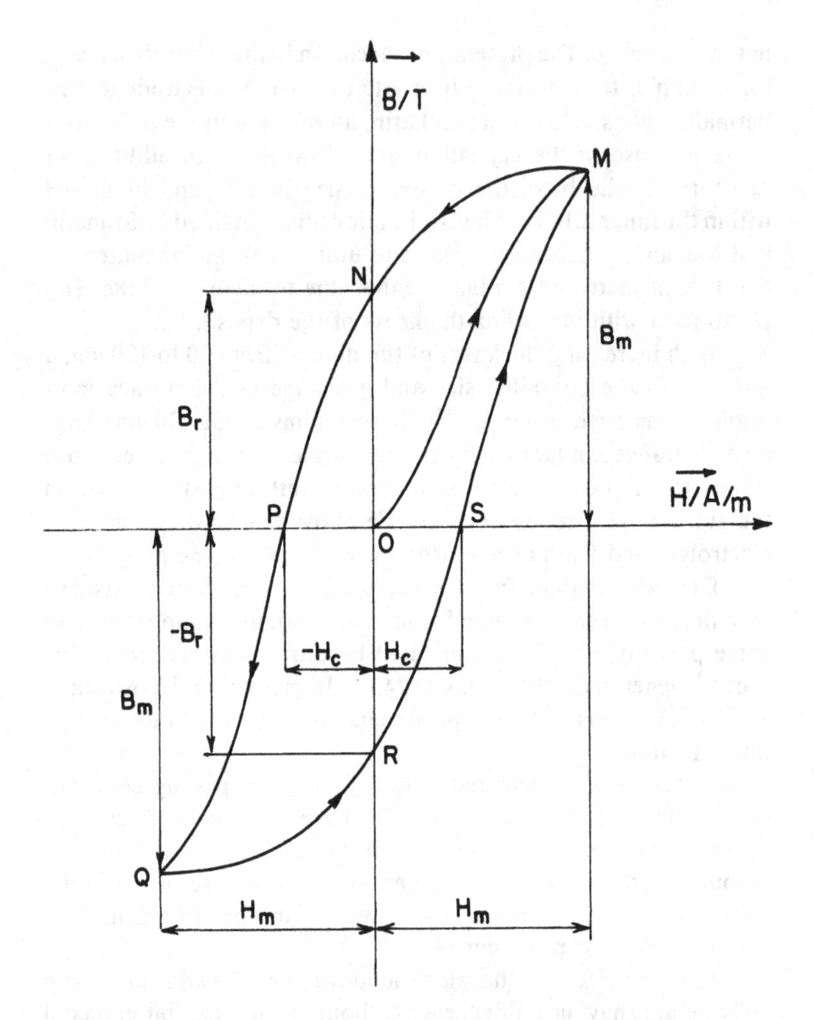

Figure 14. The hysteresis loop of ferromagnetic materials.

of the substrate, etc. As outlined in the first part of this chapter, these factors also affect the alloy composition.

(i) *The Effect of Alloy Composition and Film Thickness*

Detailed studies of the magnetic properties of electrodeposited Ni–Fe alloy films have been carried out by Wolf and co-workers.[110–114] These papers describe the methodologies employed in the electrodeposition and investigation of thin Ni–Fe alloy films in the composition range 47–100% Ni. The films of all investigated compositions had a square hysteresis loop. In the case of films with zero magnetostriction, the dependence of the rectangularity of the hysteresis loop on composition was linear. With an increase in the Ni content from 90 to 95%, deviation from this linear dependence occurred. According to many reports, the dependence of coercivity on composition passes through a minimum, and the ratio of the remanent induction to the maximum induction, B_r/B_m, reaches a maximum value at this same composition of the alloy.[115-118] This composition is characterized for Ni–Fe alloys by a zero constant of magnetostriction, which corresponds to an Fe content of about 20%.[115] The differences in the dependencies of coercivity on Ni–Fe alloy composition have been explained by the nonhomogeneous distribution of film stresses, caused by different substrate characteristics.[119]

The presence of elements such as S, P, and As in controlled, small quantities in Ni–Fe alloys affects the magnetic properties.[40-51] For instance, the presence of S in the alloy causes a decrease in coercivity, while the anisotropic field and magnetostriction remain almost constant.

Uniformity of Ni–Fe alloy composition is very important for the magnetic properties. Politycki[120] showed that films with thicknesses of 10–20 nm contained about 60% Fe, but when the thickness was increased from 20 to 100 nm, the Fe content was about 20%. With further increase of the thickness, the Fe content in the alloy remained practically constant. The films having smaller thickness and grain size had better magnetic performance. For electrodeposited films, there is great interest in the dependence of the coercivity H_c on the thickness if other parameters remain constant.[6] The theoretical dependence of the coercivity on thickness for thin magnetic films is given by[121,122]

$$H_c = Ld^{-n} \tag{35}$$

where d is the film thickness, and L and n are constants depending on the magnetization of the films, energy domain walls, structure, uniformity of the thickness, and geometry of the surface. The parameter n can be determined from the slope of the logarithmic dependence of coercivity on thickness. Based on Eq. (35), it can be concluded that, as the film thickness increases, coercivity decreases, which is in good agreement with the experimental results.[113,118] For very thin films, it is possible to use the value $\frac{4}{3}$ for n. According to Wolf[113] values between 0 and 1 can be used for n. Lloyd and Smith[123] showed that n can vary between 0 and 1.4. Uehara[119] reported values in the range 0.2-1.13 depending on the alloy composition. As film thickness increases, the coercivity increases for alloys containing 77-88% Ni. However, for alloys with 95% Ni, with increase in the thickness, the coercivity decreases.

The anisotropy of the films is also a function of deposit film thickness. With an increase in the film thickness, the anisotropy increases for alloys having high Ni content, but decreases for alloys of low Ni content.[113] The effect of the thickness on the anisotropy field is insignificant for films with zero magnetostriction.[124] The anisotropy field of thin films is affected by the roughness of the surface and the density of the crystals as well as by exposure of the films to an external magnetic field during electrolysis.[6] Politycki and Gothard[51] obtained an Ni-Fe alloy on a copper cathode with a thickness of 30 nm and with different crystallite sizes. They found that the coercivity and anisotropy field increased with an increase in crystallite size. In the presence of a superimposed magnetic field during electrodeposition, they have shown also that it is possible to obtain Ni-Fe alloy films with uniaxial anisotropy.

(ii) *The Effect of Internal Stress*

Internal stresses in the deposit have a significant influence on the magnetic properties.[125-127] The ratio of the remanent induction to the saturation induction and the anisotropy field of thin Ni-Fe alloy films depends on the mechanical compressive and tensile stresses.[128,129] An analytical expression relating the coercivity to the internal stress of ferromagnetic materials, σ, is given by the Kondorsky-Kersten equation[130,131]:

$$H_c = \frac{\sigma\lambda}{I_s} \tag{36}$$

where λ is the magnetostriction at saturation, and I_s is the saturation magnetization. Agreement of the experimental coercivity values obtained with those calculated according to Eq. (36) was found by Fisher.[132]

It is necessary to take into consideration the change of the internal stress after electrolysis, which changes the magnetic properties. Unfortunately, there are insufficient published data relating the influence of internal stresses to the magnetic properties of electrodeposited Ni–Fe alloy films.

(*iii*) *The Effect of the Substrate*

The character of the substrate has a great influence on the composition and properties of Ni–Fe alloys. Thus, a very important step in the electrodeposition of an Ni–Fe alloy is selection of the substrate.

Fisher and Haber[133] proposed the use of nickel amorphous layers chemically deposited on polymeric materials as substrates for electrodeposition of thin magnetic films. They showed that permalloy films obtained with a thickness of 200 nm on this substrate had a coercivity of 0.07 $kA\,m^{-1}$ and an anisotropy field of 0.22 $kA\,m^{-1}$. According to Luborsky,[134] gold layers are the best substrates for electrodeposition of magnetic alloys.

Ilyushenko *et al.*[135] investigated the effect of the material of the substrate on the magnetic properties and surface microstructure of electrodeposited permalloy films. As substrates, they used layers of copper, gold, and chromium with a thickness of 150 nm vaporized on glass in vacuum at 200, 150, and 60°C. In the case of the substrates obtained at lower temperatures, the coercivity of permalloy films had lower values. The coercivity of the permalloy films (70 nm) depended on the character of the substrate material, decreasing in the order Cu > Au > Cr.

By suitable control of the substrate roughness, it is possible to obtain the necessary magnetic properties for memory devices.[136] A change in the composition of electrodeposited Ni–Fe alloy on wire substrates by only 0.03% leads to a change in magnetostriction of about 5%.[137]

Venkatasetty[138] obtained permalloy films (thickness of 500 nm) with zero magnetostriction, using an electrodeposited copper substrate. He showed that, in the case of a copper substrate obtained from a citrate bath, with increasing roughness of the substrate, the coercivity and anisotropy field of permalloy films are constant up to some value, after which they suddenly increase. In the case of copper substrates obtained from a pyrophosphate bath, with increasing roughness, the coercivity and anisotropy field increase up to some value, beyond which they decrease slowly.

3. The Effect of Different Electrolysis Conditions on the Magnetic Properties of Ni–Fe Alloy Films

(*i*) *pH*

It has been shown that pH affects the alloy composition and structure and hence the magnetic properties. With increase of the pH of the electrolyte from 1 to 4, the coercivity of Ni-Fe alloy decreased, although within the pH range 2-3.5 it remained constant. This is explained by the fact that in this pH range the alloy composition remained constant.[139]

The coercivity of electrodeposited permalloy films changes less with change of pH than the coercivity of the electrodeposited pure metal films. With increase in the pH of the electrolyte above 2, the coercivity of Ni-Fe alloy with 90% Fe (500 nm) increases.[140]

Venkatasetty[4] showed that with increasing pH within the range 3.5-4.0, the coercivity decreases. Within the pH range 4.0-5.5, the coercivity and anisotropy field are practically constant; with further increase of the pH, the coercivity and anisotropy field increase (Fig. 15).

Barauskas and Guttensohn[59] found that Ni-Fe alloy films electrodeposited from an ammoniacal bath within the pH range 8.39-8.50 had better magnetic performance than those obtained in the pH range 9.20-9.25.

(*ii*) *Effect of Temperature*

An increase in the temperature of the bath leads to a decrease of cathodic polarization and to a high grain size in the deposits. Klok and White[33] showed that Ni-Fe alloy films with good magnetic performance can be electrodeposited from chloride solutions only at temperatures above 65°C.

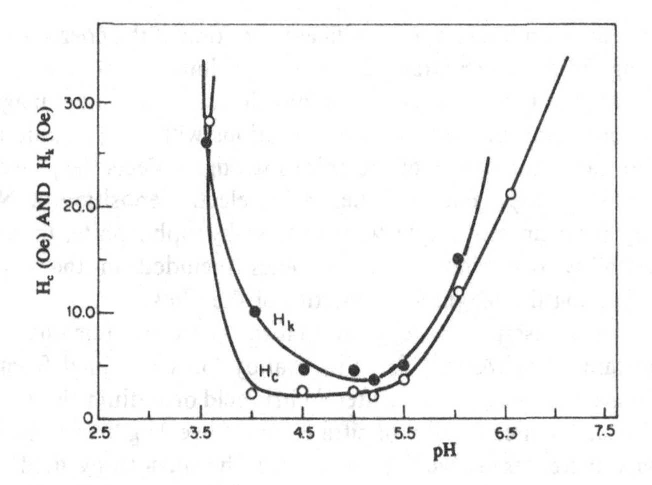

Figure 15. Variation of H_c and H_k for Ni–Fe films (300 nm thick) as a function of bath pH; $i = 10\ \mathrm{mA\ cm^{-2}}$ at 30°C.[4]

Wolf[141] found that the magnetic properties of permalloy films did not change significantly with temperature when they were electrodeposited from chloride-sulfate solution in the temperature range 10–60°C. On the other hand, electrodeposition of Ni–Fe alloys at higher temperatures (50–70°C) is generally recommended because the films obtained have better mechanical properties and less mechanical stress. For example, alloys with 97% Fe obtained at temperatures below 50°C were found to be brittle.

(iii) *Electrolyte Composition*

As outlined earlier, bath composition affects alloy composition, thus leading to different properties of the Ni–Fe alloys. However, using various solutions, it is possible to prepare alloy electrodeposits with the same composition but having different properties, which was clearly shown for the example of permalloy films. The differences were greater in the case of thin films.[26-28]

With increase of the total ion concentration in the solution ($[Me^{2+}] = [Fe^{2+}] + [Ni^{2+}]$), the coercivity and anisotropy field increase.[6] However, in the case of solutions having high metal-ion concentration, there is no significant variation of the coercivity with change in the concentration of the metal ions.

Electrodeposited Ni–Fe alloys having the same magnetic properties can be obtained from solutions with various anions. In some cases, the nature of the anions strongly affects the properties of Ni–Fe alloys. For example, in the electrodeposition of Ni–Fe alloy from an electrolyte with sodium hypophosphite, there is a possibility that phosphorus becomes included in the deposit, influencing the magnetic properties of the films.

Venkatasetty[4] has reported that the coercivity remains almost constant ($0.24\,kA\,m^{-1}$) for Ni–Fe alloy films obtained from solutions containing 20–120 g $liter^{-1}$ citric acid or sodium citrate. With increasing concentration of citrate ions above 25 g $liter^{-1}$, the coercivity increases as well, as also does the anisotropy field. With increase in the saccharin concentration from 0 to 1.25 g $liter^{-1}$, a reduction in the internal stress of permalloy films was observed.

By addition of substances containing sulfur-containing groups such as thiourea, sodium lauryl sulfate, etc., a significant reduction of the internal stress of the deposit was observed.[27,28,51,67–69,113,142] The reduction in internal stress is attributed to the inclusion of sulfur in the deposit, which influences the decrease in the anisotropic field and coercivity. The properties of the permalloy films electrodeposited from sulfate solutions containing thiourea have been investigated by several workers. Luborsky and Barber[28,69] showed that the anisotropy field is almost constant and that the coercivity of thin permalloy films reached a minimum when solutions contained thiourea at concentrations within the range 0.1–0.3 g $liter^{-1}$. The addition of thiourea caused not only a change in the Ni–Fe alloy composition (from 95 to 8% Ni), but also a decrease in the coercivity. These alloys showed a structural change from bcc to fcc.[142]

The presence of saccharin in the solutions for the permalloy electrodeposition leads to electrodeposits having small crystals and small internal stress, causing a decrease in coercivity.[143] The addition of saccharin (4 g $liter^{-1}$) to an Ni–Fe alloy plating solution reduces the coercivity, which was explained by a decrease in the crystal size within 65–30 nm. Various compounds containing sulfur have been used as additives for electrodeposition of Ni–Fe alloys from sulfate solutions.[144] Ni–Fe alloy thin films having the lowest coercivity were obtained from solutions containing thiosemicarbazide, another substance containing sulfur, at a concentration of 0.05 g $liter^{-1}$. Thus, it is important to determine the quantity of sulfur in the deposit as well as to explain the mechanism of inclusion of S into the deposit. The following questions arise: Does S become included in the alloy crystal lattice from a solid solution? Does S exist in the form of some chemical compound with the metal elements or as elementary S? Further investigation to answer them are required.

It was found that inclusion in the deposit of some nonmetallic substances such as P or As leads to improvement in the magnetic properties.[145] The possibility of improving the magnetic properties through the inclusion of some additives represents a very important advantage of electrodeposition with respect to other methods for production of magnetic alloys.

(iv) *The Effect of Current and Potential on the Properties of Electrodeposited Ni–Fe Alloy Films*

Current density has substantial influence on the alloy composition, as it determines, for example, the quantity of H included in the deposit and the size and the orientation of the crystallites, and thus also affects the properties of the alloy. In general, electrodeposition at a smaller current density leads to deposits with higher grain size. With increase in the current density, the rate of nucleation increases, causing the formation of deposits having smaller grain size. A sufficient increase in the current density causes depletion of the layer near the electrode surface with respect to metal ions, leading to diffusion control in electrodeposition. Under these conditions, crystal growth in the direction perpendicular to the electrode surface can occur. On the other hand, because of an increase in the discharge rate of hydrogen ions, the pH near the electrode surface can increase. Thus, conditions can arise that lead to metal-ion hydrolysis. Hydrolysis products formed under these conditions can become included in the deposit, causing deleterious effects on the magnetic properties of the electrodeposited Ni–Fe alloy.[4]

The range of current densities for Ni–Fe alloy electrodeposition is relatively wide (0.025–50 A dm^{-2}). The dependence of the coercivity of permalloy films with a thickness of 200 nm on the current density from 0.5 to 2.5 A dm^{-2} passes through a minimum at 1 A dm^{-2} (Ref. 6). In the current density region from 0.025 to 0.2 A dm^{-2}, the Fe content in the alloy decreases from 60 to 30%, while the coercivity has a minimum at a current density of about 0.1 A dm^{-2} (Fig. 16).[146]

In the electrodeposition of alloys, their compositions and properties depend very much on the character of the polarization current. Alternating current superimposed on dc in the electrodeposition of films with low coercivity influences the decrease of the composition gradient whereas pulsating current contributes to the formation of new crystallization centers, thus leading to deposits with a small grain size.[147] Using ac superimposed on dc in electrodeposition of nickel leads to an increase in the grain size of the crystallites.

With an increase in the ratio i_p/i_{dc} in the range 0–2, the coercivity shows a decrease.[6] With further increase of this current ratio, the coercivity increases. Similar results were obtained by Djokić *et al.*,[146] who showed that for i_p/i_{dc} ratios within the range 0–1.5, the cercivity has a minimum at about 1.5. For i_p/i_{dc} ratios

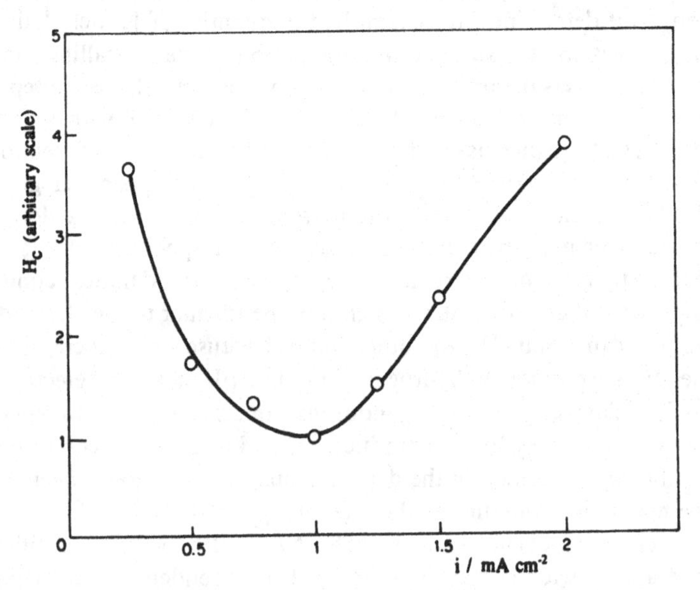

Figure 16. The effect of dc density on the coercivity.[146]

in the range 0–1, the value of the coercivity was almost constant (Fig. 17).

The periodically changing current parameters affect the Ni–Fe alloy composition only when the Q_c/Q ratio is greater than 1, which in the case of ac superimposed on dc is realized only for $i_p/i_{dc} > 1$.[97] Thus, with changes of the i_p/i_{dc} ratio within the interval 0–1.5, the coercivity decreases.[146]

Permalloy films electrodeposited at frequencies between 5 and 500 Hz exhibit a minimum in coercivity at frequencies within the range 25–50 Hz.[6] Ni–Fe alloy thin films with a thickness of 300 nm and with various compositions were electrodeposited under ac superimposed on dc conditions at a ratio $i_p/i_{dc} = 1.5$ and within the frequency range 1–5000 Hz.[146] The coercivity passes through a minimum at a frequency of about 50 Hz. It was noticed also that the coercivity of Ni–Fe alloy films electrodeposited at frequencies greater than 100 Hz was practically constant. The constant part of the coercivity dependence on the frequency above 100 Hz is explained as follows. When the frequency is increased, a damped

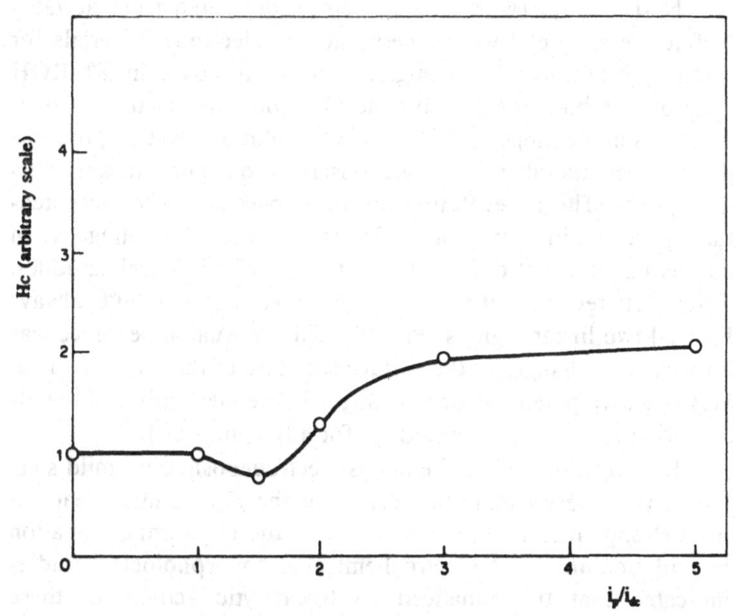

Figure 17. The effect of the i_p/i_{dc} ratio on the coercivity.[146]

faradaic current appears so that, in such a case, the behavior characteristic of dc itself is dominant.[100] Thus, coatings obtained at frequencies above 100 Hz have almost the same quality as those obtained by electrolysis carried out under dc conditions at a corresponding current density. Similar trends of the coercivity dependence on the corresponding parameters were found with Ni-Fe alloy films obtained by electrolysis carried out under constant or pulsating potential (potentiostatic plating conditions).[146]

VI. ELECTROCHEMICAL BEHAVIOR OF ELECTRODEPOSITED Ni-Fe ALLOYS

The development of new electrode materials with high electrolytic activity for the hydrogen and oxygen evolution reactions in unipolar water electrolyzers has been directed to ward systems in which various active coatings could be electrodeposited. The electrodeposited Ni-Fe alloys thus have a very significant importance in this field.

Ni-Fe alloys (with 65% Fe) electrodeposited from acetate-sulfate electrolyte[148] were investigated as electrode materials for electrolysis of water.[149] Investigations were carried out in 28% KOH solution for both the H_2 and the O_2 evolution reaction at temperatures in the range 25-80°C. The O_2 evolution reaction proceeds at rates comparable to those observed on pure nickel electrodeposits. The H_2 evolution reaction arises at lower overpotentials, resulting in a drop of ca. 250 mV for the total voltage when compared with that observed with the use of mild-steel cathodes. Tafel plots recorded at temperatures between 25 and 80°C always showed two linear regions (Fig. 18). This unusual appearance was attributed to changes in the surface structure of the Ni-Fe alloy at very negative potentials or to changes in the mechanism of the H_2 evolution reaction and conditions for adsorption of H.

Investigations of Ni-Fe alloys electrodeposited on mild steel have shown very enhanced activity for the H_2 evolution reaction from strong alkaline solutions after activation by partial oxidation in acid medium.[150-152] Electrochemical and morphological studies indicated that the enhanced electrocatalytic activity of these materials for the H_2 evolution reaction was mainly due to the highly

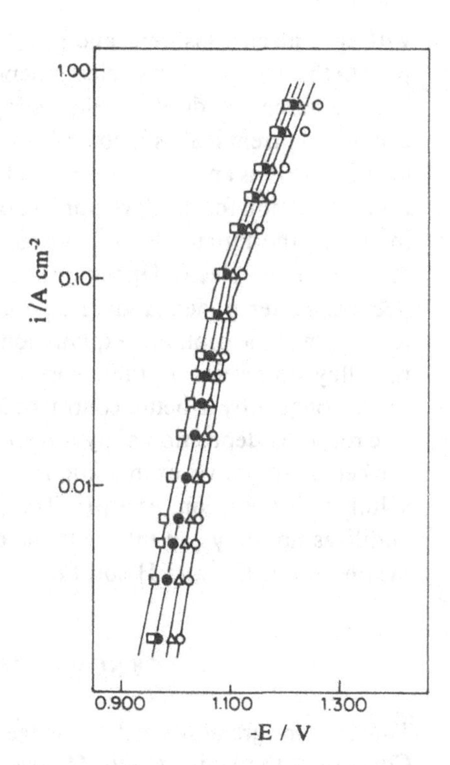

Figure 18. Tafel plots for the hydrogen evolution reaction on Ni-Fe codeposits in 28% KOH at different temperatures: ○, 25°C; △, 40°C; ●, 60°C; □, 80°C.[149]

porous nature of the deposits.[153] The electrodeposited Ni–Fe alloy for the same purpose shows an enhanced activity and good stability for long-term operation.[154]

VII. CONCLUSIONS

The composition and, consequently, the properties of Ni–Fe alloys depend on the plating conditions. In spite of intensive investigations, there is difficulty in making general conclusions about the dependencies of the alloy composition and/or properties upon various plating variables. Although Ni–Fe alloy films with desirable properties for practical applications have been obtained, the dependencies of the magnetic characteristics on chemical composition, crystallite size, and growth conditions have not yet been satisfactorily explained in a quantitative way. In order to obtain deposits with specific compositions and properties, great attention has been paid to the investigation of electrodeposition from various solutions with different additives. Electrodeposition under periodically changing current makes it possible to obtain films with a wide range of compositions and magnetic properties, and this offers a promising future direction for the development of plating technology. In order to explain the anomalous codeposition effect, as well as the kinetics of Ni-Fe alloy electrodeposition, further experimental studies are needed. Better understanding of these phenomena will probably lead to more acceptable explanations regarding dependencies of the alloy properties on the composition and/or plating variables, and consequently to better control of Ni-Fe alloy electrodeposition. The role of codeposition of hydrogen is important and also requires further quantitative examinations in relation to current density, solution composition, and pH. The presence of sulfur-containing additives not only can influence the deposit itself but modifies H_2 evolution kinetics and H sorption.

ACKNOWLEDGMENT

The authors gratefully acknowledge the help of Professor B. E. Conway of the University of Ottawa in editing the first version of this chapter and for suggesting various additions.

REFERENCES

[1] R. Walker and S. A. Halagan, *Plat. Surf. Finish.* **75**(4) (1985) 68.
[2] J. Horkans, *J. Electrochem. Soc.* **126** (1979) 1861.
[3] S. N. Srimathi, S. M. Mayanna, and B. S. Sheshadri, *Surf. Technol.* **16** (1982) 277.
[4] H. V. Venkatasetty, *J. Electrochem. Soc.* **117** (1970) 403.
[5] S. Armyanov, *Elecktrokhimiya* **24** (1986) 1011.
[6] L. F. Ilyshenko, *Elektroliticheski Osazhdenie Magnitnie Plenki*, Nauka i Tekhnika, Minsk, 1979.
[7] S. S. Djokić, Ph.D. thesis, University of Belgrade, 1988.
[8] N. Nakamura and T. Hayashi, *Plat. Surf. Finish.* **75**(8) (1985) 42.
[9] S. Harty, J. A. McGeough, and R. M. Tulloh, *Surf. Technol.* **12** (1981) 39.
[10] J. S. Hadley and J. O'Grady, *Trans. Inst. Met. Finish.* **59** (1981) 89.
[11] A. T. Vagramayan and Z. B. Solveeva, *Technology of Electrodeposition*, R. Draper Ltd., Teddington, England, 1961.
[12] A. Brenner, *Electrodeposition of Alloys*, Vol. I, Academic Press, New York, 1963.
[13] S. Glasstone and T. E. Symes, *Trans. Faraday Soc.* **23** (1927) 213.

[14] S. Glasstone and T. E. Symes, *Trans. Faraday Soc.* **24** (1928) 370.
[15] N. V. Korovin, *Zh. Neorg. Khim.* **2** (1957) 2259.
[16] A. T. Vagramyan and T. A. Fatueva, *J. Electrochem. Soc.* **110** (1963) 1030.
[17] H. Dahms, *J. Electroanal. Chem.* **8** (1964) 5.
[18] H. Dahms and I. M. Croll, *J. Electrochem. Soc.* **112** (1965) 771.
[19] L. Giuliani and M. Lazzari, *Electrochim. Metal.* **3** (1968) 45.
[20] E. Beltowska-Lehman and Riesenkampf, *Surf. Technol.* **11** (1980) 349.
[21] M. Lieder and S. Biallozor, *Surf. Technol.* **26** (1985) 23.
[22] S. Biallozor and M. Lieder, *Surf. Technol.* **21** (1984) 1.
[23] S. Hessami and C. W. Tobias, *J. Electrochem. Soc.* **136** (1989) 3611.
[24] P. Andricacos, C. Arana, J. Tabib, J. Dukovic, and L. T. Romanikiw, *J. Electrochem. Soc.* **136** (1989) 1336.
[25] S. Djokić and M. Maksimović, *Hem. Ind.* **41** (1987) 239.
[26] W. D. Doyle, *J. Appl. Phys.* **38** (1967) 1441.
[27] R. Girard, *J. Appl. Phys.* **38** (1967) 1423.
[28] F. E. Luborsky, *J. Appl. Phys.* **38** (1967) 1445.
[29] R. Burkiewich, M. Jarmolinska, and H. Koslowska, *Arch. Electrotechniki* **11** (1962) 177.
[30] A. M. Susloparov, *Fiz. Plenok* **1** (1972) 81.
[31] J. Vengris and S. Semaska, *Liet. TSR Mokslu. Akad. Darb., Ser. B* (6) (1978) 15.
[32] A. T. Vagramyan and B. S. Petrova, *Fiziko-Khimicheskie Svoystva Elektroliticheskih Splavov*, Akad. Nauk SSSR, Moscow, 1960.
[33] A. Klok and H. White, *J. Electrochem. Soc.* **110** (1963) 98.
[34] L. E. Stout and A. J. Carol, *Trans. Am. Electrochem. Soc.* **58** (1930) 357.
[35] F.K. Andryushenko, V. V. Orekhova, and J. I. Gritsenko, *Zh. Prikl. Khim.* **48** (1975) 815.
[36] V. Sree and T. Ramachar, *Bull. Indian Sect. Electrochem. Soc.* **7** (1958) 72.
[37] V. Sree and T. Ramachar, *J. Electrochem. Soc.* **108** (1961) 64.
[38] V. L. Kotov, A. K. Krivtzov, and V. A. Trostina, *Izv. Vyssh. Uchebn. Zaved., Khim. Khim. Tekhnol.* **18** (1975) 237.
[39] S. N. Srimathi and S. M. Mayanna, *Plat. Surf. Finish.* **75**(12) (1985) 76.
[40] W. O. Freitag, J. S. Mathias, and G. Di Gulio, *J. Electrochem. Soc.* **111** (1964) 35.
[41] A. W. Goldenstein, W. Rostoker, F. Schossberger, and G. Gutzeit, *J. Electrochem. Soc.* **104** (1957) 104.
[42] J. P. Marton and M. Schlesinger, *J. Electrochem. Soc.* **115** (1968) 16.
[43] M. Maeda and K. Mukasa, *Jpn. J. Appl. Phys.* **6** (1967) 895.
[44] M. Maeda and K. Mukasa, *Jpn. J. Appl. Phys.* **4** (1965) 557.
[45] P. A. Albert, Z. Kovac, and H. R. Lilientkal, *J. Appl. Phys.* **38** (1967) 258.
[46] A. F. Schmeckenbecher, *J. Electrochem. Soc.* **113** (1966) 778.
[47] H. H. Zappe, *J. Appl. Phys.* **38** (1967) 4536.
[48] K. Mukasa, M. Sato, and M. Maeda, *J. Electrochem. Soc.* **117** (1970) 22.
[49] W. O. Freitag, G. Di Guilio, and J. S. Mathias, *J. Electrochem. Soc.* **113** (1966) 441.
[50] J. L. Jostan and A. F. Bogenschutz, *IEEE Trans. Magn.* **MAG-5**(2) (1969) 112.
[51] A. Polyticki and H. Gothard, *Z. Angew. Phys.* **14** (1962) 363.
[52] V. V. Bondar, V. V. Grinina, and V. N. Pavlov, *Itogi Nauki Tekh. Ser. Elektrokhim.* **16** (1980) 175.
[53] B. Ya. Kaznachei and A. K. Rozhdestvinskaya, USSR Patent 191.981, 1967.
[54] S. S. Misra and T. L. Ramachar, *Met. Finish.* **65** (1967) 62.
[55] G. A. Sodakov, A. A. Mazin, V. V. Gordienko, V. V. Kovalov, and N. T. Kudryavtzov, *Zh. Prikl. Khim.* **53** (1980) 2038.
[56] T. R. Long, *J. Appl. Phys.* **31** (1960) 123.

[57] J. Bielinski and J. Przyluski, *Surf. Technol.* **9** (1979) 65.
[58] S. N. Srimathi and S. M. Mayanna, *J. Appl. Electrochem.* **16** (1986) 69.
[59] R. L. Barauskas and A. E. Guttensohn, *J. Electrochem. Soc.* **120** (1973) 341.
[60] S. S. Djokić, N. T. Petrović, and O. G. Fancikić, *J. Serb. Chem. Soc.* **55** (1990) 477.
[61] L. T. Romanikiw, J. V. Powers, and E. E. Castellani, paper 152 presented at the Electrochemical Society Meeting, Pittsburgh, Pennsylvania, October 15-20, 1978.
[62] J. Horkans, *J. Electrochem. Soc.* **128** (1981) 45.
[63] N. V. Korovin, *Zh. Neorg. Khim.* **2** (1957) 2259.
[64] J. Bielinski and J. Przyluski, *Surf. Technol.* **9** (1979) 53.
[65] S. N. Srimathi and S. M. Mayanna, *J. Appl. Electrochem.* **13** (1983) 679.
[66] R. Girard, *J. Appl. Chem.* **3S** (1967) 1423.
[67] R. C. Smith, L. E. Godycki, and J. C. Lloyd, *J. Electrochem. Soc.* **108**(1961) 996.
[68] F. E. Luborsky, *IEEE Trans. Magn.* **MAG-4** (1968) 19.
[69] F. E. Luborsky and W. D. Barber, *J. Appl. Phys.* **39** (1968) 1746.
[70] A. T. Vagramyan, T. A. Fedoseeva, D. V. Fedoseev, and L. A. Uvarov, *Elektrokhimiya* **6** (1970) 1773.
[71] J. Arulaj and G. Venkatachari, *Trans. Soc. Adv. Electrochem. Sci. Technol.* **13**(1) (1978) 23.
[72] V. V. Sysoeva, *Zh. Prikl. Khim.* **32** (1959) 128.
[73] V. V. Sysoeva and A. L. Rotinyan, *Zh. Prikl. Khim.* **35** (1962) 20.
[74] T. A. Fedoseeva and A. T. Vagramyan, *Elektrokhimiya* **8** (1972) 851.
[75] T. A. Fedoseeva, L. A. Uvarov, D. V. Fedoseev, and A. T. Uvarov, *Elektrokhimiya* **6** (1970) 1841.
[76] A. T. Vagramyan, M. A. Schamagorzyanc, G. F. Savtsenko, and L. A. Uvarov, *J. Phys. Chem.* **238** (1968) 67.
[77] G. H. Cocket and E. C. Spencer-Times, *J. Electrochem. Soc.* **108** (1961) 906.
[78] H. Dahms, *Electrochem. Technol.* **4** (1966) 530.
[79] F. H. Edelman, *J. Electrochem. Soc.* **109** (1962) 440.
[80] V. V. Sysoeva and A. L. Rotinyan, *Zh. Prikl. Khim.* **35** (1962) 2430.
[81] V. V. Sysoeva and A. L. Rotinyan, *Zh. Prikl. Khim.* **37** (1964) 1840.
[82] Yu. M. Polukarov and V. V. Grinina, *Itogi Nauki Tekh. Ser. Elektrokhim.* **22** (1985) 3.
[83] K. I. Popov and M. D. Maksimovic, in *Modern Aspects of Electrochemistry*, No. 19, Ed. by B. E. Conway, J. O'M. Bockris, and R. E. White, Plenum Press, New York, 1989, p. 193.
[84] A. M. Pesco and H. Y. Cheh, in *Modern Aspects of Electrochemistry*, No. 19, Ed. by B. E. Conway, J. O'M. Bockris, and R. E. White, Plenum Press, New York, 1989, p. 251.
[85] T. Cheng and H. Y. Cheh, paper presented at the 161st Meeting of the Electrochemical Society, Montreal, 1982.
[86] T. Cheng and H. Y. Cheh, paper presented at the 164th Meeting of the Electrochemical Society, Washington, D.C., 1983.
[87] M. W. Verbrugge and C. W. Tobias, *J. Electrochem. Soc.* **137** (1985) 1298.
[88] C. R. Beauchamp, paper presented at the 168th Meeting of the Electrochemical Society, Las Vegas, 1985.
[89] Z. Kovac, *J. Electrochem. Soc.* **118** (1971) 51.
[90] T. R. Roseburgh and W. Lash Miller, *J. Phys. Chem.* **14** (1910) 916.
[91] B. S. Sheshadri, Vasanta Koppa, B. S. Jai Prakash, and S. M. Mayanna, *Surf. Technol.* **13** (1981) 111.
[92] S. N. Srimathi, B. S. Sheshadri, and S. M. Mayanna, *Surf. Technol.* **17** (1982) 217.
[93] V. A. Hamaev and A. K. Krivtzov, *Izv. Vyssh. Uchebn. Zaved., Khim. Khim. Tekhnol.* **13** (1970) 240.

[94] V. A. Hamaev, V. L. Kotov, and A. K. Krivtzov, *Izv. Vyssh. Uchebn. Zaved., Khim. Khim. Tekhnol.* **16** (1973) 1247.
[95] D. L. Grimmett, M. Schwartz, and K. Nobe, *J. Electrochem. Soc.* **137** (1990) 3414.
[96] M. D. Maksimović and S. S. Djokić, *Surf. Coat. Technol.* **31** (1987) 325.
[97] M. D. Maksimović and S. S. Djokić, *Surf. Coat. Technol.* **35** (1988) 21.
[98] M. D. Maksimović, D. C. Totovski, and A. P. Ivić, *Surf. Technol.* **18** (1983) 233.
[99] N. Ibl, *Surf. Technol.* **10** (1980) 81.
[100] M. D. Maksimović and S. K. Zečević, *Surf. Coat. Technol.* **30** (1987) 405.
[101] V. Bondar, K. M. Gorbunova, and Yu. M. Polukarov, *Fiz. Met. Metalloved.* **26** (1968) 568.
[102] N. P. Fedotev, N. N. Bibikov, A. M. Vyacheslavov, and S. Ya. Grilihes, *Electrodeposited Alloys*, Mashgiz, Moscow, 1962.
[103] E. Raub, *Metalloberflaeche* **7** (1953) A17.
[104] B. P. Dzekanovskaya, N. A. Erohov, and G. V. Davidov, *Fiz. Met. Mettaloved.* **26** (1968) 729.
[105] L. F. Ilyushenko and L. P. Kostyuk-Kulgavtchuk, *Izv. Akad. Nauk BSSR, Ser. Fiz. Mat. Nauk* **5** (1972) 116.
[106] L. F. Ilyushenko, M. U. Sheleg, and A. V. Boltuskin, *Dokl. Akad. Nauk BSSR* **13** (1969) 683.
[107] J. M. Levy, *Plating* **56** (1963) 903.
[108] M. Chomakova and S. Vitkova, *J. Appl. Electrochem.* **16** (1986) 669.
[109] M. Chomakova and S. Vitkova, *J. Appl. Electrochem.* **16** (1986) 673.
[110] I. W. Wolf, *Proc. Am. Electroplaters Soc.* **44** (1957) 121.
[111] I. W. Wolf and V. P. McConnell, *Proc. Am. Electroplaters Soc.* **43** (1956) 215.
[112] I. W. Wolf, *J. Electrochem. Soc.* **108** (1961) 959.
[113] I. W. Wolf, *Electrochem. Technol.* **1** (1963) 164.
[114] I. W. Wolf and T. S. Crowther, *J. Appl. Phys.* **34** (1963) 1205.
[115] E. F. Schneider and G. Carmichael, NBS Report P.B. 151525, National Bureau of Standards, U.S. Department of Commerce, Washington, D.C., 1958.
[116] S. Bouwman, *Int. Sci. Technol.* **12** (1962) 20.
[117] N. R. Korovin and P. S. Titov, *Izv. Vyssh. Uchebn. Zaved., Tsvetn. Metall.* **1** (1958) 164.
[118] K. Seizo and T. Noboru, *J. Appl. Phys.* **34** (1963) 795.
[119] V. Uehara, *Jpn. J. Appl. Phys.* **2** (1963) 451.
[120] A. Z. Polyticki, *Angew. Phys.* **13** (1961) 465.
[121] L. Néel, *J. Phys. Radium* **13** (1956) 250.
[122] L. Néel, *Magnitnaya Struktura Feromagnetikov*, IL, Moscow, 1959.
[123] J. C. Lloyd and W. R. S. Smith *Can. J. Phys.* **40** (1962) 454.
[124] W. Freitag and J. S. Mathias, *Electroplat. Met. Finish.* **17** (1964) 42.
[125] A. T. Vagramyan and Yu. S. Petrova, *Fiziko-Mechanicheskie Svoystva Elektroliticheskih Osadkov*, Izd. Akad. Nauk SSSR, Moscow, 1967.
[126] M. Ya. Poperka, *Vnutrenie Napryazhneniya Osazhdennih Metallov*, Zapadnoe Sibirsko Knizhnoe Izdatelstro, Novosibirsk, 1966.
[127] G. S. Sotirova-Chakarova and S. A. Armyanov, *J. Electrochem. Soc.* **137** (1990) 3551.
[128] E. Kresin, *Monatsber., Dtsch. Akad. Wiss., Berlin* **4** (1961) 112.
[129] A. V. Bolwinshki, *Metody Issled. Tonkih Magnitnih Plenok* **3** (1968) 80.
[130] E. I. Kondorsky, *Sov. Phys.* **11** (1937) 597.
[131] M. Kersten, *Elekrotechnische Zeitschrift* **60** (1939) 498.
[132] R. D. Fisher, *J. Electrochem. Soc.* **109** (1962) 749.
[133] R. D. Fisher and H. E. Haber, *Nature* **199** (1963) 163.
[134] F. W. Luborsky, *J. Electrochem. Soc.* **117** (1970) 76.

[135] L. F. Ilyushenko, A. V. Bolotushkin, M. U. Shelag, and N. A. Krivenko, *Izv. Akad. Nauk BSSR, Ser. Fiz. Mat. Nauk* **1968**(5) (1968) 83.
[136] H. D. Richards, J. Humpage, and J. C. Hendy, *IEEE Trans. Magn.* **MAG-4** (1968) 243.
[137] F. E. Luborsky and B. J. Drummond, *Plating* **61** (1974) 243.
[138] H. V. Venkatasetty, *Plating* **59** (1972) 571.
[139] M. S. Blois, *J. Appl. Phys.* **26** (1955) 975.
[140] M. U. Sheleg, PhD thesis, University of Minsk, 1971.
[141] I. W. Wolf, *J. Appl. Phys.* **33S** (1962) 1152.
[142] J. Tsu, *Plating* **47** (1960) 632.
[143] R. De Mars, *J. Electrochem. Soc.* **108** (1961) 782.
[144] L. F. Ilyushenko and L. P. Kostyuk-Kulgavtchuk, *Izv. Akad. Nauk BSSR, Ser. Fiz. Mat. Nauk* **1967**(1) (1967) 114.
[145] W. O. Freitag, G. Di Gulio, and J. S. Mathias, *J. Electrochem. Soc.* **113** (1966) 441.
[146] S. S. Djokić, M. D. Maksimović, and D. Č. Stefanović, *J. Appl. Electrochem.* **19** (1989) 802.
[147] G. T. Bakhvalov, *Novaya Tekhnologiya Elektroosazhdeniya Metallov*, Metallurgizdatelstro, Moscow, 1966.
[148] D. Singh and V. B. Singh, *Indian J. Technol.* **13** (1975) 52.
[149] L. A. Avaca, E. R. Gonzalez, A. A. Tanaka, and G. Tremiliosi-Filho, in *Proceedings of the 4th World Hydrogen Energy Conference*, Pasadena, California, June 13–17, 1982, Pergamon Press, Oxford, p. 251.
[150] J. de Carvalho, G. Tremiliosi-Filho, L. A. Avaca, and E. R. Gonzalez, in *Hydrogen Energy Progress—V*, Vol. 2, Ed. by T. N. Veziroglu and J. B. Telylor, Pergamon Press, New York, 1984, p. 923.
[151] J. de Carvalho, G. Tremiliosi-Filho, L. A. Avaca, and E. R. Gonzalez, Extended Abstracts, Vol. 85-1, 167th Electrochemical Society Meeting, Toronto, Canada, 1985, p. 528.
[152] E. R. Gonzalez, L. A. Avaca, A. Carubelli, A. Tanaka, and G. Tremiliosi-Filho, *Int. J. Hydrogen Energy* **9** (1984) 689.
[153] J. de Carvalho, G. Tremiliosi-Filho, L. A. Avaca, and E. R. Gonzalez, in *Hydrogen Energy Progress—V*, Ed. by T. N. Veziroglu, N. Geroff, and P. Weizirel, Pergamon Press, New York, 1986.
[154] J. de Carvalho, G. Tremiliosi-Filho, L. A. Avaca, and E. R. Gonzalez, in *Proceedings of the Second Symposium on Electrode Materials and Processes for Energy Conversion and Storage*, Ed. by S. Srinivasan, S. Wagner, and H. W. Wroblowa, Proceedings Vol. 87-12, The Electrochemical Society, Pennington, New Jersey, 1987.

5

Microelectrode Techniques in Electrochemistry

Benjamin R. Scharifker

Departamento de Química, Universidad Simón Bolívar, Caracas 1080-A, Venezuela

I. INTRODUCTION

Microelectrode techniques have been increasingly utilized in electrochemistry since the distinct advantages of low-dimensioned electrodes began to be realized. Electrodes of micrometer dimensions were practically not used in electrochemical research or practice prior to 1980, but interest in them has increased exponentially during the past decade. Because many of the undesirable aspects of electrochemical techniques are reduced with microelectrodes and because they are easily implemented and involve relatively low costs, a growing number of laboratories are applying microelectrode techniques to investigate a wide variety of problems in diverse systems, some of them not accessible with conventional, larger electrodes. The purpose of the present chapter is to describe the most salient features of microelectrodes, reviewing the sound theoretical framework that has been woven around their use, as well as to discuss some practical aspects such as construction and handling techniques, instrumentation, and some of the problems to which they have been applied.

Three major consequences arise from the reduction in size of an electrode: (i) mass transport rates to and from the electrode are

Modern Aspects of Electrochemistry, Number 22, edited by John O'M. Bockris *et al.*
Plenum Press, New York, 1992.

increased because of nonplanar diffusion; (ii) the double-layer capacitance is reduced due to the reduction in surface area; and (iii) ohmic losses, which are the product of electrode current and solution resistance, are reduced due to the diminished current. The small electrode size may also be of advantage in special situations, such as *in vivo* analysis, where only small perturbations to the system under study are sought, or in surface electrochemical studies, where an extremely small area may allow for the study of the behavior of single catalytic centers or the formation and growth of single nuclei.

A microelectrode may be defined as an electrode with properties that depend on its size, typically with dimensions comparable to the thickness of the diffusion layer. The diffusion flux to a small electrode contains transient and steady-state contributions; if the electrode is made small enough, then the steady-state term dominates except at very short times. This property has provided a starting point for the application of microelectrodes to electrode kinetics and electroanalytical studies.[1-12]

II. FUNDAMENTALS

1. Theoretical Aspects

(i) *Mass Transport*

Consider the simple electron transfer reaction

$$O + ne^- \rightleftharpoons R$$

where both reactants and products are soluble, and assume that the reaction is diffusion-controlled. With a large, effectively infinite planar electrode, mass transport is described by semi-infinite planar diffusion,

$$\frac{\partial c_j(x, t)}{\partial t} = D_j \frac{\partial^2 c_j(x, t)}{\partial x^2}, \qquad j = O, R \tag{1}$$

with the appropriate boundary conditions. The $c_j(x, t)$ are the concentrations of species at a distance x from the electrode at a time t, and the D_j are the diffusion coefficients. The chronoamperometric response to a potential step is given by the Cottrell equation[13]

$$i = nFD^{1/2}c^{\infty}/(\pi t)^{1/2} \tag{2}$$

where i is the current density, c^{∞} is the bulk concentration of the reactant, and nF is the molar charge transferred.

With a small electrode, nonplanar contributions to mass transport arise at sufficiently long times. Even with a planar surface, the contribution of the edges of the electrode are important, so that the diffusion flux becomes convergent to the electrode, as shown in Fig. 1. A description of the nonplanar flux may be obtained from the solution of the following partial differential equation:

$$\frac{\partial c_j}{\partial t} = D_j\left[\frac{\partial^2 c_j}{\partial r^2} + \frac{1}{r}\frac{\partial c_j}{\partial r} + \frac{\partial^2 c_j}{\partial z^2}\right] \tag{3}$$

where r is the distance from the center of the electrode in the plane of its surface, and z is the perpendicular distance from the electrode surface.

In the case of a spherical electrode with radius r_0, Eq. (3) adopts a particularly simple form, and the current transient in response to a potential step is[13]

$$I = \frac{4\pi r_0^2 nFD^{1/2}c^{\infty}}{(\pi t)^{1/2}} + 4\pi nFDc^{\infty}r_0 \tag{4}$$

where I is the overall current. Two limiting situations can be identified in Eq. (4). At short times, the first, time-dependent term

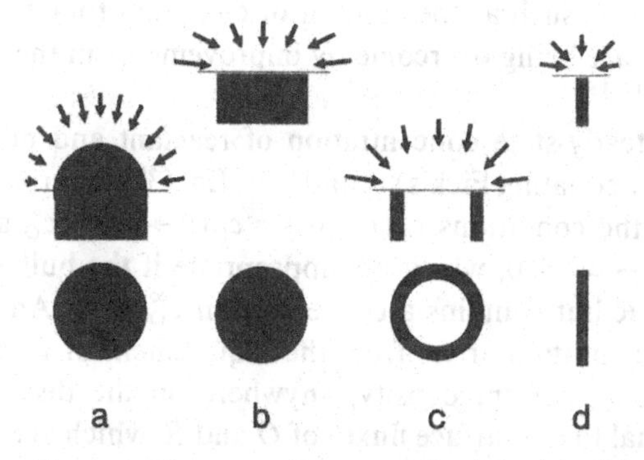

Figure 1. Convergent flux to small sphere (a), disk (b), ring (c), and band (d) electrodes. Top: side view; bottom: plane view of the microelectrodes.

dominates and the current approaches that at a planar electrode of the same area. At long times, a steady-state current is approached, as given by

$$I_s = 4\pi nFDc^{\infty} r_0 \tag{5}$$

A finite planar microelectrode embedded in an infinite coplanar insulator is a much more complex case to treat, in spite of being of great interest because of the relative ease of its fabrication. Diffusion at the edges of the planar conducting surface introduces nonplanar components into the mass transport problem and nonuniform distribution of current over the surface of the electrode. The concentration of electroactive species in the diffusion layer is therefore dependent on two spatial coordinates rather than one as with electrodes that exhibit uniform current distribution, such as the spherical electrode.

The most common geometry for planar microelectrodes is that of a flat microdisk. The description of mass transport requires the solution of Eq. (3) with the boundary conditions relevant to the electrochemical system of interest. Perhaps the most tractable approach is simulation by finite differences[14–17] or orthogonal collocation,[18] with its advantage of flexibility, allowing treatment of various boundary value problems (i.e., electrochemical techniques) and various combinations of diffusion with heterogeneous or homogeneous chemical and/or electrochemical reactions. The major drawback of digital simulation procedures is the lack of generality of the solutions obtained for each particular case. Other disadvantages, such as the amount of computer time and memory consumed, are being overcome by improvements in the algorithms employed.[14–18]

The steady-state concentration of reactant and product may be found by equating Fick's second law [Eq. (3)] to zero and solving subject to the conditions $c_O(r \to \infty) = c_O(z \to \infty) = c_O^{\infty}$ and $c_R(r \to \infty) = c_R(z \to \infty) = 0$, which are appropriate if the bulk solution is devoid of R but contains a concentration c_O^{∞} of O. An additional boundary condition arises from the requirement of Faraday's law that the local current density, anywhere on the disk surface, is proportional to the surface fluxes of O and R, which are themselves given by Fick's first law:

$$i/nF = \pm J_j(z = 0) = \mp D_j(\partial c_j/\partial r)_{z=0} \tag{6}$$

The total current is accessible from the integral

$$I = 2\pi \int_0^a ir\,dr \tag{7}$$

where a is the radius of the microdisk. The solution of Eq. (3) for a microdisk under extreme polarization in the steady state has been obtained by Saito,[19] using discontinuous integrals of Bessel functions, and has been reiterated frequently.[20-22] The steady-state diffusion current to a microdisk is

$$I_d = 4nFDc^{\infty}a \tag{8}$$

For any degree of polarization, the result describing a typical voltammetric wave is[22]

$$I(E) = K(E)I_d/[1 + K(E)] \tag{9}$$

where K is a potential-dependent quantity defined by the ratio of the surface concentrations of O and R:

$$c_O^s/c_R^s = \exp[nF/RT(E - E^0)] = D_R/D_OK \tag{10}$$

The time dependence of the current at a stationary finite disk electrode in quiescent solution has also attracted considerable attention.[1,14,23-30] A comprehensive analytic treatment of the problem has been made[25,28,30] using different approaches to determine the current at short and long times. Introducing a dimensionless time variable

$$\tau = 4Dt/a^2 \tag{11}$$

the current can be expressed as

$$I/4nFDc^{\infty}a = f(\tau) \tag{12}$$

The long-time expansion of the current, accurate for $\tau > 1$, is[25,29]

$$\lim_{\tau\to\infty} f(\tau) = 1 + 4\pi^{-3/2}\tau^{-1/2} + 32(9^{-1} - \pi^{-2})\pi^{-3/2}\tau^{-3/2} + \cdots \tag{13}$$

whereas the short-time expansion is

$$\lim_{\tau\to 0} f(\tau) = 2^{-1}\pi^{1/2}\tau^{-1/2} + \pi/4 - \cdots \tag{14}$$

The first term in Eq. (14) is the familiar Cottrell term. The first two terms of Eq. (14) have been obtained by various methods.[14,24,29,31] A useful analogy[24] is that the current flowing to a polarized inlaid disk electrode initially resembles that flowing to a quarter-sphere of the same radius, but becomes ultimately equal to that flowing to a hemisphere of radius $2a/\pi$.

One further important aspect about mass transport is that a true steady state of diffusion is achieved with a microelectrode provided that the dimensions of its surface are smaller than the characteristic length of the natural convection layer, that is, less than about 20 μm. Thus, a steady-state current is attained without resorting to forced convection, as in hydrodynamic techniques such as the rotating disk electrode. The composition of the electrolytic solution is perturbed only within a small distance from the microelectrode (a few times the radius under totally steady-state conditions), and thus the response is practically unaffected by movements in solution. In fact, the average steady-state current density (total current divided by geometrical surface area) observed at a microelectrode of 5-μm radius is equivalent to that at a rotating disk electrode operated at about 16,000 rpm and increases rapidly as the size of the microelectrode becomes smaller. Table 1 reports the equivalent rotation speed required for the same steady-state current density as observed at microdisks of different radii in aqueous solution. For small electrodes the rotation rate is beyond the usual experimental limit available.

Table 1
Rotation Speed for Rotating Disk Electrodes Required for the Same Average Steady-State Current Density as Observed at Stationary Microdisks of Different Radii[a]

a (μm)	$10^{-2}f$ (Hz)	10^3 rpm
1	67.1	403
5	2.69	16.1
10	0.671	4.03
25	0.107	0.644
50	0.0268	0.161

[a] Average current density for stationary microdisk electrodes calculated from $i = 4nFDc^\infty/\pi a$; for rotating disk electrode, from $i = 0.620nFD^{2/3}\omega^{1/2}\nu^{-1/6}c^\infty$, with $D = 1 \times 10^{-5}$ cm^2 s^{-1} and $\nu = 1 \times 10^{-2}$ cm^2 s^{-1}.

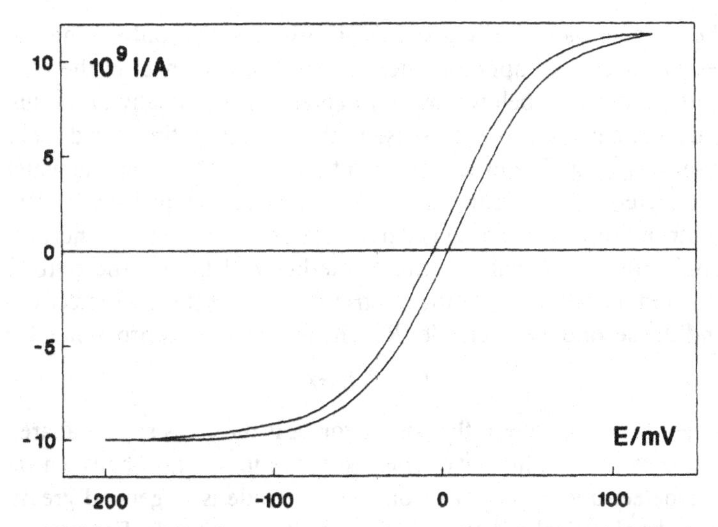

Figure 2. Cyclic voltammogram of 0.1 M $K_3Fe(CN)_6$ and 0.1 M $K_4Fe(CN)_6$ in aqueous 1 M KNO_3 at a platinum microdisk of 5-μm radius. Sweep rate: 30 mV s^{-1}. From Ref. 5.

The time to reach a new steady state of diffusion after a perturbation is on the order $a^2/4D$ [cf. Eqs. (11)-(14)], that is, usually less than 10 ms for microdisks of 5-μm radius. Thus, during voltammetry at moderate sweep rates, as shown in Fig. 2, essentially steady-state conditions are maintained, and the cyclic voltammograms do not develop the typical peaks observed with electrodes of conventional size.

(*ii*) *Electrical Properties*

The small dimensions of microelectrodes give rise to two important characteristics of their electrical properties. One of them is the reduced double-layer capacitance, which is proportional to the electrode area and, in terms of the electrochemical response of a microelectrode, determines the shortest time at which meaningful measurements of faradaic currents can be made. For a potential pulse, the charging current is

$$I_c = (\Delta E/R)\exp(-t/RC_d) \tag{15}$$

where ΔE is the amplitude of the pulse, R is the resistance of the electrochemical cell, C_d is the double-layer capacitance, and t is the time measured since the application of the pulse. The cell resistance has an approximately inverse dependence on the electrode radius (see below), but C_d depends quadratically on radius; thus, the net result is a decrease in the relaxation time for double-layer charging. The other important electrical characteristic which is affected by the reduced size is the electrical resistance. Two contributions need be considered: the resistance within the electronic conductor that constitutes the body of the microelectrode, R_m, and the resistance of the electrolyte between the microelectrode and the secondary electrode, R_s. The first may be approximated as

$$R_m = (l/\sigma A) \tag{16}$$

where l is the length of the conductor, A is the cross-sectional area, and σ is the conductivity. The resistance in solution between the microelectrode tip and a secondary electrode is in general greater than the electrical resistance within the microelectrode. For a microscopic hemispherical electrode, the cell resistance can be calculated from the (specific) conductivity of the electrolyte, κ. The contribution to the resistance dR_s by an element of area $4\pi r^2$ and thickness dr is

$$dR_s = \frac{1}{\kappa}\frac{dr}{2\pi r^2} \tag{17}$$

If the second electrode is placed at a large, effectively infinite distance from the microelectrode, the resistance is found by integration between r_0 and ∞[5,32,33]:

$$R_s = \int_{r_0}^{\infty} \frac{dr}{2\pi\kappa r^2} = \frac{1}{2\pi\kappa r_0} \tag{18}$$

Although the resistances are undoubtedly high and increase as the radius of the microelectrode is made smaller, the resistances per unit area of electroactive surface (which is the figure of merit when measuring with fast transient techniques under non-steady-state conditions of diffusion) are low and become lower with decreasing radius of the microelectrode. Values of the resistance and the product of the resistance and the geometric surface area for several electrode radii are given in Table 2.

It is evident that the resistances per unit area are in fact smaller than the uncompensated resistances between the Luggin capillary

Table 2
Resistance of Hemispherical Pt Microelectrodes Immersed in Aqueous 0.1 *M* KNO_3 Solution

r_0 (μm)	R_m [a] (Ω)	R_s [b] (kΩ)	$10^3 RA$ [c] (Ω cm^2)
∞	—	—	6.90[d]
50	0.135	2.20	0.345
25	0.542	4.39	0.172
10	3.39	11.0	0.0690
5	13.5	22.0	0.0345
1	339	110	0.00692

[a] Microwire of length 1 cm, with $\sigma_{Pt} = 9.4 \times 10^4\ \Omega^{-1}\ \text{cm}^{-1}$.
[b] For hemisphere, with $\kappa = 0.0145\ \Omega^{-1}\ \text{cm}^{-1}$.
[c] Product of R and geometric surface area.
[d] Planar electrode with Luggin tip 1 mm from the surface.

and the working electrode in conventional three-electrode cells. Studies of electrochemical kinetics invariably require either the control or the measurement of the potential of the working electrode with respect to a nonpolarized reference electrode in equilibrium with the electrolytic solution. Thus, voltage step or linear sweep techniques invariably involve (i) a potentiostat, which controls the potential of the working electrode with respect to the reference electrode, and (ii) the measurement of the current between the working electrode and a third, counter electrode. To reduce the effects of the cell resistance on the controlled potential, the electrical contact between the reference electrode and the cell is made through a Luggin capillary. For practical reasons and to avoid deformation of the electric field near the working electrode, the tip of the Luggin capillary, though close to the electrode, is nevertheless at a significant distance from it, and an uncompensated resistance is present which introduces an error into the value of the controlled potential, proportional to the current passing through the cell. If the current is maintained at a low value, the potential of the working electrode can be controlled in a two-electrode cell by passing the necessary current between the working electrode and a reference electrode of much larger area, and therefore maintained at equilibrium with the electrolytic solution. Table 2 shows that the product of the resistances and the geometric surface area in microelectrode systems are in fact smaller than the uncompensated resistances that arise in electrochemical systems with electrodes of conventional size. Thus, the use of two-electrode configurations in cells employing microelectrodes is in general justified.

The development of microelectrodes has allowed the direct measurement of fast homogeneous[12,34-42] or heterogeneous[43-46] kinetics using fast transient electrochemical techniques. These unique capabilities arise from the fact that the ohmic drop (IR) and time constants (RC_d) relative to capacitive currents are considerably reduced at microelectrodes, thus allowing the realization of, for example, cyclic voltammetry at very fast scan rates. Indeed, for a disk electrode of radius a, the resistance varies as a^{-1}, whereas the current (in the situation of planar diffusion relevant to ultra-fast cyclic voltammetry; see Refs. 47 and 48 for steady-state currents) or capacitance is proportional to electrode surface area; that is, it varies as a^2. Thus, IR and RC_d vary as a and tend to be negligibly small for radii on the order of a few micrometers or less.[49] Taking advantage of these features, the feasibility of performing meaningful cyclic voltammetry at scan rates approaching 10^6 V s^{-1} has been established, making it possible to measure rates of heterogeneous electron transfer in the range of centimeters per second or homogeneous rate constants corresponding to submicrosecond lifetimes.[12,34-46,49] This will be discussed together with instrumental aspects in Section II.2(ii).

(*iii*) *Electrode Geometry*

We have been discussing some of the fundamental properties of microspheres and microdisks that arise from the convergent nature of diffusion to their surfaces. A valuable attribute of a microelectrode is its ability to sustain a high steady-state current density. Whereas the surface of spherical electrodes (or of electrodes having the shape of segments of spheres) is uniformly accessible, the local steady-state current density for a microdisk electrode increases toward the edge. At long times, the approach to a spherical concentration field at distances far from the surface of the microdisk leads to the establishment of a stationary state.

A steady state of diffusion is also established with electrodes having the shape of a small ring (cf. Fig. 1) when the diffusion length exceeds the radius of the ring. Thin-ring microelectrodes have a number of advantages compared to microdisk electrodes.[50] The limiting current is dominated by the radius of the ring, whereas the current density depends on its thickness, so that measurements at higher current densities can be made with thin rings. For diffusion-limited currents, the average steady-state current density at a disk is about 27% larger than at a hemisphere, whereas thin rings are vastly superior in this respect. The complexities of the analysis due to nonuniform accessibility at the disk are decreased for thin rings. Ohmic potential losses in the solution are also determined mainly by the ring thickness and are therefore small for very thin rings.

The transient and steady-state behavior of microring electrodes has been described both theoretically and experimentally.[20,21,50–52] The following expression has been derived[21] for the steady-state diffusion current at a thin ring, that is, when the radius, a, is much larger than the thickness of the annulus, $a - b$:

$$I = \frac{\pi^2 nFc^\infty D(a + b)}{\ln[16(a + b)/(a - b)]} \tag{19}$$

Cylindrical, microband, or simply line electrodes have a width smaller than the diffusion length, yet the large electrode area that results due to the length of cylinder or band yields higher currents. However, because they are not smaller than the diffusion length along one of the axes that define their surfaces, the current does not reach a steady state.[53] Thus, their behavior is fundamentally different from that of spherical, disk, or ring-shaped microelectrodes. At long times the overall current at a microband electrode is given by[54]

$$I(t) = \frac{\pi nFDc^\infty l}{\ln[8(Dt)^{1/2} e^{-\gamma}/w]} \tag{20}$$

where w is the width of the band, l is its length, and γ is Euler's constant ($\gamma = 0.5772156$). Even though a true steady state of diffusion is not attained at line electrodes, the inverse logarithmic function of $t^{1/2}$ that describes the transient response decays slowly at sufficiently long times, and thus a quasi-steady-state current ensues.[55] Microbands are particularly useful when assembled in arrays (cf. Section III.6).

2. Experimental Aspects

(i) *Construction Techniques*

The progress in studies of electrochemical reactions at electrodes is dependent on the techniques available as well as on the quality and reproducibility of the electrode surfaces used, whatever their size. It is not impossible, and not even difficult, to construct microelectrodes having very small dimensions[56,57] and use them in a reproducible way. However, the construction of microelectrodes and the preparation of their surfaces is, as with solid electrodes of conventional size, an art that requires the development of a certain degree of expertise.

Microelectrodes of different geometries have been fabricated. These include regular shapes such as disks, hemispheres, cylinders, microrings, and bands, as well as a number of irregular geometries. Some of the techniques for fabrication of microelectrodes have been reviewed.[57] Here we will mainly refer to microdisks, which have been the preferred geometry for electrochemical kinetic studies because of (i) their large edge-to-surface ratio, which maximizes the average steady-state current density,[24,58,59] (ii) their smaller surface area for a given characteristic dimension, which minimizes the double-layer capacitance and hence optimizes the microelectrode response time,[49] and (iii) the relative ease of their fabrication. Microhemispheres, due to the uniform current distribution over their surfaces, are of interest for electrode kinetics studies. Nonetheless, they are mentioned only in passing, given the current difficulties in preparing them with the same degree of confidence, durability, and reproducibility achieved with microdisks of comparable size. Techniques for the fabrication of solid microelectrodes of hemispherical shape are being developed.[60]

A microdisk electrode is most conveniently fabricated by exposing the circular cross-sectional area of a conducting thin wire or fiber, elsewhere isolated from solution. As with electrodes of conventional size, the insulator used must be chemically inert and not soluble in the electrolyte. Because microelectrodes invariably show a large edge-to-surface ratio, it is particularly important to achieve a good and uniform seal between conductor and insulator; otherwise, the behavior would be dominated by penetration of the solution between these materials. This is probably the single and most important difficulty in constructing a microdisk electrode, and also the reason for the widespread use of carbon fibers[2,5,61-63] and platinum wires[5,6,35,64-66] sealed to glass or insulated with epoxy resins as microelectrodes.

Platinum microelectrodes are usually made out of platinum wire,[8,56,67] which may be coated with a layer of silver for diameters below 5 μm (known as Wollaston wire). An appropriate length (ca. 1 cm) of this wire is soldered to a fine copper helical spring, used to absorb mechanical tension, which in turn is soldered to a thicker copper rod, which may be the core of a coaxial cable. The assembly is then placed inside a glass capillary and heat-sealed under vacuum. If Wollaston wire is used, about 5 mm of wire, protruding through the end of the capillary, is immersed in HNO_3 to remove the silver layer and then pushed back into the glass capillary before sealing it into glass.

It is important to use an appropriate type of glass in manufacturing the microelectrode, in order to maintain the sealing upon cooling the assembly from the temperature at which the glass is softened to ambient temperature. The coefficient of thermal expansion for platinum is ca. 90×10^{-7} $(°C)^{-1}$, and a close match is obtained with potash soda lead, soda lime,[68] or lead borosilicate glass, the latter of which, used for diode manufacture, also possesses a high resistivity, in excess of 10^{17} Ω-cm at 25°C.[69]

Carbon microelectrodes are constructed from carbon fibers, usually with diameters in the range of 6 to 12 μm. The thermal expansion coefficient of carbon fibers is much lower than that of platinum, and thus a good seal is obtained with borosilicate glass. About 1 cm of fiber is sealed into glass under vacuum (which also serves to prevent oxidation of the carbon fiber while flame heated), and the external contact may be made with a copper wire, either with the help of Hg or by melting Pb or Woods metal inside the glass tube.

The final stage in the construction of a microdisk electrode is the exposure of the electrode surface and its preparation. The tip of the metal wire or carbon fiber is exposed by grinding with 600-grit emery paper and later polished on cloth with alumina powders in aqueous suspension down to 0.05 μm. This stage is best undertaken with frequent examination under an optical microscope, with the help of which the existence of cracks or imperfections within the seal can be clearly distinguished. Finally, the microelectrode is ultrasonically cleaned or wiped on clean, dry cloth[70] to remove alumina particles that might have adhered to the electrode surface during mechanical polishing.

Other insulating materials that have been extensively used for the manufacture of microdisk electrodes are epoxy resins,[1,35,71-75] thus avoiding the severe thermal treatment and mismatch involved in sealing the conducting material with glass. Their use, however, raises other concerns, such as the adherence of the epoxy to the electrode material, which may be increased by silanization prior to potting the epoxy, and their stability in the electrolyte, particularly in nonaqueous media.

(*ii*) *Instrumentation*

Due to the low currents usually involved in microelectrode measurements, conventional potentiostatic control with three-electrode cells is in general not necessary, and the reference electrode serves also as counter electrode. Potentiostatic control is usually achieved by applying the potential (with inverted sign) directly to the reference/counter electrode by means of a waveform generator. The currents are measured with a high-gain picoammeter or a purpose-built current follower connected in series with the cell. Experiments with microelectrodes are therefore generally simpler from the instrumental point of view than experiments with electrodes of larger sizes.

Some precautions are necessary to exclude electrical noise from the potential applied to the cell and the currents measured by the associated circuitry. The main source of noise in the measured current is due to capacitive coupling resulting from the high impedance of the microelectrode. It is thus important to electrically shield the cell and its electrical connections from external sources of ac voltage. This can be achieved by enclosing the electrochemical cell in a Faraday cage and using low-noise coaxial cable to connect it with the waveform generator and current-measuring devices. Capacitive coupling also induces stray currents due to mechanical vibrations, which should be avoided. When using a purpose-built current amplifier, considerable reduction in noise is achieved by reducing the length of wire connecting the microelectrode to the input stage of the current amplifier and enclosing the cell and the current amplifier within the same Faraday cage, supplying the necessary regulated power from the outside.

The use of disk microelectrodes considerably reduces the ohmic and capacitive contributions that limit the extraction of meaningful fast kinetics data from electrochemical systems at short times. Recent developments in this field[40-46] have allowed a gain of about three orders of magnitude in the maximum reaction rate that can be measured in the determination of rate constants of heterogeneous electron transfers and of associated homogeneous reactions as compared to that for conventional-size (millimeter diameter) electrodes.

The response time of the current amplifier while measuring low currents may be of concern, particularly with low-amplitude perturbations at high frequencies. The large impedance usually needed to attain amplification requires minimization of stray capacitance at the input, and thus again, a short electrical path connecting the microelectrode to the input of the current amplifier is beneficial. Flat response up to about 60 kHz with small-amplitude perturbation techniques or ca. 10 kV s^{-1} in cyclic voltammetry or down to 20 μs with pulse techniques can be achieved with readily available high-input impedance operational amplifiers or conventional current amplifiers,[5,76-78] using two-electrode cells.

At higher scan rates in cyclic voltammetry or shorter times in potential-step techniques, the bandpass limitations of the amplifiers, the effects of stray capacitances, and the ohmic drop in solution become more apparent, interfering with the determination of the kinetic parameters of interest.[49,79,80] Several strategies have been devised to extend the time resolution to submicroseconds and the scan rates to the 10^6 V s^{-1} range. These include *post factum* numerical simulation taking into account the effects of ohmic drop and double-layer charging,[41,43,45,81] or on-line compensation of the effects of the ohmic drop with current-measuring devices based on current feedback operational amplifiers,[80] or the use of positive voltage feedback compensation[49] of the type previously described for electrodes of conventional size,[82-84] this latter on-line compensation technique requiring the use of three-electrode cells. Some recent approaches to increasing the bandwidth of the potential control and the measurement of current include potentiostats based on voltage follower circuits[85] and current-to-photon conversion.[86]

III. APPLICATIONS

1. Kinetics under Steady-State Conditions

The large steady-state mass transport rates achievable with microelectrodes offer possibilities for the study of electrode kinetics under steady-state conditions. Simple solutions of the mass transport problem can be obtained for hemispherical or spherical microelectrodes, provided that the reaction kinetic terms are first order or pseudo-first order. An example is the study of the Hg/Hg_2^{2+} system,[3,4,87] where an ensemble of approximately 10^6 mercury droplets of radii 100 nm, electrodeposited on vitreous carbon, was used as electrode in a thin-layer cell, and values of k^0 and α were obtained from quasi-steady-state current-voltage measurements. The use of spherical or hemispherical microelectrodes is, however, somewhat restricted. They have been applied, for example, to the electrodeposition of single mercury droplets[88,89] or electrolysis with dispersions[90,91] and metallic or semiconducting colloidal particles.[92,93]

Disk and the more recently introduced ring microelectrodes[50] are of easier construction, but the quantitative description of mass transport becomes more involved. The mathematical difficulties are due to the discontinuities at the edges of the electrodes, where, in principle, the diffusion-limited flux for a reversible system becomes infinite,[24] although in a real system the combined effects of the finite rates of surface reactions and the distribution of potential and concentration will limit the rates at the edges. In spite of the more difficult theoretical description, disk and ring microelectrodes share the advantage that quasispherical diffusion fields are established at relatively short times, giving rise to high rates of mass transfer to their surfaces, so that the kinetics of fast reactions at the electrode surface or in solution can be studied under steady-state conditions.[94]

The description of the steady state of diffusion is obtained by equating Eq. (3) to zero; that is,

$$\frac{\partial^2 c}{\partial r^2} + \frac{1}{r}\frac{\partial c}{\partial r} + \frac{\partial^2 c}{\partial z^2} = 0 \qquad (21)$$

This equation must be solved with the boundary conditions corresponding to the geometry of the microelectrode and constant concentraton or constant flux at the electrode surface.

For a reversible reaction at the surface of a microdisk electrode with constant surface concentration c^s,

$$c^s = \text{constant}, \qquad 0 < r < a, z = 0$$

$$\partial c/\partial z = 0, \qquad r > a, z = 0$$

the total flux may be obtained by integration:

$$F = 2\pi D \int_0^a \left[\frac{\partial c}{\partial z}\right]_{z=0} r\,dr \tag{22}$$

Using discontinuous integrals of Bessel functions, it has been shown[51,95] that

$$F = 4D(c^\infty - c^s)a \tag{23}$$

leading to the definition of the mass transfer coefficient under constant surface concentration conditions as

$$(k_m)_c = 4D/\pi a \tag{24}$$

For irreversible reactions, there will be a constant flux Q to the surface of the disk, and thus the boundary condition

$$D(\partial c/\partial z) = Q \text{ (a constant)}, \qquad 0 < r < a, z = 0$$

holds. Evaluation of the average surface concentration leads to the definition of the mass transfer coefficient for the uniform flux conditions[51,95]:

$$(k_m)_Q = 3\pi D/8a \tag{25}$$

Mass transfer to the electrode is thus governed by Eq. (25) for irreversible and quasireversible reactions at low overpotentials and by Eq. (24) for all reactions at very high potentials.[95] The effects of the distribution of potential in the solution (also known as the tertiary current distribution; see below) cause Eq. (25) to be an appropriate description of polarization curves obtained using microdisk electrodes under most conditions.[20,51,96]

The effects of the assumptions about the magnitude of the mass transfer coefficient are small and near the detection limit for present experiments.[95,97] The solution of the mass transfer problem with the boundary conditions that arise from consideration of the concentration of reactants at the electrode surface.

$$D(\partial c/\partial z) = kc, \qquad 0 < r < a, z = 0$$

admits, for reversible first-order electrode reactions, a relatively simple solution[22]; cf. Eq. (9). Other reaction mechanisms can be treated approximately using the mass transfer coefficients given by Eqs. (24) and (25). For the particular case of a simple redox couple following Butler-Volmer kinetics, the form of the polarization curves has been obtained[98] from

$$i_0 = \left[\left(\frac{c_O^\infty - c_O(r)}{c_O^\infty}\right)\exp(-\alpha\eta F/RT) - \left(\frac{c_R^\infty + c_R(r)}{c_R^\infty}\right)\exp[(1-\alpha)\eta F/RT]\right] \tag{26}$$

where η is the overpotential and α is the charge transfer coefficient, by calculating the surface concentrations using the mass transfer coefficient in Eq. (25) for constant flux conditions:

$$\frac{i}{F(k_m)_Q c^\infty} = \frac{\exp(-\alpha\eta F/RT) - \exp[(1-\alpha)\eta F/RT]}{F(k_m)_Q c^\infty/i_0 + \exp(-\alpha\eta F/RT) + \exp[(1-\alpha)\eta F/RT]} \tag{27}$$

Equation (27) however does not describe correctly the limiting current, since at that limit it is expected that the concentration is uniform over the surface, and thus $(k_m)_c$ should be used instead of $(k_m)_Q$ at high overpotentials. The error incurred at the limiting current is $[(k_m)_c - (k_m)_Q]/(k_m)_c$, that is, ca. 7.5%. The use of the average of $(k_m)_c$ and $(k_m)_Q$, that is, assuming that $k_m = [(k_m)_c + (k_m)_Q]/2$, reduces the maximum error to ca. 3.5%, and this can be further reduced by, for example, making the assumption

$$k_m = \{(k_m)_Q[(i_l - i)/i_l] + (k_m)_c(i/i_l)\} \tag{28}$$

in which case the maximum error becomes less than 2.5%. When the mass transfer coefficient given by Eq. (28) is used, the error becomes virtually zero at potentials up to the half-wave potential, the overpotential range of greatest interest in kinetic studies, and also zero at the limiting current, the region of greatest interest for analytical applications.[98]

Further complications arise in the description of electrode reactions which are nonlinear in the concentration terms, either because of reaction orders different from one or because of adsorption of reactants.[95] Moreover, the solution of the diffusion equation, Eq. (21), for a microdisk leads to the conclusion, as mentioned earlier, that the flux at its edge becomes infinite in the limiting current region, an unreal situation as the local overpotential would also have to become infinite at the edge, revealing that the effects of the distribution of the potential in the solution (the tertiary current distribution) need to be taken into account.

Even though a rigorous and general description of the steady state of mass transport to microdisk electrodes is at the present time not available, the feasibility of obtaining meaningful kinetic information from this kind of experiments, using judicious approximations, has already been established.[8]

The kinetics of fast first-order or pseudo-first-order homogeneous chemical reactions coupled to heterogeneous electron transfer reactions has been the subject of several investigations.[6,62,99] Fleischmann *et al.*[6] studied the electroreduction of acetic acid using platinum microdisks of radii in the range 0.3–25 μm. This reaction proceeds via a chemical-electrochemical (CE) reaction mechanism:

$$\mathrm{HA} \underset{k_b}{\overset{k_f}{\rightleftharpoons}} \mathrm{H}^+ + \mathrm{A}^-$$

$$\mathrm{H}^+ + e^- \rightleftharpoons \tfrac{1}{2}H_2$$

where k_f and k_b are related to one another by the equilibrium constant for the acid dissociation. For a spherical diffusion field, using the approximation that the average current density at a microdisk of radius a is identical to that of a sphere of radius $\pi r_0/4$[24] and in the presence of a large excess of A^-, the steady-state current density is expressed by

$$\frac{1}{i_d} = \frac{(k_b c^{\infty}_{\mathrm{A}^-})^{1/2}}{nF(D_{\mathrm{H}^+})^{1/2} k_f c^{\infty}_{\mathrm{HA}}} + \frac{\pi a}{nFD_{\mathrm{HA}} c^{\infty}_{\mathrm{HA}}} \tag{29}$$

Equation (29) shows a further analogy between the behaviour of a microelectrode under steady-state conditions and that of a rotating disk electrode. Whereas with a rotating disk electrode, kinetic data are obtained by plotting the inverse of the current as a function of the inverse of the square root of the rotation rate,[13] with a microelectrode the current function is plotted with respect to the radius. A good linear relationship was obtained between i_d^{-1} and a, as shown in Fig. 3. The value of c_{HA}^{∞} was obtained from the slope, and k_b, was calculated from the intercept[6] and found to be 4×10^{10} $\mathrm{dm^3\,mol^{-1}\,s^{-1}}$, in good agreement with previously reported values.[100]

Mediated or indirect electrode reactions such as

$$R \rightleftharpoons O + ne^-$$

$$O + A \xrightarrow{k} R + B$$

where the electron transfer process is followed by a homogeneous chemical reaction regenerating the electroactive species are of increasing interest.[101] The kinetics of the coupled chemical reaction can be determined by establishing the ratio of the steady-state

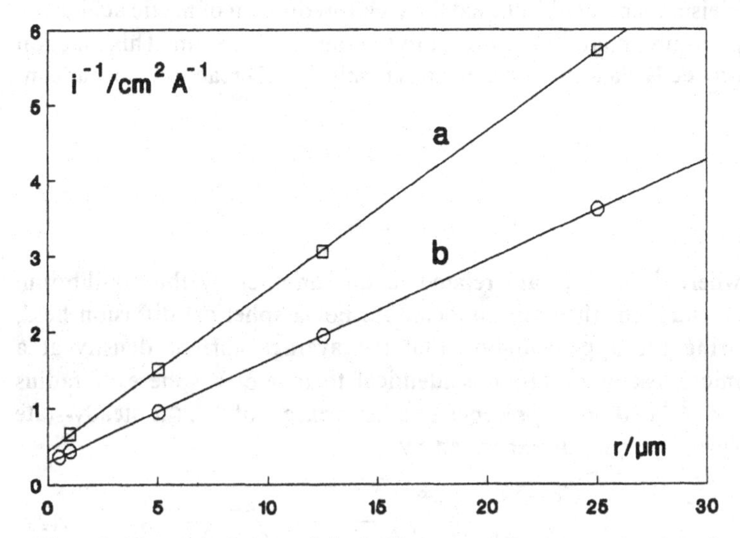

Figure 3. Plot of the inverse of the limiting current density as a function of electrode radius for the reduction of acetic acid in 1 M aqueous sodium acetate. Acetic acid concentration: (a) 40 mM; (b) 70 mM. From Ref. 6.

limiting currents with and without A in solution, I_k and I_d, respectively, as a function of the radius of the microdisk.[102] In the presence of a large excess of A, the chemical reaction becomes pseudo-first order and

$$I_k/I_d = 1 + (kc_A^\infty/D)^{1/2}(\pi a/4) \tag{30}$$

whereas under second-order kinetics conditions:

$$\frac{I_k}{I_d} = 1 - \frac{k\pi^2 a^2 c_R^\infty}{32D} + \frac{1}{2}\left[\left(\frac{k\pi^2 a^2 c_R^\infty}{16D}\right)^2 + \frac{k\pi^2 a^2 c_A^\infty}{4D}\right]^{1/2} \tag{31}$$

I_k must be measured experimentally while I_d may either be measured by experiment or calculated from $I_d = 4nFDc_R^\infty a$ [cf. Eq. (8)], provided D is known. With the help of Eq. (31), the silver(I)-mediated oxidation of chromium(III) and manganese(II) has been studied using microelectrodes with radii in the range of 0.3 μm and 62.5 μm.[78] In the oxidation of chromium(III) the rate-determining step was found to be the homogeneous electron transfer reaction

$$\text{Ag(II)} + \text{Cr(III)} \rightarrow \text{Ag(I)} + \text{Cr(IV)}$$

with a rate constant of 10^4 dm^3 mol^{-1} s^{-1}, whereas the oxidation of manganese(II) was found to be controlled by the diffusion of Mn(II) to the surface except for microelectrodes with radii below 5 μm, in which case the rate constant of the homogeneous reaction was estimated to be of the order of 10^6 dm^3 mol^{-1} s^{-1}.

A different situation arises when the homogeneous reaction proceeds at a slow rate. Under this condition the product of the preceding electrochemical reaction can diffuse away from the vicinity of the microelectrode surface before regenerating the reactant, and thus a minimal catalytic contribution is observed. This has been exploited in the determination of extracellular dopamine in the brain,[62] where it is known that ascorbic acid, present in large concentrations in the extracellular fluid, undergoes a homogeneous electron transfer reaction with the oxidation product of dopamine, regenerating it. When carbon fiber microdisks of 3-μm radius are used, only a small portion of the regenerated dopamine can return to the electrode, largely reducing the interference caused by the presence of ascorbic acid.

A common group of reactions in organic electrochemistry are described by the following general scheme:

$$A \pm n_1 e^- \rightleftharpoons B$$

$$B \underset{k_{-1}}{\overset{k_1}{\rightleftharpoons}} C$$

$$C \pm n_2 e^- \rightleftharpoons D$$

$$B + C \overset{k_2}{\rightleftharpoons} A + D$$

where the fourth step is a homogeneous electron transfer in competition with the third, heterogeneous electron transfer step. When the fourth step in the scheme is unimportant and the second, a chemical step, is rate determining, it reduces to the ECE mechanism, whereas when the third step is unimportant and the second is rate determining, it corresponds to a DISP 1 mechanism.[103]

Steady-state current measurements at microdisks have been applied to the study of such reactions.[104] The treatment of data is simplified if the apparent number of electrons involved, n_{app}, which can be obtained from the value of the plateau currents, is analyzed instead of the limiting currents themselves. A further simplification was achieved by defining $\Delta n = n_{app} - n_1$, in terms of which, and using the analogy that a microdisk behaves like a sphere of radius $\pi r_0/4$, the ECE and DISP 1 mechanisms were described by linear plots of Δn^{-1} as a function of a^{-1}, as given by Eqs. (32) and (33), respectively:

$$1/\Delta n = (4/\pi)(D/n^2 k_1)^{1/2}(1/a) + 1/n \tag{32}$$

$$1/\Delta n = (4/\pi)(2D/n^2 k_1)^{1/2}(1/a) + 1/n \tag{33}$$

As a test ECE system, the oxidation of anthracene in dry acetonitrile was investigated, whereas the oxidation of hexamethylbenzene was studied as a test DISP 1 system. Both systems produced linear relationships between Δn^{-1} and a^{-1}, as predicted, yielding values of the rate constants that were in fair agreement with previously reported data. Although for a given value of k_1 the slope for the DISP 1 mechanism is $\sqrt{2}$ times greater than that for the ECE mechanism, microelectrode data alone do not permit differentiation between these mechanisms.

Systems with second-order rate-determining steps have been also explored,[104] but these cannot be treated in the simple manner depicted by Eqs. (32) and (33). Because of the uneven distribution of reaction intermediates along the surface that arises from the nonuniform current distribution, the available simplified theories are not applicable. Thus, at the present time they require full two-dimensional simulation of the microdisk.[14,15,29,105]

Other situations in which microelectrodes are experimentally advantageous are encountered in the study of electrochemical reactions under very extreme conditions, such as those relevant to the operation of fuel cells. In the phosphoric acid fuel cell, the conditions of measurement are extremely taxing, 190°C and 2% water in phosphoric acid,[106] the solubility and diffusion coefficient of oxygen are low, and thus limiting currents for its reduction in this medium are extremely small, particularly at lower temperatures. The increased limiting currents obtained at microelectrodes made possible the determination of charge transfer coefficients, exchange current densities, and the apparent heat of activation for the oxygen reduction reaction at Pt microdisks with radii of 12.5 μm and 250 μm over a wide range of temperatures.[107]

The examples described demonstrate the usefulness of steady-state measurements with microelectrodes for the determination of electrode kinetics. With microelectrodes of radii on the order of 1 μm, heterogeneous rate constants of ca. 0.1 cm s^{-1} can be measured.[8] The rates of mass transport increase rapidly with the decreasing size of the microelectrode, and thus with radii on the order of 0.05 μm, it should be possible to observe, under steady-state conditions, electron transfer reaction rate constants of the order of 50 cm s^{-1}.[5,108]

The transition time to the steady state is on the order of $a^2/4D$. For electrodes having dimensions of 1 μm this is about 250 μs, and thus rate constants of first-order reactions up to 10^3 s^{-1} can be determined. With radii of the order of 0.05 μm, this time scale is reduced to about 600 ns, defining an upper limit for reaction rate constant measurements near 10^6 s^{-1}. Such measurements have been virtually impossible using other existing electrochemical techniques due to charging effects and mass transport limitations.[108]

2. Kinetics with Fast Transients

Whereas steady-state measurements at microelectrodes rely on the development of spherical diffusion fields near their surface at long times, planar diffusion dominates during short times after the onset of electron transfer processes at microelectrode/solution interfaces. Thus, the kinetics of electrode reactions can be studied, for example, with short measurement times in potential-step techniques[5,46,109-111] or with high-scan-rate cyclic voltammetry,[12,35,38-40,45] applying essentially the methods for analysis of data already developed for electrodes of conventional size.[13] The electrical properties of microelectrodes that make them advantageous for large-signal-amplitude, fast transient techniques have been already discussed in Section II.2(ii). An additional advantage that arises from transient kinetic determinations at short times as compared with steady-state measurements is related to the increased importance of the edges at microelectrodes. Nonuniform current distributions at the edges are less likely to develop at short times while diffusion to the electrode surface remains essentially planar.

In spite of the difficulties associated with the extraction of kinetic data from cyclic voltammograms at high sweep rates, this has been by and large the preferred transient technique for the study of fast reactions at microelectrodes.[12] The investigation of a reaction with a given rate constant requires experiments over a given range of scan rates. The combined effects of electrolyte resistance, electrode capacitance, concentration and diffusion coefficient of reactants, and the small contribution from edge diffusion defines the most convenient size of the microelectrode for that study.[81]

The reduction of anthracene in acetonitrile at gold microdisks has been used as a test system for the methods developed.[44,49,81] The rate constant of the reaction can be obtained from the variation of the anodic to the cathodic peak potential distance, ΔE_p, with the scan rate.[112] It has been shown that current techniques[49,80] can be applied to the accurate determination of large heterogeneous rate constants (≥ 3 cm s^{-1}) at scan rates up to ca. 100×10^3 V s^{-1} without having to account for distortions in the voltammograms due to the capacitive and uncompensated ohmic contributions, as shown in Fig. 4, or as high as 10^6 V s^{-1} with distorted, though readily interpretable, voltammograms.[42] In terms of the kinetics of coupled homogeneous reactions, these scan rates correspond to half-lives of less than 100 ns for electrogenerated intermediates, thus extending the range of first-order rate constants that can be determined with microelectrodes to ca. 10^7 s^{-1}.

High-scan-rate cyclic voltammetry at microelectrodes has also proven useful for the determination of thermodynamic data related to electrode processes.[34] The oxidation of most aromatic compounds, for example, exhibits chemically irreversible behavior at low sweep rates (below 100 V s^{-1}), which is manifested in the absence of the cathodic component on the return potential sweep, largely due to competition from fast follow-up reactions of the metastable arene cation radicals formed, which become unimportant at high scan rates, thus enabling the determination of the standard potential of the electrode reaction.

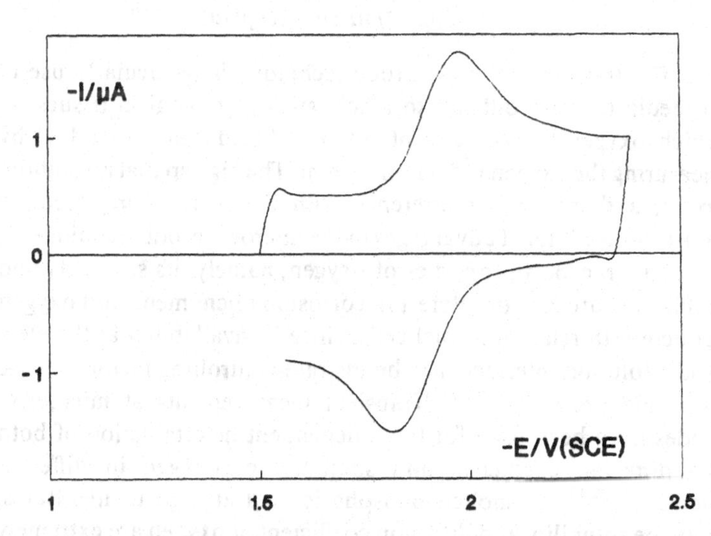

Figure 4. Cyclic voltammogram of 10 m*M* anthracene in acetonitrile with 0.6 *M* Et_4NBF_4 as supporting electrolyte at a gold microdisk electrode of 2.5-μm radius. Scan rate: 95 kV s^{-1}. From Ref. 49.

3. Transport Properties in Solution

Transport phenomena occur in almost every chemical process, and in solutions they are closely related to the structure of liquids and to the changes in this structure caused by the solute.[113] If a microelectrode is operated under conditions where the current is limited by the availability of the electroactive species in the surrounding medium, its response is determined by the diffusion coefficient of the reacting species, its bulk concentration, and the size of the microelectrode [cf. Eq. (8)]. Under such conditions the time constant of the electrode response to polarization corresponds to the time for establishing the steady state of diffusion, which is of the order of $a^2/4D$. Thus, if the time after polarization is small compared to the characteristic diffusion time, then according to Eq. (14) a plot of the measured current as a function of $1/\sqrt{t}$ is linear,[25,114] and diffusion coefficients[26,115-117] and concentrations can be determined from the slope and intercept of this plot, obtained from potential-step experiments[118]:

$$D = a^2 \,(\text{intercept})^2/\pi(\text{slope})^2 \tag{34}$$

$$c = (\text{slope})^2/nF(\text{intercept})a^3 \tag{35}$$

The transient microelectrode technique is particularly useful in media that are difficult to reach, such as biological tissues, in which oxygen transport is of interest[119] and can be studied by measuring the oxygen reduction current. The high spatial resolution, speed, and minimal interference with the surrounding medium constitute additional advantages of the microelectrode technique.[118]

The transport properties of oxygen, namely, its solubility and diffusivity, are also of interest in corrosion phenomena and oxygen reduction in relation to fuel cells, since its availability at the electrode/solution interface may be the rate-controlling factor in these electrode processes.[120,121] Transient measurements at microelectrodes have been used for the independent determination of both the diffusion coefficient and solubility of oxygen in different media.[107,122,123] In molten phosphoric acid at high temperatures, both the solubility and diffusion coefficient of oxygen are extremely low and vary anomalously with temperature because of changes in the structure of phosphoric acid, but the determination of c and D

becomes relatively easy with transient measurements using microelectrodes.[107]

Near-critical and supercritical fluids are of interest because the solvent properties are very different from those of the corresponding liquid phase and the transport properties are adjustable over a wide range by variation of temperature and pressure. Recent studies have demonstrated that electrochemical experiments can be carried out in diverse supercritical fluids,[124-128] including H_2O, NH_3, $CO_2 + H_2O$, CH_3CN, and SO_2. Transient measurements at platinum microdisks of 12.5-μm radius, for example, have been used to determine the diffusion coefficient of $Fe(bpy)_3^{2+}$ in supercritical SO_2 over a wide range of conditions,[128] and it was found to vary from 1.3×10^{-6} cm^2 s^{-1} at 190 K and 0.03 bar to 1×10^{-3} cm^2 s^{-1} as the temperature and pressure were increased to 448 K and 123 bar.

Low-temperature electrochemistry has been investigated in a number of contexts.[129-132] As conductivities are reduced with decreasing temperatures, ohmic drops in solution become more significant, but a distinct improvement is obtained with microelectrodes.[133,134] Diffusion coefficients have been obtained in different solvents at low temperatures from steady-state measurements.[81] It was found that the increase in solution resistivity at lower temperature is matched by a decrease in diffusion coefficient. The steady-state current to a microdisk is proportional to D, and the solution resistance is proportional to the resistivity, so in the steady state the ohmic drop is roughly independent of temperature, the increasing solution resistance being compensated by decreasing currents.[81] In contrast, under linear diffusion conditions, such as those that obtain at planar electrodes or at microelectrodes during transient measurements (e.g., high-scan-rate cyclic voltammetry), the current depends on the square root of the diffusion coefficient so only partial compensation is observed (the solution resistance increases more sharply than the currents decrease as the temperature is lowered), leading to larger errors due to ohmic drop.

4. Electrochemistry in Highly Resistive Media

Electrochemical techniques usually require the presence of a supporting electrolyte in order to reduce the solution resistance and to suppress migration currents whenever charged species are under study. The presence of the electrolyte necessarily alters the ionic strength and may also complicate the interpretation of the data, by, for example, ion-pairing with reactants, thus altering the thermodynamic and kinetic properties and thereby complicating or invalidating comparisons of electrochemical and spectroscopic measurements.[135] Some inert solvents, such as the saturated hydrocarbons, have not been used in electrochemical studies due to their inability to dissolve electrolytes.[136] The supporting electrolyte is additionally one of the main sources of impurities in electrochemical systems and thus may interfere with the determination of the kinetics of slow reactions, with double-layer properties, and in electroanalysis.

One of the advantages of microelectrodes is the low ohmic potential drop between the counter and working electrodes.[2,5] Thus, it might be anticipated that microelectrodes would be superior to planar electrodes for minimization of resistance effects in poorly conducting solutions. The lower currents associated with small electrodes, however, are compensated by increased resistances, and, in principle, the ohmic drop is independent of the radius of the microelectrode under steady-state conditions. In the presence of a large excess of supporting electrolyte.

$$IR = nFDc^{\infty}\kappa \tag{36}$$

Equation (36) has been shown to apply to working electrodes of any geometry, infinitely distant from infinitely large counter electrodes, with dimensions small enough to permit the establishment of a diffusional steady state unimpeded by convection,[47] requirements that are not difficult to meet when working with microelectrodes. With a 50-fold excess of supporting electrolyte, the ohmic drop reduces to a negligible amount under steady-state potentiostatic conditions. In the absence of supporting electrolyte, for example, in the electrolysis of a pure binary electrolyte, theoretical analysis[47] indicates that ohmic drops are not negligible and remain independent of the electrode radius, but become independent of the concentration of electroactive species, provided the thickness of the diffuse double layer remains small compared to the diffusion layer thickness.

Experimental studies with microelectrodes in the absence of deliberately added supporting electrolytes have shown, however, that in contrast to the theoretical expectations, ohmic drops in steady-state measurements are indeed reduced with decreasing microelectrode radius. The oxidation of ferrocene (Fc) in acetonitrile is a reversible one-electron process that produces the ferricinium cation[137]:

$$\mathrm{Fc} \rightleftharpoons \mathrm{Fc}^{+} + e^{-}$$

Using Pt microelectrodes of 25-μm radius, the sigmoidal-shaped voltammograms obtained at slow sweep rates[65], shown in Fig. 5, are consistent with the establishment of steady-state diffusion conditions. In the absence of deliberately added supporting electrolyte, a significant ohmic drop was present, as indicated by reciprocal slopes of $\log[(i_d - i)/i]$ versus E plots of the data considerably larger than 59 mV, the value obtained in the presence of a large excess of tetraethylammonium perchlorate. By contrast, with 0.5-μm radius microelectrodes the voltammetric curves obtained without adding electrolyte (Fig. 5b) were very similar to those obtained in the presence of electrolyte. Thus, the use of small microelectrodes does seem to be advantageous in highly resistive media. These studies have been extended to solutions of lower permittivity in mixed solvent systems,[138] but then large corrections are needed to account for the effects of ohmic drops, and the procedures fail for solutions of very low dielectric permittivity.

The work described on the reduction of ferrocene in acetonitrile[65] was most likely done in the presence of water as an impurity, and water can readily dissociate and be present at sufficient concentration to act as an electrolyte. The electrode reaction produces the ferricinium cation, which contributes to charge transport under the combined effects of diffusion and migration, but also introduces lack of electroneutrality, which may be partially compensated by reactions with the solvent or with traces of water.[139] There are also double-layer effects to be taken into account, since the Debye length is large for a solution with low ionic content and can be comparable to the size of the microelectrode and the diffusion layer around it.[53] It is therefore unlikely that the simple analysis that leads to Eq. (36) can be applied to microelectrodes of small radii in solutions sparsely populated with ions.

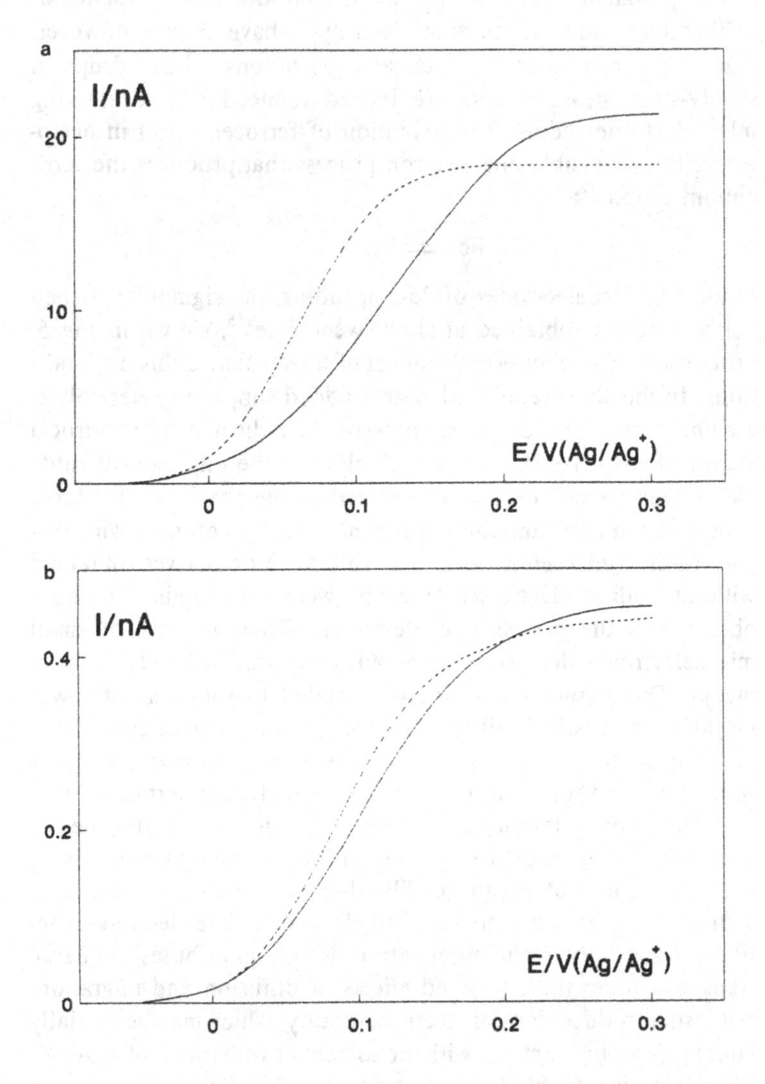

Figure 5. Voltammogram for oxidation of 1 m*M* ferrocene in acetonitrile: (a) at a Pt microelectrode of 25 μm radius, in the presence (- - -) and in the absence (——) of 0.1 *M* Et_4NClO_4; (b) at a Pt microelectrode of 0.5-μm radius, in the presence (- - -) and in the absence (——) of 1 m*M* Et_4NClO_4. From Ref. 65.

In spite of the difficulties associated with the full interpretation of results obtained in poorly conducting media, steady-state measurements at microelectrodes have proven useful for analytical determinations of charged and uncharged species in aqueous solutions in the absence of supporting electrolyte[10] or in the exploration of systems in unusual situations, such as in the gas phase,[140] or at extreme values of potential.[141,142] The positive potential limit is frequently determined by reactions of the electrolyte anion added and thus in its absence can be considerably extended. This allowed, for example, the study of the anodic oxidation of methane, butane, and other aliphatic alkanes in acetonitrile at potentials up to ca. 4.3 V versus Ag/Ag^+.[141]

5. Nucleation and Phase Formation

The first step in the formation of a new phase is the formation of stable nuclei from clusters of subcritical size. This takes place by the stepwise addition (and removal) of atoms or molecules, and therefore a distribution of clusters of different sizes occurs in the process. Nucleation consists in the propagation of clusters through this distribution and thus is a stochastic process. It is well known that the magnitude of fluctuations relative to the mean increases as the size of the system decreases.[143] It follows that if a small number of nuclei are grown on an electrode, then large fluctuations of individual transients from the mean over a number of experiments are to be expected. It further follows that additional kinetic information may be obtained from analysis of the higher moments in the statistical distribution.[144]

Most studies of electrochemical nucleation and phase formation are carried out under conditions of multiple nucleation at electrodes of conventional size.[145] The centers taking part in the phase transformation interact with each other not only as the result of their physical overlap but also as an array of sinks in a three-dimensional diffusion field (an ensemble of microelectrodes; cf. Section III.6). A distribution of nuclear birth times implies also a distribution of crystallite sizes and a time-dependent spatial distribution of centers,[146] and the consequences of these generally unknown distributions are not well understood.

If the area of the electrode is restricted to a size such that only one nucleus is allowed to develop, then under potentiostatic conditions the measurement of the nucleation rate is reduced to the determination of the distribution of times for the appearance of the first growing center.[89,147-149] The growth rate can be determined from the subsequently observed current.[5,150]

(*i*) *Phase Growth Rates*

In the deposition of mercury single nuclei onto carbon fiber and platinum microelectrodes,[5] a long induction period (which varies from experiment to experiment) is observed before the birth of a nucleus, but after it is established, the growth of mercury nuclei was found to be controlled by localized spherical diffusion under near steady-state conditions. The results agree well with the predictions of theory[151-154] and also with data obtained by computer simulation.[155] For hemispherical growth, the radius of the nucleus varies with the square root of time according to

$$r = (2DcM/\rho)^{1/2}[1 - \exp(-nF\eta/RT)]^{1/2}t^{1/2} \tag{37}$$

where M and ρ are, respectively, the molecular weight and density of the electrodeposited species. Under conditions of steady-state diffusion the current also varies with $\sqrt{t}$, as shown in Fig. 6. Thus, nuclei can be grown on microelectrodes to controllable sizes as small as 10^{-6} cm. These individual metallic centers have been used as "submicroelectrodes," with very high limits for the mass transfer flux, for the study of the kinetics of fast metal/metal-ion reactions.[89] Droplets of mercury grown on platinum microelectrodes have been used to measure by anodic stripping voltammetry the *in situ* Pb^{2+} concentration at precise locations within a lead-acid battery.[156]

Carbon, platinum, and lead microelectrodes have been used to study electrocrystallization reactions relevant to lead-acid battery electrochemistry.[150] Single centers of $\alpha-$ and β-PbO_2 were grown on carbon fiber microdisk electrodes, allowing direct measurement of the kinetics of phase growth in the absence of overlap. In this case the rate of growth is controlled by the rate of incorporation of molecules into the lattice. The current-time relationship for isotropic three-dimensional growth of a single center is

$$I(t) = HnFM^2k^3t^2/\rho^2 \tag{38}$$

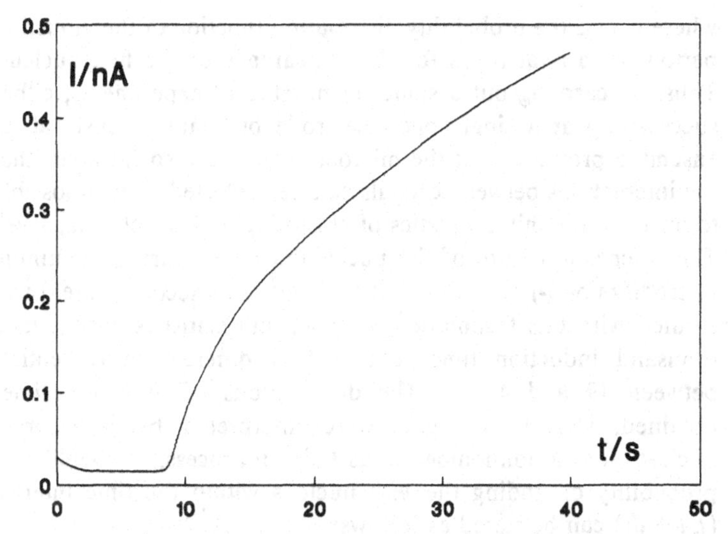

Figure 6. Potentiostatic transient for the deposition of a single nucleus of mercury on a platinum microelectrode of 5-μm radius from a 1 mM solution of $Hg_2(NO_3)_2$ in aqueous 1 M KNO_3, at 5-mV cathodic overpotential. From Ref. 5.

where k is the rate of crystal growth, and H is a shape factor which has the value 2π if the center is a hemisphere. The excellent linearity of the $i^{1/2}$-t plots obtained[150] proved the usual assumption made in studies of electrocrystallization that k, H, and ρ remain constant throughout the growth of the centers.

(ii) *Nucleation Rates*

The feasibility of confining the system to a few and, in the limit, a single growing center provides direct access to the clustering processes that lead to the formation of nuclei of the new phase. The essentially deterministic electrochemical process of phase growth therefore amplifies the stochastic event of nucleation.[147] The distribution of the induction times for the birth of the first nucleus in ensembles of experiments is determined by the kinetics of formation of the critical nuclei. The cumulative probability of nucleation up to a time t is expressed by

$$P(t) = \int_0^t g(u)\, du \tag{39}$$

where $g(t)$ is the probability distribution function of the induction periods or arrival times for the appearance of the first nucleus. Thus, by carrying out a sufficient number of experiments, either successively at a single microelectrode or simultaneously at an ensemble provided that the microelectrodes are so far apart that the interactions between crystals can be neglected,[157] it is possible to measure the initial kinetics of clustering at the molecular level. The stochastic nature of the nucleation of mercury on platinum microdisks of 5-μm radius (8×10^{-7} cm^2 cross-sectional area) was studied with this technique,[88] carrying out statistics over several thousand induction times obtained at different overpotentials between 19 and 41 mV. The distributions of induction times obtained, shown in Fig. 7, were interpreted by considering nucleation as a nonhomogeneous Poisson process,[158] whereby the probability of finding the mth nucleus within the time interval $(t, t + dt)$ can be stated as follows:

$$dP = J(t) \exp[-E(t)] \frac{[E(t)]^{m-1}}{(m-1)!} dt \tag{40}$$

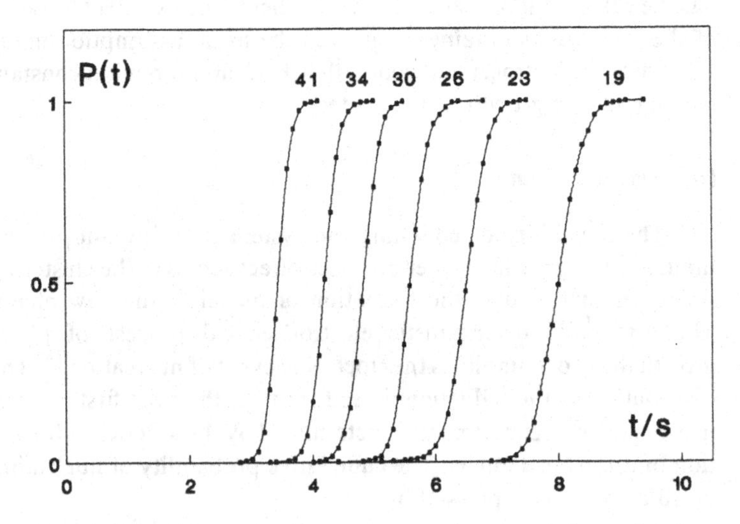

Figure 7. Probability of observing a mercury nucleus on a platinum microelectrode of 5-μm radius in aqueous 1 M KNO_3, 50 mM Hg_2NO_3, at the cathodic overpotentials indicated, in mV. From Ref. 88.

where $J(t)$ is the rate of nucleation at time t, and $E(t)$ is given by

$$E(t) = \int_0^t J(u)\, du \tag{41}$$

The distribution of the times elapsing from the beginning of the clustering process to the first nucleation event can be expressed by substituting $m = 1$ in Eq. (40):

$$g(t) = dP/dt = J(t) \exp[-E(t)] \tag{42}$$

One way of interpreting the experimental data is through the time dependence of the transient nucleation rate $J(t)$, obtained, for example, from non-steady-state classical nucleation theory,[159] from which values for the steady-state nucleation rate and the kinetic time lag for the establishment of steady-state concentrations of subcritical clusters can be obtained.[88] An alternative analysis of the same data can be carried out with fewer restrictions by direct application of the nonhomogeneous Poisson distribution to the probability of not observing the growth of a nucleus at time t,[148]

$$P_0 = 1 - \sum_1^{\infty} P_m(t) \tag{43}$$

This method provides also a direct estimation of the stoichiometry of the critical nucleus, which resulted in a value of n^* of ca. 40. The numerical values of $J(t)$ obtained by this method differ from those obtained by application of the classical theory of nucleation by several orders of magnitude. In spite of the numerical differences, both types of analysis lead to the result that the clustering process that precedes the appearance of critical nuclei of size n^* during the deposition of mercury on Pt at low overpotentials is composed of reactions of the type

$$n - 1 \rightleftharpoons n \rightleftharpoons n + 1 \rightleftharpoons \cdots \rightleftharpoons n^*$$

normally known as a birth-and-death process, while the further evolution of sufficiently developed nuclei can be considered as a pure-birth process,

$$n^* \rightarrow \cdots \rightarrow n - 1 \rightarrow n \rightarrow n + 1$$

for which the reverse reaction rates can be neglected.

Using carbon microdisk electrodes of 4-μm radius, single centers of α-PbO_2 can be grown at high overpotentials, up to 400 mV,[149] and the kinetics of its nucleation can be studied under conditions of very high supersaturation. An alternative model of interpretation of distributions of induction times was introduced, based on the concept of the coexistence of mixed populations of critical clusters of different shapes and sizes, formed through pure-birth processes, whereby the rates of the dissolution steps are neglected for clusters of sizes below critical too. A further assumption underlying the model proposed[149] is that the populations of subcritical clusters, and thus the nucleation rate, are in steady state throughout the process. At very high overpotentials the results indicated that the critical clusters were of a single kind, with a size that correlates with the known crystal structure of *α-*PbO_2.

(*ii*) *Electrocrystallization Processes*

Even for cases where the number of centers cannot be confined to a sufficiently small number so as to separate the nucleation and growth rates, microelectrodes provide advantages with respect to electrodes of larger size. An important one is the reduction in ohmic losses in solution, which has allowed the detailed study of aluminum deposition[160,161] and the kinetics of the Li/Li^+ couple in a number of solvents of low dielectric constant.[142,162-164] Relevant to batteries with extremely high power density are studies of this couple in the LiCl : KCl eutectic, where the concentration of Li^+ is 17 *M* and that of Cl^- is 30 *M*, at 400°C. In this aggressive environment, microelectrodes become necessary to overcome ohmic drops that arise not because of low conductivity but because of extremely high current densities: lithium stripping from a tungsten microdisk of 12.5-μm radius can give current densities in excess of 64 A cm^{-2}, and possibly as high as 1300 A cm^{-2} during short periods, with little distortion from *IR* drop in solution.[165]

The interconversion of lead, lead sulfate, and lead dioxide in the lead–acid battery system have been also investigated[150] using lead microdisk electrodes, under conditions such that the entire surface is involved in phase growth. For this complex system, the details of the reaction mechanisms obtained from measurements at microelectrodes cannot be derived from electrochemical measurements at electrodes of conventional size.

6. Ensembles of Microelectrodes

In spite of the various advantages of microelectrodes, one obvious consequence of their small surface area is that the current flow is correspondingly small. This may be mitigated by the construction of ensembles of microelectrodes, arranged so that their individual diffusion fields remain isolated within the time scale of the experiment, thus effectively amplifying the processes occurring with high mass transport rates at individual microelectrodes.

Different geometries have been considered for the construction of ensembles of microelectrodes. These include arrays of band, circular, or irregularly shaped microelectrodes, assembled in a number of possible ordered or disordered patterns. They may be arranged so that their conducting surfaces are on the same plane or otherwise protrude or recess from the insulator that separates them. Some examples are shown in Fig. 8. In spite of their differences, common features appear in their behavior with respect to mass transport.

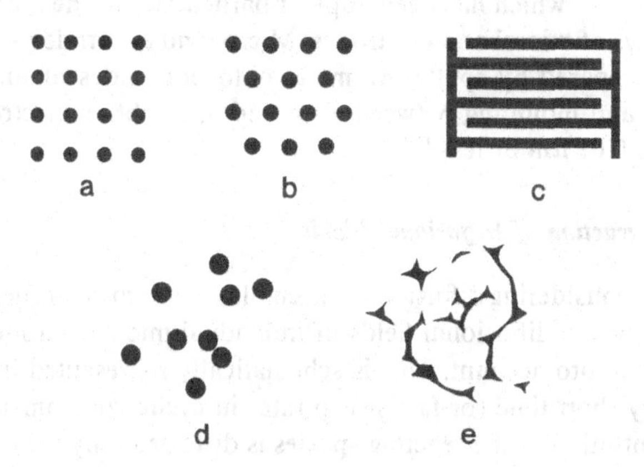

Figure 8. Ensembles of microelectrodes in ordered patterns as square (a) and hexagonal (b) arrays of microdisks or interdigitated microband electrodes (c) and in disordered patterns as randomly arrayed microdisks (d) or disorderly arrayed and irregularly shaped microelectrode ensembles (e).

Perhaps the simplest geometry from the standpoint of fabrication is the random array of irregularly shaped microelectrodes, which can be made by combining powders or chips of electrode material with an insulator.[166,167] Another type of random array is made by using a reticulated electrode material, for example, reticulated vitreous carbon, with an insulator to fill the pores. Although difficult to characterize geometrically, these random arrays are fairly simple to construct from readily available materials.[168,169]

Arrays of electrodes based on disks have also been constructed,[157,170] by embedding carbon fibers or thin wires in an epoxy matrix.[171-173] A technique to construct ensembles of very small disk-shaped elements involves the electrodeposition of platinum into the pores of a microporous host membrane[174] or the filling of the pores with a conducting medium such as carbon paste.[175,176]

Arrays of Hg microspheres have been used to measure the kinetics of the mercurous ion/mercury system.[3,4,87] Arrays of microspheres or oblate microspheroids may be fabricated by electrodepositing a metal onto a substrate which is electroinactive for the reaction of interest or by depositing controlled amounts of a metal onto a previously prepared array of microdisks.

The most common technique for the fabrication of arrays with well-defined regular geometry is the photolithographic technique,[177-184] which has been applied particularly to the fabrication of arrays of microband electrodes. Microband electrode pairs have been prepared by sputtering metal onto both sides of thin mica sheets and mounting between glass slides, to obtain electrodes as thin as 0.01 μm or less.[185]

(*i*) *Interaction of Diffusional Fields*

In considering diffusion to ensembles of microelectrodes, interaction of the diffusional fields of individual microelectrodes must be taken into account. This is schematically represented in Fig. 9. At very short time (or fast sweep rates in cyclic voltammetry), the concentration of the reacting species is distorted only very close to the electrochemically active part of the surface and diffusion is planar to each one of these segments. At intermediate times, the concentration distortion extends over longer distances, comparable to or larger than the characteristic size of the electrochemically

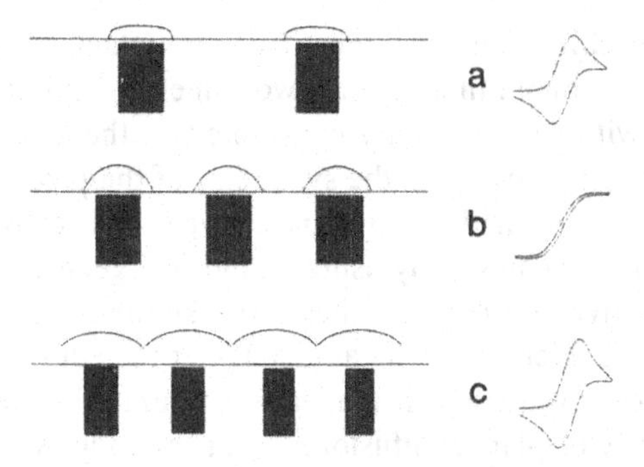

Figure 9. Development of diffusion concentration profiles in ensembles of microelectrodes. Concentration distortions at very short times during chronoamperometry or fast sweep rates during cyclic voltammetry (a), intermediate times or sweep rates (b), and long times or slow seep rates (c). Voltammetric responses are schematically shown.

active elements of the array, and the diffusion profiles become hemispherical. Finally, at long times, the concentration distortion extends over a distance comparable to the intermicroelectrode separation, the diffusional fields of adjacent microelectrodes overlap, and planar diffusion becomes again dominating.

The quantitative description of mass transport to ensembles of microelectrodes has been realized both analytically[177,186] and by computer simulation, using finite differences[187,188] or collocation techniques.[189] These descriptions all agree in regard to the short- and long-time behaviors: at very short times the current is Cottrellian to the electrochemically active surface area, whereas at very long times it is also Cottrellian but to the entire surface of the array. Some differences appear in the description of the mass flux at intermediate times, at which the diffusion fields of individual microelectrodes start to interact and eventually coalesce. It is only during this interval that the currents depend on the way the microelectrodes are arranged in the ensemble.

A major complication arises from the spreading of the concentration profiles originating from each microelectrode in three dimensions, extending to the bulk of the solution, coupled with their essentially two-dimensional distribution over the plane of the array. The problem has been most frequently tackled with microelectrodes distributed in a regular two-dimensional lattice, usually hexagonal, with each hexagon considered as the base of a semi-infinite unit cell. Because of the symmetry of the problem, there is no flux of reactant at the boundary between unit cells. A further simplification is achieved by using cylindrical geometry.

An intuitive approach has been to take the nonplanar flux to a free, isolated microelectrode and consider the equivalent area of plane surface toward which the same amount of material would diffuse by way of planar diffusion.[190] For microdisks,

$$\pi r_d^2 J_p(t) = \pi a^2 J(t) \tag{44}$$

where $J(t)$ is the nonplanar diffusional flux at the microdisk, and

$$J_p(t) = Dc^\infty/(\pi Dt)^{1/2} \tag{45}$$

is the planar flux. The area of the equivalent electrode functioning under planar diffusion conditions is πr_d^2, and thus its radius is given by

$$r_d^2 = a^2 J(t)/J_p(t) \tag{46}$$

The nonplanar flux corresponding to the real situation may be sustained only if the equivalent electrode expands. In an ensemble, the spread of the segments arising from each microelectrode extends until they meet. Consideration of the overlap of the equivalent segments spreading from each microelectrode then leads to the description of mass transport to the ensemble. This greatly simplifies the treatment of the interaction of the diffusional fields of individual microelectrodes by transforming them into equivalent two-dimensional entities.

Thus, if an expression is available for $J(t)$ at a single microelectrode, then the transient response to an ensemble can be obtained. Taking, for example, the short-times solution for the nonplanar flux to an isolated microdisk of radius a[24,25,29,30]:

$$\pi a^2 J(t) = -\pi Dc^\infty a[1 + a/(\pi Dt)^{1/2}] \tag{47}$$

it follows from Eqs. (45) and (46) that

$$r_d^2 = a^2 + a(\pi Dt)^{1/2} \tag{48}$$

The current density to an ensemble composed of ν microelectrodes per unit area may be expressed by the planar diffusion current to the fraction of the area that has become covered by equivalent disks. At short times, when these do not overlap, the current density to the ensemble is given by

$$i = \pi\nu anFDc^{\infty}[1 + a/(\pi Dt)^{1/2}] \tag{49}$$

Expressing the time in nondimensional units as

$$u = \nu a(\pi Dt)^{1/2} \tag{50}$$

and normalizing with respect to the overall current of independent microelectrodes, then

$$i/\pi\nu anFDc^{\infty} = 1 + \nu a^2/u \tag{51}$$

At longer times, the current density is affected by the overlap of equivalent disks. The extent of overlap and the time at which it starts influencing the current depend on the geometry of the ensemble and on its packing density. Equations that describe the current density for microelectrodes arrays according to several patterns have been derived[190]; for example, for an ensemble of randomly arrayed microdisks:

$$i/\pi\nu anFDc^{\infty} = \{1 - \exp[-\pi(1 + \nu a^2/u)]\} \tag{52}$$

Finally, at even longer times, the ensemble of microelectrodes becomes totally covered by the equivalent disks, and the current density is fully given by planar diffusion to its entire area,

$$i/\pi\nu anFDc^{\infty} = 1/\pi u \tag{53}$$

Figure 10 shows plots of the normalized current as a function of the inverse of the square root of time for hexagonal, square, and random arrays of microdisks of uniform size. At short times, the slopes of the i-$t^{-1/2}$ plots are proportional to the electroactive area, whereas at long times they are proportional to the total area of the ensemble.

For hexagonal arrays, both experimental[177] and simulated[187-189] data are available. Good agreement has been obtained with the response given by the equivalent expanding disks description.[190]

A transition between nonplanar and planar diffusion regimes occurs between two limiting behaviors. Its characteristic time, t_{sp},

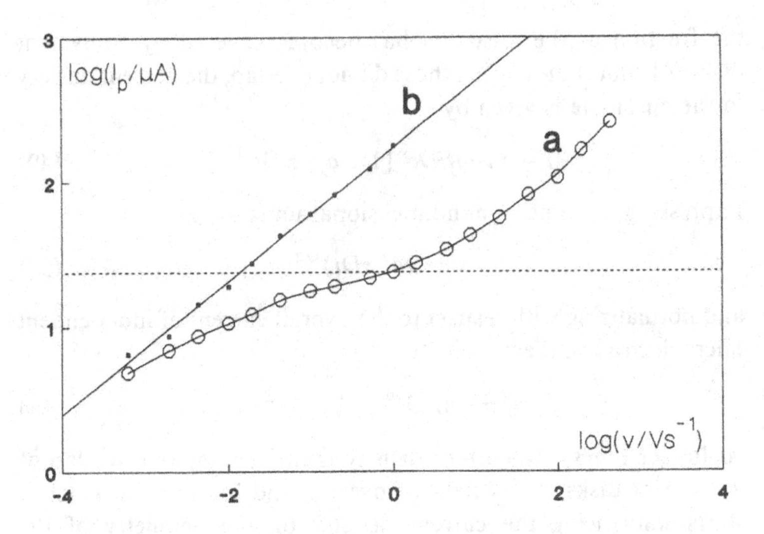

Figure 11. Dependence of peak currents on sweep rate for cyclic voltammetry of 1 mM $Fe(bpy)_3^{2+}$ on a random array of microdisks with $\nu = 1.36 \times 10^6$ cm^{-2} and $a = 4$ μm (a) and on a macrosized electrode with the same area as the array (b). The broken line is the pure radial diffusion current according to Eq. (59). From Ref. 175.

The voltammetric behavior of microdisk electrode ensembles has also been studied by digital simulation, through finite differences[192] and orthogonal collocation[193] techniques, with results consistent with the behavior outlined above.

(iii) *Arrays of Microbands*

Arrays of closely spaced microband electrodes present important differences with respect to microdisk ensembles, above all in that they may be individually addressed.[194]

The close electrode spacing has profound effects on the response of these arrays.[179] In linear sweep voltammetry at moderate scan rates, each microelectrode displays essentially a sigmoidal current response due to nonplanar diffusion at the edges, even though a steady-state current is not truly attained with the cylindrical symmetry characteristic of microband [54,195] A consequence of the individual addressability and the close spacing of microbands in an array is the possibility of detecting electrogenerated products definition of r_d:

$$r_d^2 = a^2 + (4/\pi)a(\pi Dt)^{1/2} \tag{57}$$

from which

$$t_{sp} = \pi\theta^2/4DP^2 \tag{58}$$

Thus, for all θ, interaction of individual diffusion fields becomes important when t is of the order of θ^2/DP^2.

(ii) *Cyclic Voltammetry*

The voltammetric response of an ensemble of microdisks depends on the scan rate of the experiment. Three limiting cases can be identified,[175,192,193] as shown schematically in Fig. 9. The first occurs at very high scan rates, where the diffusion layers are thin and extend linearly from the individual ensemble elements. A peak-shaped, planar diffusion voltammogram typical of macrosized electrodes is observed, with currents proportional to the overall active element area. At lower scan rates, radial diffusion fields develop at the microdisks. The voltammogram becomes sigmoidal with a limiting current density given by [viz. Eq. (8)]

$$i_l = 4\nu nFDc^{\infty}a \tag{59}$$

Finally, at very low scan rates, the individual diffusion layers merge to yield a net planar diffusion field, and the voltammogram becomes peak-shaped again, but with currents proportional to the total geometric area of the ensemble. This is shown in Fig. 11, where the peak current is represented as a function of the scan rate in a logarithmic plot[175] for cyclic voltammograms on a random array of microdisks and on a macrosized electrode with the same area as the array. At low scan rates the currents converge to that corresponding to a totally active electrode, whereas at very high sweep rates the currents approach the response expected for individual microelectrodes behaving independently. An inflection is observed at intermediate sweep rates, at a value corresponding to the current given by Eq. (59).

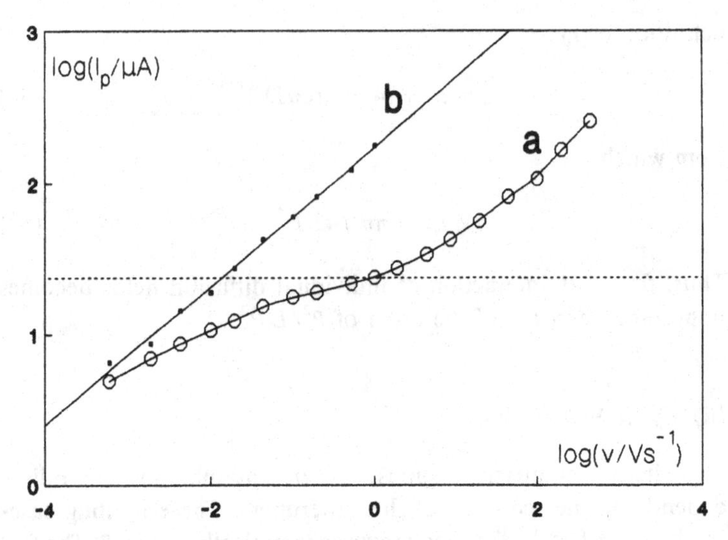

Figure 11. Dependence of peak currents on sweep rate for cyclic voltammetry of 1 mM Fe(bpy)$_3^{2+}$ on a random array of microdisks with $\nu = 1.36 \times 10^6$ cm^{-2} and $a = 4$ μm (a) and on a macrosized electrode with the same area as the array (b). The broken line is the pure radial diffusion current according to Eq. (59). From Ref. 175.

The voltammetric behavior of microdisk electrode ensembles has also been studied by digital simulation, through finite differences[192] and orthogonal collocation[193] techniques, with results consistent with the behavior outlined above.

(*iii*) *Arrays of Microbands*

Arrays of closely spaced microband electrodes present important differences with respect to microdisk ensembles, above all in that they may be individually addressed.[194]

The close electrode spacing has profound effects on the response of these arrays.[179] In linear sweep voltammetry at moderate scan rates, each microelectrode displays essentially a sigmoidal current response due to nonplanar diffusion at the edges, even though a steady-state current is not truly attained with the cylindrical symmetry characteristic of microband [54,195] A consequence of the individual addressability and the close spacing of microbands in an array is the possibility of detecting electrogenerated products at the adjacent electrodes. For example, the reduced form of a solution species generated at one microelectrode can be collected at adjacent microelectrodes, held at a potential where oxidation can occur, in analogy with collection experiments using conventional rotating ring-disk electrodes of macroscopic dimension.[196] However, as opposed to rotating ring-disk experiments, where hydrodynamics imposes flow from the generating disk to the collecting ring, and thus the current at the disk is unaffected by the reactions taking place at the ring, for closely spaced stationary electrodes products at the collector can diffuse back to the generator, where they may be electrolyzed. This produces an additional feedback current at the generator. If the microbands are all held at the same potential, then the current at one electrode can reduce, or shield, that at its neighbor due to overlap of individual diffusion layers, as discussed in Section III.6(i) for ensembles of microdisks.

Experimental and simulated results[179] demonstrate that collection efficiencies for close-spaced microband electrodes are very high: more than half of the product generated at one microband is collected at another separated by an insulating gap of 1 μm. With two collector electrodes, one on either side of a centrally positioned generating electrode, the collection efficiency increases to nearly 80%. As may be expected, the intermicroelectrode gap strongly affects the collection efficiencies and the shielding and feedback currents.

(*iv*) *Advantages of Ensembles over Single Microelectrodes*

While microelectrodes exhibit enhanced mass transfer rates and reduced ohmic drops per electroactive surface area, the magnitude of the current is diminished due to the reduced dimension. In the limit of very small electrodes, an extremely low current would have to be measured, imposing serious constraints on the instrumentation needed, and its response time, for determination of the kinetics of fast electrode reactions. A higher overall current is recovered with ensembles of microelectrodes, thus relieving the need for extraordinary current amplification, without compromising the advantages of single microelectrodes for fast transient kinetic determinations. The amplification effect of ensembles of microelectrodes makes them also useful for electroanalytical applications by improving the signal-to-noise ratio since, under optimal conditions, the signal controlled by diffusion of the electroactive species is proportional to the total area, while the noise is proportional to the active area of the electrode.[192]

Ensembles of microelectrodes may be used advantageously to carry out diverse electrochemical processes. In the recovery of metals from dilute solutions, the mass transport enhancement due to nonplanar diffusion increases both the rate of metal recovery as well as the current efficiency,[171] since the competing reactions would not usually be limited by mass transport. Other areas of application include electrochemical energy conversion (gas diffusion porous electrodes such as those used in fuel cells are, in fact, ensembles of microelectrodes) and possibly electrosynthesis on practical scales, where substantial savings of electrocatalysts and increased efficiencies arise from the use of ensembles of microelectrodes.

Consideration must also be given to three-dimensional ensembles of microelectrodes, where electrochemical reactions take place at the surface of small conducting particles or even clusters of molecules[92] in solution. Even though the microelectrodes may not be electrically connected to an external source, electrochemical reactions are driven by a mixture potential.[197] Small semiconductor particles also act like microelectrodes, with redox reactions initiated by electrons and positive holes created in the particles upon illumination.[93] In poorly conducting or nonconducting media, dispersions of microelectrodes may be operated in the form of bipolar fluidized bed reactors[91] to carry out electrolysis.

IV. CONCLUSION

The response of small electrodes is substantially different from that of larger electrodes. Convergent diffusion establishes a steady-state mass transport flux which increases sharply with the decrease in size of the microelectrode, while the transient diffusion response is drastically reduced. Thus, the conventional cyclic voltammogram transforms into a more easily tractable polarographic-type wave. The decreased surface area reduces the effects of the electrode capacitance. Transient measurements for the study of fast processes can then be made at higher perturbation speeds. In addition, electrochemical phenomena can be studied in highly resistive media without additional supporting electrolytes, in frozen solutions[133] and with the electrode surface in contact with the gas[140] or solid[198] phases. This opens up possibilities for studies in new physiochemical systems as well as for novel applications in analytical techniques and the design of sensors and microelectrochemical devices.[199-201]

Electrodes of very small size can be used to probe the structure and dynamics of systems at levels not accessible with electrodes of larger size. Microelectrodes can be used to measure the local concentration of chemical species within the diffusion layer of another electrode with space and time resolutions on the order of microns and milliseconds, respectively.[202,203] When electroactive species spring from a very localized point, as in local or pit corrosion, spatial control of the microprobe allows the location of the corrosion site.[204] Due to the improved electrical properties of microelectrodes, investigations under conditions closer to those of natural corrosion environments can be carried out.[205]

With microelectrodes having at least one dimension smaller than the Debye length, double-layer effects on fluid properties and diffusional transport have been observed.[55,206,207]

The reduction in size when using microelectrodes magnifies fluctuations occurring at the molecular level that trigger large-amplitude signals. Thus, stochastic measurements on diverse electrochemical systems can be carried out, such as two-dimensional nucleation on screw-dislocation-free single-crystal surfaces,[208] three-dimensional nucleation on single crystals[209] or polycrystalline surfaces,[88] and the formation of pores in lipid bilayers.[210] Large changes in capacitance occur during the reorientation and two-dimensional condensation of organic molecules on mercury and other metals. Transitions of this type can be studied by stochastic methods with sufficiently small electrodes.[211]

It is thus clear that the use of microelectrodes provides new and improved ways of studying traditional electrochemical systems, as well as the possibility of experimenting with others hitherto not accessible. Further and still not foreseen advances in our understanding of the metal/solution interface and the dynamics of electrochemical processes may be realized once the size of electrodes becomes comparable to the molecular dimensions.

REFERENCES

[1] M. A. Dayton, J. C. Brown, K. J. Stutts, and R. M. Wightman, *Anal. Chem.* **52** (1980) 946.
[2] R. M. Wightman, *Anal. Chem.* **53** (1981) 1125A.
[3] P. Bindra, A. P. Brown, M. Fleischmann, and D. Pletcher, *J. Electroanal. Chem.* **58** (1975) 31.
[4] P. Bindra, A. P. Brown, M. Fleischmann, and D. Pletcher, *J. Electroanal. Chem.* **58** (1975) 39.
[5] B. Scharifker and G. Hills, *J. Electroanal. Chem.* **130** (1981) 81.
[6] M. Fleischmann, F. Laserre, J. Robinson, and D. Swan, *J. Electroanal. Chem.* **177** (1984) 97.
[7] K. Aoki, K. Akimoto, K. Tokuda, H. Matsuda, and J. Osteryoung, *J. Electroanal. Chem.* **171** (1984) 219.
[8] M. I. Montenegro, *Port. Electrochim. Acta* **3** (1985) 165.
[9] T. E. Edmonds, *Anal. Chim. Acta* **175** (1985) 1.
[10] M. Ciszkowska and Z. Stojek, *J. Electroanal. Chem.* **213** (1986) 189.
[11] M. Fleischmann, S. Pons, D. R. Rolison, and P. P. Schmidt, eds., *Ultramicroelectrodes*, Datatech, Morganton, North Carolina, 1987.
[12] R. M. Wightman and D. O. Wipf, in *Electroanalytical Chemistry*, Vol. 15, Ed. by A. J. Bard, Marcel Dekker, New York, 1989, p. 267.
[13] A. J. Bard and L. R. Faulkner, *Electrochemical Methods*, Wiley, New York, 1980.
[14] J. Heinze, *J. Electroanal. Chem.* **124** (1981) 73.
[15] J. Heinze and M. Störzbach, *Ber. Bunsenges. Phys. Chem.* **90** (1986) 1043.
[16] D. Shoup and A. Szabo, *J. Electroanal. Chem.* **160** (1984) 1.
[17] A. C. Michael, R. M. Wightman, and C. Amatore, *J. Electroanal. Chem.* **267** (1989) 33.
[18] J. F. Cassidy, S. Pons, A. S. Hinman, and B. Speiser, *Can. J. Chem.* **62** (1984) 716.
[19] Y. Saito, *Rev. Polarogr. (Japan)* **15** (1968) 177.
[20] M. Fleischmann and S. Pons, *J. Electroanal. Chem.* **222** (1987) 107.
[21] A. Szabo, *J. Phys. Chem.* **91** (1987) 3108.
[22] A. M. Bond, K. B. Oldham, and C. G. Zoski, *J. Electroanal. Chem.* **245** (1988) 71.
[23] Z. G. Soos and P. G. Lingane, *J. Phys. Chem.* **68** (1964) 3821.
[24] K. B. Oldham, *J. Electroanal. Chem.* **122** (1981) 1.
[25] K. Aoki and J. Osteryoung, *J. Electroanal. Chem.* **122** (1981) 19.
[26] J. B. Flanagan and L. Marcoux, *J. Phys. Chem.* **77** (1973) 1051.
[27] M. Kakihana, H. Ikeuchi, G. P. Sato, and K. Tokuda, *J. Electroanal. Chem.* **117** (1981) 201.
[28] T. Hepel and J. Osteryoung, *J. Phys. Chem.* **86** (1982) 1406.
[29] D. Shoup and A. Szabo, *J. Electroanal. Chem.* **140** (1982) 237.
[30] T. Hepel, W. Plot, and J. Osteryoung, *J. Phys. Chem.* **87** (1983) 1278.
[31] D. K. Cope and D. E. Tallman, *J. Electroanal. Chem.* **235** (1987) 97.
[32] D. Ilkovic, *Collect. Czech, Chem. Commun.* **8** (1936) 13.
[33] J. Newman, *J. Electrochem. Soc.* **113** (1966) 501.
[34] J. O. Howell, J. Goncalves, C. Amatore, L. Klasinc, R. M. Wightman, and J. K. Kochi, *J. Am. Chem. Soc.* **106** (1984) 3968.
[35] J. O. Howell and R. M. Wightman, *Anal. Chem.* **56** (1984) 524.
[36] K. R. Wehmeyer and R. M. Wightman, *Anal. Chem.* **57** (1985) 1989.
[37] J. O. Howell, W. G. Kuhr, R. E. Ensman, and R. M. Wightman, *J. Electroanal. Chem.* **209** (1986) 77.
[38] M. I. Montenegro and D. Pletcher, *J. Electroanal. Chem.* **200** (1986) 371.
[39] A. Fitch and D. H. Evans, *J. Electroanal. Chem.* **202** (1986) 83.

[40] C. Amatore, A. Jutand, and F. Pflüger, *J. Electroanal. Chem.* **218** (1987) 361.
[41] C. P. Andrieux, P. Hapiot, and J. M. Savéant, *J. Phys. Chem.* **92** (1988) 5987.
[42] D. O. Wipf and R. M. Wightman, *Anal. Chem.* **60** (1988) 2460.
[43] D. O. Wipf, E. W. Kristensen, M. R. Deakin, and R. M. Wightman, *Anal. Chem.* **60** (1988) 306.
[44] C. P. Andrieux, D. Garreau, P. Hapiot, J. Pinson, and J. M. Savéant, *J. Electroanal. Chem.* **243** (1988) 321.
[45] C. P. Andrieux, D. Garreau, P. Hapiot, and J. M. Savéant, *J. Electroanal. Chem.* **248** (1988) 447.
[46] C. P. Andrieux, P. Hapiot, and J. M. Savéant, *J. Phys. Chem.* 92 (1988) 5992.
[47] S. Bruckenstein, *Anal. Chem.* **59** (1987) 2098.
[48] K. B. Oldham, *J. Electroanal. Chem.* **237** (1987) 808.
[49] C. Amatore, C. Lefrou, and F. Pflüger, *J. Electroanal. Chem.* 270 (1989) 43.
[50] M. Fleischmann, S. Bandyopadhyay, and S. Pons, *J. Phys. Chem.* **89** (1985) 5537.
[51] M. Fleischmann and S. Pons, in *Ultramicroelectrodes*, Ed. by M. Fleischmann, S. Pons, D. R. Rolison, and P. P. Schmidt, Datatech, Morganton, North Carolina, 1987, p. 17.
[52] L. C. R. Alfred, J. C. Myland, and K. B. Oldham, *J. Electroanal. Chem.* **280** (1990) 1.
[53] K. B. Oldham, in *Ultramicroelectrodes*, Ed. by M. Fleischmann, S. Pons, D. R. Rolison, and P. P. Schmidt, Datatech, Morganton, North Carolina, 1987, p.276.
[54] A. Szabo, D. K. Cope, D. E. Tallman, P. M. Kovach, and R. M. Wightman, *J. Electroanal. Chem.* **217** (1987) 417.
[55] K. R. Wehmeyer, M. R. Deakin, and R. M. Wightman, *Anal. Chem.* **57** (1985) 1913.
[56] A. M. Bond, M. Fleischmann, S. B. Khoo, S. Pons, and J. Robinson, *Indian J. Technol.* **24** (1986) 492.
[57] D. R. Rolison, in *Ultramicroelectrodes*, Ed. by M. Fleischmann, S. Pons, D. R. Rolison, and P. P. Schmidt, Datatech, Morganton, North Carolina, 1987, p. 65.
[58] K. B. Oldham, *J. Electroanal. Chem.* **260** (1989) 461.
[59] R. L. Birke, *J. Electroanal. Chem.* **274** (1989) 297.
[60] R. M. Penner, M. J. Heben, and N. S. Lewis, *Anal. Chem.* **61** (1989) 1630.
[61] J. L. Ponchon, R. Cespuglio, F. Gonon, M. Jouvet, and J. F. Pujol, *Anal. Chem.* **51** (1979) 1483.
[62] M. A. Dayton, A. G. Ewing, and R. M. Wightman, *Anal. Chem.* **52** (1980) 2392.
[63] P. M. Kovach, M. R. Deakin, and R. M. Wightman, *J. Phys. Chem.* **90** (1986) 4612.
[64] P. M. Kovach, W. L. Caudill, D. G. Peters, and R. M. Wightman, *J. Electroanal. Chem.* **185** (1985) 285.
[65] A. M. Bond, M. Fleischmann, and J. Robinson, *J. Electroanal. Chem.* **168** (1984) 299.
[66] B. J. Feldman, A. G. Ewing, and R. W. Murray, *J. Electroanal. Chem.* **194** (1985) 63.
[67] D. Swan, Ph.D. thesis, Southampton, 1981.
[68] D. C. Boyd and D. A. Thompson, in *Kirk-Othmeer Encyclopedia of Chemical Technology*, Vol. 11, 3rd. ed., Wiley, New York, 1980, p. 807.
[69] J. Fong, personal communication, 1986.
[70] G. Denuault, M. Fleischmann, and D. Pletcher, *J. Electroanal. Chem.* **280** (1990) 255.
[71] Y. T. Kim, D. M. Scarnulio, and A. G. Ewing, *Anal. Chem.* **58** (1986) 1782.
[72] R. C. Engstrom, *Anal. Chem.* **56** (1984) 890.
[73] G. Schulze and W. Frenzel, *Anal. Chim. Acta* **159** (1984) 95.
[74] T. E. Mallouk, V. Cammarata, J. A. Crayston, and M. S. Wrighton, *J. Phys. Chem.* **90** (1986) 2150.
[75] J. W. Bixler and A. M. Bond, *Anal. Chem.* **58** (1986) 2859.

[76] H. L. S. Maia, M. J. Medeiros, M. I. Montenegro, D. Court, and D. Pletcher, *J. Electroanal. Chem.* **164** (1984) 347.
[77] A. S. Baranski, *J. Electrochem. Soc.* **133** (1986) 93.
[78] L. M. Abrantes, M. Fleischmann, L. M. Peter, S. Pons, and B. R. Scharifker, *J. Electroanal. Chem.* **256** (1988) 229.
[79] D. Garreau, P. Hapiot, and J. M. Savéant, *J. Electroanal. Chem.* **272** (1989) 1.
[80] D. Garreau, P. Hapiot, and J. M. Savéant, *J. Electroanal. Chem.* **281** (1990) 73.
[81] W. J. Bowyer, E. E. Engelman, and D. H. Evans, *J. Electroanal. Chem.* **262** (1989) 67.
[82] D. Garreau and J. M. Savéant, *J. Electroanal. Chem.* **35** (1972) 309.
[83] C. Lamy and C. C. Herrmann, *J. Electroanal. Chem.* **59** (1975) 113.
[84] D. Britz, *J. Electroanal. Chem.* **88** (1978) 309.
[85] D. E. Tallman, G. Shepherd, and W. J. MacKellar, *J. Electroanal. Chem.* **280** (1990) 327.
[86] L. R. Faulkner, in *Ultramicroelectrodes*, Ed. by M. Fleischmann, S. Pons, D. R. Rolison, and P. P. Schmidt, Datatech, Morganton, North Carolina, 1987, p. 225.
[87] P. Bindra and J. Ulstrup, *J. Electroanal. Chem.* **140** (1982) 131.
[88] G. A. Gunawardena, G. J. Hills, and B. R. Scharifker, *J. Electroanal. Chem.* **130** (1981) 99.
[89] G. Hills, A. Kaveh Pour, and B. Scharifker, *Electrochim. Acta* **28** (1983) 891.
[90] M. Fleischmann, J. Ghoroghchian, and S. Pons, *J. Phys. Chem.* **89** (1985) 5530.
[91] M. Fleischmann, J. Ghoroghchian, D. Rolison, and S. Pons, *J. Phys. Chem.* **90** (1986) 6392.
[92] A. Henglein, *Top. Curr. Chem.* **143** (1988) 113.
[93] M. Grätzel, *Faraday Dis. Chem. Soc.* **70** (1980) 359.
[94] A. Russell, K. Repka, T. Dibble, J. Ghoroghchian, J. Smith, M. Fleischmann, and S. Pons, *Anal. Chem.* **58** (1986) 2961.
[95] M. Fleischmann, J. Dashbach, and S. Pons, *J. Electroanal. Chem.* **263** (1989) 189.
[96] M. Fleischmann and S. Pons, *J. Electroanal. Chem.* **250** (1988) 257.
[97] M. Fleischmann, J. Dashbach, and S. Pons, *J. Electroanal. Chem.* **250** (1988) 269.
[98] M. Fleischmann, J. Daschbach, and S. Pons, *J. Electroanal. Chem.* **263** (1989) 205.
[99] M. Fleischmann, D. Pletcher, G. Denuault, J. Daschbach, and S. Pons, *J. Electroanal. Chem.* **263** (1989) 225.
[100] W. J. Albery and R. P. Bell, *Proc. Chem. Soc.* **1963** 169.
[101] E. Steckhan, *Top. Curr. Chem.* **142** (1987) 1.
[102] G. Denuault, M. Fleischmann, D. Pletcher, and O. R. Tutty, *J. Electroanal. Chem.* **280** (1990) 243.
[103] C. Amatore and J. M. Savéant, *J. Electroanal. Chem.* **85** (1977) 27.
[104] M. Fleischmann, F. Laserre, and J. Robinson, *J. Electrocnal. Chem.* **177** (1984) 115.
[105] B. Speiser and B. S. Pons, *Can. J. Chem.* **60** (1982) 1352.
[106] A. J. Appleby, in *Assessment of Research Needs for Advanced Fuel Cells*, Ed. by S. S. Penner, Department of Energy, Washington, D.C., 1985, p.. 13.
[107] B. R. Scharifker, P. Zelenay, and J. O'M. Bockris, *J. Electrochem. Soc.* **134** (1987) 2714.
[108] S. Pons and M. Fleischmann, in *Ultramicroelectrodes*, Ed. by M. Fleischmann, S. Pons, D. R. Rolison, and P. P. Schmidt, Datatech, Morganton, North Carolina, 1987, p. 1.
[109] R. S. Robinson and R. L. McCreery, *J. Electroanal. Chem.* **182** (1985) 61.
[110] K. Aoki and Y. Tezuka, *J. Electroanal. Chem.* **267** (1989) 55.
[111] M. Kalaji, L. M. Peter, L. M. Abrantes, and J. C. Mesquita, *J. Electroanal. Chem.* **274** (1989) 289.
[112] R. S. Nicholson, *Anal. Chem.* **37** (1965) 1351.

[113] T. Erdey-Grúz, *Transport Phenomena in Aqueous Solutions*, A. Hilger, London, 1974.
[114] K. Aoki and J. Osteryoung, *J. Electroanal. Chem.* **125** (1981) 315.
[115] P. J. Lingane, *Anal. Chem.* **36** (1964) 1723.
[116] M. Kakihana, H. Ikeuchi, G. P. Satô, and K. Tokuda, *J. Electroanal. Chem.* **108** (1980) 381.
[117] J. C. Myland and K. B. Oldham, *J. Electroanal. Chem.* **147** (1983) 295.
[118] C. P. Winlove, K. H. Parker, and R. K. C. Oxenham, *J. Electroanal. Chem.* **170** (1984) 293.
[119] J. Klinowski, S. E. Korsner, and C. P. Winlove, *Cardiovasc. Res.* **16** (1982) 448.
[120] J. Kruger and J. P. Calvert, *J. Electrochem. Soc.* **114** (1967) 43.
[121] N. Sato, K. Kudo, and T. Noda, *Corros. Sci.* **10** (1970) 785.
[122] P. Zelenay, B. R. Scharifker, J. O'M. Bockris, and D. Gervasio, *J. Electrochem. Soc.* **133** (1986) 2262.
[123] V. Jovancicevic, P. Zelenay, and B. R. Scharifker, *Electrochim. Acta* **32** (1987) 1553.
[124] W. M. Flarsheim, K. P. Johnson, and A. J. Bard, *J. Electrochem. Soc.* **135** (1988) 1939.
[125] R. M. Crooks and A. J. Bard, *J. Phys. Chem.* **91** (1987) 1274.
[126] M. E. Philips, M. R. Deakin, M. V. Novotny, and R. M. Wightman, *J. Phys. Chem.* **91** (1987) 3934.
[127] R. M. Crooks and A. J. Bard, *J. Electroanal. Chem.* **243** (1988) 117.
[128] C. R. Cabrera, E. Garcia, and A. J. Bard, *J. Electroanal. Chem.* **260** (1989) 457.
[129] J. O'M. Bockris, R. Parsons, and H. Rosenberg, *Trans. Faraday Soc.* **47** (1951) 766.
[130] B. E. Conway and M. Salomon, *J. Chem. Phys.* **41** (1964) 3169.
[131] U. Stimming and W. Schmickler, *J. Electroanal. Chem.* **150** (1983) 125.
[132] T. Dinan and U. Stimming, *J. Electrochem. Soc.* **133** (1986) 2662.
[133] A. M. Bond, M. Fleischmann, and J. Robinson, *J. Electroanal. Chem.* **180** (1984) 257.
[134] W. J. Bowyer and D. H. Evans, *J. Electroanal. Chem.* **240** (1988) 227.
[135] D. C. Bradley and M. Ahmed, *Polyhedron* **2** (1983) 87.
[136] R. Lines and V. D. Parker, *Acta Chem. Scand., Ser. B***31** (1977) 369.
[137] I. M. Kolthoff and F. G. Thomas, *J. Phys. Chem.* **69** (1965) 3049.
[138] M. J. Peña, M. Fleischmann, and N. Garrard, *J. Electroanal. Chem.* **220** (1987) 31.
[139] A. Bond, M. Fleischmann, and J. Robinson, *J. Electroanal. Chem.* **172** (1984) 11.
[140] J. Ghoroghchian, F. Sarfarazi, T. Dibble, J. Cassidy, J. J. Smith, A. Russell, G. Dunmore, M. Fleischmann, and S. Pons, *Anal. Chem.* **58** (1986) 2278.
[141] J. Cassidy, S. B. Khoo, S. Pons, and M. Fleischmann, *J. Phys. Chem.* **89** (1985) 3933.
[142] J. D. Genders, W. M. Hedges, and D. Pletcher, *J. Chem. Soc., Faraday Trans. 1* **80** (1984) 3399.
[143] G. Nicolis and I. Prigogine, *Self-Organization in Nonequilibrium Systems*, Wiley, New York, 1977, p. 223.
[144] P. Bindra, M. Fleischmann, J. W. Oldfield, and D. Singleton, *Faraday Disc. Chem. Soc.* **56** (1973) 180.
[145] B. Scharifker and G. Hills, *Electrochim. Acta* **28** (1983) 879.
[146] B. R. Scharifker, *Acta Cient. Venez.* **35** (1984) 211.
[147] R. de Levie, in *Advances in Electrochemistry and Electrochemical Engineering*, Vol. 13, Ed. by H. Gerischer and C. W. Tobias, Wiley, New York, 1984, p. 1.
[148] R. Sridharan and R. de Levie, *J. Electroanal. Chem.* **169** (1984) 59.
[149] M. Fleischmann, L. J. Li, and L. M. Peter, *Electrochim. Acta* **34** (1989) 475.
[150] L. J. Li, M. Fleischmann, and L. M. Peter, *Electrochim. Acta* **34** (1989) 459.

[151] G. J. Hills, D. J. Schiffrin, and J. Thompson, *Electrochim. Acta* **19** (1974) 657.
[152] S. Fletcher, *J. Chem. Soc., Faraday Trans. 1* **79** (1983) 467.
[153] P. A. Bobbert, M. M. Wind, and J. Vlieger, *Physica A* **141** (1987) 58.
[154] C. L. Colyer, D. Luscombe, and K. B. Oldham, *J. Electroanal. Chem.* **283** (1990) 379.
[155] G. A. Gunawardena, G. J. Hills, and I. Montenegro, *Electrochim. Acta* **23** (1978) 693.
[156] L. J. Li, M. Fleischmann, and L. M. Peter, *Electrochim. Acta* **32** (1987) 1585.
[157] R. L. Deutscher and S. Fletcher, *J. Electroanal. Chem.* **239** (1988) 17.
[158] E. Parzen, *Stochastic Processes*, Holden-Day, San Francisco, 1962.
[159] D. Kashchiev, *Surf. Sci.* **14** (1969) 209.
[160] J. N. Howarth and D. Pletcher, *J. Chem. Soc., Faraday Trans. 1* **83** (1987) 2787.
[161] J. N. Howarth and D. Pletcher, *J. Chem. Soc., Faraday Trans. 1* **83** (1987) 2795.
[162] W. M. Hedges and D. Pletcher, *J. Chem. Soc., Faraday Trans. 1* **82** (1986) 179.
[163] W. M. Hedges, D. Pletcher, and C. Gosden, *J. Electrochem. Soc.* **134** (1987) 1334.
[164] K. S. Aojula, J. D. Genders, A. D. Holding, and D. Pletcher, *Electrochim. Acta* **34** (1989) 1535.
[165] R. T. Carlin and R. A. Osteryoung, *J. Electrochem. Soc.* **136** (1989) 1249.
[166] D. E. Weisshaar, D. E. Tallman, and J. L. Anderson, *Anal. Chem.* **53** (1981) 1809.
[167] L. Falat and H. Y. Cheng, *Anal. Chem.* **54** (1982) 2109.
[168] J. Wang and B. A. Freiha, *J. Chromatogr.* **298** (1984) 79.
[169] N. Sleszynski, J. Osteryoung, and M. Carter, *Anal. Chem.* **56** (1984) 130.
[170] W. L. Caudill, J. O. Howell, and R. M. Wightman, *Anal. Chem.* **54** (1982) 2531.
[171] R. C. Paciej, G. L. Cahen, G. E. Stoner, and E. Gileadi, *J. Electroanal. Chem.* **132** (1985) 1307.
[172] M. Ciszkowska and Z. Stojek, *J. Electroanal. Chem.* **191** (1985) 101.
[173] S. M. Lipka, G. L. Cahen, G. E. Stoner, L. L. Scribner, and E. Gileadi, *J. Electrochem. Soc.* **135** (1988) 368.
[174] R. M. Penner and C. R. Martin, *Anal. Chem.* **59** (1987) 2625.
[175] I. F. Cheng, L. D. Whiteley, and C. R. Martin, *Anal. Chem.* **61** (1989) 762.
[176] I. F. Cheng and C. R. Martin, *Anal. Chem.* **60** (1988) 2163.
[177] T. Gueshi, K. Tokuda, and H. Matsuda, *J. Electroanal. Chem.* **89** (1978) 247.
[178] B. J. Seddon, H. H. Girault, and M. J. Eddowes, *J. Electroanal. Chem.* **266** (1989) 227.
[179] A. J. Bard, J. A. Crayston, G. P. Kittlesen, T. Varco Shea, and M. S. Wrighton, *Anal. Chem.* **58** (1986) 2321.
[180] C. E. Chidsey, B. J. Feldman, C. Lungren, and R. W. Murray, *Anal. Chem.* **58** (1986) 601.
[181] L. E. Fosdick and J. L. Anderson, *Anal. Chem.* **58** (1986) 2431.
[182] E. W. Paul, A. J. Ricco, and M. S. Wrighton, *J. Phys. Chem.* **89** (1985) 1441.
[183] W. Thormann, P. van den Bosch, and A. M. Bond, *Anal. Chem.* **57** (1985) 2764.
[184] K. Aoki, M. Morita, O. Niwa, and H. Tabei, *J. Electroanal. Chem.* **256** (1988) 269.
[185] T. Varco Shea and A. J. Bard, *Anal. Chem.* **59** (1987) 2102.
[186] J. Lindemann and R. Landsberg, *J. Electroanal. Chem.* **30** (1971) 79.
[187] H. Reller, E. Kirowa-Eisner, and E. Gileadi, *J. Electroanal. Chem.* **138** (1982) 65.
[188] D. Shoup and A. Szabo, *J. Electroanal. Chem.* **160** (1984) 19.
[189] J. Cassidy, J. Ghoroghchian, F. Sarfarazi, and S. Pons, *Can. J. Chem.* **63** (1985) 3577.
[190] B. R. Scharifker, *J. Electroanal. Chem.* **240** (1988) 61.
[191] B. R. Scharifker, *Microelectrodes: Theory and Applications*, Ed. by M. I. Montenegro, M. A. Queirós, and J. L. Daschbach, NATO Advanced Study Institutes Series, Series E, Vol. 197, Kluwer, Dordrecht, 1991, p. 227.

[192] H. Reller, E. Kirowa-Eisner, and E. Gileadi, *J. Electroanal. Chem.* **161** (1984) 247.
[193] J. Cassidy, J. Ghoroghchian, F. Sarfarazi, J. J. Smith, and C. Pons, *Electrochim. Acta* **31** (1986) 629.
[194] H. S. White, G. P. Kittlesen, and M. S. Wrighton, *J. Am. Chem. Soc.* **106** (1984) 5375.
[195] D. K. Cope, C. H. Scott, U. Kalapathy, and D. E. Tallman, *J. Electroanal. Chem.* **280** (1990) 27.
[196] W. J. Albery and M. L. Hitchman, *Ring Disc Electrodes*, Clarendon, Oxford, 1971.
[197] M. Spiro and P. L. Freund, *J. Chem. Soc., Faraday Trans. 1* **79** (1983) 1649.
[198] B. J. Feldman and R. Murray, *Anal. Chem.* **58** (1986) 2844.
[199] M. Morita, M. L. Longmire, and R. W. Murray, *Anal. Chem.* **60** (1988) 2770.
[200] M. S. Wrighton, J. W. Thackeray, M. J. Natan, D. K. Smith, G. A. Lane, and D. Belanger, *Phil. Trans. R. Soc. London B*, **316** (1987) 13.
[201] S. Chao and M. S. Wrighton, *J. Am. Chem. Soc.* **109** (1987) 2197.
[202] R. C. Engstrom, T. Meaney, R. Tople, and R. M. Wightman, *Anal. Chem.* **59** (1987) 2005.
[203] R. C. Engstrom, R. M. Wightman, and E. W. Kristensen, *Anal. Chem.* **60** (1988) 652.
[204] K. Aoki and M. Sakai, *J. Electroanal. Chem.* **267** (1989) 47.
[205] K. Wikiel and J. Osteryoung, *J. Electrochem. Soc.* **135** (1988) 1915.
[206] R. B. Morris, D. J. Franta, and H. S. White, *J. Phys. Chem.* **91** (1987) 3559.
[207] J. D. Seibold, E. R. Scott, and H. S. White, *J. Electroanal. Chem.* **264** (1989) 281.
[208] W. Obretenov, V. Bostanov, and V. Popov, *J. Electroanal. Chem.* **132** (1982) 273.
[209] V. Tsakova and A. Milchev, *Electrochim. Acta* **35** (1990) 339.
[210] H. G. Schindler and U. Quast, *Proc. Natl. Acad. Sci. U.S.A.* **77** (1980) 3052.
[211] R. Sridharan and R. de Levie, *J. Phys. Chem.* **86** (1982) 4489.

Cumulative Author Index for Numbers 1–22

Author	Title	Number
Frumkin, A. A. N.	Adsorption of Organic Compounds at Electrodes	3
Fuoss, R. M.	Physical Chemistry of Synthetic Polyelectrolytes	1
Gileadi, E.	The Behavior of Intermediates in Electrochemical Catalysis	3
Gileadi, E.	The Mechanism of Oxidation of Organic Fuels	4
Goddard, E. D.	Electrochemical Aspects of Adsorption on Mineral Solids	13
Goodisman, J.	Theories for the Metal in the Metal-Electrolyte Interface	20
Gores, H.-J.	Temperature Dependence of Conductance of Electrolytes in Nonaqueous Solutions	13
Goruk, W. S.	Anodic and Electronic Currents at High Fields in Oxide Films	4
Grätzel, M.	Interfacial Charge Transfer Reactions in Colloidal Dispersions and Their Application to Water Cleavage by Visible Light	15
Green, M.	Electrochemistry of the Semiconductor-Electrolyte Interface	2
Gregory, D. P.	Electrochemistry and the Hydrogen Economy	10
Gurevich, Y. Y.	Electrochemistry of Semiconductors: New Problems and Prospects	16
Gutmann, F.	Charge-Transfer Complexes in Electrochemistry	12
Gutmann, F.	The Electrochemical Splitting of Water	15
Gutmann, F.	An Introduction to the Electrochemistry of Charge Transfer Complexes II	13
Habib, M. A.	Solvent Dipoles at the Electrode-Solution Interface	12
Haering, R. R.	Physical Mechanisms of Intercalation	15
Hamann, S. D.	Electrolyte Solutions at High Pressure	9
Hamelin, A.	Double-Layer Properties at sp and sd Metal Single-Crystal Electrodes	16
Hamnett, A.	Surface States on Semiconductors	22
Hansma, P. K.	Scanning Tunneling Microscopy: A Natural for Electrochemistry	21
Heiland, W.	The Structure of the Metal–Vacuum Interface	11

Cumulative Title Index for Numbers 1–22

Index

Printed in the United States
by Baker & Taylor Publisher Services